Emerging and Re-emerging Infectious Diseases of Livestock

Jagadeesh Bayry
Editor

Emerging and Re-emerging Infectious Diseases of Livestock

Springer

Editor
Jagadeesh Bayry
INSERM
Paris
France

ISBN 978-3-319-47424-3 ISBN 978-3-319-47426-7 (eBook)
DOI 10.1007/978-3-319-47426-7

Library of Congress Control Number: 2017931062

Printed on acid-free paper

This Springer imprint is published by Springer Nature
The registered company is Springer International Publishing AG
The registered company address is: Gewerbestrasse 11, 6330 Cham, Switzerland

Preface

Emerging and reemerging infectious diseases caused by virus, bacteria, fungi and parasites are causing significant morbidity and mortality not only in humans but also in various livestock including cattle, horses, birds, pigs, sheep, camels and others. In addition, these diseases are instigating significant economy and trade losses and disruption of global travel. Many of these diseases, including influenza, Middle East respiratory syndrome and Hanta, are of public health importance. The reasons for alarmingly raising prevalence of emerging infectious diseases are multifactorial such as deforestation and increased contact with wild animals and birds, climate changes, increase in global travel and altered life cycle of vectors.

In veterinary science, an appropriate referencing book on emerging and reemerging infectious diseases is lacking. Therefore, Springer has recently taken initiatives to start a book programme in this field. This book of *Emerging and Re-emerging Infectious Diseases of Livestock* focuses on various aspects of emerging and reemerging infectious diseases such as details on etiological agent, host range, epidemiology, pathogenesis, diagnosis, therapy and preventive measures including vaccines. The Chaps. 1, 2, 3, 4, 5, 6, 7, 8, 9, 10, 11, 12, 13, 14, 15 and 16 mainly present emerging viral diseases of livestock. Chapter 17 provides details on rickettsial disease. Chapters 18 and 19 describe parasitic and mycotic diseases, while Chap. 20 outlines emerging infectious diseases of camelids.

I hope that this book will serve as good reference for veterinary scientists, field veterinarians, general public and policy makers. I am also confident that this book will inspire new investigations on pathogenesis, diagnosis, therapies and preventive measures for these infectious diseases and might prove useful in the event of emergence of new infectious diseases. I am indebted to all the contributors for writing excellent and detailed chapters on individual diseases, to my family and to Silvia Herold, editor of Biomedicine/Life Sciences, Springer, for her assistance and support.

Paris, France Jagadeesh Bayry

Contents

Part I

Emerging Viral Diseases of Livestock

Bluetongue: Aetiology, Epidemiology, Pathogenesis, Diagnosis and Control

1

Pavuluri Panduranga Rao, Nagendra R. Hegde, Karam Pal Singh, Kalyani Putty, Divakar Hemadri, Narender S. Maan, Yella Narasimha Reddy, Sushila Maan, and Peter P.C. Mertens

1.1 Bluetongue Virus (BTV) and Its Biology

1.1.1 BTV Structure and Proteins

Bluetongue virus (BTV) is the type species of genus *Orbivirus*, subfamily *Sedoreovirinae*, family *Reoviridae*. The virus particle contains seven distinct proteins, comprising three concentric capsid layers that encase the ten linear segments of the dsRNA genome. The innermost 'sub-core' layer is composed of viral protein 3 [VP3(T2)], which encloses the ribonucleoprotein 'transcriptase complexes'

P.P. Rao • N.R. Hegde (✉)
Ella Foundation, Genome Valley, Turkapally, Shameerpet Mandal, Hyderabad 500078, India
e-mail: hegden@ellafoundation.org

K.P. Singh
Centre for Animal Disease Research and Diagnosis, Indian Veterinary Research Institute, Izatnagar, Uttar Pradesh 243122, India

K. Putty • Y.N. Reddy
College of Veterinary Science, P.V. Narasimha Rao Telangana Veterinary University, Hyderabad 500030, India

D. Hemadri
National Institute of Veterinary Epidemiology and Disease Informatics, Yelahanka, Bengaluru 560064, India

N.S. Maan • S. Maan
College of Veterinary Science, Lala Lajpat Rai University of Veterinary and Animal Sciences, Hisar, Haryana, India

P.P.C. Mertens (✉)
Vector-borne Viral Diseases Programme, The Pirbright Institute, Ash Road, Pirbright, Woking, Surrey GU24 0NF, UK

School of Veterinary Medicine and Science, University of Nottingham, Sutton Bonington Campus, Leicestershire, LE12 5RD, UK
e-mail: peter.mertens@pirbright.ac.uk

J. Bayry (ed.), *Emerging and Re-emerging Infectious Diseases of Livestock*, DOI 10.1007/978-3-319-47426-7_1

(TC), each of which comprises of an individual genome segment closely associated with the viral RNA polymerase VP1(Pol), the RNA capping enzyme and transmethylase VP4(CaP) and the viral helicase VP6(Hel) (Mertens and Diprose 2004). The outer surface of the sub-core provides a base for the attachment of the 'outer core' layer, composed of VP7(T13), which provides added strength and rigidity to the sub-core layer. The outer core is surrounded by an 'outer capsid' composed of VP2 (outer capsid protein-1) [VP2(OC1)] and VP5 (outer capsid protein-2) [VP5(OC2)]. Each virus particle encapsidates one copy of each of the ten dsRNA segments (identified as Seg-1 to Seg-10 in order of decreasing molecular weight) (Sung and Roy 2014).

Besides the typical fully intact non-enveloped particles, BTV can exist as other structural variants. The virus can bud out of infected cells to produce membrane-enveloped virus particles (MEVP). Protease treatment of BTV particles cleaves VP2(OC1), although the cleavage products are still associated with the surface of the resulting 'infectious subviral particles' (ISVP) (Mertens et al. 2008). In addition, 'core particles' lacking the outer capsid proteins can also be observed.

Each of BTV's seven 'structural' proteins as well as two nonstructural (NS) proteins [tubule protein NS1(TuP) and viral inclusion body matrix protein NS2(ViP)] is encoded by different genome segments (Roy 2005). However, VP6 (Hel) and NS4 are both translated from different reading frames of Seg-9 (Belhouchet et al. 2011; Ratinier et al. 2011), while NS3 and NS3a are produced from alternate initiation sites within Seg-10 (Wu et al. 1992). Seg-10 has also recently been shown to encode the putative protein NS5 from an alternate reading frame (Stewart et al. 2015). The structure of the BTV particle is shown in Fig. 1.1, and characteristics of various proteins encoded by the different genome segments of BTV are shown in Table 1.1.

1.1.2 BTV Entry, Transcription, Genome Replication, Assembly, Egress and Release

The BTV infectious particle (MEVP, 'intact' virus particle, ISVP, or core) can enter host cells by clathrin-mediated endocytosis, micropinocytosis or via other as yet undetermined mechanisms (Hassan and Roy 1999; Hassan et al. 2001; Gold et al. 2010). Attachment of intact virus particles, or ISVP, to mammalian cells occurs through the binding of VP2(OC1) to an as yet unknown sialoglycoprotein and/or possibly to other receptors or co-receptors (Zhang et al. 2010). The BTV core particles that have lost the outer capsid proteins have a surface composed entirely of the VP7(T13) and have reduced infectivity for mammalian cells (e.g. baby hamster kidney (BHK)-21 fibroblast cells) but are highly infectious to adult *Culicoides* midges or *Culicoides* cell lines (KC cells). The BTV core particle interacts with unknown receptors on the cell surface and can be neutralized by antibodies to VP7(T13). The core particles can bind to glycosaminoglycans, and VP7(T13) contains a surface-exposed conserved Asp-Gly-Glu (RGD) motif, suggesting the involvement of integrins in attachment to cells (Xu et al. 1997).

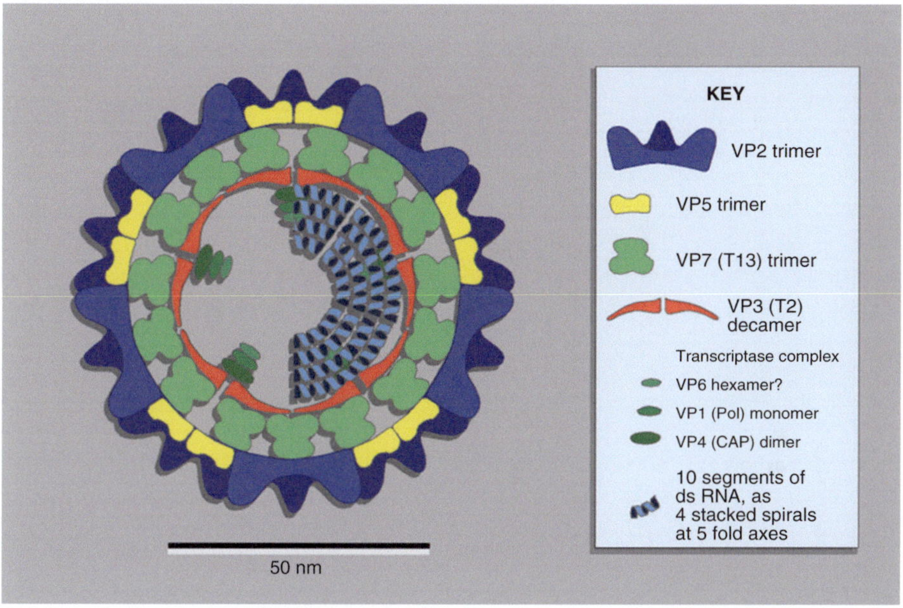

Fig. 1.1 Relative positions and organization of the major structural proteins and genomic RNAs of BTV particle. The virus particle has a triple-layered structure from inside to outside of *VP3(T2)*, *VP7(T13)* and *VP2(OC1)/VP5(OC2)* which enclose the transcription complex of *VP1(Pol)/VP4(CaP)/VP6(Hel)* which closely associates with the ten segmented dsRNA genome. Further descriptions can be found in the text

During endocytosis of the BTV particle, the low pH of endosomes triggers conformational changes in the outer capsid proteins, facilitating uncoating of the virus core, and is essential for the intact virion to infect mammalian cells. Uncoating is followed by penetration of the core particle through the endosomal membrane, which is mediated by the acid-dependent fusion activity activated by conformational changes in VP5(OC2). Removal of the outer capsid proteins from the core particle activates the TCs, initiating viral mRNA synthesis as soon as the core is released into the cytoplasm, without further disassembly (Mertens et al. 2004; Noad et al. 2009). Since BTV cores can infect *Culicoides* cells in the absence of either VP2(OC1) or VP5(OC2), there must be alternative mechanism(s) to translocate the core particle across the cell membrane in these cells.

Viral RNA synthesis is mediated by the TCs. The polymerase VP1(Pol) transcribes from the negative strand of the dsRNA genome to produce positive sense ssRNA (transcripts). Its function is aided by the VP6(Hel), which separates the dsRNA into its component strands as well as helping to release the nascent transcript from the parental negative strand. Evidence suggests that each of the genome segments is associated with a single TC, allowing all ten mRNAs to be transcribed simultaneously and for the reinitiation of transcription of each segment as soon as the previous round is completed. Consequently, more mRNA copies are generated for the smaller genome segments. The rate of transcription can be enhanced by NS1,

Table 1.1 BTV genome segments, the encoded proteins and their characteristics and functions

Genome segment	Segment size (bp)	Protein(s) encoded	Protein length (aa)	Protein molecular weight (kDa)	Estimated subunits per virion	Location in the virion	Characteristics/functions
Seg-1	3954	VP1(Pol)	1302	150.6	10–12	Transcriptase complex (TC) within the sub-core	RNA-dependent RNA polymerase (Pol; transcriptase and replicase)
Seg-2	2926	VP2(OC1)	956	111.1	180 (60 trimers)	Outer shell	Cell attachment, entry
							Haemagglutination
							Virus neutralization antigen, determines serotype
Seg-3	2770	VP3(T2)	901	103.3	120 (12 decamers)	Sub-core capsid shell	*T*=2 icosahedral symmetry
							Scaffold for addition of core surface layer, outer coat proteins
							Localizes TCs on internal surface of the capsid, at 5-fold axes
Seg-4	1981	VP4(CaP)	644	76.4	5–24	TC within the sub-core	Capping enzyme (CaP) – nucleoside phosphohydrolase, guanylyltransferase, transmethylase
							Covalently binds GMP from GTP
Seg-5	1769	NS1(TuP)	552	64.4	NA	Nonstructural	Abundant in infected cell cytoplasm
							Forms tubules (TuP), with unknown function
							Aids virus release from insect cells
							Viral protein translation enhancer
							Co-localizes with the centrosome – may play a role in disruption and blocking of cell division in mammalian cells

Seg-6	1638	VP5(OC2)	526	59.1	360	Outer capsids	Has coiled coil domain – induces membrane permeabilization during initiation of infection and can cause syncytia formation
							Affects specificity of virus neutralization
Seg-7	1156	VP7(T13)	349	38.5	780 (260 trimers)	Core surface	Core structural protein
							T=13 icosahedral symmetry
							Receptor binding for *Culicoides* cells
							Group-/virus-species-specific antigen
Seg-8	1124	NS2(ViP)	357	41.0	NA	Nonstructural	Forms inclusion bodies (ViP)
							Aids in early morphogenesis
							Binds mRNA
							Phosphorylated by protein kinase 1
							Blocks spindle formation and cell division
Seg-9	1046	VP6(Hel)	329	35.7	37 / 72	Inner core	Helicase (Hel)
							Binds ss/dsRNA
							ATPase
		NS4	78	17	NA	Nonstructural	Imparts viral fitness to IFN response

(continued)

Table 1.1 (continued)

Genome segment	Segment size (bp)	Protein(s) encoded	Protein length (aa)	Protein molecular weight (kDa)	Estimated subunits per virion	Location in the virion	Characteristics/functions
Seg-10	822	NS3	229	25.6	NA	Nonstructural	Glycoprotein
							Binds cellular exocytotic components
							Aids virus release
							Contains coiled coil domain
							Localizes to cell membrane late in infection – interacts with lipid droplets in infected cell
		NS3A	216	24.0	NA	Nonstructural	Glycoprotein
							Binds cellular exocytotic components
							Viroporin – aids virus release from insect cells
		NS5	50–59	7.6	NA	Nonstructural	Localizes to nucleus Interferon antagonist?

Information available at http://www.reoviridae.org/dsRNA_virus_proteins/BTV.htm
NA not applicable

although it is not required for the function of the TCs (Patel and Roy 2014). The polymerase is fully conservative and consequently both strands of the dsRNA genome segments are retained within the BTV core, while the newly synthesized mRNA strands are exported out of the core via pores through the VP3(T2) and VP7(T13) layers. The viral transcripts are capped and methylated (addition of m^7GpppG at the 5′ end) by VP4(CaP), a requirement for efficient translation, but they are not poly-adenylated (Patel and Roy 2014).

The assembly of virus particles takes place in the host-cell cytoplasm, within viral inclusion bodies (VIBs). The NS proteins play essential roles in virus replication, assembly and egress. NS1(TuP) forms characteristic 'tubules' (Patel and Roy 2014), although its contribution to BTV assembly is unknown. NS2(ViP) forms VIBs at perinuclear locations; associates intimately with VP1(Pol), VP3(T2), VP4(CaP) and VP6(Hel) within the VIB; and facilitates their assembly in the vicinity of one another (Patel and Roy 2014). Preassembled VP1(Pol)-VP4(CaP)-VP6(Hel)-dsRNA complexes are rapidly recruited by VIB-resident VP3(T2) structures which are immediately stabilized by VP7(T13) (Patel and Roy 2014). Packaging of all ten dsRNA segments is essential for virus replication (Feenstra et al. 2014b), and the process requires the RNA cap structure, the secondary structure at the 3′ non-coding region, conformation formed by the interaction of the 5′ and 3′ termini of the RNA segments as well as specific interactions between short complimentary oligonucleotide regions on different viral mRNAs (Boyce and McCrae 2015). NS2 also co-localizes with the centrosomes and the condensed chromosomes during mitosis and appears to block attachment of the spindle fibres, leading to cell cycle arrest in mammalian cells (Shaw et al. 2013a). The incorporation of ssRNAs into nascent progeny subviral particles proceeds from small to large segments (Sung and Roy 2014), and NS2(ViP) is required for the recruitment of ssRNA (Patel and Roy 2014). At a late phase of assembly of nascent virus particles, VP1(Pol) synthesizes the negative sense RNA strand, reforming the dsRNA genome segments, a process that is thought to accompany the importation of the positive strand ssRNAs into the sub-core.

The final assembly of the outer capsid layer occurs at the periphery of the VIB, as progeny core particles are released into the cell cytoplasm. The sub-core proteins, VP1(Pol), VP4(CaP), VP6(Hel), VP3(T2), along with NS2(ViP) and the ten ssRNAs are sufficient to produce infectious particles, but the inclusion of VP7(T13) and NS1 greatly enhances infectivity (Patel and Roy 2014). Assembly of the outer capsid requires membrane microdomains (lipid rafts) of exocytic vesicles where VP2(OC1), VP5(OC2) and NS3 co-localize (Patel and Roy 2014). The virion co-opts cellular secretory machinery via VP5(OC2) and VP2(OC1), as well as NS3/NS3a, in order to be transported from the VIB to the cell membrane in mammalian cells. NS3 redirects virions from the periphery of the VIBs to a secretory process, leading to the release of BTV particles from insect cells, with little apparent damage to the cell membrane. This, along with the ability of NS3 to control the extent of BTV replication, may allow persistent infection of insect, but not mammalian cells (Patel and Roy 2014). The NS4 protein does not take part in virus replication or assembly per se, but is involved in counteracting host cell's antiviral functions (see below). The presence of NS4 in the cell membrane late in infection suggests that it may be present in the envelope of MEVP budding out of infected mammalian cells. The role of NS5 has not yet been determined, although it appears to be an interferon antagonist. The possibility of expression of additional NS proteins has not been excluded.

The assembled particles are released through direct membrane penetration, cell lysis or budding in mammalian cells and by budding or pore formation in insect cells, the divergent pathways being regulated by NS3/NS3a, and possibly also by NS1(TuP) (Patel and Roy 2014). The MEVP that are released from BTV-infected mammalian cells are thought to be unstable and have not been fully characterized. It is uncertain if they are infectious on their own or whether they play any significant role in virus infectivity and dissemination. The ISVP, on the other hand, can enter mammalian cells as efficiently as intact virion, but are >100-fold more infectious for *Culicoides* cells or adult *Culicoides* insects, suggesting that they play an important role in initiation of infection in the insect vector (Darpel et al. 2011; Drolet et al. 2015).

Although all seven structural proteins, along with the genome segments, are part of the fully intact and infectious BTV particle, removal of the outer capsid proteins to release the virus core does not completely abrogate infectivity (Mertens et al. 2008). Non-infectious virus-like particles (VLP), lacking the genome and the associated TC proteins, can be generated by co-expression of just VP2(OC1), VP3(T2), VP5(OC2) and VP7(T13) (Hewat et al. 1994).

1.2 Epidemiology of Bluetongue

1.2.1 BTV Serotypes, Topotypes and Nucleotypes

Twenty-seven distinct serotypes of BTV have been recognized based on the ability of antibodies generated during infection of the mammalian host to neutralize only the homologous virus type (Hofmann et al. 2008; Maan et al. 2011; Zientara et al. 2014). There are also reports of two other putative serotypes (Nomikou et al. 2015a; Wright et al. 2012). Variations in the sequence of Seg-2 and of its translated protein VP2(OC1) correlate with BTV serotype. Although the sequence of Seg-6 and its product VP5(OC2) can also show a partial correlation with serotype, large variations are sometimes detected in different isolates of the same serotype, and isolates of different serotypes may contain Seg-6/VP5(OC2) with almost identical sequences (Maan et al. 2009). Identifying BTV serotypes, and analysing their appearance and reappearance, provides important epidemiological data and is essential for the design and implementation of effective control and prevention strategies for BT (particularly vaccine matching).

Genome segments can be exchanged between different serotypes, and all segments can reassort independently (Shaw et al. 2013b), potentially giving rise to a complex mosaic of viruses. Many different BTV serotypes exist in multiple but separated geographic regions, but show sequence variations in different segments

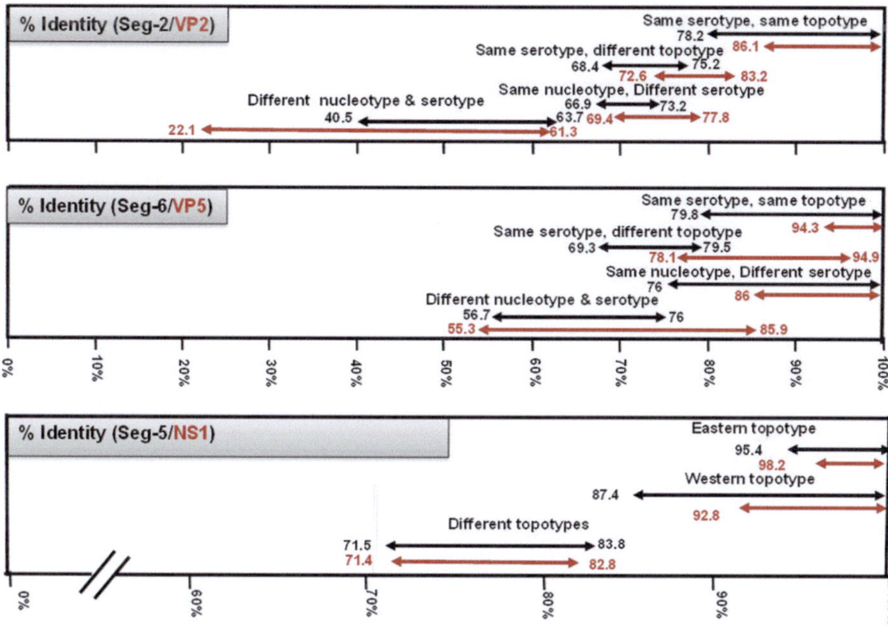

Fig. 1.2 Sequence identities in Seg-2/VP2(OC1), Seg-6/VP5(OC2) and Seg-5/NS1(TuP) within and between BTV serotypes and topotypes. The levels of sequence identity detected within the major eastern and western topotypes are given for Seg-2, Seg-5 and Seg-6 (*in black*) and proteins VP2(OC1), VP5 (OC2) and NS1(ViP) (*in red*)

that reflect their geographic separation (topotypes) (Maan et al. 2009, 2010). The genome segments of most BTV isolates can be grouped into either the major 'eastern' or 'western' topotypes (Maan et al. 2009, 2010). The major eastern group includes isolates from Australasia, the Middle East and the Mediterranean, while the western group includes viruses from Africa, the Mediterranean and the Americas. In addition, there is evidence for several further diverse groups including the recently discovered BTV-25 (SWI2008/01), BTV-26 (KUW2010/02) and BTV-27. The nucleotide/amino acid identity levels are shown in Fig. 1.2.

The eastern topotypes for Seg-2 and Seg-6 (the serotype-determining segments) of BTV-1, BTV-2, BTV-3, BTV-4, BTV-9 and BTV-16 have 25–30 % nucleotide sequence variation with their western topotype counterparts. Similar levels of sequence variation have also been observed for all non-serotype-determining segments. However, multiple clusters have been identified for both eastern and western topotypes of Seg-7 and Seg-10, suggesting that factors other than simple geographic separation and sequence drift have shaped the evolution of these two segments. These additional groups may reflect different insect populations/species that act as vectors within the separated eastern and western geographic regions, since both VP7(T13) and NS3 (encoded by Seg-7 and Seg-10, respectively) are believed to play important roles in the infection of, and viral exit from, insect cells (Tan et al. 2001; Celma and Roy 2009).

Within each major topotype, lower levels of variation (subgroups) can also be detected in the different genome segments, indicating further separation into more closely related local topotypes (Pritchard et al. 2004). For example, within the western topotype, a separate American subgroup has been observed for most of the segments. The level of divergence between the eastern and western topotype viruses indicates that they have been evolving in geographically separated regions for a long time. The levels of divergence between the American and African viruses indicate that they separated more recently than the eastern and western (African) viruses.

The presence of the same BTV serotypes within different regions (topotypes) suggests that these serotypes initially emerged from a common ancestor, before spreading out to different regions, gradually acquiring point mutations through successive rounds of virus replication in the vertebrate and invertebrate hosts, as well as adaptation to local ecosystems (and vector populations) over long periods of separation. The processes of random mutation, genome segment reassortment and selection may have contributed to the divergence of BTV into topotypes in most of the BTV genome segments. However, new strains of BTV, representing exotic topotypes or serotypes, have occasionally entered naïve or endemic areas via natural or anthropogenic routes, potentially changing the genome segment 'mix' and local genotype, and are frequently associated with severe clinical disease outbreaks (Nomikou et al. 2015b).

By considering the major topotypes separately, it is possible to set levels of sequence variation that may be used as a guide to identify different BTV serotypes. Overall, BTV strains within the same serotype can show up to 31.6 % nucleotide (nt) and 27.4 % amino acid (aa) variation in Seg-2 and VP2(OC1) (Maan et al. 2009, 2010). Viruses belonging to different serotypes can show up to 26.8 % nt and 22.2 % aa identity, making the delineation of clear limits for identification of distinct serotypes difficult. However, the maximum level of variation in Seg-2/VP2(OC1) for isolates of the same serotype and the same major topotype drops to a 21.8 % nt and 13.9 % aa sequences, respectively, 5.0 % and 8.3 % lower than the minimum levels of nucleotide and amino acid variation detected between different BTV serotypes. These levels, based on Seg-2 sequences, can therefore be used as a guide for the identification of existing serotypes, regardless of their geographic origin. If a novel isolate falls into the 'gap' [(between 26.8 % and 21.8 % nt and 22.2 % and13.9 % aa variation in Seg-2/VP2(OC1) compared to known BTV isolates], it could represent either an entirely novel serotype that is related to but distinct from other established serotypes or an isolate of an existing serotype but belonging to a different and distinct 'major' topotype (Maan et al., in preparation). In either case, additional serological analyses would be required. Such 'new viruses' would then become 'reference strains' for the new serotype or the topotype. It is therefore important to accurately identify and curate reference strains. A collection of well-documented BTV isolates, as well as sequence data and other available information, can be accessed at http://www.reoviridae.org/dsRNA_virus_proteins/ReoID/virus-nos-by-country.htm.

There is a clear correlation between the low levels of serological cross-reactivity observed between different BTV serotypes and nucleotide sequence similarities in Seg-2 and Seg-6. These clusters of related virus serotypes are said to belong to the

Table 1.2 Seg-2- and Seg-6-based nucleotypes and their relationship to serotypes

Seg-2 nucleotype	Serotype	Seg-6 nucleotype
A	BTV-4, BTV-10, BTV-11, BTV-17, BTV-20 and BTV-24	A
B	BTV-3, BTV-13 and BTV-16	B
C	BTV-6, BTV-14 and BTV-21	B
E	BTV-5, BTV-9 (eastern)	B
	BTV-9 (western)	C
H	BTV-1	C
I	BTV-2	C
D	BTV-23	C
	BTV-8, BTV-18	G
F	BTV-7 and BTV-19	D
G	BTV-12 and BTV-22	E
J	BTV-15	F
K	BTV-25 and BTV-27	H
L	BTV-26	I

same nucleotype (for Seg-2 and Seg-6 separately). Nucleotypes may represent a node of divergence among different serotypes during evolution. The distribution of various serotypes among the different Seg-2 and Seg-6 nucleotypes is shown in Table 1.2. As expected, serotypes from different Seg-2 nucleotypes are more distinct. However, Seg-6 from different serotypes within a particular 'Seg-6 nucleotype' can show very high levels of identity, approaching 100 % (Fig. 1.2), suggesting that they contain an essentially similar Seg-6/VP5(OC2) as a result of genome segment reassortment, but this does not result in an identical serotype, because of the more dominant variation in Seg-2/VP2(OC1).

1.2.2 The Incidence and Prevalence of Disease and BTV Serotypes

Bluetongue (BT) was regarded as an African disease until 1943, when an outbreak was recorded in Cyprus. Since then, the disease has been reported from the Americas, Australia, Europe, the Indian subcontinent and parts of the Far East (Walton 2004). The increasing distribution and devastation of outbreaks suggest that major and worldwide changes are still ongoing in the epidemiology of BT. Table 1.3 summarizes the BTV serotypes reported from the various regions.

1.2.2.1 Sub-Saharan Africa

Although BT is endemic in sub-Saharan Africa, most of the information concerning BT from this region is from South Africa, where BTV circulation is year-round in the north and seasonal in the south. South Africa presents the highest diversity of BTV serotypes with only BTV-20 and BTV-21 of the 24 'classical' serotypes not

Table 1.3 BTV serotype distribution in different geographic regions

Episystem	Western topotype	Eastern topotype
Sub-Saharan Africa	1–19, 22, 24, (1 putative new serotype)	
Americas	1–6, 8–14, 17–19, 22, 24	
Australasia	2, 5, 7, 10, 12, 24	1–4, 9, 15, 16, 20–21, 23
Middle East, Mediterranean and Europe	1, 2, 4, 5, 6, 8, 10, 11, 12, 14, 15–24, 25, 26, 27, (1 putative new serotype)	1, 9, 16,

reported from there. A putative new serotype has also been reported from the region (Wright et al. 2012). Besides the circulation of multiple BTV serotypes, more than one serotype is also commonly detected during an outbreak. In general, three to five serotypes dominate every season in South Africa, possibly reflecting dynamic changes in herd immunity to specific serotypes (Coetzee et al. 2012).

1.2.2.2 The Mediterranean, Middle East and Europe

North Africa, west Asia and Southern European countries surrounding the Mediterranean form an endemic area for BT. This region is separated from the rest of Africa by the Sahara and the rest of Asia by the Arabian Desert and other west Asian arid areas. BTV-2, BTV-4, BTV-6, BTV-10 and BTV-16 are considered endemic in Israel, along with active circulation in 2008 of BTV-5, BTV-8 and BTV-24 and, more recently, BTV-12 and BTV-15 (Brenner et al. 2010; Shimshony 2004). In Lebanon, BTV-1, BTV-4, BTV-6, BTV-8, BTV-16 and BTV-24 were identified in 2011 (El Hage et al. 2013). In other west Asian countries, BT has been reported from Iraq, Iran, Jordan, Oman, Saudi Arabia, Syria, Turkey and Yemen; the newer serotype, BTV-26, has been isolated from Kuwait (Maan et al. 2011). In North Africa, BTV-1, BTV-4, BTV-10 and BTV-12 have been reported from Egypt (eubtnet.izs.it/btnet/inFocus/pdf/bluetoungeDisease_mod_1.pps), while BTV-1, BTV-2 and BTV-4 have been reported from Algeria, Libya, Morocco and Tunisia since 2002 (WAHIS 2015).

In southern Europe, BTV-3 was first reported in Cyprus in 1924, followed by BTV-10 in Spain and Portugal during 1956–1960 and BTV-4 again in Cyprus in 1969. Since 1998, 11 different serotypes (BTV-1, BTV-2, BTV-4, BTV-6, BTV-8, BTV-9, BTV-11, BTV-14, BTV-16, BTV-25 and BTV-27) have been reported in Europe (Wilson and Mellor 2009), including strains of BTV-6, BTV-11 and BTV-14 with very high identity levels to modified live virus (MLV) vaccines (http://ec.europa.eu/food/animal/diseases/controlmeasures/bluetongue_en.htm#serotypes). BTV-25 and BTV-27 have also been isolated from this region (Hofmann et al. 2008; Zientara et al. 2014). Regular incursions of BTV-1 and BTV-4 into southern Europe have been observed between 2006 and 2015. During 2006, BTV was detected for the first time in northern Europe, beyond the traditional BTV boundary. This outbreak, caused by BTV-8 (Maan et al. 2008), spread across most parts of Europe and was successfully controlled through the use of inactivated vaccines, but it reappeared in France during 2015 after a gap of 5 years (Sailleau et al. 2015).

The incursions of BT into Europe have occurred mainly via three routes: (i) from Morocco to Spain through the Straits of Gibraltar, (ii) from Tunisia to Italy through Sicily or Sardinia, and (iii) from Turkey to Greece and Bulgaria through the Aegean islands or the land borders between these countries (Gomez-Trejedor 2004; Wilson and Mellor 2009). However, BTV-8, which is thought to have originated from sub-Saharan Africa (Maan et al. 2008), arrived directly into northern Europe via an unknown route. It is also unclear how vaccine strains of BTV-6, BTV-11 and BTV-14 arrived in Europe, although it could be due to illegal use of MLV vaccines. Genome analysis of viruses that have circulated in West Asia, North Africa and Europe since 1998 indicates that some of BTV-1, BTV-9 and BTV-16 may have originated from an 'eastern' ecosystem; however, the origin of BTV-16 is less certain due to the use of MLV vaccines in southern Europe and the Middle East (Mellor and Wittmann 2002; Gomez-Trejedor 2004; Savini et al. 2008). Strains of BTV-25, BTV-26 and BTV-27 show high levels of diversity compared to BTV-1 to BTV-24, and these may represent ancient lineages of BTV that have existed in Europe, the Middle East and the Mediterranean for a long period.

1.2.2.3 Australasia

In Australasia, the incidence of BT extends from northern China, Mongolia, Kazakhstan and the Asian part of southern Russia, through to northern Australia (Kirkland 2004; Koltsov et al. 2014). Tropical southern and Southeast Asia forms a major endemic area, with several serotypes in circulation (Boyle et al. 2012). Phylogenetic analyses of Seg-2 and Seg-6 sequences of viruses circulating in Australasia indicate that BTV-1, BTV-2, BTV-3, BTV-4, BTV-9 and BTV-16 belong to the eastern topotype (Boyle et al. 2012; Yang et al. 2012; Maan et al. 2012b, 2015b; Rao et al. 2012), indicating the spread of these serotypes from Southeast Asia to Australia, while BTV-20, BTV-21 and BTV-23 are unique to this region (Maan et al. 2009; Boyle et al. 2012; Susmitha et al. 2012). Western topotype strains of BTV-5, BTV-7, BTV-12 and BTV-24 have also recently entered the area (Lee et al. 2011; Boyle et al. 2012; Rao et al. 2015a; Yang et al. 2012, 2015; WAHIS 2015; Krishnajyothi et al., 2016; Yang et al. 2016; unpublished data) and have become established in more than one country, although western topotype MLV-like strains of BTV-2 and BTV-10 (Gollapalli et al. 2012; Maan et al. 2012c) did not become widely established.

1.2.2.4 The Americas

Serological evidence exists for BTV circulation in most parts of the USA and Central and South America, except the southern parts of the Pampas and Patagonia, and most of Canada and Alaska (Tabachnick 2004; Legisa et al. 2014). The serotypes reported in South America include (a) BTV-1, BTV-2, BTV-6, BTV-10, BTV-12, BTV-13, BTV-14, BTV-17 and BTV-24 from French Guyana and BTV-4 from Argentina by virus isolation (Legisa et al. 2014), and (b) BTV-4, BTV-6, BTV-14, BTV-17, BTV-18 and BTV-20 from Brazil; BTV-12, BTV-14 and BTV-17 from Colombia; BTV-6, BTV-14 and BTV-17 from Suriname and BTV-14 and BTV-17 from Guyana by serology (Wilson et al. 2009b). BTV-4 and BTV-12 are reported to be isolated from animals imported from Brazil and are under quarantine in the USA (Groocock and Campbell 1982).

In the USA and Mexico, BTV-2, BTV-10, BTV-11, BTV-13 and BTV-17 are considered endemic; several other serotypes (BTV-1, BTV-3, BTV-5, BTV-6, BTV-9, BTV-12, BTV-14, BTV-19, BTV-22 and BTV-24) have also been isolated from the USA (Johnson et al. 2007). Occasional incursions of BTV from the USA into the Okanagan Valley, British Columbia, Canada (Clavijo et al. 2000b; Dulac et al. 1989) have been observed, and the virus has been recently reported for the first time in southwestern Ontario (http://www.oie.int/wahis_2/public/wahid.php/Reviewreport/Review?reportid=18593). Limited studies of the isolates from this region indicate genetic relatedness among the isolates and suggest to and fro movement of viruses from the Caribbean Basin, Central America and North America (Wilson et al. 2000; MacLachlan et al. 2007; Balasuriya et al. 2008; Legisa et al. 2013; Viarouge et al. 2014).

1.2.3 BTV Transmission and Movement

Traditionally, BT has been restricted to areas between the latitudes 40°S and 53°N, coinciding with the distribution of the transmission-competent *Culicoides* vector species. Both natural and anthropogenic factors that contribute to movement of vertebrate and invertebrate hosts of BTV can increase the spread and transmission of the virus. Adult *Culicoides* can fly 2–5 km in a few days, leading to the local spread of BTV. However, long-distance spread can also occur as a result of dispersal of the insects by wind, particularly over water (Eagles et al. 2014). Anthropogenic activities including the use of incompletely attenuated/inactivated vaccines, import of infected ruminants, semen and embryos, and transportation of infected midges via planes, ships and containers may also contribute to the long-distance and even transcontinental spread of BT.

Once introduced, the survival and transmission of BTV in a new area depends on the availability and density of susceptible vertebrate hosts, their collective herd immunity as well as the transmission competency, seasonal activity and the density of adult *Culicoides* populations. Once the virus has been introduced, domestic ruminant population densities are often sufficient to sustain BTV. Consequently, the presence and activity of competent *Culicoides* is thought to be a limiting factor for BTV transmission and spread (Mellor and Boorman 1995).

The geographical range of *Culicoides* is dictated by climatic conditions. Models predicting significant correlations between climatic zones and geographical range of different *Culicoides* species (Brugger and Rubel 2013; Guichard et al. 2014) have indicated their spread to and within temperate areas and towards the poles, associated with global warming. This therefore represents an increased risk for the incidence and spread of BT in these regions. On the other hand, BTV infection of animals at temperate latitudes is distinctly seasonal (typically July to November in the northern hemisphere). However, BTV can survive from one "vector season" to the next during the winter months (overwintering) when cold temperatures restrict the numbers, activity and reproduction of adult *Culicoides*, effectively preventing vector transmission. The mechanisms involved in overwintering are poorly understood, although vertical transmission of BTV to the ruminant foetus, resulting in release of the virus at birth and horizontal transmission to other naïve hosts, and persistent infection of adult vectors that manage to survive throughout the winter months have been suggested (Wilson et al. 2008; Darpel et al. 2009; Mayo et al. 2014). Viral RNA has been detected in midge larvae (White et al. 2005), suggesting that at least some component of the virus can be vertically transmitted in the insect vector. However, attempts to recover infectious virus have been unsuccessful. Trans-stadial and trans-ovarial passage of BTV in hard and soft ticks, respectively, suggests that they could also play a role in virus transmission and overwintering (Bouwknegt et al. 2010). Persistent infection of ovine T cells may provide another potential mechanism (Takamatsu et al. 2003).

Some BTV serotypes (e.g. BTV-1, BTV-8, BTV-26) can be transmitted horizontally in the absence of adult *Culicoides* (Menzies et al. 2008; van der Sluijs et al. 2011; Batten et al. 2014), and their transmission may therefore be independent of climatic conditions.

1.2.4 Vector Distribution and Bluetongue Epidemiology

Variations in the distribution, activity and virus transmission competence of adult *Culicoides* populations can determine if an area is endemic, disease-free or seasonally disease-free. Despite the existence of more than 2000 species of *Culicoides*, only a few are known to be competent vectors for BTV. Natural barriers such as oceans and deserts affect the free movement of insects and BTV between the different geographical areas where competent vector species are present.

The most widespread of the known BTV vector is *C. imicola*, with habitat extending from most parts of Africa, the Mediterranean, southern Europe, west and south Asia and east Asian countries, including Laos, Vietnam and China. Recently, *C. imicola* has been found in new areas of southern Europe, coinciding with the spread of BTV to these regions (Wilson and Mellor 2009; Carpenter et al. 2013). In endemic areas where *C. imicola* has not been detected, including the Americas, northern Europe, northern China, southeast Asia and Australia, other *Culicoides* species can act as vectors for BTV transmission (see Table 1.4) (Mellor et al. 2009).

1.2.5 BTV Episystems, Climatic Zones, 'Source' and 'Sink' Populations

Climatic and geographical barriers restrict the movement of both vectors and hosts between different ecosystems, limiting the spread of BTV strains and even individual BTV genome segments, leading to their independent evolution and genetic divergence over long periods of time (Gibbs and Greiner 1994; Nomikou et al. 2015b). It has been suggested that the distribution of different BTV serotypes in

Table 1.4 Vector species implicated in the transmission of BTV in different geographic regions

	Principle vector	Other known and potential species
Sub-Saharan Africa	*C. imicola*	*C. bolitinos, C. magnus, C. bedfordi, C. leucostictus, C. pycnostictus, C. gulbenkiani, C. milnei, C. tororoensis*
Mediterranean	*C. imicola*	
North Europe	*C. obsoletus, C. pulicaris*	
West Asia	*C. imicola*	
South Asia	*C. imicola*	*C. peregrinus, C. oxystoma*
East Asia	Unknown	*C. imicola, C. schultzei, C. gemellus, C. peregrinus, C. arakawae, C. circumscriptus, C. actoni, C. homotomus*
Southeast Asia	*C. fulvus, C. peregrinus*	*C. actoni, C. brevitarsis, C. fulvus, C. wadai, C. brevipalpis, C. peregrinus, C. oxystoma, C. nudipalpis, C. orientalis*
Australia	*C fulvus, C. wadai, C. actoni, C. brevitarsis*	*C. brevipalpis, C. peregrinus*
North America	*C. sonorensis*	*C. variipennis, C. stellifer, C. insignis*
Central America	*C. insignis*	*C. pusillus*
South America	*C. insignis*	*C. pusillus*

different geographical areas may be linked to the *Culicoides* species inhabiting those areas (episystems) (Tabachnick 2004). However, there is little unequivocal evidence to support this hypothesis. Indeed, recent incursions of exotic BTV serotypes and/or topotypes into the Americas, Europe and Australasia (Johnson et al. 2007; MacLachlan et al. 2007; Lee et al. 2011; Boyle et al. 2012; Legisa et al. 2014; Viarouge et al. 2014; Maan et al. 2015c; Nomikou et al. 2015b; Rao et al. 2016; Yang et al. 2015) demonstrate the abilities of different regional populations of *Culicoides* to transmit exotic BTV strains/genotypes.

Rather, the presence and spread of BTV into different climatic regions fits well into 'source' and 'sink' population dynamics. Tropical regions, with large vertebrate and invertebrate populations and conducive climate, support the circulation of multiple BTV serotypes throughout the year, acting as virus 'source' areas. In contrast, in temperate areas, BTV is often not sustained over long periods due to increasing herd immunity over time, and the massive reduction in adult vector populations during winters, favouring seasonal incursions caused by one or a few serotypes (Sellers 1980; Rao et al. 2016). Phylogenetic and evolutionary analyses show that the seasonal outbreaks in these virus 'sink areas', which may be interspersed with disease-free periods, reflect novel virus introductions from 'source areas' (Carpi et al. 2010; Nomikou et al. 2015b). The extent of source and sink areas depends on the availability of susceptible vertebrate host populations and competent vectors. South and Central America, North Africa and the Mediterranean and southeast Asia represent source areas, while most parts of North America, northern Europe and parts of Australia represent the associated sink areas, respectively (Sellers and Maarouf 1989; Mellor and Wittmann 2002; Johnson et al. 2007; Purse et al. 2008; Daniels et al. 2009; Boyle et al. 2012). The 'sink' area is especially evident in northern Europe where different exotic serotypes have entered from neighbouring or distant endemic areas and have been sustained for short periods, followed by the re-emergence or entry of the same or different strains or serotypes (Nomikou et al. 2015b).

Combining source and sink systems with climatic zones, we propose the division of tropical areas into four episystems: sub-Saharan Africa, the Mediterranean, Australasia and the Americas. Each episystem contains its own source and sink areas overlapping with climatic zones (Fig. 1.3). Evidence exists for sharing and circulation of virus populations within these proposed episystems, and more specifically, each source and sink area within each episystem has closely related populations of circulating viruses (Wilson et al. 2000; Pritchard et al. 2004; Potgieter et al.

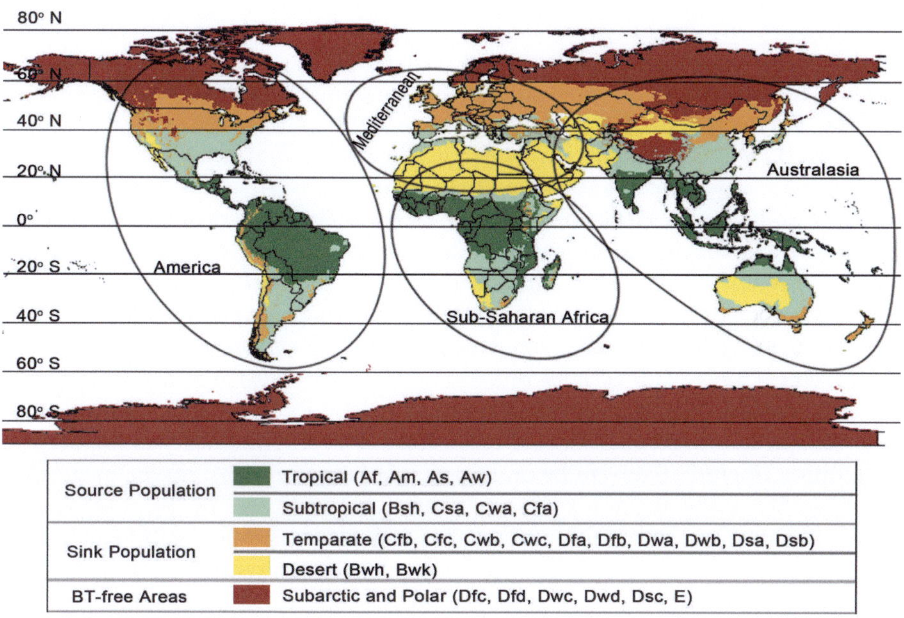

Fig. 1.3 Climatic zones and proposed BT episystems. Climatic zone map was prepared using FAO GeoNetwork (FAO 2013), as per Köppen's classification (Peel et al. 2007), and BTV sink and source population areas are mapped based on the conditions suitable for propagation of *Culicoides* and occurrence of BT.
Source areas include tropical (*dark green*) and subtropical areas including the Mediterranean (*light green*), which are most congenial for *Culicoides* propagation. Sink areas include hot and cold deserts (*yellow*) and temperate areas (*orange*), where only part of the year is suitable for the sustenance of *Culicoides* populations. Subarctic and polar regions (*red*) are considered as BTV-free zones, where harsh weather hampers the survival of *Culicoides*. Four distinct episystems (sub-Saharan Africa, the Mediterranean, the Americas and Australasia) are shown with each system having sink and source areas. The boundaries of both climatic zones and episystems are indicative and not definitive and may overlap

2005; Breard et al. 2007; Johnson et al. 2007; MacLachlan et al. 2007; Balasuriya et al. 2008; Daniels et al. 2009; Boyle et al. 2012; Legisa et al. 2013; Viarouge et al. 2014; Maan et al. 2015c; Nomikou et al. 2015b; Rao et al. 2015a). The *Culicoides* species and populations in different regions may have different vector competencies and capacity for different BTV strains, serotypes and topotypes (MacLachlan et al. 2007; Purse et al. 2008; Mellor et al. 2009), and this may redefine source and sink areas within an episystem. The definition of episystems helps predict movement of viruses from source to neighbouring sink areas and preparedness for surveillance and control.

1.3 Bluetongue Disease, Pathology and Pathogenesis

1.3.1 Host Tropism and Range of Disease Manifestations

All ruminants are susceptible to BTV infection, but clinical disease is most often observed in sheep and in some species of deer. There is significant variation in the severity of disease in sheep from different geographic regions. Native tropical and subtropical breeds usually remain asymptomatic after infection with endemic BTV strains/topotypes, whereas fine wool European breeds such as Merino typically manifest severe disease. Disease severity also depends on the virus strain involved. For example, MLV strains, which are presumably safe for South African sheep, caused severe and even lethal disease in naïve European sheep (Veronesi et al. 2010). Similarly, local sheep breeds of India were also severely affected, possibly after the introduction and emergence of BTV strains belonging to exotic 'western' topotypes (Prasad et al. 2009; Gollapalli et al. 2012). These observations may relate in part to an inherent resistance against endemic local BTV strains, which is overcome by the significant variations in gene and protein sequences that are observed in viruses from different topotypes. In addition, there is likely to be an important level of cross-serotype protection between different viruses belonging to the same topotype, by T-cell-mediated immune responses, which are much less effective across topotype. Other factors which influence disease severity include age, environmental stress and nutritional and immune status.

Bluetongue in sheep is characterized by one or more of the following: (a) hyperaemia, oedema, ulceration and/or haemorrhage of the face and oral cavity (muzzle, lips, gums, tongue), (b) salivation and lacrimation, (c) oedema of the ears and neck, (d) respiratory distress, (e) congestion of other hairless parts of the body (axilla, inner thighs, flank, perineum), (f) coronitis and (g) occasional enteritis (MacLachlan et al. 2009; Worwa et al. 2010; Umeshappa et al. 2011). The severity of disease and case fatality levels depend on the immune status of the individual animal and of the population as a whole. Mortality levels in naïve sheep populations may reach up to 70 % when susceptible animals are introduced into endemic areas or when incursions occur. Convalescence is protracted, but some recovering animals may die suddenly due to severe and rapidly progressing pulmonary oedema (Verwoerd and

Erasmus 2004, MacLachlan et al. 2009). Animals developing chronic disease show extreme weakness, prostration and torticollis, as well as wool-break, and some may succumb to bacterial complications, mainly pasteurellosis (Parsonson 1990; Verwoerd and Erasmus 2004, MacLachlan et al. 2009). Infection during early pregnancy (5–6 weeks of gestation) may lead to abortion, foetal death or birth of underdeveloped "dummy" lambs with or without malformations such as cerebral and/or skeletal anomalies (MacLachlan et al. 2009).

Infection of cattle and goats is mostly asymptomatic or subclinical, despite prolonged viraemia (Tweddle and Mellor 2002). However, clinical disease similar to that in sheep was reported in cattle during the recent epidemics caused by BTV-8 in West and Central Europe (Thiry et al. 2006; Darpel et al. 2007; Elbers et al. 2008). Infection of dams in early pregnancy can result in death and resorption of the embryo or abortion or the birth of malformed or weak calves (MacLachlan et al. 2000; Desmecht et al. 2008; Elbers et al. 2008). Foetuses infected at 70–130 days of gestation can develop serious cerebral defects, leading to 'dummy-calf' syndrome, while those infected a few weeks before delivery usually develop only mild encephalitis (Waldvogel et al. 1992; MacLachlan et al. 2000; MacLachlan et al. 2009).

Bluetongue has also been described in a variety of wild ruminants, although most remain asymptomatic. Cervid deer, especially red deer, play an important role in BTV epidemiology in Europe, serving to sustain the sylvatic cycle (Falconi et al. 2011; Lorca-Oro et al. 2014; Rossi et al. 2014; Ruiz-Fons et al. 2014), possibly because of the long-lasting viraemia in them (Lopez-Olvera et al. 2010). In the USA, capreoline deer, especially white-tailed deer, have been shown to be particularly susceptible and are used as sentinels in surveillance (Parsonson 1990; Stallknecht and Howerth 2004). Wild antelopes may also play an important role in the spread of BT in South Africa (Coetzee et al. 2012). Severe, lethal BTV infection with acute respiratory distress has also been described in llamas and alpacas in Europe, although South American camelids are thought to be resistant (Henrich et al. 2007; Meyer et al. 2009). The disease has also been recorded in a number of other wild, free-ranging and captive zoo ruminants.

A fatal BTV infection was reported in dogs immunized with a BTV-contaminated vaccine, and the disease syndrome was also reproduced experimentally in pregnant bitches which aborted and succumbed to severe respiratory distress (Levings et al. 1996). Bluetongue was also reported in Eurasian lynx held at Belgian zoo and fed on potentially infected ruminant foetuses (Jauniaux et al. 2008). BTV-specific antibodies have also been reported in several wild and domestic carnivores, including dogs (Alexander et al. 1994; Howerth et al. 1995; Dunbar et al. 1998), suggesting infection without disease manifestation.

1.3.2 Pathology of Bluetongue

Pathological changes due to BT in sheep include one or more of the following: (a) hyperaemia; ulceration and/or haemorrhage of the upper gastrointestinal tract, including ruminal pillars and reticular folds; conjunctiva; and coronary bands around the hooves; (b) inflammation of the upper respiratory tract, pulmonary oedema and bronchopneumonia, emphysema and atelectasis with petechial to ecchymotic haemorrhages in the apical and diaphragmatic lobe of the lungs; (c) pericardial, pleural and peritoneal haemorrhages and effusion; (d) petechiae and ecchymoses, necrosis and oedema of skeletal muscle and muscles of the abdominal wall; (e) haemorrhage and necrosis of cardiac muscles, especially the papillary muscle of the left ventricle; (f) petechial and ecchymotic subintimal haemorrhage in the pulmonary artery near the junction of the heart and sometimes in the aorta; (g) oedema and congestion or haemorrhage of the lymph nodes, spleen and tonsils; and (h) occasional petechial haemorrhage of the kidney and the gut, particularly the ileocecal junction (Parsonson 1990; MacLachlan et al. 2009; Worwa et al. 2010; Umeshappa et al. 2011). Histological lesions are characterized by oedema, haemorrhage, congestion and microvascular thrombosis, with subsequent tissue infarction in the affected sites and hypertrophy of the capillary endothelium. Skeletal and cardiac muscles show haemorrhage and necrosis in acute cases and fibrosis in chronic cases (MacLachlan et al. 2009). Other pathology includes degenerative changes in the trapezius muscle, engorgement of pulmonary capillaries, inter- and intra-alveolar haemorrhage and oedema and hyperaemia of the corium of the hooves (Umeshappa et al. 2011; Channappanavar et al. 2012).

Natural vertical BTV transmission had been considered rare until the recent BTV-8 epizootic in Europe, where transplacental transmission was frequently observed in different ruminants (MacLachlan et al. 2009; Saegerman et al. 2011; van der Sluijs et al. 2016). Experimentally, only egg- or cell culture-adapted (e.g. modified live virus (MLV) strains, and not unadapted viruses have been shown to cross the placenta and result in reproductive failure and congenital abnormalities (MacLachlan et al. 2009). The brain of affected foetuses shows acute necrotizing meningoencephalitis progressing to hydrocephaly, subcortical cysts and cavitations, with the involvement of the cerebellum in severe cases (MacLachlan et al. 2009; Coetzee et al. 2013).

In cattle, in the majority of the cases, no or minimal pathological changes are observed upon BTV infection (Parsonson 1990). On the other hand, the pathology, including occasional severe pulmonary oedema, in the case of the recent BTV-8 outbreak in Europe resembled those of sheep, although effusions in the pleura, pericardium and peritoneum were not as characteristic as in sheep (MacLachlan et al. 2009). White-tailed deer develop disseminated intravascular coagulopathy and haemorrhagic diathesis (Howerth et al. 1988). Llamas and alpacas may develop severe pulmonary oedema, hydrothorax and pericardial effusions (Henrich et al. 2007; Meyer et al. 2009).

1.3.3 Pathogenesis of Bluetongue

It is evident that (a) the outcome of infection and the severity of disease are influenced by the virus serotype, topotype and strain; certain genome segments and their encoded proteins; host species, breed and their native or exotic origin, relative to the virus strain; and the immune status of the host, including both humoral and cell-mediated responses, and (b) a combination of virus- and host response-mediated damages is ultimately responsible for the pathogenesis of BT (Huismans et al. 2004; Caporale et al. 2014; Coetzee et al. 2014; Janowicz et al. 2015). Although viral virulence determinants are not well understood, both serotype-dependent and independent phenomena have been described (Groocock et al. 1982; Jeggo et al. 1987), and both structural [VP1(Pol), VP2(OC1), VP5(OC2)] and nonstructural [NS2(ViP), NS3/NS3a] proteins have been implicated to contribute to virulence (Gould and Eaton 1990; Carr et al. 1994; Bernard et al. 1997; Huismans et al. 2004; Owens et al. 2004).

Insect bite is the predominant route of infection of BTV, although some strains can be transmitted horizontally, transplacentally as well as through the colostrum (Menzies et al. 2008; Backx et al. 2009; Mayo et al. 2010; van der Sluijs et al. 2011; Coetzee et al. 2013; Batten et al. 2014). The virus, which is deposited subdermally by infected *Culicoides*, establishes a primary infection in dermal fibroblasts, mononuclear phagocytes, dendritic cells (DCs), lymphocytes and endothelial cells (MacLachlan et al. 2009; Darpel et al. 2012). The virus is transported to the draining lymph nodes where it replicates in phagocytes and DCs (Hemati et al. 2009; MacLachlan et al. 2009; Drew et al. 2010b). Secondary replication occurs in other lymphatic tissues such as the spleen, thymus, tonsils, etc., before the virus disseminates haematogenously via the efferent lymphatics (MacLachlan et al. 2009). In early viraemia, the virus is associated with all blood cells, while at later stages, it mostly associates with erythrocytes, being sequestered in membrane invaginations, allowing for prolonged survival of the virus in the presence of neutralizing antibodies (NAbs) (MacLachlan 2004; MacLachlan et al. 2009). Free virus in low titres is found in blood plasma only at the initial stages of infection, but infectious BTV may be detected for 35–60 days (MacLachlan et al. 1990; Barratt-Boyes and MacLachlan 1994; MacLachlan 1994). At late stages of infection, viral genome can be detected by reverse transcription polymerase chain reaction (RT-PCR) for up to 200 days in cattle and 150 days in sheep, in the absence of infectious BTV (Katz et al. 1994). Following viraemia, the virus spreads to secondary organs (Parsonson 1990; MacLachlan et al. 2009), before spreading to the peripheral sites such as the oral mucosa and the hooves.

The pathology of BT is due mainly to injury to microvasculature in target tissues, resulting in breach of permeability, fluid extravasation, vascular occlusion and tissue infarction. Besides direct damage due to virus-induced apoptosis and necrosis, vasoactive mediators produced by thrombocytes, DCs, macrophages and infected endothelial cells damage the endothelium, interfere with its function and increase vascular permeability, leading to the development of oedema and effusions (MacLachlan et al. 2009; Drew et al. 2010a, b; Darpel et al. 2012). In addition, BTV induces the production of several pro-inflammatory mediators by plasmacytoid DCs (pDCs) and macrophages, leading to haemorrhagic fever-like syndrome (Schwartz-Cornil et al. 2008; MacLachlan et al. 2009; Drew et al. 2010a, 2010b; Sanchez-Cordon et al. 2013).

Infection of sheep and cattle with BTV increases plasma levels of prostacyclin (a potent vasodilator and inhibitor of platelet aggregation) and thromboxane (a strong procoagulant) (DeMaula et al. 2002). A direct correlation between the concentration of serum acute-phase proteins and severity of vascular lesions has also been noted with BTV-1 and BTV-8 (Sanchez-Cordon et al. 2013). The lower sensitivity of cattle to BTV-induced microvascular injury and thrombosis may be due to a lower susceptibility of their endothelial cells to BTV infection, higher levels of inflammatory and vasoactive mediators and/or a higher prostacyclin/thromboxane ratio (DeMaula et al. 2001, 2002).

Bluetongue virus is a strong inducer of type I interferon (IFN) (MacLachlan and Thompson 1985; Foster et al. 1991), and this would be expected to counter virus replication. However, the functional effect of IFN can be subverted by BTV. One of the major inducers of type I IFN response is dsRNA, but the dsRNA genome of BTV is retained within the virus core throughout the replication cycle, reducing the potential for induction of the cytokine. The exposed VP7(T13) structure of the core surface layer can also bind dsRNA molecules, potentially sequestering them further (Diprose et al. 2002). In addition, BTV actively counteracts the IFN response in cells by pre-transcriptional attenuation of IFN-induced gene expression by NS3(Chauveau et al. 2012); enhancement of BTV-specific translation by NS1, possibly countering the IFN-induced translation inhibition in infected cells (Boyce et al. 2012); countering IFN response via NS4 (Ratinier et al. 2011), possibly also by NS5 (Stewart et al. 2015); and the 2′-O methylase function of VP4(CaP), contributing to viral fitness in infected cells (Stewart and Roy 2015).

The mechanism by which the foetal-maternal barrier is breached during BTV infection is not understood, although migrating infected macrophages have been implicated in the process (Osburn 1994). The susceptibility of the ruminant foetus to BTV infection-associated neuropathology is related to the age of the developing brain, as well as the immune status of the foetus at the time of infection. Neural and glial progenitor cells in the cerebral subependymal region are susceptible to BTV-mediated cytolysis during early gestation, but become progressively less susceptible later on (MacLachlan et al. 2009). On the other hand, competency in producing NAbs by the developing immune system (gestatory day 95 in lambs and 175 in calves) can curtail the spread of the virus to the foetus (Osburn 1994).

1.4 Surveillance and Laboratory Diagnosis of Bluetongue

Isolation and identification of BTV, and rapid and accurate typing of circulating viruses, are essential for the rapid confirmation of a clinical diagnosis, and this allows the implementation of control measures to limit the spread of disease.

1.4.1 Surveillance

The focus of BTV surveillance in endemic zones is the detection and identification of the circulating serotypes, in order to develop and target appropriate control measures (such as vaccination). In seasonally free temperate regions, the focus is on the initial detection of BTV infection and circulation and identification of BTV serotype, in order to select an appropriate vaccine and implement control programmes. In countries or zones that are free of BT (e.g. Iceland, Patagonia, Hawaii and New Zealand), periodic surveillance is needed to maintain the 'disease-free' status for trade purposes (Melville 2004; Caporale and Giovannini 2010; EFSA Panel on Animal Health and Welfare 2011; OIE 2015b).

Surveillance for BT can be achieved by various methods. Clinical surveillance is useful at the flock level and has frequently detected the start of an outbreak in disease-free (e.g. BTV-8 outbreaks in the Netherlands in 2006 and in France in 2015) and seasonally free areas. Clinical surveillance can also provide information on losses (especially in endemic areas), the extent of the infected area and the expansion, contraction or persistence of an outbreak. However, clinical surveillance requires confirmation by laboratory diagnosis. Laboratory-based surveillance can be accomplished using serum-based assays to detect BTV-specific antibodies. Serological surveillance can be used to identify the boundary of the infected zone, to identify the circulating serotypes, to monitor vaccination coverage and herd immunity, to monitor the spread of exotic serotypes after a recent incursion, to establish or retain disease-free status and for regulation of exports.

An important aspect of surveillance is the ability to detect and predict outbreaks using sentinel animals, usually cattle. Seronegative animals located in strategic places can be periodically monitored to detect virus activity and identify the circulating serotype(s). Since infected ruminants remain seropositive for long periods, possibly for life, serosurveys can be used to analyse the history of BTV circulation in a region (seroprevalence). Sentinel monitoring can be used to determine the extent of the infected zone, identify the circulating or newly invading serotypes in endemic areas and determine the seasonality of transmission in seasonally free zones. It can also be used to alert authorities in areas that are infection-free, but considered 'at risk', or in areas that have previously experienced only a limited range of serotypes. If used to monitor a herd, sentinel surveillance can pinpoint the precise date of infection and therefore the dates of virus circulation.

In endemic areas, virological surveillance either by virus isolation or by detection of viral RNA in blood or tissue samples (e.g. by RT-PCR) is more important than serological surveillance, since most of the animals will be seropositive. All of the currently available BTV vaccines will induce antibodies to all of the BTV proteins, and there is no accompanying assay to *d*ifferentiate *i*nfected from *v*accinated *a*nimals (DIVA), invalidating BTV serogroup-specific serological surveillance for the same BTV serotype in vaccinated animals. Virological surveillance is useful to confirm clinical cases of BT, in order to declare outbreaks, to confirm serological data and to identify and characterize circulating serotypes and topotypes.

Virus isolation from wild caught *Culicoides* can also be used for the identification of circulating BTV serotypes and to support sentinel studies. However, the number of infected insects can represent only a very small percentage of the entire insect population, making detection and isolation of the virus difficult. Studies on vector abundance can be useful for developing models of disease spread in both endemic and seasonally free areas (Eagles et al. 2012; Sedda et al. 2012; Burgin et al. 2013; Eagles et al. 2013; Kelso and Milne 2014).

The suitability of various tests for surveillance and diagnosis of BT under various circumstances is shown in Table 1.5 (OIE 2015a).

1.4.2 Detection of Anti-BTV Antibodies

Previous circulation of BTV can be effectively identified by the detection of anti-BTV antibodies in animal sera. Identification of group-specific antibodies is sufficient to confirm virus circulation, while serum neutralization assays can be used to identify serotypes. However, sequential infections with different serotypes can elicit a cross-serotype immunity, making it difficult to identify individual serotypes in endemic regions where multiple strains/serotypes co-circulate. Under these circumstances, serotype-specific molecular assays can positively identify RNA from multiple different serotypes in a single blood sample or virus isolate (Maan et al. 2012a; Reddy et al. 2016).

Group-specific 'anti-BTV' antibodies can be detected using assays such as agar gel immunodiffusion (AGID), competitive enzyme-linked immunosorbent assay (cELISA), indirect ELISA, immunoperoxidase or immunofluorescence assays. These assays primarily target the highly immunodominant and serotype-cross-reactive protein VP7(T13). One potential problem with these tests is reactivity with other orbiviruses, such as epizootic haemorrhagic disease virus (EHDV), although such cross-reactivities are usually very low to be of major consequence. VP7(T13)-specific monoclonal antibodies can be used in cELISA, which is more sensitive and specific than AGID, and can differentiate even closely related orbiviruses (Afshar et al. 1992). Several other variations of ELISA have also been reported.

1.4.3 Virus Isolation and Identification

Isolation of virus from diagnostic samples is regarded as the 'gold standard' for confirmation of BTV circulation. The preferred samples for isolation are uncoagulated blood from live animals, or spleen, lymph node or bone marrow from carcasses. Erythrocytes can be washed to remove any circulating antibodies and used as inoculum for virus isolation. Traditionally, BTV isolation has been achieved by inoculation of embryonated chicken eggs (ECE), either via the yolk sac or more effectively via the slightly more skill-demanding intravenous route (Clavijo et al. 2000a). Cells of insect origin (e.g. C6/36 or KC cells) can also be used for the initial isolation of BTV (Sawyer and Osburn 1988; Wechsler and McHolland 1988; Wechsler et al. 1989). These cells, especially KC cells (derived from *C. sonorensis*), are often more sensitive than mammalian cells or ECE for the isolation of field strains (Mertens et al. 1996). However, some serotypes/strains (e.g. BTV-25 and BTV-26) may not infect or replicate in some insect cell lines or in some cases in most mammalian cell lines (Maan et al. 2011; Planzer et al. 2011). Inoculation of

Table 1.5 Suitability of tests for BT diagnosis and surveillance

Test	Serogroup/ serotype-specific	Test	Use for confirmation of clinical cases	Use for surveillance and to determine prevalence	Use to determine population freedom from infection and confirm eradication	Use to confirm an animal's freedom from infection (for trade)	Use to confirm immune status postvaccination	Use to identify topotype (origin) and genetic relationships (may be coupled with sequence analyses)	Remarks
Detection of virus	Serogroup-specific	1. Real-time RT-PCR	+++	++	+++	+++	–	+++	Nucleotide diversity among serotype/ topotypes may need to be evaluated continuously
		2. RT-PCR	+++	++	+	+++	–	++	
		3. Isothermal amplification techniques[a]	++	+	++	++	–	++	
		4. Virus isolation and confirmation[b]	+++	++	+	+++	–	+++	Time-consuming, laborious, less sensitive, but definitive
	Serotype-specific	RT-PCR	+++	++	+	+	+	+++	Nucleotide diversity among serotype/ topotypes may need to be evaluated continuously
		Serum neutralization (SN)	+++	++	+	–	–	–	Virus isolation is a prerequisite. Mixed infections need special attention. Useful for surveillance of circulating serotypes. Definitive test for novel serotypes.
		Sequencing of genome segment 2	+++	+++	–	–	–	+++	Expensive, suitable for confirmation of RT-PCR and neutralization negative viruses (new serotypes, variants and mixed infections)

(continued)

Table 1.5 (continued)

Test	Serogroup/ serotype-specific	Test	Use for confirmation of clinical cases	Use for surveillance and to determine prevalence	Use to determine population freedom from infection and confirm eradication	Use to confirm an animal's freedom from infection (for trade)	Use to confirm immune status postvaccination	Use to identify topotype (origin) and genetic relationships (may be coupled with sequence analyses)	Remarks
Detection of antibodies	Serogroup-specific	cELISA[c]	+	++	++	+++[d]	++	–	Population freedom from infection (surveillance)
		AGID[c]	±	+	+	–	+	–	Population freedom from infection (surveillance)
		CFT	+	+	+	–	+	–	No longer in use
	Serotype-specific	Virus neutralization	++	++	++	+++[d]	++	–	Provides definitive identification of serotype, including novel serotypes

Adapted from OIE (2015a)
[a]Under development; not recommended by OIE
[b]Virus confirmation by RNA gel electrophoresis, RT-PCR, ELISA, immunofluorescence/immunoperoxidase, Western blot
[c]Surveillance, disease-free status, control and eradication programmes, postvaccination immune status
[d]Quarantines in vector-free conditions for 60 days

sheep (intravenous, intradermal or subcutaneous) and suckling mice (intracerebral) has also been employed, especially when the virus titres are low (Afshar 1994; OIE 2015a). Initial passages in ECE, insect cells or animals are often followed by passage in mammalian cells, usually BHK-21, BSR, Vero or cattle pulmonary artery endothelial (CPAE) cell lines. While adaptation to BHK-21 cells may require prior passage in ECE or KC cells (Sawyer and Osburn 1988; Mertens et al. 1996), CPAE can be directly inoculated (Wechsler and Luedke 1991).

Immunological or molecular assays or sequence analyses are normally used to confirm the identity of BTV isolates and to differentiate them from other orbiviruses (see below). One simple method is the observation of characteristic, size-dependent migration pattern of BTV dsRNA segments (electropherotype) during agarose gel electrophoresis (AGE). This methodology can even distinguish mosquito-borne, tick-borne or *Culicoides*-borne orbiviruses (Belaganahalli et al. 2015). In contrast, the migration of genome segments during polyacrylamide gel electrophoresis (PAGE) is determined by both their size and primary nucleotide sequence, thus allowing differentiation of even closely related BTV strains (Squire et al. 1983; Samal et al. 1987). Virus isolation is increasingly being replaced by either nucleic acid-based or immunological techniques, because it is laborious and time-consuming. However, virus isolation is still important for the genetic and pathological characterization of the circulating strain(s).

Besides the immunological assays described above, the identity of BTV isolates can also be confirmed by conventional or real-time RT-PCR to detect RNA of one of the group-specific segments. Several real-time RT-PCR assays (based on Seg-1, Seg-5, Seg-7, Seg-9 or Seg-10) targeting all the 24 'classical' serotypes have been developed and show high levels of sensitivity (Wade-Evans et al. 1990; Katz et al. 1993; Jimenez-Clavero et al. 2006; Orru et al. 2006; Anthony et al. 2007; Shaw et al. 2007; Wilson et al. 2009a; Yin et al. 2010). The recently developed Seg-9-based TaqMan RT-PCR can detect all the 27 known serotypes (Maan et al. 2015a). Assays targeting Seg-5 and Seg-10 have also been used to differentiate between co-circulating strains of BTV (de Santis et al. 2004; Monaco et al. 2006). In addition, rapid pen-side assays based on isothermal amplification methods have also been developed (Mohandas et al. 2015; Ambagala et al. 2015). However, the existence of significant variability among virus isolates belonging to the different topotypes suggests that the use of more than one assay targeting different genome segments or more than one topotype in parallel or as multiplexed assays could provide some safeguards against false-negative results.

1.4.4 Serotyping

1.4.4.1 Neutralization Tests

The gold standard for serotyping BTV is the neutralization test, where reference viruses can be used to determine the presence of specific antibodies in test sera (serum neutralization test), or reference antisera can be used to identify unknown virus isolates (virus neutralization test). Micro-neutralization assays employing indicator systems such as ECE or mammalian cells (BHK-21, Vero, etc.) are the most commonly used (Hamblin 2004; OIE 2015a). Plaque reduction assays can be used to detect lower levels of NAbs and may be more suited to explore the low levels of cross-neutralization between serotypes. However, neutralization tests are laborious, time-consuming and expensive and may fail to identify different viruses in isolates containing multiple serotypes, or the specificity of antibodies if the animal has previously been infected or vaccinated with multiple serotypes (as in endemic areas), leading to a cross-serotype reaction (Jeggo et al. 1986). The assays also require reference strains of all known BTV serotypes and reference monospecific antisera, which are not always readily available. It is also essential to maintain strict quality control and biosafety measures, as these assays are based on live viruses.

An alternative to neutralization tests is inhibition of agglutination of sheep or chicken erythrocytes by BTV (Hubschle 1980; van der Walt 1980). Haemagglutination inhibition (HI) can be serotype-specific, but the assay requires purification of the virus to remove non-specific inhibitors. In addition, this assay also cannot identify mixed serotype infections.

Although neutralization-based typing cannot distinguish different strains or topotypes of the same serotype, it can provide some indication of the epidemiology of BTV in a specific region, particularly, identifying the introduction and emergence of previously exotic serotypes.

1.4.4.2 Nucleic Acid-Based Methods

Sequence comparisons of Seg-2 and Seg-6 from isolates of the 27 recognized BTV serotypes show that variations in Seg-2 correlate well with serotypes, although they can also show differences within each topotype (Maan et al. 2009). Conventional RT-PCR assays targeting Seg-2 can be used to identify serotypes (Johnson et al. 2000; Zientara et al. 2006; Mertens et al. 2007; Maan et al. 2012a; Reddy et al. 2016). These techniques have been and continue to be used to identify incursions by multiple different BTV types into Europe, Brazil, India, Australia and the USA.

The development of geographic region-specific multiplex RT-PCR assays to detect all of the serotypes present in a region may be suitable for initial rapid screening to identify and differentiate circulating serotypes, but may not detect incursions caused by exotic serotypes or topotypes (Johnson et al. 2000; Reddy et al. 2016). Other nucleic acid-based techniques such as oligonucleotide chip or bead-based systems (e.g. Luminex, Magpix) can also be used for the positive and sensitive detection and discrimination of individual BTV serotypes (Frost et al. 2014). The more precise identification of BTV serotypes by RT-PCR and Seg-2-sequence-based techniques have contributed to the recent identification of several novel BTV types (BTV-25 to BTV-27 and two new putative types). However, RT-PCR-based assays may fail to detect variants of existing serotypes that can still be identified by antibody-based assays. The significant sequence variations in Seg-2 between different topotypes (20–30 %) of the same serotype (Maan et al. 2009) make it necessary to update the primer and probe sequences for typing, in order to ensure that they can detect the virus types circulating in a region. This is essential and inevitable, as the virus continues to evolve in different parts of the world and new strains are discovered or emerge. It is also essential that RT-PCR assays are performed with proper care and controls to avoid false negatives/positives due to cross-contamination.

1.5 Genome Analysis and Molecular Epidemiology

1.5.1 Genome Sequencing and Analyses

Sequencing of individual segments or the entire genome of novel BTV isolates, directly from clinical samples (e.g. blood), can provide a basis for molecular epidemiological studies. Minor variations in RNA sequence can be detected in multiple genome segments, even within isolates from a single BT outbreak, and it is possible to track the spatial and temporal spread of individual viruses from farm to farm in a manner that is impossible using serological assays. By phylogenetic comparisons to existing sequence databases, it is possible to identify newly emerging variants, novel genotypes and reassortants and to explore the movement and spread of individual segments in the virus population within a region (Rao et al. 2012; Nomikou et al. 2015b; Maan et al. 2015c). In addition, Seg-2 sequencing can be used to unambiguously 'type' isolates where either conventional or real-time RT-PCR have failed (Rao et al. 2013).

Because of the segmented nature of the BTV genome and the absence of conserved regions that would provide reliable binding sites for primers to amplify and sequence the segments of all BTV isolates, novel methods have been designed to permit full-length analyses of individual genome segments, without prior knowledge of the sequence. Exploiting the dsRNA nature of the genome and conserved terminal hexanucleotides, 'single primer amplification technique' (SPAT) (Potgieter et al. 2002) and 'full-length amplification of cDNAs' (FLAC) (Maan et al. 2007) were developed. With the advent of high-throughput parallel sequencing methods, full genomes can now be sequenced with relative ease, and these can be combined with SPAT/FLAC to generate data directly from suspected blood samples (Potgieter et al. 2009; Yang et al. 2012, 2015; Rao et al. 2013).

Full genome sequence analyses often show significant changes in the consensus sequences for one or more genome segment. This suggests that BTV infections by more than one strain or genotype are not uncommon. It also indicates that selection of specific variants from the virus population, during isolation/adaptation to tissue culture, can generate strains that contain only a subset of the original genetic variants in the blood sample (Maan et al. 2008). In cases where multiple serotypes are present in a single clinical sample (e.g. in endemic areas), it may be essential to plaque purify the viruses (e.g. in Vero or BHK-21 cells). However, it is important to recognize that such samples may contain a wide range of different reassortant viruses, containing individual genome segments derived from different parental strains (Nomikou et al. 2015b).

1.6 Experimental Models of Bluetongue

1.6.1 Experimental Infection of Natural Hosts

Experimental infection with BTV has been accomplished by a variety or a combination of different inoculation routes using ruminant- or cell culture-passaged virus (Eschbaumer et al. 2010; Umeshappa et al. 2011). However, experimental infections often result in mild disease, although severe disease can be reproduced under certain conditions (Verwoerd and Erasmus 2004; Dal Pozzo et al. 2009). Intravenous inoculation of BTV-4-infected sheep blood to sheep has been shown to produce fulminant disease and pathology similar to that of natural infection (MacLachlan et al. 2008). Similar observations have been made with cell culture-adapted BTV-8 inoculated intravenously and subcutaneously to calves (Dal Pozzo et al. 2009). A recent study has shown that intradermal and subcutaneous inoculation of BTV-23 can produce early onset of clinical signs, severe pathological lesions, and simulate the natural dissemination of BTV from the skin to secondary target organs more effectively than intravenous inoculation in sheep (Umeshappa et al. 2011). Reproductive dysfunction, transplacental infection and teratogenicity can also be reproduced by experimental inoculation of sheep, goats and cattle dams or via the intrauterine route (MacLachlan et al. 2009; Worwa et al. 2009; van der Sluijs et al. 2011; Belbis et al. 2013; Coetzee et al. 2013).

In wild ruminants, BT disease can be experimentally induced in white-tailed deer, with characteristic haemorrhagic manifestations (Parsonson 1990; MacLachlan et al. 2009; Drolet et al. 2013). In contrast, black-tailed and European red deer remain asymptomatic, although viraemia may be observed and viral RNA can be detected for prolonged periods (Work et al. 1992; Lopez-Olvera et al. 2010). The virus can also infect and replicate in American bison although clinical signs and transmission to other hosts have not been observed (Tessaro and Clavijo 2001). An inefficient infection is also observed following experimental inoculation of camels, and virus can be isolated from the blood of these animals (Batten et al. 2011).

In relation to experimental inoculation of BTV, one needs to bear in mind that passage of wild-type viruses in cell culture (particularly in mammalian cell lines) can lead to selection of a subpopulation of the virus (as demonstrated by sequence analyses) and may lead to a significant reduction in virulence. This effect is less pronounced on isolation in KC cells, where the virus usually remains virulent for at least the first few passages (Caporale et al. 2014).

1.6.2 Laboratory Animal Models of Bluetongue

Suckling mice are highly susceptible to BTV infection, especially when inoculated intracranially. An early study showed that BTV causes necrotizing encephalitis and cavitations, similar to lesions observed in ovine and/or bovine foetuses infected with BTV during early gestation. It was also found that the susceptibility to BTV decreases rapidly with age, as even two-week-old mice are largely refractory to infection (Narayan and Johnson 1972). On the other hand, α/β-interferon receptor-deficient ($IFNAR^{-/-}$) mice, when inoculated through various routes, are highly susceptible to infection with several serotypes of BTV at any age and demonstrate similar BTV tissue tropism and gross lesions similar to that in the ruminant host (Ortego et al. 2014). This animal model recapitulates serotype-dependent virus replication and disease severity of ruminants as well as attenuation following laboratory passaging of the virus (Ortego et al. 2014). The model also reflects the haemorrhagic disease-like pathogenesis and pathology of BT, including heightened inflammatory cytokine and vasoactive mediator responses (Ortego et al. 2014). Finally, the model greatly facilitates preliminary studies on immune responses to BTV vaccines (Ortego et al. 2014). However, the results of experiments using $IFNAR^{-/-}$ mice should be interpreted with caution since BTV is a very potent inducer of type I IFNs, which are important controllers of early BTV replication.

1.7 Vaccination and Control Measures

1.7.1 Immune Responses to BTV

Interferons are important barriers to BTV infection. BTV is a strong inducer of type I IFN in sheep (Foster et al. 1991), cattle (MacLachlan and Thompson 1985) and mice (Jameson et al. 1978). The importance of IFN response *in vivo* is exemplified by the fact that blocking type I IFN or knocking out IFN receptor allows BTV replication, and reproduction of disease in mice (Ortego et al. 2014). Experimental infection of bovine foetus (4-month old), as well as calves, leads to detectable levels of IFN in blood for 3–4 days postinfection (DPI), and declining levels coincide with viraemia. In sheep, IFN is detected in serum at 5–6 DPI, partially overlapping with primary viraemia, but much before secondary viraemia at 10 DPI. Plasmacytoid but not classical DCs of the blood are the main producers of type I IFN in response to BTV infection or killed virus in sheep (Chauveau et al. 2012; Ruscanu et al. 2012).

Neutralizing antibody responses to BTV play a crucial role in controlling the virus. Although NAbs are by and large serotype-specific, targeting and determined by VP2(OC1), some amount of cross-protection to closely related heterologous serotypes can be observed (Cowley and Gorman 1989; Erasmus 1990; Dungu et al. 2004; Schwartz-Cornil et al. 2008; Zulu and Venter 2014). To a lesser extent, NAbs are also elicited against or may be influenced by VP5(OC2) (Cowley and Gorman 1989; Mertens et al. 1989; Roy 1992). Antibodies are also generated against other structural and NS proteins, including group-specific antibodies directed against the immunodominant antigen VP7(T13), but such antibodies do not usually correlate with protection (Huismans and Erasmus 1981; Richards et al. 1988; MacLachlan 2004).

Host cell-mediated immunity (CMI), mediated by both $CD4^+$ and $CD8^+$ T cells, plays an important role in the control and elimination of many viruses. The role of CMI against BTV is not fully understood. Partial protection can be afforded by primed T-cell-enriched lymph fractions (Jeggo et al. 1984), and T cells commonly show cross-reactivity to heterologous serotypes (Takamatsu and Jeggo 1989). In sheep, $CD4^+$ cells increase initially and then decrease, while $CD8^+$ cells decrease initially and increase during later stages of infection (Channappanavar et al. 2012), suggesting their importance during early and late stages of disease, respectively. At least one protein, VP7(T13), has been shown to be a target for both $CD4^+$ and $CD8^+$ T cells, with the involvement of multiple epitopes in each case (Rojas et al. 2011). Virus-specific $CD8^+$ cytotoxic T lymphocytes (CTL) have been demonstrated in sheep (Jeggo et al. 1985), and these may be important *in vivo* (MacLachlan 1994; Schwartz-Cornil et al. 2008). Multiple CTL epitopes have been identified in NS1(TuP) and VP2(OC1): most sheep appear to elicit protective CTL responses to NS1(TuP), whereas only some animals produce such responses to VP2(OC1), VP3(T2), VP5(OC2) and VP7(T13) (Janardhana et al. 1999). However, there is as yet no comprehensive study of the importance of cell-mediated protection across BTV serotypes and between topotypes.

A major cytokine produced in specific response to BTV is IFN-γ. Its expression following BTV infection of sheep is low during the initial stages but increases significantly by 15 DPI. The increased IFN-γ production in the lymph nodes and spleen is associated with enhanced $CD8^+/CD4^+$ ratio. The expression of another cytokine, TNF-α, is high at early stages and remains high till 15 DPI in lymph nodes, whereas in peripheral blood mononuclear cells and spleen, its levels are initially low, but upregulated slowly to peak around 15 DPI. Incidentally, TNF-α activity in the BTV-infected animals correlates positively with $CD8^+$ T-cell frequency (Channappanavar et al. 2012). However, the role of either IFN-γ or TNF-α in protecting sheep or whether differences in levels or kinetics exist between infection with different serotypes are unclear.

Several soluble mediators, such as interleukin-1 (IL-1), IL-6, IL-8, IL-12, chemokine (C-C) motif ligand 2 (CCL2), CCL4, inducible nitric oxide synthase (iNOS) and cyclooxygenase 2 (Cox-2), can be induced by BTV infection of pDCs and macrophages (DeMaula et al. 2001, 2002; Schwartz-Cornil et al. 2008; Drew et al. 2010b). *In vivo,* different cytokines are upregulated at different times following infection (Channappanavar et al. 2012). Although some of these may be involved in coordinating immune responses, they may also be responsible for pathology as described above.

1.7.2 Vaccines Against Bluetongue

In spite of the widespread distribution of BT in sheep-rearing areas, vaccination against BT is not as popular as it is for other diseases. BT is one of the first animal diseases for which vaccines were developed. In fact, vaccine research in BT led to the identification of plurality of serotypes and development of serotype-specific diagnostic assays (Coetzee et al. 2012). Antigenic plurality, lack of epidemiological data, lack of facilities for differential diagnosis with other important diseases (such as *Peste des petits ruminants* and foot-and-mouth disease) and the inaccessibility of vaccines may be some of the reasons for this.

All of the current BTV vaccines are serotype-specific, and NAbs are used as correlates of protection, with VP2 (OC1) being the major target. The other more conserved BTV proteins also contribute to the efficacy of BTV vaccines by acting as targets for cross-reactive and at least partially protective CMI responses (Jeggo et al. 1985, 1986; Janardhana et al. 1999). However, there have been no comprehensive studies examining the intra- and inter-serotype efficacy of BTV vaccines, to explore their effects on the severity of clinical signs and potential to reduce viraemia, cross-topotype.

1.7.2.1 Modified Live Virus (MLV) Vaccines

The first BT vaccine was an attenuated virus generated by serially passaging infected sheep blood in sheep (Coetzee et al. 2012). However, this vaccine failed repeatedly due to the plurality of serotypes. Later, trivalent MLV vaccines were developed by serial passage of the viruses in ECE. Subsequently, polyvalent vaccines for up to 15 serotypes were developed by serial passage of ECE-attenuated MLVs in cell cultures (Coetzee et al. 2012). Attenuation of the virus was verified by assessing animals for clinical score and level of viraemia following 3–5 serial passages of inoculated sheep blood. Traditionally, MLVs are considered safe if the viraemia in the inoculated sheep is $\leq 10^3$ tissue culture infective dose 50 ($TCID_{50}$)/ml, a level at which infection of *Culicoides* is thought to be inefficient, although evidence to the contrary has been reported (Venter et al. 2004; Savini et al. 2008). The use of MLV vaccines may result in transient fever and reduced milk yield, although they have been effective in controlling clinical disease in South Africa, especially when used during the vector-free season. MLV vaccines have also been used for more than five decades in the USA and Israel, and have been briefly used in Italy and parts of Europe (Dungu et al. 2004; Shimshony 2004; Savini et al. 2008; McVey and MacLachlan 2015). However, safety issues include severe clinical signs in naïve European sheep, with a high associated viraemia, leading to transmission to insect vectors and then to unvaccinated hosts (Veronesi et al. 2010), reassortment with field strains, along with abortions, embryonic deaths, teratogenicity and secretion of the virus in semen, leading to resistance to the use of MLV in many countries. There is also concern over their suitability for use to combat different BTV topotypes in other geographic regions.

1.7.2.2 Inactivated Vaccines

Until recently, interest in the development of inactivated vaccines to combat was lacking due to the higher cost of production as compared to MLV vaccines. However, safety, efficacy and expanded markets due to endemicity and frequent incursions into Europe led to the wider acceptance of these vaccines. Monovalent (BTV-1, BTV-4, BTV-8 and BTV-9) and bivalent (BTV-1/BTV-4, BTV-1/BTV-8, BTV-2/BTV-4) vaccines are being manufactured and marketed in Europe, while a pentavalent vaccine (BTV-1, BTV-2, BTV-10, BTV-16 and BTV-23) is being used in India (Savini et al. 2008; Reddy et al. 2010). Inactivated vaccines are safe, with limited adverse reactions that may include a local reaction, fever and a slight reduction in fertility immediately after vaccination (Feyer et al. 2011; Nusinovici et al. 2011). However, the large quantities of antigen required for each serotype, the limited number of serotypes that can be accommodated and variable immunity against different serotypes are drawbacks of inactivated vaccines. Postvaccination detection of viral RNA may also be a cause of concern as it may interfere in epidemiological studies (Steinrigl et al. 2010), although recent studies suggest that this is both low level and limited to a few days (de Leeuw et al. 2015). Although inactivated vaccines may not entirely prevent viraemia post-challenge, they appear to be effective in reducing viraemia, thus reducing infection of adult *Culicoides* and breaking the transmission cycle (Savini et al. 2008; Szmaragd et al. 2010). Vaccination with an inactivated vaccine does not pose a risk of transplacental transmission of the virus and subsequent foetal infection (van der Sluijs et al. 2012). NAbs have been shown to persist for at least 3 years post-vaccination, although yearly vaccination is recommended to control the disease. Inactivated vaccines have been effectively used to control and eradicate BTV from some of the European countries (Gubbins et al. 2012; Baetza 2013).

1.7.2.3 Recombinant Viral Proteins and Virus-Like Particles (VLPs)

Individual or a combination of viral proteins has been evaluated as vaccines for BT. The protein VP2(OC1) released from the virion (Huismans et al. 1987) or expressed through baculovirus (Roy et al. 1990), bacteria (Mohd Jaafar et al. 2014) or plants (Thuenemann et al. 2013) has been shown to elicit NAbs and protect sheep against challenge with homotypic virus. Addition of VP5(OC2) to VP2(OC1) or the use of VLPs containing VP3(T2), VP7(T13), VP5(OC2) and VP2(OC1) (produced in insect cells, through recombinant baculovirus vectors, or in plants) can reduce the dose required for protection (Roy et al. 1990; Mohd Jaafar et al. 2014). Baculovirus-derived VLPs can elicit NAbs against several BTV serotypes in sheep and cattle and protect against homotypic virus challenge (Perez de Diego et al. 2011). However, higher production and purification costs and stability issues are matters of concern for their commercialization. On the other hand, VLPs produced by transient expression in plants (Thuenemann et al. 2013) have the potential to be commercialized. Serotype-independent vaccines based on inner capsid proteins VP3(T2) and VP7(T13) have been proposed as vaccines, but they fail to protect animals against disease or infection when challenged (Stewart et al. 2012). Recently, NS1(TuP) has been shown to generate a cross-serotype-protective immune response (Marin-Lopez et al. 2014).

A major advantage of subunit or VLP vaccines is the possibility of using proteins that are not included in the vaccine as a target for DIVA compliance. MLV or inactivated vaccines elicit antibodies against all of the viral proteins, invalidating serological assays for surveillance purposes. VLP or subunit vaccines based on structural proteins and/or NS1(TuP) could use detection of antibodies to one of the other NS proteins, e.g. NS2(ViP), for DIVA assays (Perez de Diego et al. 2011).

1.7.2.4 Vectored and DNA Vaccines

Different viral vectors expressing different proteins of BTV have been tested for their ability to protect sheep or IFNAR$^{-/-}$ mice challenged with homo- or heterotypic BTV. Vaccinia virus expressing VP2(OC1) and VP5(OC2) alone or in combination could protect sheep, with better protection when both proteins were expressed together (Lobato et al. 1997). Similarly, a canarypox virus co-expressing both the proteins produced high levels of NAbs and protected sheep against homotypic challenge. In contrast, a mixture of capripox viruses individually expressing VP2(OC1), VP7(T13), NS1(TuP) and NS2(ViP) provided only partial protection (Boone et al. 2007), whereas a myxoma virus expressing VP2(OC1) and VP5(OC2) failed to protect sheep against virulent virus challenge, whereas that expressing VP2(OC1) alone could provide partial protection (Top et al. 2012). Herpesvirus vectors [bovine herpesvirus 4, equine herpesvirus 1(EHV-1)] expressing VP2(OC1) alone elicited NAbs, but could not completely protect IFNAR$^{-/-}$ mice from clinical symptoms and viraemia (Ma et al. 2012). However, EHV-1 expressing both VP2(OC1) and VP5(OC2) induced a protective response (Franceschi et al. 2011; Ma et al. 2012). A replication-defective human adenovirus serotype 5 expressing either VP7(T13), VP2(OC1) or NS3 proteins could protect IFNAR$^{-/-}$ mice against lethal challenge, but failed to completely protect sheep against challenge; protection was attributed to T-cell response rather than NAbs (Martin et al. 2015). A single-cycle recombinant vesicular stomatitis virus expressing VP2(OC1) alone could protect sheep from developing clinical disease and viraemia when challenged with a virulent virus (Kochinger et al. 2014). A DNA prime and a modified vaccinia virus Ankara (MVA) boost, both expressing VP2(OC1), VP5(OC2) and VP7(T13), protected IFNAR$^{-/-}$ mice against lethal challenge, while the same regimen with DNA/MVA expressing VP2(OC1) and VP5(OC2) failed to protect them completely (Calvo-Pinilla et al. 2009; Jabbar et al. 2013). Similarly, it has been reported that DNA vaccine prime followed by a recombinant fowlpox virus expressing VP2(OC1) and VP5(OC2) induced a high titre of NAbs in sheep (Li et al. 2015). Available data for vectored vaccines suggest that both VP2 and VP5 are required to elicit higher levels of NAbs, which is similar to the response observed with protein immunization.

1.7.2.5 Single-Cycle Infectious Virus Vaccines

With the advent of reverse genetics system, it is now possible to generate infectious BTV from plasmids. Disabled infectious single-cycle (DISC) viruses, in which a large portion of Seg-8 is deleted, and the virus rescued by providing VP6 (Hel) in a complementing cell line (Matsuo et al. 2011), can undergo productive infection only in complementing cell lines, but replication is aborted after a single cycle in non-complementing lines. Immunization with DISC viruses can protect sheep from clinical disease as well as viraemia following challenge with a virulent homotypic BTV. No viraemia has been observed after vaccination with DISC virus, substantially reducing the chances of reassortment and reversion to virulence. However, large quantity of virus (5×10^7plaque forming units per animal) was used for vaccination while *in vitro* yield was low (1.3×10^4 to 1.7×10^6 tissue culture infective dose-50/mL for different serotypes) (Matsuo et al. 2011; Celma et al. 2013), necessitating significant improvement in yield in order for this approach to be cost-effective.

One of the advantages of DISC vaccines is their amenability for DIVA. NS3 (Seg-10) is not essential for BTV replication in mammalian cells, but is essential for replication in insect cells. This principle has been utilized to develop a disabled infectious single animal (DISA) virus vaccine (van Rijn et al. 2013; Feenstra et al. 2014a, 2015).

This vaccine is reported to be safer than MLV, as it cannot be transmitted from vaccinated to unvaccinated animals by *Culicoides* and hence could potentially replace MLVs in the future. The accompanying NS3 ELISA as well as RT-PCR can be applied for DIVA (Tacken et al. 2015), paving the way for relaxation of restrictions on animal movement during outbreaks. However, other concerns of MLVs such as duration of viraemia (using a PCR targeting other than Seg-10 and/or virus isolation), genome segment reassortment in the vaccinated animal upon superinfection with a field strain, the effect of the vaccine strain on pregnancy, transplacental transmission and other non-vector-mediated transmissions of BTV need to be ascertained for DISA vaccines to replace the inactivated vaccines.

One of the major issues in the development of BT vaccines is that laboratory-adapted virus strains may cause inapparent infection in experimental animals, posing problems in evaluating protection against challenge. In addition, not all breeds of sheep are equally susceptible and not all serotypes/strains of viruses cause clinical symptoms in the same sheep. Furthermore, although it is recommended that challenge virus be prepared by passaging only in ruminant animals and with no or limited ECE or cell culture passages, elevated body temperature may be the only sign that is observed reliably. Prevention of viraemia after challenge may therefore be considered as the gold standard, rather than depending on a more unpredictable clinical score (Martinelle et al. 2011). In this regard, IFNAR$^{-/-}$ mice may be more reliable both for the identification of avirulent and virulent strains of BTV and for vaccine efficacy testing (Caporale et al. 2011). In view of identification of different topotypes from different geographical areas, it is important to plan the experiments using challenge virus sourced from different geographical regions (topotypes) and species of animals.

1.7.3 Control Measures

Control measures for BT include surveillance, identification of affected animals, vector control to reduce the transmission, control of animal movement to reduce the spread to new areas and vaccination of susceptible animals to reduce transmission and economical losses. In the absence of a fully cross-reactive/cross-serotype vaccine, BTV eradication may be difficult to achieve in source areas where multiple serotypes of BTV co-circulate and an abundant vector population is available throughout the year. Vaccination strategies for endemic areas may therefore be more focused on reducing the economic losses and could be directed primarily at sheep or susceptible species of deer. It is also essential to use multivalent vaccines to control the disease in endemic areas, and selection of relevant serotypes as well as inclusion of relevant topotypes may both be important.

Countries such as South Africa, Israel and the USA, where good surveillance system for identification of circulating serotypes is available, have included the majority of serotypes circulating in their country in the vaccine (Shimshony 2004; Coetzee et al. 2012; McVey and MacLachlan 2015). Although the use of MLV containing more than five serotypes is feasible, it would be difficult to prepare and use multivalent inactivated vaccines in endemic countries. In India, which is an endemic source area, a pentavalent inactivated vaccine has been recently introduced for use in sheep on a voluntary basis. Inclusion of all the circulating strains in an inactivated vaccine was not considered economically feasible. It will be interesting to monitor the effects of this vaccine since at least six more serotypes (BTV-4, BTV-5, BTV-9, BTV-12, BTV-21, BTV-24) have been reported to be circulating in India during the last decade (Rao et al. 2015a, 2016; Krishnajyothi et al. 2016; unpublished data). It is uncertain if CMI against cross-reactive structural and NS proteins will be sufficient to provide a broadly protective response against clinical disease, even if they do not prevent infection by other serotypes. It will therefore be important to have adequate surveillance systems to monitor circulating serotypes and herd immunity against different serotypes to evaluate if the strategy used has been effective and to decide which strains should be included in future vaccination campaigns.

In contrast, control and eradication of the disease in 'sink' areas, where only a limited number of BTV serotypes circulate in a season, could be achieved using vaccination along with other control measures, but would need to target both sheep and other susceptible species, including cattle. Indeed, BTV-8 was eradicated successfully from most of the northern European countries in this way (Baetza 2013).

Another potential, but difficult, control measure is vector control, which can be at least partially achieved by housing susceptible animals during dawn and dusk when midges bite (Calvete et al. 2010). Alternate approaches involve elimination of *Culicoides* breeding sites by draining and drying wet soils rich in organic matter. The control of adult midges can be carried out by using approved insecticides (Carpenter et al. 2013). A more advanced vector control approach could involve the replacement of vector population with those having lower or zero vector competence, generated by genetic manipulation of the vector, although long-term ecological consequences of this approach will need to be carefully evaluated.

Conclusions

Bluetongue is one of the most important viral diseases of livestock, causing massive economic losses mainly due to its effect on trade. Although the disease is relatively easy to control in seasonally and completely free areas, highly concerted efforts are required to control it in endemic areas. The situation is complicated by the existence of multiple serotypes, topotypes and genotypes of BTV, involvement of asymptomatic hosts and transmission by insect vectors. Because different geographical regions present unique situations in terms of viruses, hosts, vectors and climatic conditions, it is essential that continuous monitoring of the circulation of BTV be carried out, not only to identify the different strains of BTV circulating in specific regions but also to develop relevant vaccines and diagnostics. In view of the threat of the virus entering new areas, epidemiology and surveillance in endemic areas, which act as the source of BTV to neighbouring sink areas, is of major importance.

Acknowledgements We are thankful to innumerable students, postdocs, technicians, other laboratory personnel, collaborators, field veterinarians and para-veterinarians, farmers, personnel and officials of the departments of animal husbandry and others who have directly or indirectly helped past and ongoing studies in the laboratories of all the authors. This chapter was a concerted effort of the Indo-UK collaborative consortium to study important arboviruses of livestock in India and is funded by the Governments of India (Department of Biotechnology; BT/IN/Indo-UK/FADH/46/SM/2013) and the UK (Biotechnology and Biological Sciences Research Council; BB/L004690/1). We apologize to those whose original work could not be cited due to space constraints.

References

Afshar A (1994) Bluetongue: laboratory diagnosis. Comp Immunol Microbiol Infect Dis 17(3–4):221–242

Afshar A, Eaton BT, Wright PF, Pearson JE, Anderson J, Jeggo M, Trotter HC (1992) Competitive ELISA for serodiagnosis of bluetongue: evaluation of group-specific monoclonal antibodies and expressed VP7 antigen. J Vet Diagn Invest 4(3):231–237

Alexander KA, MacLachlan NJ, Kat PW, House C, O'Brien SJ, Lerche NW, Sawyer M, Frank LG, Holekamp K, Smale L et al (1994) Evidence of natural bluetongue virus infection among African carnivores. Am J Trop Med Hyg 51(5):568–576

Ambagala A, Pahari S, Fisher M, Lee PA, Pasick J, Ostlund EN, Johnson DJ, Lung O (2015) A rapid field-deployable reverse transcription-insulated isothermal polymerase chain reaction assay for sensitive and specific detection of bluetongue virus. Transbound Emerg Dis. doi:10.1111/tbed.12388

Anthony S, Jones H, Darpel KE, Elliott H, Maan S, Samuel A, Mellor PS, Mertens PP (2007) A duplex RT-PCR assay for detection of genome segment 7 (VP7 gene) from 24 BTV serotypes. J Virol Methods 141(2):188–197

Backx A, Heutink R, van Rooij E, van Rijn P (2009) Transplacental and oral transmission of wild-type bluetongue virus serotype 8 in cattle after experimental infection. Vet Microbiol 138(3–4):235–243

Baetza HJ (2013) Eradication of bluetongue disease in Germany by vaccination. Vet Immunol Immunopathol 158(1–2):116–119

Balasuriya UB, Nadler SA, Wilson WC, Pritchard LI, Smythe AB, Savini G, Monaco F, De Santis P, Zhang N, Tabachnick WJ, Maclachlan NJ (2008) The NS3 proteins of global strains of bluetongue virus evolve into regional topotypes through negative (purifying) selection. Vet Microbiol 126(1–3):91–100

Barratt-Boyes SM, MacLachlan NJ (1994) Dynamics of viral spread in bluetongue virus infected calves. Vet Microbiol 40(3–4):361–371

Batten CA, Harif B, Henstock MR, Ghizlane S, Edwards L, Loutfi C, Oura CA, El Harrak M (2011) Experimental infection of camels with bluetongue virus. Res Vet Sci 90(3):533–535

Batten C, Darpel K, Henstock M, Fay P, Veronesi E, Gubbins S, Graves S, Frost L, Oura C (2014) Evidence for transmission of bluetongue virus serotype 26 through direct contact. PLoS One 9(5):e96049

Belaganahalli MN, Maan S, Maan NS, Brownlie J, Tesh R, Attoui H, Mertens PP (2015) Genetic characterization of the tick-borne orbiviruses. Viruses 7(5):2185–2209

Belbis G, Breard E, Cordonnier N, Moulin V, Desprat A, Sailleau C, Viarouge C, Doceul V, Zientara S, Millemann Y (2013) Evidence of transplacental transmission of bluetongue virus serotype 8 in goats. Vet Microbiol 166(3–4):394–404

Belhouchet M, Mohd Jaafar F, Firth AE, Grimes JM, Mertens PP, Attoui H (2011) Detection of a fourth orbivirus non-structural protein. PLoS One 6(10):e25697

Bernard KA, Israel BA, Wilson WC, Schultz KT (1997) Sequence and cognitive analyses of two virulence-associated markers of bluetongue virus serotype 17. Intervirology 40(4):226–231

Boone JD, Balasuriya UB, Karaca K, Audonnet JC, Yao J, He L, Nordgren R, Monaco F, Savini G, Gardner IA, Maclachlan NJ (2007) Recombinant canarypox virus vaccine co-expressing genes encoding the VP2 and VP5 outer capsid proteins of bluetongue virus induces high level protection in sheep. Vaccine 25(4):672–678

Bouwknegt C, van Rijn PA, Schipper JJ, Holzel D, Boonstra J, Nijhof AM, van Rooij EM, Jongejan F (2010) Potential role of ticks as vectors of bluetongue virus. Exp Appl Acarol 52(2):183–192

Boyce M, McCrae MA (2015) Rapid mapping of functional cis-acting RNA elements by recovery of virus from a degenerate RNA population: application to genome segment 10 of bluetongue virus. J Gen Virol 96(10):3072–3082

Boyce M, Celma CC, Roy P (2012) Bluetongue virus non-structural protein 1 is a positive regulator of viral protein synthesis. Virol J 9:178

Boyle DB, Bulach DM, Amos-Ritchie R, Adams MM, Walker PJ, Weir R (2012) Genomic sequences of Australian bluetongue virus prototype serotypes reveal global relationships and possible routes of entry into Australia. J Virol 86(12):6724–6731

Breard E, Sailleau C, Nomikou K, Hamblin C, Mertens PP, Mellor PS, El Harrak M, Zientara S (2007) Molecular epidemiology of bluetongue virus serotype 4 isolated in the Mediterranean Basin between 1979 and 2004. Virus Res 125(2):191–197

Brenner J, Oura C, Asis I, Maan S, Elad D, Maan N, Friedgut O, Nomikou K, Rotenberg D, Bumbarov V, Mertens P, Yadin H, Batten C (2010) Multiple serotypes of bluetongue virus in sheep and cattle, Israel. Emerg Infect Dis 16(12):2003–2004

Brugger K, Rubel F (2013) Characterizing the species composition of European Culicoides vectors by means of the Koppen-Geiger climate classification. Parasit Vectors 6(1):333

Burgin LE, Gloster J, Sanders C, Mellor PS, Gubbins S, Carpenter S (2013) Investigating incursions of bluetongue virus using a model of long-distance Culicoides biting midge dispersal. Transbound Emerg Dis 60(3):263–272

Calvete C, Estrada R, Miranda MA, Del Rio R, Borras D, Beldron FJ, Martinez A, Calvo AJ, Lucientes J (2010) Protection of livestock against bluetongue virus vector Culicoides imicola using insecticide-treated netting in open areas. Med Vet Entomol 24(2):169–175

Calvo-Pinilla E, Rodriguez-Calvo T, Sevilla N, Ortego J (2009) Heterologous prime boost vaccination with DNA and recombinant modified vaccinia virus Ankara protects IFNAR(−/−) mice against lethal bluetongue infection. Vaccine 28(2):437–445

Caporale V, Giovannini A (2010) Bluetongue control strategy, including recourse to vaccine: a critical review. Rev Sci Tech 29(3):573–591

Caporale M, Wash R, Pini A, Savini G, Franchi P, Golder M, Patterson-Kane J, Mertens P, Di Gialleonardo L, Armillotta G, Lelli R, Kellam P, Palmarini M (2011) Determinants of bluetongue virus virulence in murine models of disease. J Virol 85(21):11479–11489

Caporale M, Di Gialleonorado L, Janowicz A, Wilkie G, Shaw A, Savini G, Van Rijn PA, Mertens P, Di Ventura M, Palmarini M (2014) Virus and host factors affecting the clinical outcome of bluetongue virus infection. J Virol 88(18):10399–10411

Carpenter S, Groschup MH, Garros C, Felippe-Bauer ML, Purse BV (2013) Culicoides biting midges, arboviruses and public health in Europe. Antiviral Res 100(1):102–113

Carpi G, Holmes EC, Kitchen A (2010) The evolutionary dynamics of bluetongue virus. J Mol Evol 70(6):583–592

Carr MA, de Mattos CC, de Mattos CA, Osburn BI (1994) Association of bluetongue virus gene segment 5 with neuroinvasiveness. J Virol 68(2):1255–1257

Celma CC, Roy P (2009) A viral nonstructural protein regulates bluetongue virus trafficking and release. J Virol 83(13):6806–6816

Celma CC, Boyce M, van Rijn PA, Eschbaumer M, Wernike K, Hoffmann B, Beer M, Haegeman A, De Clercq K, Roy P (2013) Rapid generation of replication-deficient monovalent and multivalent vaccines for bluetongue virus: protection against virulent virus challenge in cattle and sheep. J Virol 87(17):9856–9864

Channappanavar R, Singh KP, Singh R, Umeshappa CS, Ingale SL, Pandey AB (2012) Enhanced proinflammatory cytokine activity during experimental bluetongue virus-1 infection in Indian native sheep. Vet Immunol Immunopathol 145(1–2):485–492

Chauveau E, Doceul V, Lara E, Adam M, Breard E, Sailleau C, Viarouge C, Desprat A, Meyer G, Schwartz-Cornil I, Ruscanu S, Charley B, Zientara S, Vitour D (2012) Sensing and control of bluetongue virus infection in epithelial cells via RIG-I and MDA5 helicases. J Virol 86(21):11789–11799

Clavijo A, Heckert RA, Dulac GC, Afshar A (2000a) Isolation and identification of bluetongue virus. J Virol Methods 87(1–2):13–23

Clavijo A, Munroe F, Zhou EM, Booth TF, Roblesky K (2000b) Incursion of bluetongue virus into the Okanagan Valley, British Columbia. Can Vet J 41(4):312–314

Coetzee P, Stokstad M, Venter EH, Myrmel M, Van Vuuren M (2012) Bluetongue: a historical and epidemiological perspective with the emphasis on South Africa. Virol J 9:198

Coetzee P, Stokstad M, Myrmel M, Mutowembwa P, Loken T, Venter EH, Van Vuuren M (2013) Transplacental infection in goats experimentally infected with a European strain of bluetongue virus serotype 8. Vet J 197(2):335–341

Coetzee P, van Vuuren M, Venter EH, Stokstad M (2014) A review of experimental infections with bluetongue virus in the mammalian host. Virus Res 182:21–34

Cowley JA, Gorman BM (1989) Cross-neutralization of genetic reassortants of bluetongue virus serotypes 20 and 21. Vet Microbiol 19(1):37–51

Dal Pozzo F, Saegerman C, Thiry E (2009) Bovine infection with bluetongue virus with special emphasis on European serotype 8. Vet J 182(2):142–151

Daniels P, Lunt R, Prichard I (2009) Bluetongue viruses in Australasia and East Asia. In: Mertens PPC, Baylis M, Mellor PS (eds) Bluetongue. Elsevier Academic Press, London, pp 223–234

Darpel KE, Batten CA, Veronesi E, Shaw AE, Anthony S, Bachanek-Bankowska K, Kgosana L, bin-Tarif A, Carpenter S, Muller-Doblies UU, Takamatsu HH, Mellor PS, Mertens PP, Oura CA (2007) Clinical signs and pathology shown by British sheep and cattle infected with bluetongue virus serotype 8 derived from the 2006 outbreak in northern Europe. Vet Rec 161(8):253–261

Darpel KE, Batten CA, Veronesi E, Williamson S, Anderson P, Dennison M, Clifford S, Smith C, Philips L, Bidewell C, Bachanek-Bankowska K, Sanders A, Bin-Tarif A, Wilson AJ, Gubbins S, Mertens PP, Oura CA, Mellor PS (2009) Transplacental transmission of bluetongue virus 8 in cattle, UK. Emerg Infect Dis 15(12):2025–2028

Darpel KE, Langner KF, Nimtz M, Anthony SJ, Brownlie J, Takamatsu HH, Mellor PS, Mertens PP (2011) Saliva proteins of vector Culicoides modify structure and infectivity of bluetongue virus particles. PLoS One 6(3):e17545

Darpel KE, Monaghan P, Simpson J, Anthony SJ, Veronesi E, Brooks HW, Elliott H, Brownlie J, Takamatsu HH, Mellor PS, Mertens PP (2012) Involvement of the skin during bluetongue virus infection and replication in the ruminant host. Vet Res 43:40

De Leeuw I, Garigliany M, Bertels G, Willems T, Desmecht D, De Clercq K (2015) Bluetongue virus RNA detection by real-time RT-PCR in post-vaccination samples from cattle. Transbound Emerg Dis 62(2):157–162

de Santis P, Orru G, Solinas F, Piras V, Savini G, Caporale V (2004) Molecular differentiation of field and vaccine strains of bluetongue virus serotype 2 using the real-time polymerase chain reaction and fluorescence resonance energy transfer hybridisation probes. Vet Ital 40(4):572–576

DeMaula CD, Jutila MA, Wilson DW, MacLachlan NJ (2001) Infection kinetics, prostacyclin release and cytokine-mediated modulation of the mechanism of cell death during bluetongue virus infection of cultured ovine and bovine pulmonary artery and lung microvascular endothelial cells. J Gen Virol 82(Pt 4):787–794

DeMaula CD, Leutenegger CM, Bonneau KR, MacLachlan NJ (2002) The role of endothelial cell-derived inflammatory and vasoactive mediators in the pathogenesis of bluetongue. Virology 296(2):330–337

Desmecht D, Bergh RV, Sartelet A, Leclerc M, Mignot C, Misse F, Sudraud C, Berthemin S, Jolly S, Mousset B (2008) Evidence for transplacental transmission of the current wild-type strain of bluetongue virus serotype 8 in cattle. Vet Rec 163(2):50–52

Diprose JM, Grimes JM, Sutton GC, Burroughs JN, Meyer A, Maan S, Mertens PP, Stuart DI (2002) The core of bluetongue virus binds double-stranded RNA. J Virol 76(18):9533–9536

Drew CP, Gardner IA, Mayo CE, Matsuo E, Roy P, MacLachlan NJ (2010a) Bluetongue virus infection alters the impedance of monolayers of bovine endothelial cells as a result of cell death. Vet Immunol Immunopathol 136(1–2):108–115

Drew CP, Heller MC, Mayo C, Watson JL, Maclachlan NJ (2010b) Bluetongue virus infection activates bovine monocyte-derived macrophages and pulmonary artery endothelial cells. Vet Immunol Immunopathol 136(3–4):292–296

Drolet BS, Reister LM, Rigg TD, Nol P, Podell BK, Mecham JO, VerCauteren KC, van Rijn PA, Wilson WC, Bowen RA (2013) Experimental infection of white-tailed deer (Odocoileus virginianus) with Northern European bluetongue virus serotype 8. Vet Microbiol 166(3–4):347–355

Drolet BS, Reister LM, Lehiy CJ, Van Rijn PA, Bowen RA (2015) Effect of Culicoides sonorensis salivary proteins on clinical disease outcome in experimental bluetongue virus serotype 8 infection of Dorset sheep. Vet Ital 51(4):379–384

Dulac G, Dubuc C, Myers D, Afshar A, Taylor E, Ward D, Sterritt W (1989) British Columbia. Incursion of bluetongue virus type 11 and epizootic hemorrhagic disease of deer type 2 for two consecutive years in the Okanagan Valley. Can Vet J 30(4):351

Dunbar MR, Cunningham MW, Roof JC (1998) Seroprevalence of selected disease agents from free-ranging black bears in Florida. J Wildl Dis 34(3):612–619

Dungu B, Gerdes T, Smit T (2004) The use of vaccination in the control of bluetongue in southern Africa. Vet Ital 40:616–622

Eagles D, Deveson T, Walker PJ, Zalucki MP, Durr P (2012) Evaluation of long-distance dispersal of Culicoides midges into northern Australia using a migration model. Med Vet Entomol 26(3):334–340

Eagles D, Walker PJ, Zalucki MP, Durr PA (2013) Modelling spatio-temporal patterns of long-distance Culicoides dispersal into northern Australia. Prev Vet Med 110(3–4):312–322

Eagles D, Melville L, Weir R, Davis S, Bellis G, Zalucki MP, Walker PJ, Durr PA (2014) Long-distance aerial dispersal modelling of Culicoides biting midges: case studies of incursions into Australia. BMC Vet Res 10:135

EFSA Panel on Animal Health and Welfare (2011) Scientific opinion on bluetongue monitoring and surveillance. EFSA J 9(6):1–61

El Hage J, Lorusso A, Carmine I, Di Gennaro A, Portanti O, Olivieri S, Casaccia C, Pisciella M, Teodori L, Sghaier S, Savini G (2013) Bluetongue virus in Lebanon. Transbound Emerg Dis 60(5):390–394

Elbers AR, Backx A, Meroc E, Gerbier G, Staubach C, Hendrickx G, van der Spek A, Mintiens K (2008) Field observations during the bluetongue serotype 8 epidemic in 2006. I. Detection of first outbreaks and clinical signs in sheep and cattle in Belgium, France and the Netherlands. Prev Vet Med 87(1–2):21–30

Erasmus BJ 1990 Bluetongue virus. In: Dinter Z, Morein B (eds) Virus infections of ruminants. Elsevier Science, New York pp 227–237

Eschbaumer M, Wackerlin R, Rudolf M, Keller M, Konig P, Zemke J, Hoffmann B, Beer M (2010) Infectious blood or culture-grown virus: a comparison of bluetongue virus challenge models. Vet Microbiol 146(1–2):150–154

Falconi C, Lopez-Olvera JR, Gortazar C (2011) BTV infection in wild ruminants, with emphasis on red deer: a review. Vet Microbiol 151(3–4):209–219

FAO (2013) GeoNetwork. http://www.fao.org/geonetwork/srv/en/main.home. Accessed 24th Oct 2015

Feenstra F, Maris-Veldhuis M, Daus FJ, Tacken MG, Moormann RJ, van Gennip RG, van Rijn PA (2014a) VP2-serotyped live-attenuated bluetongue virus without NS3/NS3a expression provides serotype-specific protection and enables DIVA. Vaccine 32(52):7108–7114

Feenstra F, van Gennip RG, van de Water SG, van Rijn PA (2014b) RNA elements in open reading frames of the bluetongue virus genome are essential for virus replication. PLoS One 9(3):e92377

Feenstra F, Drolet BS, Boonstra J, van Rijn PA (2015) Non-structural protein NS3/NS3a is required for propagation of bluetongue virus in Culicoides sonorensis. Parasit Vectors 8:476
Feyer D, Kemper D, Reist M, Kaufmann T (2011) Bluetongue disease: impact of the 2008 vaccination on fertility in supervised dairy herds. Schweiz Arch Tierheilkd 153(6):257–262
Foster NM, Luedke AJ, Parsonson IM, Walton TE (1991) Temporal relationships of viremia, interferon activity, and antibody responses of sheep infected with several bluetongue virus strains. Am J Vet Res 52(2):192–196
Franceschi V, Capocefalo A, Calvo-Pinilla E, Redaelli M, Mucignat-Caretta C, Mertens P, Ortego J, Donofrio G (2011) Immunization of knock-out alpha/beta interferon receptor mice against lethal bluetongue infection with a BoHV-4-based vector expressing BTV-8 VP2 antigen. Vaccine 29(16):3074–3082
Frost M, Gu X, Van der Sag M, Kirkland P (2014) Development of multiplex serotype specific microsphere assays for the detection and typing of bluetongue viruses. IV international conference on bluetongue and other orbiviruses, Rome, 5–7 Nov 2014, p 158
Gibbs EP, Greiner EC (1994) The epidemiology of bluetongue. Comp Immunol Microbiol Infect Dis 17(3–4):207–220
Gold S, Monaghan P, Mertens P, Jackson T (2010) A clathrin independent macropinocytosis-like entry mechanism used by bluetongue virus-1 during infection of BHK cells. PLoS One 5(6):e11e360
Gollapalli SR, Mallavarapu S, Uma M, Rao PP, Susmitha B, Prasad PU, Chaitanya P, Prasad G, Hegde NR, Reddy YN (2012) Sequences of genes encoding type-specific and group-specific antigens of an Indian isolate of bluetongue virus serotype 10 (BTV-10) and implications for their origin. Transbound Emerg Dis 59(2):165–172
Gomez-Trejedor C (2004) Brief overview of bluetongue situation in Mediterranean Europe. Vet Ital 40(3):57
Gould AR, Eaton BT (1990) The amino acid sequence of the outer coat protein VP2 of neutralizing monoclonal antibody-resistant, virulent and attenuated bluetongue viruses. Virus Res 17(3):161–172
Groocock CM, Campbell CH (1982) Isolation of an exotic serotype of bluetongue virus from imported cattle in quarantine. Can J Comp Med 46(2):160–164
Groocock CM, Parsonson IM, Campbell CH (1982) Bluetongue virus serotypes 20 and 17 infections in sheep: comparison of clinical and serological responses. Vet Microbiol 7(3):189–196
Gubbins S, Hartemink NA, Wilson AJ, Moulin V, Vonk Noordegraaf CA, van der Sluijs MT, de Smit AJ, Sumner T, Klinkenberg D (2012) Scaling from challenge experiments to the field: Quantifying the impact of vaccination on the transmission of bluetongue virus serotype 8. Prev Vet Med 105(4):297–308
Guichard S, Guis H, Tran A, Garros C, Balenghien T, Kriticos DJ (2014) Worldwide niche and future potential distribution of Culicoides imicola, a major vector of bluetongue and African horse sickness viruses. PLoS One 9(11):e112491
Hamblin C (2004) Bluetongue virus antigen and antibody detection, and the application of laboratory diagnostic techniques. Vet Ital 40(4):538–545
Hassan SS, Roy P (1999) Expression and functional characterization of bluetongue virus VP2 protein: role in cell entry. J Virol 73(12):9832–9842
Hassan SH, Wirblich C, Forzan M, Roy P (2001) Expression and functional characterization of bluetongue virus VP5 protein: role in cellular permeabilization. J Virol 75(18):8356–8367
Hemati B, Contreras V, Urien C, Bonneau M, Takamatsu HH, Mertens PP, Breard E, Sailleau C, Zientara S, Schwartz-Cornil I (2009) Bluetongue virus targets conventional dendritic cells in skin lymph. J Virol 83(17):8789–8799
Henrich M, Reinacher M, Hamann HP (2007) Lethal bluetongue virus infection in an alpaca. Vet Rec 161(22):764
Hewat EA, Booth TF, Roy P (1994) Structure of correctly self-assembled bluetongue virus-like particles. J Struct Biol 112(3):183–191
Hofmann MA, Renzullo S, Mader M, Chaignat V, Worwa G, Thuer B (2008) Genetic characterization of toggenburg orbivirus, a new bluetongue virus, from goats, Switzerland. Emerg Infect Dis 14(12):1855–1861

Howerth EW, Greene CE, Prestwood AK (1988) Experimentally induced bluetongue virus infection in white-tailed deer: coagulation, clinical pathologic, and gross pathologic changes. Am J Vet Res 49(11):1906–1913

Howerth EW, Dorminy M, Dreesen DW, Spires EA, Stallknecht DE (1995) Low prevalence of antibodies to bluetongue and epizootic hemorrhagic disease viruses in dogs from southern Georgia. J Vet Diagn Invest 7(3):393–394

Hubschle OJ (1980) Bluetongue virus hemagglutination and its inhibition by specific sera. Arch Virol 64(2):133–140

Huismans H, Erasmus BJ (1981) Identification of the serotype-specific and group-specific antigens of bluetongue virus. Onderstepoort J Vet Res 48(2):51

Huismans H, van der Walt NT, Cloete M, Erasmus BJ (1987) Isolation of a capsid protein of bluetongue virus that induces a protective immune response in sheep. Virology 157(1):172–179

Huismans H, van Staden V, Fick WC, van Niekerk M, Meiring TL (2004) A comparison of different orbivirus proteins that could affect virulence and pathogenesis. Vet Ital 40(4):417–425

Jabbar TK, Calvo-Pinilla E, Mateos F, Gubbins S, Bin-Tarif A, Bachanek-Bankowska K, Alpar O, Ortego J, Takamatsu HH, Mertens PP, Castillo-Olivares J (2013) Protection of IFNAR (−/−) mice against bluetongue virus serotype 8, by heterologous (DNA/rMVA) and homologous (rMVA/rMVA) vaccination, expressing outer-capsid protein VP2. PLoS One 8(4):e60574

Jameson P, Schoenherr CK, Grossberg SE (1978) Bluetongue virus, an exceptionally potent interferon inducer in mice. Infect Immun 20(1):321–323

Janardhana V, Andrew ME, Lobato ZI, Coupar BE (1999) The ovine cytotoxic T lymphocyte responses to bluetongue virus. Res Vet Sci 67(3):213–221

Janowicz A, Caporale M, Shaw A, Gulletta S, Di Gialleonardo L, Ratinier M, Palmarini M (2015) Multiple genome segments determine virulence of bluetongue virus serotype 8. J Virol 89(10):5238–5249

Jauniaux TP, De Clercq KE, Cassart DE, Kennedy S, Vandenbussche FE, Vandemeulebroucke EL, Vanbinst TM, Verheyden BI, Goris NE, Coignoul FL (2008) Bluetongue in Eurasian lynx. Emerg Infect Dis 14(9):1496–1498

Jeggo MH, Wardley RC, Brownlie J (1984) A study of the role of cell-mediated immunity in bluetongue virus infection in sheep, using cellular adoptive transfer techniques. Immunology 52(3):403–410

Jeggo MH, Wardley RC, Brownlie J (1985) Importance of ovine cytotoxic T cells in protection against bluetongue virus infection. Prog Clin Biol Res 178:477–487

Jeggo MH, Wardley RC, Brownlie J, Corteyn AH (1986) Serial inoculation of sheep with two bluetongue virus types. Res Vet Sci 40(3):386–392

Jeggo MJ, Corteyn AH, Taylor WP, Davidson WL, Gorman BM (1987) Virulence of bluetongue virus for British sheep. Res Vet Sci 42(1):24–28

Jimenez-Clavero MA, Aguero M, San Miguel E, Mayoral T, Lopez MC, Ruano MJ, Romero E, Monaco F, Polci A, Savini G, Gomez-Tejedor C (2006) High throughput detection of bluetongue virus by a new real-time fluorogenic reverse transcription-polymerase chain reaction: application on clinical samples from current Mediterranean outbreaks. J Vet Diagn Invest 18(1):7–17

Johnson DJ, Wilson WC, Paul PS (2000) Validation of a reverse transcriptase multiplex PCR test for the serotype determination of U.S. isolates of bluetongue virus. Vet Microbiol 76(2):105–115

Johnson D, Mertens P, Maan S, Ostlund E (eds) (2007) Exotic bluetongue viruses identified from ruminants in the Southeastern US from 1999–2006. Proceedings of annual conference of American Association of Veterinary Laboratory Diagnosticians (AAVLD). Reno, Oct 2007

Katz JB, Alstad AD, Gustafson GA, Moser KM (1993) Sensitive identification of bluetongue virus serogroup by a colorimetric dual oligonucleotide sorbent assay of amplified viral nucleic acid. J Clin Microbiol 31(11):3028–3030

Katz J, Alstad D, Gustafson G, Evermann J (1994) Diagnostic analysis of the prolonged bluetongue virus RNA presence found in the blood of naturally infected cattle and experimentally infected sheep. J Vet Diagn Invest 6(2):139–142

Kelso JK, Milne GJ (2014) A spatial simulation model for the dispersal of the bluetongue vector Culicoides brevitarsis in Australia. PLoS One 9(8):e104646

Kirkland PD (2004) Bluetongue viruses, vectors and surveillance in Australia – the current situation and unique features. Vet Ital 40(3):47–50

Kochinger S, Renevey N, Hofmann MA, Zimmer G (2014) Vesicular stomatitis virus replicon expressing the VP2 outer capsid protein of bluetongue virus serotype 8 induces complete protection of sheep against challenge infection. Vet Res 45:64

Koltsov A, Salnikov N, Novikhova M, Kurinnov V, Lunitsyn A, Tsybanov S, Kolbasov D (2014) Investigation of Bluetongue virus persistence in Russia. IV international conference on bluetongue and other orbiviruses, Rome, 5–7 Nov 2014

Krishnajyothi Y, Maan S, Kandimalla K, Maan NS, Tutika RB, Reddy YV, Kumar A, Mrunalini N, Reddy GH, Putty K, Ahmed SM, Reddy YN, Hemadri D, Singh KP, Mertens PP, Hegde NR, Rao PP (2016) Isolation of bluetongue virus 24 from India – an exotic serotype to Australasia. Transbound Emerg Dis 63(4):360–364

Lee F, Ting LJ, Lee MS, Chang WM, Wang FI (2011) Genetic analysis of two Taiwanese bluetongue viruses. Vet Microbiol 148(2–4):140–149

Legisa D, Gonzalez F, De Stefano G, Pereda A, Dus Santos MJ (2013) Phylogenetic analysis of bluetongue virus serotype 4 field isolates from Argentina. J Gen Virol 94(Pt 3):652–662

Legisa DM, Gonzalez FN, Dus Santos MJ (2014) Bluetongue virus in South America, Central America and the Caribbean. Virus Res 182:87–94

Levings RL, Wilbur LA, Evermann JF, Stoll IR, Starling DE, Spillers CA, Gustafson GA, McKeiman AJ, Rhyan JC, Halverson DH, Rosenbusch RF (1996) Abortion and death in pregnant bitches associated with a canine vaccine contaminated with bluetongue virus. Dev Biol Stand 88:219–220

Li J, Yang T, Xu Q, Sun E, Feng Y, Lv S, Zhang Q, Wang H, Wu D (2015) DNA vaccine prime and recombinant FPV vaccine boost: an important candidate immunization strategy to control bluetongue virus type 1. Appl Microbiol Biotechnol 99(20):8643–8652

Lobato ZI, Coupar BE, Gray CP, Lunt R, Andrew ME (1997) Antibody responses and protective immunity to recombinant vaccinia virus-expressed bluetongue virus antigens. Vet Immunol Immunopathol 59(3–4):293–309

Lopez-Olvera JR, Falconi C, Fernandez-Pacheco P, Fernandez-Pinero J, Sanchez MA, Palma A, Herruzo I, Vicente J, Jimenez-Clavero MA, Arias M, Sanchez-Vizcaino JM, Gortazar C (2010) Experimental infection of European red deer (Cervus elaphus) with bluetongue virus serotypes 1 and 8. Vet Microbiol 145(1–2):148–152

Lorca-Oro C, Lopez-Olvera JR, Ruiz-Fons F, Acevedo P, Garcia-Bocanegra I, Oleaga A, Gortazar C, Pujols J (2014) Long-term dynamics of bluetongue virus in wild ruminants: relationship with outbreaks in livestock in Spain, 2006-2011. PLoS One 9(6):e100027

Ma G, Eschbaumer M, Said A, Hoffmann B, Beer M, Osterrieder N (2012) An equine herpesvirus type 1 (EHV-1) expressing VP2 and VP5 of serotype 8 bluetongue virus (BTV-8) induces protection in a murine infection model. PLoS One 7(4):e34425

Maan S, Rao S, Maan NS, Anthony SJ, Attoui H, Samuel AR, Mertens PP (2007) Rapid cDNA synthesis and sequencing techniques for the genetic study of bluetongue and other dsRNA viruses. J Virol Methods 143(2):132–139

Maan S, Maan NS, Ross-smith N, Batten CA, Shaw AE, Anthony SJ, Samuel AR, Darpel KE, Veronesi E, Oura CA, Singh KP, Nomikou K, Potgieter AC, Attoui H, van Rooij E, van Rijn P, De Clercq K, Vandenbussche F, Zientara S, Breard E, Sailleau C, Beer M, Hoffman B, Mellor PS, Mertens PP (2008) Sequence analysis of bluetongue virus serotype 8 from the Netherlands 2006 and comparison to other European strains. Virology 377(2):308–318

Maan S, Maan N, Nomikou K, Anthony S, Ross-Smith N, Singh K, Samuel A, Shaw A, Mertens P (2009) Molecular epidemiology studies of bluetongue virus. In: Mertens PPC, Baylis M, Mellor PS (eds) Bluetongue, vol 197–221. Elsevier Academic Press, London, pp 23–52

Maan S, Maan NS, van Rijn PA, van Gennip RGP, Sanders A (2010) Full genome characterisation of bluetongue virus serotype 6 from the Netherlands 2008 and comparison to other field and vaccine strains. PLoS One 5(4):e10323

Maan S, Maan NS, Nomikou K, Batten C, Antony F, Belaganahalli MN, Samy AM, Reda AA, Al-Rashid SA, El Batel M, Oura CA, Mertens PP (2011) Novel bluetongue virus serotype from Kuwait. Emerg Infect Dis 17(5):886–889
Maan NS, Maan S, Belaganahalli MN, Ostlund EN, Johnson DJ, Nomikou K, Mertens PP (2012a) Identification and differentiation of the twenty six bluetongue virus serotypes by RT-PCR amplification of the serotype-specific genome segment 2. PLoS One 7(2):e32601
Maan NS, Maan S, Guimera M, Nomikou K, Morecroft E, Pullinger G, Belaganahalli MN, Mertens PP (2012b) The genome sequence of a reassortant bluetongue virus serotype 3 from India. J Virol 86(11):6375–6376
Maan NS, Maan S, Guimera M, Pullinger G, Singh KP, Nomikou K, Belaganahalli MN, Mertens PP (2012c) Complete genome sequence of an isolate of bluetongue virus serotype 2, demonstrating circulation of a Western topotype in southern India. J Virol 86(9):5404–5405
Maan NS, Maan S, Belaganahalli M, Pullinger G, Montes AJ, Gasparini MR, Guimera M, Nomikou K, Mertens PP (2015a) A quantitative real-time reverse transcription PCR (qRT-PCR) assay to detect genome segment 9 of all 26 bluetongue virus serotypes. J Virol Methods 213:118–126
Maan S, Maan NS, Belaganahalli MN, Kumar A, Batra K, Rao PP, Hemadri D, Reddy YN, Putty K, Krishnajyothi Y, Reddy GH, Singh KP, Hegde NR, Nomikou K, Sreenivasulu D, Mertens PP (2015b) Genome sequence of bluetongue virus type 2 from India: evidence for reassortment between outer capsid protein genes. Genome Announc 3(2):e00045–e00015
Maan S, Maan NS, Belaganahalli MN, Rao PP, Singh KP, Hemadri D, Putty K, Kumar A, Batra K, Krishnajyothi Y, Chandel BS, Reddy GH, Nomikou K, Reddy YN, Attoui H, Hegde NR, Mertens PP (2015c) Full-Genome sequencing as a basis for molecular epidemiology studies of bluetongue virus in India. PLoS One 10(6):e0131257
MacLachlan NJ (1994) The pathogenesis and immunology of bluetongue virus infection of ruminants. Comp Immunol Microbiol Infect Dis 17(3–4):197–206
MacLachlan NJ (2004) Bluetongue: pathogenesis and duration of viraemia. Vet Ital 40(4):462–467
MacLachlan NJ, Thompson J (1985) Bluetongue virus-induced interferon in cattle. Am J Vet Res 46(6):1238–1241
MacLachlan NJ, Jagels G, Rossitto PV, Moore PF, Heidner HW (1990) The pathogenesis of experimental bluetongue virus infection of calves. Vet Pathol 27(4):223–229
MacLachlan N, Conley A, Kennedy P (2000) Bluetongue and equine viral arteritis viruses as models of virus-induced fetal injury and abortion. Anim Reprod Sci 60:643–651
MacLachlan NJ, Zientara S, Stallknecht DE, Boone JD, Goekjian VH, Sailleau C, Balasuriya UB (2007) Phylogenetic comparison of the S10 genes of recent isolates of bluetongue virus from the United States and French Martinique Island. Virus Res 129(2):236–240
MacLachlan NJ, Crafford JE, Vernau W, Gardner IA, Goddard A, Guthrie AJ, Venter EH (2008) Experimental reproduction of severe bluetongue in sheep. Vet Pathol 45(3):310–315
MacLachlan NJ, Drew CP, Darpel KE, Worwa G (2009) The pathology and pathogenesis of bluetongue. J Comp Pathol 141(1):1–16
Marin-Lopez A, Otero-Romero I, de la Poza F, Menaya-Vargas R, Calvo-Pinilla E, Benavente J, Martinez-Costas JM, Ortego J (2014) VP2, VP7, and NS1 proteins of bluetongue virus targeted in avian reovirus muNS-Mi microspheres elicit a protective immune response in IFNAR(−/−) mice. Antiviral Res 110:42–51
Martin V, Pascual E, Avia M, Pena L, Valcarcel F, Sevilla N (2015) Protective efficacy in sheep of adenovirus-vectored vaccines against bluetongue virus is associated with specific T cell responses. PLoS One 10(11):e0143273
Martinelle L, Dal Pozzo F, Sarradin P, De Leeuw I, De Clercq K, Thys C, Ziant D, Thiry E, Saegerman C (2011) Two alternative inocula to reproduce bluetongue virus serotype 8 disease in calves. Vaccine 29(19):3600–3609
Matsuo E, Celma CC, Boyce M, Viarouge C, Sailleau C, Dubois E, Breard E, Thiery R, Zientara S, Roy P (2011) Generation of replication-defective virus-based vaccines that confer full protection in sheep against virulent bluetongue virus challenge. J Virol 85(19):10213–10221

Mayo CE, Crossley BM, Hietala SK, Gardner IA, Breitmeyer RE, Maclachlan NJ (2010) Colostral transmission of bluetongue virus nucleic acid among newborn dairy calves in California. Transbound Emerg Dis 57(4):277–281

Mayo CE, Mullens BA, Reisen WK, Osborne CJ, Gibbs EP, Gardner IA, MacLachlan NJ (2014) Seasonal and interseasonal dynamics of bluetongue virus infection of dairy cattle and Culicoides sonorensis midges in northern California–implications for virus overwintering in temperate zones. PLoS One 9(9):e106975

McVey DS, MacLachlan NJ (2015) Vaccines for prevention of bluetongue and epizootic hemorrhagic disease in livestock: a North American perspective. Vector Borne Zoonotic Dis 15(6):385–396. doi:10.1089/vbz.2014.1698

Mellor PS, Boorman J (1995) The transmission and geographical spread of African horse sickness and bluetongue viruses. Ann Trop Med Parasitol 89(1):1–15

Mellor PS, Wittmann EJ (2002) Bluetongue virus in the mediterranean basin 1998-2001. Vet J 164(1):20–37

Mellor PS, Carpenter S, White DM, Mellor P, Baylis M, Mertens P (2009) Bluetongue virus in the insect host. In: Mellor P, Baylis M, Mertens P (eds) Bluetongue. Academic Press, Elsevier, London, United Kingdom, pp 295–320

Melville LF (2004) Bluetongue surveillance methods in an endemic area: Australia. Vet Ital 40(3):184–187

Menzies FD, McCullough SJ, McKeown IM, Forster JL, Jess S, Batten C, Murchie AK, Gloster J, Fallows JG, Pelgrim W, Mellor PS, Oura CA (2008) Evidence for transplacental and contact transmission of bluetongue virus in cattle. Vet Rec 163(7):203–209

Mertens PP, Diprose J (2004) The bluetongue virus core: a nano-scale transcription machine. Virus Res 101(1):29–43

Mertens P, Pedley S, Cowley J, Burroughs J, Corteyn A, Jeggo M, Jennings D, Gorman B (1989) Analysis of the roles of bluetongue virus outer capsid proteins VP2 and VP5 in determination of virus serotype. Virology 170(2):561–565

Mertens PP, Burroughs JN, Walton A, Wellby MP, Fu H, O'Hara RS, Brookes SM, Mellor PS (1996) Enhanced infectivity of modified bluetongue virus particles for two insect cell lines and for two Culicoides vector species. Virology 217(2):582–593

Mertens PP, Diprose J, Maan S, Singh KP, Attoui H, Samuel AR (2004) Bluetongue virus replication, molecular and structural biology. Vet Ital 40(4):426–437

Mertens PP, Maan NS, Prasad G, Samuel AR, Shaw AE, Potgieter AC, Anthony SJ, Maan S (2007) Design of primers and use of RT-PCR assays for typing European bluetongue virus isolates: differentiation of field and vaccine strains. J Gen Virol 88(Pt 10):2811–2823

Mertens PP, Ross-Smith N, Diprose J, Attoui H (2008) The structure of bluetongue virus. In: Mellor P, Baylis M, Mertens PPC (eds) Bluetongue. Academic Press, Elsevier, London, United Kingdom, pp 101–134

Meyer G, Lacroux C, Leger S, Top S, Goyeau K, Deplanche M, Lemaire M (2009) Lethal bluetongue virus serotype 1 infection in llamas. Emerg Infect Dis 15(4):608–610

Mohandas SS, Muthuchelvan D, Pandey AB, Biswas SK, Chand K, Venkatesan G, Choudhary D, Ramakrishnan MA, Mondal B (2015) Development of reverse transcription loop mediated isothermal amplification assay for rapid detection of bluetongue viruses. J Virol Methods 222:103–105

Mohd Jaafar F, Belhouchet M, Vitour D, Adam M, Breard E, Zientara S, Mertens PP, Attoui H (2014) Immunisation with bacterial expressed VP2 and VP5 of bluetongue virus (BTV) protect alpha/beta interferon-receptor knock-out (IFNAR(−/−)) mice from homologous lethal challenge. Vaccine 32(32):4059–4067

Monaco F, Camma C, Serini S, Savini G (2006) Differentiation between field and vaccine strain of bluetongue virus serotype 16. Vet Microbiol 116(1–3):45–52

Narayan O, Johnson RT (1972) Effects of viral infection on nervous system development. I. Pathogenesis of bluetongue virus infection in mice. Am J Pathol 68(1):1–14

Noad R, Roy P, Mellor P, Baylis M, Mertens P (2009) Bluetongue virus replication and assembly. Elsevier Academic Press, London

Nomikou K, Guimera M, Matos ACD, Bumbarov V, Maan S, Pullinger G, Mertens PPC (2015a) Detection, molecular characterization and vector competence of a novel Bluetongue virus type. 12th International Symposium of Double-Stranded RNA Viruses, Goa, India, 6–10 Oct 2015

Nomikou K, Hughes J, Wash R, Kellam P, Breard E, Zientara S, Palmarini M, Biek R, Mertens P (2015b) Widespread reassortment shapes the evolution and epidemiology of bluetongue virus following European invasion. PLoS Pathog 11(8):e1005056

Nusinovici S, Seegers H, Joly A, Beaudeau F, Fourichon C (2011) A side effect of decreased fertility associated with vaccination against bluetongue virus serotype 8 in Holstein dairy cows. Prev Vet Med 101(1–2):42–50

OIE (2015a) Bluetongue, vol 2. Manual of diagnostic tests and vaccines for terrestrial animals, 7th edn. World Organization for Animal Health, Paris

OIE (2015b) Terrestrial animal health code, 24 edn. World Organization for Animal Health, Paris

Orru G, Ferrando ML, Meloni M, Liciardi M, Savini G, De Santis P (2006) Rapid detection and quantitation of bluetongue virus (BTV) using a molecular beacon fluorescent probe assay. J Virol Methods 137(1):34–42

Ortego J, de la Poza F, Marin-Lopez A (2014) Interferon alpha/beta receptor knockout mice as a model to study bluetongue virus infection. Virus Res 182:35–42

Osburn BI (1994) The impact of bluetongue virus on reproduction. Comp Immunol Microbiol Infect Dis 17(3–4):189–196

Owens RJ, Limn C, Roy P (2004) Role of an arbovirus nonstructural protein in cellular pathogenesis and virus release. J Virol 78(12):6649–6656

Parsonson IM (1990) Pathology and pathogenesis of bluetongue infections. Curr Top Microbiol Immunol 162:119–141

Patel A, Roy P (2014) The molecular biology of bluetongue virus replication. Virus Res 182:5–20

Peel MC, Finlayson BL, McMahon TA (2007) Updated world map of the Köppen-Geiger climate classification. Hydrol Earth Syst Sci Discuss 4(2):1633–1644

Perez de Diego AC, Athmaram TN, Stewart M, Rodriguez-Sanchez B, Sanchez-Vizcaino JM, Noad R, Roy P (2011) Characterization of protection afforded by a bivalent virus-like particle vaccine against bluetongue virus serotypes 1 and 4 in sheep. PLoS One 6(10):e26666

Planzer J, Kaufmann C, Worwa G, Gavier-Widen D, Hofmann MA, Chaignat V, Thur B (2011) In vivo and in vitro propagation and transmission of Toggenburg orbivirus. Res Vet Sci 91(3):e163–e168

Potgieter AC, Steele AD, van Dijk AA (2002) Cloning of complete genome sets of six dsRNA viruses using an improved cloning method for large dsRNA genes. J Gen Virol 83(Pt 9):2215–2223

Potgieter AC, Monaco F, Mangana O, Nomikou K, Yadin H, Savini G (2005) VP2-segment sequence analysis of some isolates of bluetongue virus recovered in the Mediterranean basin during the 1998-2003 outbreak. J Vet Med B Infect Dis Vet Public Health 52(9):372–379

Potgieter AC, Page NA, Liebenberg J, Wright IM, Landt O, van Dijk AA (2009) Improved strategies for sequence-independent amplification and sequencing of viral double-stranded RNA genomes. J Gen Virol 90(Pt 6):1423–1432

Prasad G, Sreenivasulu D, Sigh KP, Metens PPC, Maan S (2009) Bluetongue in the Indian subcontinent. In: Mertens PPC, Baylis M, Mellor PS (eds) Bluetongue. Elsevier Academic Press, London, pp 167–195

Pritchard LI, Daniels PW, Melville LF, Kirkland PD, Johnson SJ, Lunt R, Eaton BT (2004) Genetic diversity of bluetongue viruses in Australasia. Vet Ital 40(4):438–445

Purse BV, Brown HE, Harrup L, Mertens PP, Rogers DJ (2008) Invasion of bluetongue and other orbivirus infections into Europe: the role of biological and climatic processes. Rev Sci Tech 27(2):427–442

Rao PP, Reddy YV, Meena K, Karunasree N, Susmitha B, Uma M, Prasad PUVS, Chaitanya P, Reddy YN, Hegde NR (2012) Genetic characterization of bluetongue virus serotype 9 isolates from India. Virus Genes 44(2):1–9

Rao PP, Reddy YN, Ganesh K, Nair SG, Niranjan V, Hegde NR (2013) Deep sequencing as a method of typing bluetongue virus isolates. J Virol Methods 193(2):314–319
Rao P, Reddy Y, Hegde N (2015a) Isolation and complete genome sequencing of bluetongue virus serotype 12 from India. Transbound Emerg Dis 62(5):e52–e59
Rao PP, Hegde NR, Reddy YN, Krishnajyothi Y, Reddy YV, Susmitha B, Gollapalli SR, Putty K, Reddy GH (2016) Epidemiology of bluetongue in India. Transbound Emerg Dis 63(2):e151–e164. doi:10.1111/tbed.12258
Ratinier M, Caporale M, Golder M, Franzoni G, Allan K, Nunes SF, Armezzani A, Bayoumy A, Rixon F, Shaw A, Palmarini M (2011) Identification and characterization of a novel non-structural protein of bluetongue virus. PLoS Pathog 7(12):e1002477
Reddy YKM, Manohar BM, Pandey AB, Reddy YN, Prasad G, Chauhan RS (2010) Development and evaluation of inactivated pentavalent adjuvanted vaccine for bluetongue. Indian Vet J 87(5):434–436
Reddy YV, Krishnajyothi Y, Susmitha B, Devi BV, Brundavanam Y, Gollapalli SR, Karunasri N, Sonali B, Kavitha K, Patil SR, Sunitha G, Putty K, Reddy GH, Reddy YN, Hegde NR, Rao PP (2016) Molecular typing of bluetongue viruses isolated over a decade in South India. Transbound Emerg Dis 63:e412–e418. doi:10.1111/tbed.12320
Richards RG, MacLachlan NJ, Heidner HW, Fuller FJ (1988) Comparison of virologic and serologic responses of lambs and calves infected with bluetongue virus serotype 10. Vet Microbiol 18(3–4):233–242
Rojas JM, Rodriguez-Calvo T, Pena L, Sevilla N (2011) T cell responses to bluetongue virus are directed against multiple and identical CD4+ and CD8+ T cell epitopes from the VP7 core protein in mouse and sheep. Vaccine 29(40):6848–6857
Rossi S, Pioz M, Beard E, Durand B, Gibert P, Gauthier D, Klein F, Maillard D, Saint-Andrieux C, Saubusse T, Hars J (2014) Bluetongue dynamics in French wildlife: exploring the driving forces. Transbound Emerg Dis 61(6):e12–e24
Roy P (1992) Bluetongue virus proteins. J Gen Virol 73(Pt 12):3051–3064
Roy P (2005) Bluetongue virus proteins and particles and their role in virus entry, assembly, and release. Virus Struct Assem 64:69–114
Roy P, Urakawa T, Van Dijk AA, Erasmus BJ (1990) Recombinant virus vaccine for bluetongue disease in sheep. J Virol 64(5):1998–2003
Ruiz-Fons F, Sanchez-Matamoros A, Gortazar C, Sanchez-Vizcaino JM (2014) The role of wildlife in bluetongue virus maintenance in Europe: lessons learned after the natural infection in Spain. Virus Res 182:50–58
Ruscanu S, Pascale F, Bourge M, Hemati B, Elhmouzi-Younes J, Urien C, Bonneau M, Takamatsu H, Hope J, Mertens P, Meyer G, Stewart M, Roy P, Meurs EF, Dabo S, Zientara S, Breard E, Sailleau C, Chauveau E, Vitour D, Charley B, Schwartz-Cornil I (2012) The double-stranded RNA bluetongue virus induces type I interferon in plasmacytoid dendritic cells via a MYD88-dependent TLR7/8-independent signaling pathway. J Virol 86(10):5817–5828
Saegerman C, Bolkaerts B, Baricalla C, Raes M, Wiggers L, de Leeuw I, Vandenbussche F, Zimmer JY, Haubruge E, Cassart D, De Clercq K, Kirschvink N (2011) The impact of naturally-occurring, trans-placental bluetongue virus serotype-8 infection on reproductive performance in sheep. Vet J 187(1):72–80
Sailleau C, Breard E, Viarouge C, Vitour D, Romey A, Garnier A, Fablet A, Lowenski S, Gorna K, Caignard G, Pagneux C, Zientara S (2015) Re-emergence of bluetongue virus serotype 8 in France, 2015. Transbound Emerg Dis. doi:10.1111/tbed.12453
Samal SK, Livingston CW Jr, McConnell S, Ramig RF (1987) Analysis of mixed infection of sheep with bluetongue virus serotypes 10 and 17: evidence for genetic reassortment in the vertebrate host. J Virol 61(4):1086–1091
Sanchez-Cordon PJ, Pedrera M, Risalde MA, Molina V, Rodriguez-Sanchez B, Nunez A, Sanchez-Vizcaino JM, Gomez-Villamandos JC (2013) Potential role of proinflammatory cytokines in the pathogenetic mechanisms of vascular lesions in goats naturally infected with bluetongue virus serotype 1. Transbound Emerg Dis 60(3):252–262

Savini G, MacLachlan NJ, Sanchez-Vizcaino JM, Zientara S (2008) Vaccines against bluetongue in Europe. Comp Immunol Microbiol Infect Dis 31(2–3):101–120
Sawyer M, Osburn BI (1988) Processing of wild caught Culicoides and isolation of bluetongue virus using a mosquito cell line. J Tissue Cult Methods 11(3):171–173
Schwartz-Cornil I, Mertens PP, Contreras V, Hemati B, Pascale F, Breard E, Mellor PS, MacLachlan NJ, Zientara S (2008) Bluetongue virus: virology, pathogenesis and immunity. Vet Res 39(5):46
Sedda L, Brown HE, Purse BV, Burgin L, Gloster J, Rogers DJ (2012) A new algorithm quantifies the roles of wind and midge flight activity in the bluetongue epizootic in northwest Europe. Proc Biol Sci 279(1737):2354–2362
Sellers RF (1980) Weather, host and vector – their interplay in the spread of insect-borne animal virus diseases. Epidemiol Infect 85(01):65–102
Sellers RF, Maarouf AR (1989) Trajectory analysis and bluetongue virus serotype 2 in Florida 1982. Can J Vet Res 53(1):100–102
Shaw AE, Monaghan P, Alpar HO, Anthony S, Darpel KE, Batten CA, Guercio A, Alimena G, Vitale M, Bankowska K (2007) Development and initial evaluation of a real-time RT-PCR assay to detect bluetongue virus genome segment 1. J Virol Methods 145(2):115–126
Shaw AE, Bruning-Richardson A, Morrison EE, Bond J, Simpson J, Ross-Smith N, Alpar O, Mertens PP, Monaghan P (2013a) Bluetongue virus infection induces aberrant mitosis in mammalian cells. Virol J 10:319
Shaw AE, Ratinier M, Nunes SF, Nomikou K, Caporale M, Golder M, Allan K, Hamers C, Hudelet P, Zientara S (2013b) Reassortment between two serologically unrelated bluetongue virus strains is flexible and can involve any genome segment. J Virol 87(1):543–557
Shimshony A (2004) Bluetongue in Israel–a brief historical overview. Vet Ital 40(3):116–118
Squire KR, Osburn BI, Chuang RY, Doi RH (1983) A survey of electropherotype relationships of bluetongue virus isolates from the western United States. J Gen Virol 64(Pt 10):2103–2115
Stallknecht DE, Howerth EW (2004) Epidemiology of bluetongue and epizootic haemorrhagic disease in wildlife: surveillance methods. Vet Ital 40(3):203–207
Steinrigl A, Revilla-Fernandez S, Eichinger M, Koefer J, Winter P (2010) Bluetongue virus RNA detection by RT-qPCR in blood samples of sheep vaccinated with a commercially available inactivated BTV-8 vaccine. Vaccine 28(34):5573–5581
Stewart ME, Roy P (2015) Structure-based identification of functional residues in the nucleoside-2'-O-methylase domain of Bluetongue virus VP4 capping enzyme. FEBS Open Bio 5:138–146
Stewart M, Dovas CI, Chatzinasiou E, Athmaram TN, Papanastassopoulou M, Papadopoulos O, Roy P (2012) Protective efficacy of bluetongue virus-like and subvirus-like particles in sheep: presence of the serotype-specific VP2, independent of its geographic lineage, is essential for protection. Vaccine 30(12):2131–2139
Stewart M, Hardy A, Barry G, Pinto RM, Caporale M, Melzi E, Hughes J, Taggart A, Janowicz A, Varela M, Ratinier M, Palmarini M (2015) Characterization of a second open reading frame in genome segment 10 of bluetongue virus. J Gen Virol 96(11):3280–3293
Sung PY, Roy P (2014) Sequential packaging of RNA genomic segments during the assembly of bluetongue virus. Nucleic Acids Res 42(22):13824–13838
Susmitha B, Sudheer D, Rao PP, Uma M, Prasad G, Minakshi P, Hegde NR, Reddy YN (2012) Evidence of bluetongue virus serotype 21 (BTV-21) divergence. Virus Genes 44(3):466–469
Szmaragd C, Wilson AJ, Carpenter S, Wood JL, Mellor PS, Gubbins S (2010) The spread of bluetongue virus serotype 8 in Great Britain and its control by vaccination. PLoS One 5(2):e9353
Tabachnick W (2004) Culicoides and the global epidemiology of bluetongue virus infection. Vet Ital 40(3):144–150
Tacken MG, Daus FJ, Feenstra F, van Gennip RG, van Rijn PA (2015) Development of a competitive ELISA for NS3 antibodies as DIVA test accompanying the novel Disabled Infectious Single Animal (DISA) vaccine for Bluetongue. Vaccine 33(42):5539–5545
Takamatsu H, Jeggo MH (1989) Cultivation of bluetongue virus-specific ovine T cells and their cross-reactivity with different serotype viruses. Immunology 66(2):258–263

Takamatsu H, Mellor PS, Mertens PP, Kirkham PA, Burroughs JN, Parkhouse RM (2003) A possible overwintering mechanism for bluetongue virus in the absence of the insect vector. J Gen Virol 84(Pt 1):227–235
Tan BH, Nason E, Staeuber N, Jiang W, Monastryrskaya K, Roy P (2001) RGD tripeptide of bluetongue virus VP7 protein is responsible for core attachment to Culicoides cells. J Virol 75(8):3937–3947
Tessaro SV, Clavijo A (2001) Duration of bluetongue viremia in experimentally infected American bison. J Wildl Dis 37(4):722–729
Thiry E, Saegerman C, Guyot H, Kirten P, Losson B, Rollin F, Bodmer M, Czaplicki G, Toussaint JF, De Clercq K, Dochy JM, Dufey J, Gilleman JL, Messeman K (2006) Bluetongue in northern Europe. Vet Rec 159(10):327
Thuenemann EC, Meyers AE, Verwey J, Rybicki EP, Lomonossoff GP (2013) A method for rapid production of heteromultimeric protein complexes in plants: assembly of protective bluetongue virus-like particles. Plant Biotechnol J 11(7):839–846
Top S, Foucras G, Deplanche M, Rives G, Calvalido J, Comtet L, Bertagnoli S, Meyer G (2012) Myxomavirus as a vector for the immunisation of sheep: protection study against challenge with bluetongue virus. Vaccine 30(9):1609–1616
Tweddle N, Mellor P (2002) Technical review – bluetongue: the virus, hosts and vectors. http://webarchive.nationalarchives.gov.uk/20130822084033/http://www.defra.gov.uk/animalh/diseases/notifiable/disease/bluetongue_technical.PDF. Accessed 23 Jan 2016
Umeshappa CS, Singh KP, Channappanavar R, Sharma K, Nanjundappa RH, Saxena M, Singh R, Sharma AK (2011) A comparison of intradermal and intravenous inoculation of bluetongue virus serotype 23 in sheep for clinico-pathology, and viral and immune responses. Vet Immunol Immunopathol 141(3–4):230–238
van der Sluijs M, Timmermans M, Moulin V, Noordegraaf CV, Vrijenhoek M, Debyser I, de Smit AJ, Moormann R (2011) Transplacental transmission of Bluetongue virus serotype 8 in ewes in early and mid gestation. Vet Microbiol 149(1–2):113–125
van der Sluijs MT, Schroer-Joosten DP, Fid-Fourkour A, Vrijenhoek MP, Debyser I, Gregg DA, Dufe DM, Moulin V, Moormann RJ, de Smit AJ (2012) Effect of vaccination with an inactivated vaccine on transplacental transmission of BTV-8 in mid term pregnant ewes and heifers. Vaccine 30(3):647–655
van der Sluijs MT, de Smit AJ, Moormann RJ (2016) Vector independent transmission of the vector-borne bluetongue virus. Crit Rev Microbiol 42:57–64. doi:10.3109/1040841X.2013.879850
van der Walt NT (1980) A haemagglutination and haemagglutination inhibition test for bluetongue virus. Onderstepoort J Vet Res 47(2):113–117
van Rijn PA, van de Water SG, van Gennip HG (2013) Bluetongue virus with mutated genome segment 10 to differentiate infected from vaccinated animals: a genetic DIVA approach. Vaccine 31(44):5005–5008
Venter GJ, Gerdes GH, Mellor PS, Paweska JT (2004) Transmission potential of South African Culicoides species for live-attenuated bluetongue virus. Vet Ital 40(3):198–202
Veronesi E, Darpel KE, Hamblin C, Carpenter S, Takamatsu HH, Anthony SJ, Elliott H, Mertens PP, Mellor PS (2010) Viraemia and clinical disease in Dorset Poll sheep following vaccination with live attenuated bluetongue virus vaccines serotypes 16 and 4. Vaccine 28(5):1397–1403
Verwoerd DW, Erasmus BJ (2004) Bluetongue. In: Coetzer JA, Tustin RC (eds) Infectious diseases of livestock. Oxford University Press Cape Town, pp 1201–1230
Viarouge C, Lancelot R, Rives G, Breard E, Miller M, Baudrimont X, Doceul V, Vitour D, Zientara S, Sailleau C (2014) Identification of bluetongue virus and epizootic hemorrhagic disease virus serotypes in French Guiana in 2011 and 2012. Vet Microbiol 174(1–2):78–85
Wade-Evans AM, Mertens PP, Bostock CJ (1990) Development of the polymerase chain reaction for the detection of bluetongue virus in tissue samples. J Virol Methods 30(1):15–24
WAHIS (2015) OIE. http://www.oie.int/wahis_2/public/wahid.php/Diseaseinformation/statusdetail/. Accessed 23 Jan 2016

Waldvogel A, Anderson G, Phillips D, Osburn B (1992) Association of virulent and avirulent strains of bluetongue virus serotype 11 with premature births of late-term bovine fetuses. J Comp Pathol 106(4):333–340

Walton TE (2004) The history of bluetongue and a current global overview. Vet Ital 40(3):31–38

Wechsler SJ, Luedke AJ (1991) Detection of bluetongue virus by using bovine endothelial cells and embryonated chicken eggs. J Clin Microbiol 29(1):212–214

Wechsler SJ, McHolland LE (1988) Susceptibilities of 14 cell lines to bluetongue virus infection. J Clin Microbiol 26(11):2324–2327

Wechsler SJ, McHolland LE, Tabachnick WJ (1989) Cell lines from Culicoides variipennis (Diptera: Ceratopogonidae) support replication of bluetongue virus. J Invertebr Pathol 54(3):385–393

White DM, Wilson WC, Blair CD, Beaty BJ (2005) Studies on overwintering of bluetongue viruses in insects. J Gen Virol 86(2):453–462

Wilson AJ, Mellor PS (2009) Bluetongue in Europe: past, present and future. Philos Trans R Soc Lond B Biol Sci 364(1530):2669–2681

Wilson WC, Ma HC, Venter EH, van Djik AA, Seal BS, Mecham JO (2000) Phylogenetic relationships of bluetongue viruses based on gene S7. Virus Res 67(2):141–151

Wilson A, Darpel K, Mellor PS (2008) Where does bluetongue virus sleep in the winter? PLoS Biol 6(8):e210

Wilson WC, Hindson BJ, O'Hearn ES, Hall S, Tellgren-Roth C, Torres C, Naraghi-Arani P, Mecham JO, Lenhoff RJ (2009a) A multiplex real-time reverse transcription polymerase chain reaction assay for detection and differentiation of bluetongue virus and epizootic hemorrhagic disease virus serogroups. J Vet Diagn Invest 21(6):760–770

Wilson WC, Mecham JO, Schmidtmann E, Jimenezsanchez C, Herrero M, Lager I (2009b) Current status of bluetongue virus in the Americas. In: Mellor P, Baylis M, Mertens P (eds) Bluetongue. Academic Press, Elsevier, London, United Kingdom, pp 197–221

Work TM, Jessup DA, Sawyer MM (1992) Experimental bluetongue and epizootic hemorrhagic disease virus infection in California black-tailed deer. J Wildl Dis 28(4):623–628

Worwa G, Hilbe M, Ehrensperger F, Chaignat V, Hofmann MA, Griot C, Maclachlan NJ, Thuer B (2009) Experimental transplacental infection of sheep with bluetongue virus serotype 8. Vet Rec 164(16):499–500

Worwa G, Hilbe M, Chaignat V, Hofmann MA, Griot C, Ehrensperger F, Doherr MG, Thur B (2010) Virological and pathological findings in Bluetongue virus serotype 8 infected sheep. Vet Microbiol 144(3–4):264–273

Wright I, Maartens L, Gers S, Gerdes G, van Dijk A, AC P (2012) Serological and genetic characterization of new serotypes of BTV and EHDV isolated from a single Alpaca in South Africa. Paper presented at the annual conference of the South African association of veterinary technicians, Berg-en-dal, Kruger National Park, 30–31 Oct 2012

Wu X, Chen SY, Iwata H, Compans RW, Roy P (1992) Multiple glycoproteins synthesized by the smallest RNA segment (S10) of bluetongue virus. J Virol 66(12):7104–7112

Xu G, Wilson W, Mecham J, Murphy K, Zhou EM, Tabachnick W (1997) VP7: an attachment protein of bluetongue virus for cellular receptors in Culicoides variipennis. J Gen Virol 78(Pt 7):1617–1623

Yang H, Zhu J, Li H, Xiao L, Wang J, Li N, Zhang N, Kirkland PD (2012) Full genome sequence of bluetongue virus serotype 4 from China. J Virol 86(23):13122–13123

Yang H, Lv M, Sun M, Lin L, Kou M, Gao L, Liao D, Xiong H, He Y, Li H (2015) Complete genome sequence of the first bluetongue virus serotype 7 isolate from China: evidence for entry of African-lineage strains and reassortment between the introduced and native strains. Arch Virol 161:223–227. doi:10.1007/s00705-015-2624-7

Yang H, Xiao L, Wang J, Meng J, Lv M, Liao D, Song J, Gao L, Xiong H, He Y, Niu B, Chuang X, Li H (2016) Phylogenetic characterization genome segment 2 of bluetongue virus strains belonging to serotypes 5, 7 and 24 isolated for the first time in China during 2012 to 2014. Transbound Emerg Dis. doi:10.1111/tbed.12479

Yin HQ, Zhang H, Shi LJ, Yang S, Zhang GP, Wang SQ, Zhang JG (2010) Detection and quantitation of bluetongue virus serotypes by a TaqMan probe-based real-time RT-PCR and differentiation from epizootic hemorrhagic disease virus. J Virol Methods 168(1–2):237–241

Zhang X, Boyce M, Bhattacharya B, Schein S, Roy P, Zhou ZH (2010) Bluetongue virus coat protein VP2 contains sialic acid-binding domains, and VP5 resembles enveloped virus fusion proteins. Proc Natl Acad Sci U S A 107(14):6292–6297

Zientara S, Breard E, Sailleau C (2006) Bluetongue: characterization of virus types by reverse transcription-polymerase chain reaction. Dev Biol (Basel) 126:187–196; discussion 326–327

Zientara S, Sailleau C, Viarouge C, Hoper D, Beer M, Jenckel M, Hoffmann B, Romey A, Bakkali-Kassimi L, Fablet A, Vitour D, Breard E (2014) Novel bluetongue virus in goats, Corsica, France, 2014. Emerg Infect Dis 20(12):2123–2125

Zulu GB, Venter EH (2014) Evaluation of cross-protection of bluetongue virus serotype 4 with other serotypes in sheep. J S Afr Vet Assoc 85(1):1041

Peste des Petits Ruminants 2

Balamurugan Vinayagamurthy

2.1 Introduction

Small ruminants play a vital role in maintaining the sustainable livelihood of rural poor farmers in developing and underdeveloped countries, and they are the anytime money (ATM) of poor landless farmers. PPR stands for "peste des petits ruminants" that literally means "plague of small ruminants" and is associated with high morbidity and mortality in susceptible hosts (sheep and goats) and poses heavy threat to the national economy of the countries where the disease is endemic. PPR is a highly contagious viral disease of domestic and wild small ruminants and is currently emerging to cause infections in camels. Clinically, disease is characterized by pyrexia, necrotic stomatitis, catarrhal inflammation of the ocular and nasal mucosa, enteritis and bronchopneumonia, which leads to either recovery or death of the affected animal. The disease was first described in 1942 in Côte d'Ivoire (Ivory Coast), West Africa (Gargadennec and Lalanne 1942); since then the disease has spread to different regions in sub-Saharan Africa, the Arabian Peninsula, the Middle East, Southwest Asia, Indian subcontinent and other countries (Balamurugan et al. 2014a; Kumar et al. 2014b; Muthuchelvan et al. 2015).

The causative agent, PPR virus (PPRV), is a member of the genus *Morbillivirus* of the family *Paramyxoviridae* (Gibbs et al. 1979). The PPRV is genetically grouped into four lineages (I, II, III and IV) based on the fusion (F) and nucleocapsid (N) gene sequence analyses (Shaila et al. 1996; Dhar et al. 2002; Kerur et al. 2008; Balamurugan et al. 2010b). In the past, lineages I to III circulate in

B. Vinayagamurthy
Indian Council of Agricultural Research-National Institute of Veterinary Epidemiology and Disease Informatics (ICAR-NIVEDI), Post Box No. 6450, Ramagondanahalli, Yelahanka, Bengaluru 560064, Karnataka, India
e-mail: balavirol@gmail.com; b.vinayagamurthy@icar.gov.in

J. Bayry (ed.), *Emerging and Re-emerging Infectious Diseases of Livestock*,
DOI 10.1007/978-3-319-47426-7_2

Africa, while lineage IV is generally found in Asia (Shaila et al. 1996; Dhar et al. 2002). However, a recent appearance of lineage IV, which was associated with a large epizootic in Morocco, undermines the probable risk of PPR introduction to Europe through Spain and other parts of the world (Kwiatek et al. 2011). Spread of disease to a number of new countries in Africa and Asia with involvement of various lineage of PPRV is a cause of global concern (Kwiatek et al. 2011; Balamurugan et al. 2014a) as recently, Asian lineage was introduced in some African countries (Banyard et al. 2010; Kwiatek et al. 2011). This transboundary nature of the disease is one of the main constraints in improving the productivity of small ruminants in enzootic regions. The virus is considered to be a major obstacle to the development of sustainable agriculture across the developing world and has recently been targeted by the World Organisation for Animal Health (WOAH-OIE) and the Food and Agriculture Organisation (FAO) for eradication with the aim of global elimination (Parida et al. 2015a, b). PPR recently become a major international target for improved control, marked by the adoption in 2014 of a resolution by the OIE to establish a control programme with a view to eventually eradication of the disease (OIE 2014).

As of now, a total of 76 countries with approximately 1.7 billion (80 % – of the global population) of sheep and goats in Africa, the Middle East and the Indian subcontinent have confirmed PPR within their borders, and many countries are at risk of the disease being introduced. Considering the importance of small ruminants in food security and socio-economic growth mainly in Africa, Asia and in many other parts of the world, the complete control and eradication of PPR becomes essential. The campaign will make PPR only the second animal disease ever to be eradicated, after rinderpest. Recently, FAO and OIE convened the International Conference in Abidjan, Côte d'Ivoire, and the high-level authorities from PPR-affected countries agreed on a global plan to control and eradicate PPR by 2030. This chapter provides current comprehensive aspects of PPR with special reference to pathogen, pathogenesis and epidemiology, progress in diagnosis and management of disease and prophylaxis measures with future perspectives.

2.2 Peste Des Petits Ruminants Virus

PPR is caused by an enveloped RNA virus known as PPRV, which belongs to the genus *Morbillivirus* of the family *Paramyxoviridae* (subfamily *Paramyxovirinae*) under the order *Mononegavirales* (Gibbs et al. 1979; http://www.ictvonline.org/virusTaxonomy.asp). Other members of the genus are immunologically cross-reactive viruses, which include rinderpest virus (RPV), measles virus (MV), canine distemper virus (CDV), phocine distemper virus (PDV) and dolphin and porpoise morbillivirus (DMV) (Barrett et al. 1993). The virus is a pleomorphic particle with a lipoprotein membrane enveloping a ribonucleoprotein core, which contains RNA genome (Haffar et al. 1999).

2.2.1 Genome

The PPRV genome is most similar at the nucleotide (nt) level to that of RPV. The genome is a single-stranded RNA, approximately 16 kilobases (kb) long (Chard et al. 2008) with negative polarity (Haas et al. 1995), and the genes are arranged in the order of 3′ N–P/C/V–M–F–H–L 5′ separated by inter-genic region (Diallo 1990; Bailey et al. 2005) and follow the "rule of six" (Baron and Barrett 1995). It is divided into six transcriptional units encoding eight proteins, namely, two nonstructural proteins (V, C) and six structural proteins: the surface glycoproteins which include F and haemagglutinin (H) proteins, the matrix protein (M), the nucleocapsid protein (N) and the phosphoprotein (P), which forms the polymerase complex in association with large (L) protein (Crowley et al. 1988; Sidhu et al. 1993; Diallo et al. 1994). Diallo (1990) classified the morbilliviruses into two groups (CDV and PDV formed one group and RPV, MV and PPRV formed the other) based on nt and amino acid (aa) sequences of N protein. The full-length genome sequence showed that the N, V and H protein of PPRV had close similarity with DMV, indicating a close antigenic relationship between the two viruses (Bailey et al. 2005).

The N protein is the most abundant viral protein and is the major component of nucleocapsid core. It is believed to play a major role in virus transcription and replication (Kingsbury 1990). The P gene of the morbilliviruses is responsible for coding three proteins, viz. phosphoprotein (P) and two nonstructural proteins (V and C), by overlapping ORFs (Baron and Barrett 1995) and was found to be the most poorly conserved protein of the morbilliviruses (Mahapatra et al. 2003). The V protein is translated from an mRNA produced from the P gene by addition of one or more non-template G residues, depending on the paramyxovirus, during the transcription process. The absence of V protein enhances the viral replication in MV, while overexpression attenuates the RNA synthesis (Tober et al. 1998) indicating a regulatory role of V protein in the transcription process. V protein is also a strong inhibitor of interferon actions and thereby contributes to the immunosuppression in morbillivirus infections (Nanda and Baron 2006). However, no studies have yet been carried out to establish the role of V protein in pathogenesis. However, V protein of PPRV plays a pivotal role in interfering with host innate immunity by blocking interferon (IFN) signalling through interacting with signal transducer and activator of transcription 1 (STAT1) and STAT2. Identification of amino acid residues in the V protein of PPRV is essential for interference and suppression of STAT-mediated interferon signalling (Ma et al. 2015). These findings give a new sight for the further studies to understand the delicate mechanism of PPRV to escape the IFN signalling. The C protein is a small basic non-phosphorylated protein. High degree of conservation between the different morbillivirus C proteins at the C-terminus was observed with sequence analysis (Mahapatra et al. 2003).

The PPRV matrix (M) protein possesses an intrinsic ability to bind to lipid membranes and plays a crucial role in viral assembly and further budding. Haffar et al. (1999) reported that the M protein is located inside the viral envelope and is the most conserved protein within the group. It plays a pivotal role in the formation and release of paramyxoviruses particles as it serves as a bridge between the external surface viral glycoproteins (H and F) and the nucleocapsid-ribonucleoprotein core and plays a central role in the formation of new virion, which is liberated from the infected cells by budding. The defected or mutated M protein prevents the virus from completing its infectious cycle leading to a persistent non-productive virus infection. Such a phenomenon seems to be involved in some cases of persistent MV infection causing subacute sclerosing panencephalitis (SSPE) in man and old dog encephalitis (ODE) in persistently CDV-infected dog (Carter et al. 1983; Baczko et al. 1986; Cattaneo et al. 1989). No such a similar disease has been reported so far with persistence of PPRV in animals. Small interfering RNAs targeting PPRV M mRNA increase virus-mediated fusogenicity and inhibit viral replication *in vitro* (Liu et al. 2015a).

The surface glycoproteins (H and F proteins) mediate virus attachment and penetration of the host cells (Scheid et al. 1972). Antibodies to these surface proteins probably play a key role in the development of protective immunity against the morbilliviruses (Giraudon and Wild 1985; Norrby et al. 1986). PPRV fusion (F) gene is one of the most conserved genes (Bailey et al. 2005 and Dhar et al. 2006) and codes for the F protein, which is highly immunogenic and involved in inducing protective immunity (Diallo et al. 2007). The paramyxovirus F protein is normally synthesized as precursor F0, which is cleaved in to two subunits, F1 and F2, linked by a disulphide bond. This cleavage is required for the virus to become fusogenic and thus infective. It helps in fusion of the viral envelope with cell membrane, which leads to the delivery of the nucleocapsid into the cell cytoplasm, where viral multiplication takes place. In PPRV, F protein is essential for virus penetration and cell-to-cell spread and plays a vital role in virus-induced cytopathology. Besides being biologically active and involved in cell fusion, PPRV F protein is also responsible for induced haemolysis as shown by Devireddy et al. (1999). The H protein enables the virus to bind to the cell receptors especially signalling lymphocyte activation molecule (SLAM/CD 150), a cell membrane glycoprotein of the immunoglobulin (Ig) superfamily (Tatsuo et al. 2001). H and F mutually involve in the fusion activity of the later protein (Wild et al. 1991; Das et al. 2000). The L protein (RdRp) is the largest virus protein, is least abundant and surprisingly conserved among morbilliviruses (Bailey et al. 2005), migrates with an apparent MW of 200 kDa and is 2183 aa long in MV, RPV and PPRV (Baron and Barrett 1995; McIlhatton et al. 1997; Muthuchelvan et al. 2005; Bailey et al. 2005).

2.2.2 Biology

PPRV is a recent addition to the other common morbillivirus-causing diseases. The major determinants of the host range and tissue tropism of a PPRV are cellular receptors. Interaction of the host and virus is initiated by specific receptor binding, which is mediated by the H protein of PPRV and sialic acid on the host cell membrane (Munir et al. 2013). Studies on these receptors with PPRV are meagre. However, the only established host receptor for this virus is SLAM (Tatsuo et al. 2000). Like other morbilliviruses, the target selective infection and destruction of SLAM-positive cells for PPRV are epithelial cells, activated lymphocytes and macrophages (Rey Nores et al. 1995). The marmoset B cell line (B95a) has been shown to be sensitive to both virulent and vaccine virus of rinderpest virus (RPV) and PPRV (Lund and Barrett 2000; Sreenivasa et al. 2006). siRNA approaches confirmed SLAM could be a putative co-receptor for PPRV. Pawar et al. (2008) showed relationship between the level of SLAM mRNA and replication of PPRV in peripheral blood mononuclear cells (PBMCs) of cattle, buffalo, sheep and goats. RPV and PPRV strains could use SLAMs of non-host species receptors, albeit at reduced efficiencies (Tatsuo et al. 2001). The other putative receptor, namely, ovine Nectin-4 protein, when overexpressed in epithelial cells, permits efficient replication of PPRV. This gene was predominantly expressed in epithelial tissues and encoded by multiple haplotypes in sheep breeds from around the world (Birch et al. 2013). Tumour-associated marker, PVRL4 (nectin-4), was identified as the epithelial receptor for members of the *Morbillivirus* genus, including MV, CDV and PPRV. Delpeut et al. (2014) described the role of PVRL4 in morbillivirus pathogenesis and its promising use in cancer therapies. Chaudhary et al. (2015) observed receptor tyrosine kinase signalling regulates replication of the PPRV. They found that blocking the receptor tyrosine kinase (RTK) signalling in Vero cells by tryphostin AG879 impairs the *in vitro* replication of the PPRV. These data represent the first evidence that the RTK signalling regulates replication of a morbillivirus.

The isolation of PPRV in cell culture (Gilbert and Monier 1962) was possible at least 10 years after the three (MV, CDV and RPV) other morbilliviruses were cultured in the early 1950s. For several decades, the PPRV even after its isolation was considered a variant of RPV that was adapted to goats and sheep and had lost its virulence for cattle (Laurent 1968). Besides clinical resemblance, serological cross-reactivity between morbilliviruses exists, and diagnostic tests based on polyclonal antibody with notable exception of virus neutralization test (VNT) are incapable of distinguishing PPRV and RPV. PPRV infection of PBMCs in suspension did not produce syncytia, and the CPE was characterized by marked cellular changes consisting of rounding, ballooning, clumping and degeneration of cells (Mondal et al. 2001). Moreover, PPRV produced characteristic syncytia between 48 and 72 h postinfection (pi) in B95a cells, and the virus titre increases gradually up to 5 days postinfection (dpi) and then declined (Sreenivasa et al. 2006). Meng et al. (2014) studied ultrastructural features of PPRV infection in Vero cells. Moreover, Khandelwal et al. (2014) showed the first document evidence that SNPs are capable of inhibiting a *Morbillivirus* replication *in vitro*. The leaf extract of the *Argemone maxicana* was used as a reducing agent for biological synthesis of the SNPs from silver nitrate, and silver nanoparticles impair PPRV replication by exerting a blocking effect on viral entry into the target cells. Balamurugan et al. (2008) assessed the potential antiviral effect of the Babul (*Acacia arabica*) plant aqueous extract on PPR virus replication *in vitro* using Vero cells and PPR vaccine virus.

Like other morbilliviruses, PPR virus is fragile and cannot survive for long time outside the host, but survives for long periods in chilled and frozen tissues. Its half-life has been estimated to be 2.2 min at 56 °C and 3.3 h at 37 °C. The virus is inactivated at 50 °C for 60 min. PPRV has been found to be sensitive to ether or similar lipid solvent agents. However, the virus is relatively stable between pH 5.8 and 10.0 (OIE 2013). PPRV H protein agglutinates a variety of mammalian and avian erythrocytes (Wosu 1991), and PPRV-infected cell cultures readily haemadsorb chicken erythrocytes. The transiently expressed PPRV HN protein in mammalian CV-1 cells was found to be biologically active in possessing haemadsorption and neuraminidase activities (Seth and Shaila 2001). Furthermore it was found that the transiently expressed PPRV F protein could bring about both fusion and hemifusion, whereas the RPV F protein could only bring about hemifusion, and fusion required the presence of an attachment HN protein (Seth and Shaila 2001).

2.3 Epidemiology

2.3.1 Geographical Distribution

The first authentic and scientific description of disease was reported during 1942. At that time Gargadennec and Lalanne (1942) reported on an epidemic disease in Ivory Coast (Cote d'Ivoire) of West Africa, which was clinically similar to RP but was affecting only small ruminants, while in-contact cattle remained apparently healthy. The disease was called as "kata", "psuedorinderpest", "pneumoenteritis complex" and "stomatitis-pneumoenteritis syndrome" (Braide 1981) in French-speaking countries (francophone) of West Africa. Based on outbreaks of disease in Senegal during 1871 and in French Guinea in 1927 [quoted by Curasson (1986)], it was believed that PPR might be much more historical than it has been thought of. The disease soon spread to neighbouring African countries, namely, Nigeria, Senegal and Ghana. Till early 1980s, definite outbreaks of the disease were reported from different parts of West Africa (Mornet et al. 1956; Provost et al. 1971; Hamdy et al. 1976; Bonniwell 1980). Until 1984, PPR was regarded as a disease of West African countries; however, its presence in Sudan that year (El Hag and Taylor 1984) marked beginning of spread of PPR. Since the later part of 1980s, the PPR further spread to countries of Central and Eastern Africa and parts of Asia. In India, the first confirmed outbreak of PPR in sheep was reported in Arasur village, Villupuram district of Tamil Nadu, during 1987 (Shaila et al. 1989). The disease was thought to be restricted in Southern India until severe epidemics swept through the rest of India in 1994 and onwards. Since then the disease became enzootic in many northern states of India (Mondal et al. 1995; Nanda et al. 1996), and now disease is enzootic to India. (Balamurugan et al. 2014a). Evidence of PPR infection based on either serological tests or clinical investigations has been reported from several other countries. Spread of disease to a number of new countries in Southern Africa, Central Asia, Southeast Asia, China, Southern Europe and Western Turkey with involvement of various lineages of PPRV is a cause of global animal health concern (Wu et al. 2016) especially recent introduction of Asian lineage in some African countries and presence of PPR in Europe through Western Turkey (Banyard et al. 2010; Kwiatek et al. 2011; Balamurugan et al. 2014a). Recently, many outbreaks of PPR are being reported frequently from a number of new countries in Africa (Tanzania, Democratic Republic of Congo, Angola, Ismailia Province of Egypt, Morocco and Algeria) and Asia (China). PPR is endemic in many countries with expansion of the range in recent years including across China (Wang et al. 2015a), Kazakhstan (Kock et al. 2015), Tunisia (Sghaier et al. 2015), Ismailia Province, Egypt (Soltan and Abd-Eldaim 2014), Tanzania (Chazya et al. 2014), Mauritania (El Arbi et al. 2014) and Eastern Asia (Banyard et al. 2014).

2.3.2 Transmission

PPRV needs close contact between infected and susceptible animals (Braide 1981) to spread because of either the lability of the virus outside the host or low resistance of the virus in the environment. PPRV is transmitted mainly by inhalation of infectious aerosol materials between animals living in close contact. Quantities of virus are excreted in the secretions or the discharges from eyes, nose and mouth, as well as the excretion through faeces of affected animals during the course of infection especially at least 7 days after onset of the disease and are the important source of virus infection (Abegunde and Abu 1977; Singh et al. 2004b; Balamurugan et al. 2006a). These discharges form fine infectious droplets in the air particularly when the affected animals cough and sneeze (Bundza et al. 1988; Taylor 1984a, b). Animals in close contact inhale the infectious droplets and are likely to become infected. The infectious aerosols can also contaminate water, feed troughs and bedding, turning them into additional sources of infection, but this indirect mode seems to be less important since the PPRV is not expected to survive for long time outside the host and is also sensitive to lipid solvent (Lefevre and Diallo 1990).

Trade of small ruminants at markets where animals from different sources are brought into close contact with one another provides increased opportunities for PPRV transmission. Although cattle and pigs have been shown to seroconvert following contact with the sick sheep and goats, development of clinical disease in these animals has not been reported (Dardiri et al. 1976; Gibbs et al. 1979; Nawthane and Tayler 1979; Sen et al. 2014). Transmission of PPRV either directly or indirectly from sheep or goats to cattle provides a mechanism for the virus to survive outside of the environment in the unnatural host (Abraham et al. 2005). The presence of PPRV antibodies in camels, cattle and wild ruminants besides sheep and goats suggests the natural transmission of PPRV infection among these animals under field condition (Abraham et al. 2005; Balamurugan et al. 2012a, 2014b). Nomadic animals will often come into contact with local sheep and goat populations from whom they may contact the virus (Singh et al. 2004a); subsequently, infected migratory animals may transmit the virus to another susceptible local sheep and goats. Therefore, the movement of animals plays an important role in the transmission and maintenance of PPR virus in nature. The incursion, persistence and spread of PPR in Tanzania with epidemiological patterns and predictions have also been reported (Kivaria et al. 2013).

2.3.3 Host Susceptibility

The PPRV primarily affects small ruminants such as sheep and goats, which are the main natural hosts and occasionally some other artiodactyls including camels and small ruminant wildlife. Many authors believe that although PPRV infects sheep and goats, severity of the clinical symptoms are more predominant in goats than sheep (Lefevre 1980; Wosu 1994; Tripathi et al. 1996a; Singh et al. 2004a). The details of host/species susceptibilities to PPRV have been extensively discussed in earlier book chapter (Balamurugan et al. 2015b). The data on the seroprevalence of antibodies to PPRV in small ruminants is available from a number of countries including India from low to high based on the endemicity of the disease (Singh et al. 2004a; Khan et al. 2008a, b; Balamurugan et al. 2011, 2012b, 2015b). Further, PPRV antibodies in small and large ruminants were also observed in field conditions (Abubakar et al. 2015). Recently, serological evidence of camel exposure to PPRV in Nigeria has also been reported (Woma et al. 2015).

In general, the prevalence of PPRV antibodies in young and adult sheep and goats indicates subclinical or in apparent or non-lethal infections, as vaccinations against the disease are limited and irregular in the endemic developing/underdeveloped countries. However, prevalence of anti-PPR antibodies in adult sheep or goats is not always indicative of PPR infection as there is always a high probability of these animals receiving vaccination once during a lifetime. Seroprevalence of PPR in cattle, buffaloes (Anderson and Mckay 1994; Govindarajan et al. 1997; Balamurugan et al. 2012a, 2014b), camels (Roger et al. 2000; Khalafallaa et al. 2010), bharals (*Pseudois nayaur*) (Bao et al. 2011) and other wild ruminants or in zoological collections (Furley et al. 1987) has been used to study the natural transmission of PPRV among these animals under field conditions (Abraham et al. 2005). However, the presence of infectious virus in these cases has not yet been reported except in a few hosts like gazelles (Abu-Elzein et al. 2004), camels (Khalafallaa et al. 2010) and wild bharal (Bao et al. 2011). Further, Balamurugan et al. (2012c) detected nucleic acid of PPRV in the lion's tissue samples that died of trypanosomiasis and provided new insight in the PPR epidemiology (Balamurugan et al. 2015b). There have been several reports of PPR occurring in wild ruminant species, namely, wild ungulate from three families: Gazellinae (*Dorcas gazelle*), Caprinae (Nubian ibex and Laristan sheep) and Hippotraginae (Gemsbok) (Fentahun and Woldie 2012). Initially the role of wild animals in the epizootiology of PPR was realized by some of the leading investigators (Taylor 1984a, b). Outbreaks of PPR in wild animals (Taylor 1984b) or in zoological collections in the Arabian Gulf (Furley et al. 1987) could be of considerable significance for virus perpetuation. The PPRV also caused severe disease with high mortality in dorcas gazelles (*Gazella dorcas*), gazelle and deer (Abu Elzein et al. 1990), Nubian ibex (*Capra ibex nubiana*), Laristan sheep (*Ovis orientalis laristani*) and gemsbok (*Oryx gazella*) and subclinical infection in Nilgai (*Tragelaphinae*). In general, the wildlife, by sharing same grazing field or areas and water bodies with domestic animals, may potentially be part of the epidemiology of the disease. Wild ruminants have been suspected to play a role in spreading of the disease. However, the actual role of the wildlife on the epizootiology of disease remains unclear for the moments and to be investigated as described earlier (Banyard et al. 2010; Balamurugan et al. 2015b).

The occurrence of PPR in alternate animal hosts is still a matter of debate. There have been several reports of PPR subclinical infection occurring in other ruminant species of domestic animals. Cattle and pigs are susceptible to infection with PPRV, but they do not exhibit clinical signs and do not transmit disease to other animals. There is no evidence of carrier state for PPRV. However, they may play a role in the epizootiology of PPR because they are apparently unable to transmit the disease to other animals (Furley et al. 1987). Moreover, the infected animals excrete or secrete the virus in the natural secretions up to 21 days or even more days without showing clinical signs after recovery from infection (Balamurugan et al. 2006a, 2010a). Liu et al. (2014d) investigated the dynamics of virus excretion and antibody production in animals and stated that virus particles could be detected as early as 3 dpi and virus excretion lasted for up to 26 dpi. PPRV was also suspected to have been involved in a respiratory disease, which affected camels in Ethiopia (Roger et al. 2000). There was also a report of a rinderpest-like disease in Indian buffalo, which was caused by PPRV (Govindarajan et al. 1997). The circulation of PPRV in unnatural host(s) may have a positive role or help in the control or restrict the spread of PPR in small ruminants in a particular geographical area (Balamurugan et al. 2012a, 2014b). This assumption may be due to the possibility of adaption and change in virulence of the virus where small and large ruminants are reared together in integrated farming systems (Balamurugan et al. 2014b). It is believed that, in the situation, where large and small ruminants coexist, seroconversion in cattle of cross-reacting PPRV antibodies might have also helped to eradicate dreaded rinderpest disease.

Subsequent upon development of specific diagnostics, the presence of PPR antibodies has now also been reported in cattle and buffaloes in a number of reports in India and in other countries, and also potential of camel to act as a reservoir was suggested (Hinshu et al. 2001; Haque et al. 2004; Abraham et al. 2005; Balamurugan et al. 2012a, 2014b). Cattle, goats and in some cases camels grassing alongside sheep did not show any signs of the disease. Earlier study on subclinical PPR infection in experimental cattle showed that PPRV antibody and antigen could be detected over a period of one year (Sen et al. 2014). All these reports showed that PPR can also be transmitted directly or indirectly from sheep or goats to cattle, providing a mechanism for the virus to survive outside of the environment in the unnatural host. The list of species of animals in which either PPRV antibodies or virus infection was detected by different researchers is described in earlier book chapter (Balamurugan et al. 2015b).

Besides species, the breed of the animal may also have effect on the outcome of PPRV infection and disease epidemiology. Differences in susceptibility to goat plague among different breeds and species exist. Some breeds are more susceptible to disease than others (Lefevre and Diallo 1990). However, much study on breed has not been carried out except one or few *in vitro* studies based on molecular-based approaches on receptors and their expression levels. The host innate immune system discriminates between pathogen-associated molecular patterns and self-antigens through surveillance receptors known as toll-like receptors (TLR). Dhanasekaran et al. (2015) investigated the role of TLR and cytokines in differential susceptibility of goat breeds (Barbari, Tellichery, Kanni and Salem Black) and water buffalo to PPRV. These studies and other host genetic factor analyses might provide further insights on susceptibility to PPRV and genetic polymorphisms in the host. Association between age and severity of the PPR has also been reported by Obi et al. (1983) and Taylor (1984a, b). According to these researchers, young animals aged from six months to one year old are more susceptible than adult animals. Various studies by different investigators, from different parts of the world, showed various percentages of the mortality and morbidity with the involvement of the different strains of PPRV in both sheep and goats populations. Some isolates of the virus are reported to cause as high as 90–100 % morbidity and mortality in affected animals. However, this scenario is likely to change drastically once intensive vaccinations are carried in the sheep and goats population through control programme. PPR is still a poorly recognized disease, particularly with regard to epidemiological features such as transmission dynamics under different production systems. Several PPR outbreaks go unrecorded due to inadequate animal disease reporting and surveillance systems in India and other developing countries.

2.3.4 Molecular Epidemiology

Sequences and phylogenetic analyses based on F and N gene of the different PPRV isolates/strains from different parts of the world have defined the presence of four (I, II, II and IV) different lineages of virus (Shaila et al. 1996; Dhar et al. 2002; Kerur et al. 2008; Balamurugan et al. 2010b). Lineage I is represented by viruses isolated in West Africa in the 1970s (Nigerian isolates and Senegalese strain) and later from other countries, viz. Guinea, Senegal, Cote d'Ivoire and Burkina Faso. Lineage II represented the viruses isolated during the 1980s in West Africa [Ivory Coast (Cote d'Ivoire), Guinea, Ghana] and is the only African lineage that did not cross the Red Sea to the Asian countries. Lineage III comprised isolates from Eastern African countries, viz. Meilig, Sudan (Diallo 1988), Ethiopia (Roeder et al. 1994), Yemen and Oman. The lineage IV (Asian lineage) is represented by isolates of PPRV from the Arabian Peninsula, the Middle East and Asia including India (Shaila et al. 1996). Recently, lineage IV virus was also reported in Turkey (Ozkul et al. 2002) and some other African countries (Morocco, Cameroon, Gabon). Sequence and phylogenetic analyses of structural protein (F, N, M and H) genes of different Indian isolates and vaccine strains of PPRV by different researchers showed that all virus isolates/strains belong to lineage IV (Shaila et al. 1996; Dhar et al. 2002; Kaul 2004; Kerur et al. 2008; Balamurugan et al. 2010b) except one isolate (PPR TN/92) which belongs to lineage III (Shaila et al. 1996). Further, molecular characterization of PPRV isolated from an outbreak in the Indo-Bangladesh border of Tripura state of North-East India suggests the transboundary circulation of virus between India and Bangladesh border, which warrants immediate vaccination across the international border to create an immune belt (Muthuchelvan et al. 2014; Balamurugan et al. 2015a). Recently, co-circulation of PPPV Asian lineage IV with lineage II in Nigeria has also been reported (Woma et al. 2016) with emergence of lineage IV virus in Ethiopia (Muniraju et al. 2014a). Further, molecular evolution, emergence and re-emergence and characterization of PPRV across Asia and Africa from different countries have been reported (Kumar et al. 2014a; Su et al. 2015; Muniraju et al. 2014b; Kgotlele et al. 2014; Padhi and Ma 2014; Dundon et al. 2014; Şevik and Sait 2015).

2.3.5 Seasonal Occurrence

Seasonal differences in PPR outbreaks have also been recorded in different parts of the world. Sheep and goat animal husbandry practices, agroclimatic conditions and geographical locations may have some effect on the seasonal distribution of the disease. Obi et al. (1983) and Wosu (1994) have reviewed seasonal pattern of PPR outbreaks. PPR outbreaks can occur any time during the year, but was encountered most frequently during the wet season (April to September or October.) or during the cold dry season (January and February) (Taylor 1984b). Most authors have linked the outbreak of PPR with introduction of new animals to the flocks (Taylor 1984b; Asmar et al. 1980; Tripathi et al. 1996a, b; Kumar et al. 2002; Singh et al. 2004a). PPR has been found to occur throughout the year but was encountered most frequently during the lean period either in wet season or in rainy season or during the cold dry season (December to February) (Balamurugan et al. 2011, 2012b). This seasonal occurrence of the disease was correlated with the animal movements (migratory pattern) and climate factors. Climatic factors favourable for the survival and spread of the virus may also contribute to the seasonal distribution of PPR outbreaks. Further, the majority of farmers in developing countries like India have small and marginal land holdings. Small ruminants are farmed on free-range pasture land, shrubs and forest. Due to an ongoing decrease in available pastureland and forest area, these animals will often travel long distances during the dry season in search of fodder and water. With the commencement of monsoon, nutrition improves and migration of animals also restricts due to availability of local fodder resulting in substantial decrease in the frequency of outbreaks.

2.4 Pathogenesis

Evidence from the experimental studies on PPRV infection and records based on accurate field observations in sheep, goats and cattle clearly suggested that natural PPRV infection occurs by entry of the virus through the upper respiratory tract epithelium (Taylor et al. 1965). Small ruminants infected with PPRV exhibit lesions typical of epithelial infection and necrosis. The virus multiplication and pathogenicity of the virus is proportional to that of the host resistance or innate resistance, host's immune response, parasitic infection in the host, the nutritional level of host, the breed, sex and age of the animal, etc. as mentioned earlier (Munir et al. 2013). Further, host genetics and non-genetic factors may play a significant role in variation to disease susceptibility, innate resistance and immune response to virus.

The virus is disseminated from initial replication sites by way of lymph and blood (primary viraemia), and the clinical signs emerge synchronously with replication of PPRV in target tissues. Viral tropism of both RPV and PPRV has been observed to the lymphoid and epithelial tissues especially in the mucosa of the digestive tract (Brown et al. 1991; Wohlsein et al. 1993; Brown and Torres 1994). As mucosal phase of disease progresses, the dramatic cytolytic changes occur in the primary lymphoid organs and epithelial tissues. Both thymus-dependent and thymus-independent areas were affected during pathogenesis. Lymphocytolysis is prominent in the periartiolar area of Malpighian bodies and in the germinal centres of lymph nodes. Lymphocyte destruction is widespread in Peyer's patches and the mesenteric lymph nodes (MLN). The PPRV differs from RPV in its significant affinity to the parenchymatous cells in lungs. Brown et al. (1991) found that the intensity of immunostaining increased with the increase in the inflammatory process of pneumonia. The presence of the viral antigen was detected in the cytoplasm of the bronchial and bronchiolar epithelium with highest concentration in the airway epithelium including the trachea, bronchi and bronchioles and also in the desquamated bronchial and bronchiolar epithelial cells, pneumocytes, macrophages and other leucocytes present in the lumen. Syncytial cells in most cases revealed intense viral staining in their cytoplasm and rarely in their nuclei (Saliki et al. 1994; Eligulashvili et al. 1999; Kumar et al. 2004). The PPRV was detected as dark brown granules in the nuclei and cytoplasm of the ileal epithelial cells, which were found to be nuclear and cytoplasmic viral aggregation of fibrillar strands on electron microscope (Bundza et al. 1988). Further, Eligulashvili et al. (1999) reported the presence of viral antigens in the cytoplasm of epithelial intestinal crypts and within macrophages of mucosal lamina propria. Kumar et al. (2004) demonstrated PPRV antigens in experimentally infected goats in various tissues and organs, namely, oral cavity, sebaceous glands, cryptal and villous epithelium of the small and large intestines, Peyer's patches, MLNs, lymphoid follicles and others.

In PPR infection, immune activation coincides with immune suppression, and these two activities continue for several days during convalescence stage of disease. Activation of blood mononuclear cells and tissue macrophages presumably occurs in the later stages of the incubation period in the primary replication sites. Most sensitized lymphocytes and reticular cells may undergo cytolysis, and the surviving populations are subjected to immune activation and immune suppression. The transient leucocytosis during incubation period in PPR was regarded as a response to stress (Olaleye et al. 1989), but it may also represent immune activation. It is reasonable to speculate that there is polyclonal activation of surviving B cells in circulation and in lymphoid organs and tissues during secondary viraemia as reported in measles. The resistance to challenge was obviously not mediated by humoral antibody and may be from interferon, and other cytokines provide protection. In RP and PPR, amelioration of clinical signs and tissue damage synchronizes with the appearance of neutralizing antibodies in the terminal stages of the mucosal phase (Scott

1990). Provost (1970) observed the presence of IgA in nasal secretions only in those recovering from natural or experimental RP and in those vaccinated intranasally. Kinetics of early antibody response and long-term immune response has been extensively studied in PPR-vaccinated animals (Singh et al. 2004b, c; Rajak et al. 2005; Balamurugan et al. 2007), and the finding indicated protective antibody response persists even up to 6 years after vaccination (Saravanan et al. 2010a, b). However, Zahur et al. (2014) reported that PPR vaccine (Nigerian strain 75/1) confers protection for at least 3 years in sheep and goats. The prolonged immunosuppression in RP and also in PPR involves both humoral and cell-mediated immune functions (Yamannouchi et al. 1974; Rajak et al. 2005). However, the actual mechanism of immunosuppression in PPRV infection is largely unknown barring a few reports (Rajak et al. 2005). But the demonstration of ability of PPRV to induce apoptosis in goat PBMCs (*in vitro*) could be an important mechanism (Mondal et al. 2001). Similarly, greater frequency of apoptotic lymphoid cells (laddering pattern of DNA) was demonstrated in ethidium bromide/acridine dye-stained PBMCs isolated from six goats naturally infected with PPRV (Kumar et al. 2002). Jagtap et al. (2012) showed that immunosuppressed goats had a short period of viraemia and more extensive and severe disease advancement with high mortality rate than the non-immunosuppressed goats while studying the effect of immunosuppression during pathogenesis of PPR. Further, genomic analysis of host and PPR vaccine viral transcriptome uncovers transcription factors modulating immune regulatory pathways under PPRV infection with key gene involvement in immune system regulation and spliceosomal and apoptotic pathways, and on comparison with control, it revealed 985 differentially expressed genes (Manjunath et al. 2015). Baron and Baron (2015) studied the early changes in cytokine expression during PPR pathogenesis in goats and stated that consistent reduction in $CD4^+$ T cells was observed at 4 dpi. Measurement of the expression of various cytokines showed elements of a classic inflammatory response but also a relatively early induction of interleukin 10 (Baron et al. 2014a). Truong et al. (2014) stated lymph nodes, lymphoid tissue and digestive tract organs were the predominant sites of virus replication while studying PPRV tissue tropism and pathogenesis in sheep and goats following experimental infection using a quantitative time-course study.

Several research workers reported detection of PPRV antigen and nucleic acid at various periods of infection by different diagnostic assays/tests. Libeau et al. (1994) detected viral antigen in ocular and nasal secretions after 6 dpi. Further, PPRV antigen was detected between 7 and 13 dpi in goats and 5 and 11 dpi in sheep using immunocapture ELISA (Nanda et al. 1996). Infected goats secreted/excreted viral antigen coincided with the rise in the body temperature (6–7 dpi) (Singh et al. 2004b) with maximum concentration level between 9–11 dpi and then showed a decline after 12dpi. Conventional RT-PCR detected PPRV nucleic acid in the experimental samples as early as from 4 to17 dpi. Earlier one-step multiplex RT-PCR (Balamurugan et al. 2006a) had detected the virus nucleic acid in swab materials from 5 to17 dpi. Using sensitive assays, it was possible to detect PPRV nucleic acid in preclinical swabs from an earlier stage and for a longer time during infection, which indicates that incubating animals might play a role in the transmission of

PPRV among susceptible animals. Study on subclinical PPR infection in experimental cattle showed that PPRV antibody and antigen could be detected over a period of one year (Sen et al. 2014).

Post-mortem confirmation of PPRV infection in sheep and goats was routinely carried out using the lung, MLN, spleen, internal tissue, etc., but the inflammation on comparative assessment of the concentration of the virus in different organs was hardly available. However, Balamurugan et al. (2012b) reported, besides the regular ante- (blood and swabs) and post-mortem (spleen and lymph nodes) samples, tissues from the lung, liver and heart also showed high positivity for PPRV antigen, showing that the lung, liver and heart samples from infected animals are also useful for diagnosing PPR.

2.5 Pathology

The pathology of PPR is characterized and dominated by retrogressive and necrotic changes in lymphoid tissues and epithelial cells of gastrointestinal and respiratory systems. Various reports have described gross pathological changes in naturally occurring PPR in goats and sheep (Tripathi et al. 1996a, b; Aruni et al. 1998, Kumar et al. 2002, 2004). At necropsy, the carcass is emaciated and dehydrated, and animals have soiled hindquarters. Pathology of PPR has been reported in different goat and sheep breeds from natural and experimental infections. The severity of the disease varies with species, breed and immune status of the host. Chowdhury et al. (2014) studied the natural PPRV infection in Black Bengal goats and carried out detailed virological, pathological and immunohistochemical investigation. The striking histomorphologic diagnosis of PPR was acute pneumonia, and severe gastroenteritis was observed. The gross pathological changes most commonly seen in different parts or tissues or organs are described below.

The prominent lesions of PPR-infected animals include consolidation and changes in colour of lungs, and sometimes frothy mucus is observed in cut pieces of the lung on squeezing, antero-ventral areas of the right lung are frequently involved and areas of lungs become dark red or purple and firm to touch mainly in the anterior and cardiac lobes (Kumar et al. 2004). Consolidation of lobes of lungs (Fig. 2.1) and occlusion of airway caused by secondary bacterial pneumonia are common. Congested alveolar border was found to be one of the most characteristic clinical and pathological changes in infected goats (Tripathi et al. 1996a, b; Kumar et al. 2002).

The involvement of the respiratory system is remarkable, and pneumonia is a predominant sign in PPR. Most reports found lung involvement in almost more than 90 % of cases dying during outbreak of PPR (Obi et al. 1983; Tripathi et al. 1996a, b; Kumar et al. 2002, Aruni et al. 1998). Trachea may contain mucopurulent or frothy exudates with multifocal bronchiolitis and bronchitis, and tracheal rings may show haemorrhages. Bronchopneumonia is a constant lesion, with possibility of pleuritis and hydrothorax. Lymph nodes associated with the lung (mediastinal) and intestine (mesenteric) are most commonly affected which are generally enlarged,

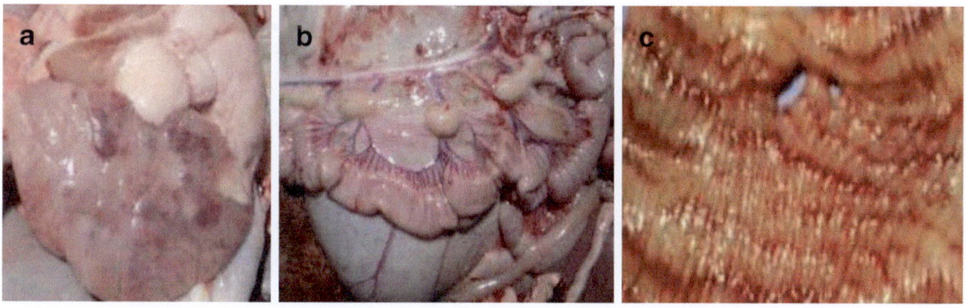

Fig. 2.1 Post-mortem lesions of PPR-affected sheep. (**a**) Congestion and consolidation of lobes of lungs. (**b**) Enlarged, oedematous and congested intestinal mesenteric lymph nodes. (**c**) Colon showing discontinuous streaks of congestion and haemorrhages (Zebra marking) on the mucosal folds

oedematous and congested. The mesenteric lymph nodes were usually oedematous and congested (Fig. 2.1) (Obi et al. 1983; Tripathi et al. 1996a, b; Kumar et al. 2002). The spleen was congested and engorged and at times showed petechiae on the capsular surface. Severe congestion and necrotic lesions in the gastrointestinal tract including the rumen, reticulum and abomasum are normal features. The small intestine (duodenum, jejunum or ileum) may show black colouration of the mucosal lining with haemorrhages and sometimes erosions in the terminal ileum. There may be lesions on the liver and other internal organs also. Necrotic or haemorrhagic enteritis with extensive necrosis and sometimes severe ulceration of Peyer's patches to be seen, which may slough off from the intestinal wall. Congestions around the ileo-caecal valve especially at the caeco-colic junction and in the rectum were observed. In the posterior part of the colon and rectum, discontinuous streaks of congestion (also known as "zebra" stripes or "zebra markings") on the mucosal folds are observed, which are typical of PPR (Fig. 2.1).

The other lesions in PPR-infected animals include oedema and excoriation of the lips; discrete or extensive areas of erosions, ulceration and necrosis on the various parts of the oral cavity; deposits of white bran-like necrotic materials on the palates, tongue and oral mucosa; and erosions on the gums, soft and hard palates, tongue, cheeks and into the oesophagus; lips become swollen and extensive erosions are found on the lips and commissures; extensive areas of erosions/ulcerations are found in the nasal mucosa, and mucosal lining is congested with clear or creamy yellow exudates. Small erosions and petechiae on the nasal mucosa, turbinates, larynx and trachea were also a feature. Further, emaciation, conjunctivitis, and erosive stomatitis involving the inside of the lower lips and the adjacent gum near the commissures and the free portion of the tongue were also characteristics to PPR. Lesions on the hard palate, pharynx and upper third of the oesophagus in severe cases may be observed.

In general, the PPRV causes the appearance of multinucleate giant cells and cytoplasmic and nuclear inclusion bodies in the permissive cell cultures and target tissues. By electron microscopy, viral nucleocapsids have been demonstrated in these inclusions (Tajima and Ushijima 1970; Hamdy et al. 1976; Bundza et al. 1988). However, the significance of the intranuclear nucleocapsids and their temporal relationship to viral assembly in the cytoplasm remain unclear. Moreover, viral antigen in the nucleus occurred in the later stages of cellular infection (Tajima and Ushijima 1970).

Most remarkable pathological changes are found in the lymphoid tissues, which have been described in a number of naturally and experimentally infected sheep and goats (Tripathi et al. 1996a, b; Kumar et al. 2002, 2004). The MLN showed a variable degree of lymphoid cell depletion particularly in the cortical lymphoid follicles. The depleted areas appeared pale, devoid of mature lymphocytes, and showed the presence of large reticular cells. In the medullary region, blood vessels engorgement, oedema and infiltration with large number of polymorphs were observed. In the liver, hepatocytes were swollen having pale and finely granular cytoplasm. Sinusoids were variably dilated and hepatic cords were disrupted. Bile ducts in the portal areas showed epithelial hyperplasia. In kidneys, the epithelial lining cells of proximal convoluted tubules were swollen and at times occluded the lumen. The cytoplasm showed more granularity than usual (Obi et al. 1983; Tripathi et al. 1996a, b; Kumar et al. 2002, 2004). In acute phase, stratified squamous epithelium of the lip showed focal areas of degeneration and necrosis. The cells in the stratum spinosum layer were swollen, pale and variably vacuolated with loosened intercellular junctions. Some cells showed the intranuclear eosinophilic inclusions. Subepithelial tissues were oedematous, congested and revealed infiltration with mononuclear cells. In some cases, sebaceous glands showed degenerative changes and neutrophil infiltration forming micro abscess. Trachea did not show much significant changes except for mucosal desquamation and necrotic debris in the lumen. There was vacuolation and hyperplasia of the tracheal epithelium and presence of eosinophilic intranuclear and intracytoplasmic inclusions. Typical broncho-interstitial pneumonia was often complicated with purulent and serofibrinous exudates (Tripathi et al. 1996a, b; Kumar et al. 2002). Hyperplasia and desquamation of bronchial and bronchiolar epithelium and the presence of inflammatory cells and debris in the lumen are characteristics. Hypertrophy and hyperplasia of alveolar lining cells were the most consistent finding. Inflammatory cells consisted of macrophages, lymphocytes and syncytia of varying sizes; intranuclear and intracytoplasmic inclusions were frequently seen in alveolar macrophages, syncytial cells and bronchiolar epithelial cells. The complication of viral pneumonia with serofibrinous and purulent type was due to *Pasteurella* pyogenic organisms, and fungus has also been reported (Obi et al. 1983; Tripathi et al. 1996a, b). Intestinal lamina propria in the majority of cases revealed mild to moderate infiltration with mononuclear cells including plasma cells and a moderate to severe degree of vascular congestion. The goblet cells were prominent and filled with mucin. Villi were reduced in height and blunt, and Peyer's patches consistently exhibited showed areas of lymphoid depletion, which was similar to those seen in lymph nodes. Occasionally, typical syncytia were seen in Peyer's patches of the ileum. Submucosal oedema of variable degree was also a consistent feature in these cases. The large intestine showed congestion of blood vessels in the mucosa and submucosa and infiltration of mononuclear cells

(Tripathi et al. 1996a, b; Kumar et al. 2002, 2004). Jagtap et al. (2012) the histopathological changes and PPRV antigen distribution in the gastrointestinal tract were more extensive and diffused in immunosuppressed goats infected with PPRV with more load of antigen in the atypical organ(s)/tissues like the liver, kidney and heart.

A consistent haematological event in PPR is leucopenia (lymphocytopenia), due to lymphoid necrosis in the reticuloendothelial organs, which generally synchronizes with the onset of pyrexia in the prodromal phase and progresses considerably and tissues account for the lymphopenia (Rajak et al. 2005). Further, serum analysis of infected goat revealed a non-significant decrease in the total protein and the serum enzymes, whereas levels of total protein and globulin were increased significantly in sheep. Raghvendra et al. (2000) reported no appreciable changes in the levels of enzyme alanine amino transferase (ALT) and electrolytes, sodium (Na^+) and potassium (K^+) between pre- and post inoculation periods. Rajak et al. (2005) observed a marked immunosuppression evidenced by lymphocytic leucopenia and reduced antibody response to specific and nonspecific antigens. Sreenivasa et al. (2000) reported the PPR vaccine also induced a transient lymphopenia without affecting the immune response.

2.6 Clinical Manifestations

Clinically, the disease is manifested by high fever (pyrexia), oculonasal discharges, necrotizing and erosive stomatitis, gastroenteritis, diarrhoea and bronchopneumonia, followed by either death of the animal or recovery from the disease. Clinical picture of PPR could be characterized by "3Ds", i.e. discharge, diarrhoea and death, with additional fourth component, the *bronchopneumonia.* Generally, the incubation period is 4–6 days, but may range from 3 to 10 days. The incubation period ranged from 2 to 6 days (Nanda et al. 1996; Tripathi et al. 1996a, b; Kumar et al. 2002, 2004). The disease exhibits different stages, viz. (i) incubation period (short) 5–7 days, (ii) prodromal phase (febrile reaction), (iii) mucosal phase (pyrexia, ocular and nasal discharges, hyperaemia of conjunctiva and mucosa of anterior nares and erosions on the tongue, palate, lips and other parts of the oral mucosa), (iv) diarrhoeal stage (diarrhoea, pneumonia, dehydration, death) and, (v) in nonfatal cases, "recovery stage" in which sheep and goats that recover from PPR develop an active lifelong immunity. Several studies have described clinical signs and subsequent complications in the spontaneous occurrence of PPR in goats and sheep (Tripathi et al. 1996a, b). PPR may manifest in different forms like per acute, acute and mild depending on the severity of the disease (OIE, 2013). Peracute form is seen in young goats with an incubation period of 2 days followed by hyperthermia (41–42 °C), prostration, piloerection, anorexia, oral and ocular discharges and constipation and later leads to profuse diarrhoea. Death occurs within 5–6 days of infection. The acute form is characterized by fever followed by oculonasal discharges (watery to mucopurulent (Fig. 2.2) with eyelids sticking together) and partial blockage of the nose due to

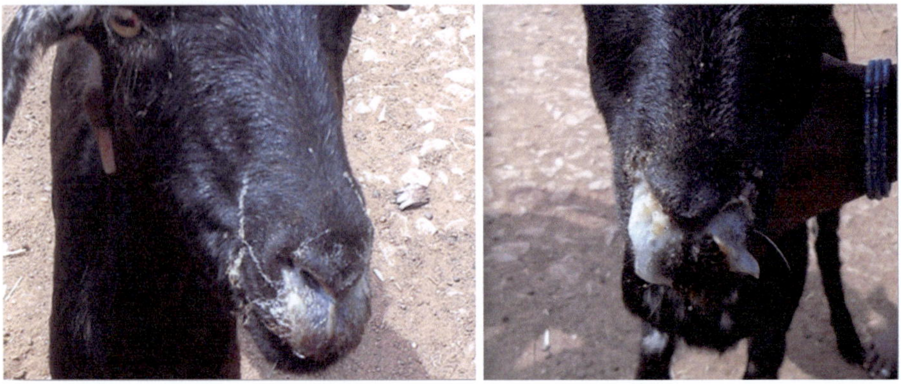

Fig. 2.2 Frothy mucopurulent nasal discharges from PPRV-infected goats

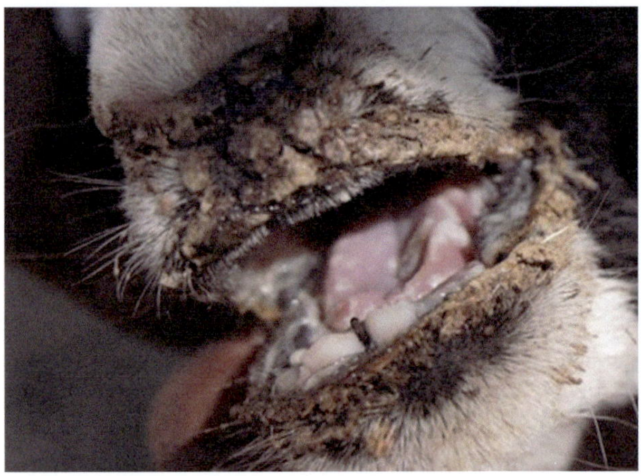

Fig. 2.3 Necrotic spots-pinprick lesions in buccal cavity and tongue of PPR-infected sheep

nasal discharges. Later, diarrhoea followed by necrotic lesions appears in the oral cavity (Fig. 2.3). The mortality rate is 70–80 % and death occurs within 10–12 days after the onset of disease. The severity of disease depends on various factors, namely, the PPRV virulence/lineage (Couacy-Hymann et al. 2007), species, breed, immune status and age of the animals.

PPR is frequently confused with other diseases that present fever and grossly similar clinical signs, especially when it is newly introduced. The other diseases with similar signs for differential diagnosis are bluetongue (BT), contagious ecthyma (Orf), foot and mouth disease (FMD), contagious caprine pleuropneumonia (CCPP), pasteurellosis, etc., Sometimes, mixed infection of PPR either with goat pox, sheep pox, Orf, BT, border disease virus (BDV) or adenovirus (Gibbs et al. 1977; Saravanan et al. 2007; Mondal et al. 2009; Malik et al. 2011; Toplu et al. 2012; Mbyuzi et al. 2014) has also been reported. Mixed infection of PPRV and other respiratory viruses {parainfluenza virus 3 (PIV 3), respiratory syncytial virus (RSV), bovine herpes virus 1 (BHV-1), bovine viral diarrhoea (BVD) and adenovirus} were also reported in dromedary camels in Sudan in an abattoir study (Saeed et al. 2015).

2.7 Diagnosis and Management

Tentatively PPR was diagnosed by characteristic clinical signs and post-mortem lesions and laboratory confirmation by using various serological and molecular techniques. A plethora of serological tests and molecular assays are available to detect and identify PPRV antigen, nucleic acid and antibodies. Serological tests such as agar gel immunodiffusion test (AGID), counter-immuno-electrophoresis (CIE) and indirect ELISA were popular in earlier days for tentative diagnosis (Obi and Pattrick 1984; Obi et al. 1990). Advent of cell culture and molecular biological technique has allowed development of specific, rapid and sensitive diagnostics. Virus isolation, nucleic acid hybridization (Diallo et al. 1989a; Shaila et al. 1989; Taylor et al. 1990; Pandey et al. 1992), haemagglutination using piglet (Wosu 1985) or chicken (Shaila et al. 1996) red blood cells, immunohistochemical detection (Saliki et al. 1994), serum or virus neutralization test (VNT) (Rossiter et al. 1985; Taylor et al. 1990; Chandran et al. 1995) and ELISA using PPRV-specific polyclonal or monoclonal antibodies (MAbs) are used for diagnosis of PPR (Libeau et al. 1994, Singh et al. 2004b, c; Balamurugan et al. 2007). Most of the conventional tests are time-consuming, labour intensive, less sensitive and not rapid and, therefore, not suitable for primary diagnosis but useful in secondary confirmatory testing and retrospective epidemiological studies. To overcome the drawbacks associated with these tests, the recent molecular biology tools and techniques like RT-PCR (Forsyth and Barrett 1995; Balamurugan et al. 2006a), real-time RT-PCR (Bao et al. 2008; Balamurugan et al. 2010a, 2012e) and loop-mediated isothermal amplification (LAMP) assays (Li et al. 2010; Dadas et al. 2012) have been used for the rapid and sensitive detection of PPRV nucleic acid from clinical samples.

Among all the methods, isolation of the virus remains the "gold standard". Virus isolation cannot always be done as routine diagnostic assays because they are time-consuming and cumbersome and require cell culture facilities, and they are not as sensitive as RT-PCR (Brindha et al. 2001). The PPRV can be isolated and grown *in vitro* in primary bovine and sheep cells (Taylor et al. 1990) as well as established cell lines such as Vero (African green monkey kidney) cells (Lefevre and Diallo 1990) and marmoset B-lymphoblastoid-B95a cells (Sreenivasa et al. 2006). The virus manifests specific cytopathic effect (CPE) after 3–5 days of infection, which includes initial rounding of the infected cells in grape-bunch-like clusters, followed by vacuolation, granulation of the cell cytoplasm, fusion of the monolayer cells and formation of syncytia, which are characteristics of PPRV. VNT was performed for confirmation of PPR viruses (Furley et al. 1987; Lefevre and Diallo 1990), either in tubes or in microtiter plates (micro-VNT) for the detection of PPRV antibodies in field serum samples (Rossiter et al. 1985; Taylor et al. 1990).

As a rapid, simple and sensitive assay, ELISA has been widely used in serological profiling of PPRV in mass screening of samples for seromonitoring/

serosurveillance or clinical prevalence. Various research workers have used MAb produced against PPRV for detection of antibodies and antigens in ELISA (Saliki et al. 1993; Anderson and McKay 1994; Libeau et al. 1995). Saliki et al. (1993) and Anderson and McKay (1994) used neutralizing MAbs against H protein of PPRV for specific detection of antibodies in competitive ELISA (c-ELISA) and blocking ELISA (B-ELISA). MAb-based PPR c-ELISA for the detection of PPRV antibodies assay is currently being employed extensively throughout India for seromonitoring and serosurveillance of PPR (Singh et al. 2004c). Further, a polyclonal antibody-based indirect ELISA was also developed for detection of antibodies to PPRV in the serum samples of goats and sheep, which may be a good alternative tool to c-ELISA for sero-epidemiological surveys (Balamurugan et al. 2007). Recently, the combined ELISA for detection of PPRV and bluetongue virus (BTV) has also been developed (Yousuf et al. 2015).

Further, MAb-based immunocapture ELISA (Libeau et al. 1994) and sandwich ELISA (s-ELISA) (Singh et al. 2004b) have been used extensively for detection of PPRV in field clinical specimens. The immunocapture ELISA using a biotinylated anti-'N′ MAb was developed in the World Reference Laboratory (CIRAD-EMVT, France) and is an internationally accepted assay for PPRV antigen detection in clinical samples (Libeau et al. 1994, 1995). Similarly, In India, the s-ELISA kit routinely is being used for clinical prevalence or detection of PPRV antigen in clinical specimens (Singh et al. 2004b). Recently, Berguido et al. (2016) developed luciferase immunoprecipitation system (LIPS) for the rapid detection of antibodies against PPRV in serum samples and for specific differentiation from antibodies against RPV.

Due to improved understanding of the viral genome and molecular biological techniques, nucleic acid-based specific and sensitive assays were also developed. RT-PCR, simple and aqueous phase ELISA (SNAP-ELISA) and nucleic acid hybridization have also been used for the detection of PPRV (Shaila et al. 1989; Pandey et al. 1992; Forsyth and Barrett 1995; Diallo et al. 1995; Couacy-Hymann et al. 2002; Forsyth et al. 2003). RT-PCR and cDNA hybridization techniques are the two sensitive means to diagnose the PPRV infection, but are time-consuming and cumbersome for routine diagnosis with a large sample size. Nucleic acid hybridization is also suitable for providing diagnosis utilizing field materials, which are either collected from the putrefied or semi-putrefied carcases or get putrefied during transit period (Pandey et al. 1992). Despite the best sensitivity, radiolabelled probes were not widely used because of short half-life of ^{32}P and the requirements of fresh specimens and isotopes handling facility. This led to the development of non-radioactive probes using biotinylated DNA or digoxigenin (DIG)-labelled oligonucleotides (Pandey et al. 1992; Diallo et al. 1995).

Further, the nucleic acid hybridization using radiolabelled or biotinylated cDNA probes, competitive and immunocapture ELISAs and differential immunohistochemical staining using MAbs are less sensitive than PCR for the detection of viruses. RT-PCRs have been reported for detection and differential diagnosis of PPR in clinical specimens. Forsyth and Barrett (1995) developed genus-specific and universal morbillivirus primers to distinguish among known morbilliviruses. The

PCR techniques have been developed targeting F gene (Forsyth and Barrett 1995), N gene (Couacy-Hymann et al. 2002; George et al. 2006), M gene (George et al. 2006; Balamurugan et al. 2006a) and H gene (Kaul 2004; Balamurugan et al. 2010b) and used for specific detection of PPRV from clinical samples. A two-step RT-PCR has been shown to be useful for the rapid detection of virus-specific RNA in the samples submitted for laboratory diagnosis (Shaila et al. 1996; Couacy-Hymann et al. 2002; George et al. 2006). Further, PCR strategies targeting M and N gene either two steps (George et al. 2006) or one step (Balamurugan et al. 2006a) have been developed for detection and differentiation of PPRV in sheep and goats using the clinical samples (Balamurugan et al. 2006a). A highly sensitive N gene-based RT-PCR-ELISA for the detection and differentiation of PPRV has been developed using DIG-labelled RT-PCR product with a detection limit of viral RNA in the infected tissue culture fluid with a titre as low as 0.01 $TCID_{50}$/100 μl (Saravanan et al. 2004).

Over conventional PCR approaches, the real-time PCR techniques targeting either N or M gene using TaqMan hydrolysis probe and SYBR Green with melting curve analysis have been in use for rapid, highly sensitive and specific detection and quantitation of PPRV. Earlier, N gene-based one-step TaqMan real-time RT-PCR (Bao et al. 2008), M gene-based two-step TaqMan hydrolysis probe (Balamurugan et al. 2010a), one-step real-time RT-PCR based on SYBR Green chemistry (Balamurugan et al. 2012e) and two-step SYBR Green I-based real-time RT-PCR assay (Abera and Thangavelu 2014a, b) have been optimized for specific detection and quantification of PPRV in clinical samples. Polci et al. (2015) developed N gene-based duplex real-time RT-PCR assay for a simple and rapid diagnosis of PPR. Kihu et al. (2015) detected PPRV in formalin-fixed tissues by qRT-PCR assay and stated that field pathological samples of PPR-suspected cases, collected and stored in 10 % formalin for 2 years, could be used for PPRV RNA extraction for confirmation. In the recent past, as a field diagnostic tool, a simple, rapid, specific and highly sensitive novel technique called LAMP has also been developed targeting N gene (Li et al. 2010; Dadas et al. 2012) and evaluated using clinical samples and suitable diagnostic tool in less equipped rural diagnostics laboratory settings. Moreover, Bhuiyan et al. (2014) established dried fluid spots in the filter papers for archiving RNA from local PPRV field isolates for molecular detection and genotyping of PPRV. This clearly reveals the excellent capacity of filter papers to store genetic material that can be sampled using a non-invasive approach.

A simple dot-ELISA has also been developed using either anti-M protein MAb (Obi and Ojeh 1989) or anti-N protein MAb (Saravanan et al. 2006) for the detection of PPRV antigen in tissue homogenate/swab materials of sheep and goats origin and could be used as a penside test for diagnosis of PPR. Lateral flow test for detection of PPRV antigen and antibody has been developed, but its usefulness in field diagnosis has not been satisfactory. However, Raj et al. (2008) developed user-friendly screening (immunofiltration) and confirmatory (antigen-competition ELISA) assays/tests for detection of PPRV antigen in the diagnosis of PPR in sheep with advantages of being quick, easy to perform and not requiring technical skill or expertise. Baron et al. (2014b) developed an immunochromatographic test for the diagnosis of PPR under field conditions in Ivory Coast, Pakistan, Ethiopia and Uganda and suggesting that this diagnostic tool may be useful for current efforts to control the spread of infection.

Due to the advancement in rDNA technology, gene expression technology, production of recombinant viral proteins has become easier and more efficient, and application of these recombinant proteins for disease diagnosis will be of beneficial. Several PPRV N, M, F and H proteins were expressed in different heterologous systems and successfully used as diagnostic antigen in ELISA for serodiagnosis and surveillance (Yadav et al. 2009; Ismail et al. 1995: Choi et al. 2005; Liu et al. 2013; Balamurugan et al. 2006b). Balamurugan et al. (2006b) expressed PPRV H protein in Vero cell and evaluated expressed protein as potential antigen source in ELISA for serosurveillance of PPR. Similarly, Yadav et al. (2009) expressed truncated and full-length N protein of PPRV in *E. coli* and showed reactivity in s-ELISA and tested as a coating antigen in c-ELISA for serological diagnosis of PPR. Wang et al. (2013a) expressed the truncated F protein of PPRV in *E. coli* and assessed its immunoreactivity. Similarly, in our laboratory, we have expressed truncated PPRV N, F and H proteins individually in *E.coli* system (Apsana et al. 2016; Balamurugan et al. 2016). Generally, these assays are safe and better alternatives to live PPRV antigen in ELISA for clinical or serosurveillance of PPR in enzootic or non-enzootic countries. Recently, Baron and Baron (2015) created RNA polymerase (L) gene deleted PPRV, which is helper cell-dependent form of virus, useful in a system for production of biosafe antigen. Different diagnostic techniques, methods and assays/tests for diagnosis of PPR with their advantages are summarized in Table 2.1.

There is no specific treatment for PPR, so management of the PPR is highly essential for control of infection. However, treatment of affected animals by administration of antibiotics (long-acting oxytetracycline, chlortetracycline) to prevent secondary bacterial infections and anti-diarrhoeal medicines has been practised with supportive therapy (B complex and dextrose saline) for 5–7 days, which may be useful to reduce the severity of the disease. Treatment and management of clinical cases of PPR in the event of outbreaks in sheep and goats are also necessary in order to minimize the economic losses to farmers (Balamurugan et al. 2014a)

2.8 Prophylaxis and Control Measures

2.8.1 Vaccine

Considering the close antigenic relationship between RPV and PPRV, the Plowright's tissue culture rinderpest (TCRP) (Plowright and Ferris 1961), vaccine was earlier tested in goats for vaccination against PPR (Taylor 1979). Therefore, earlier PPR was controlled in different parts of the world by using TCRP (heterologous) vaccine. But the use of TCRP vaccine to control PPR was later banned in all animal species worldwide so as to achieve the status of RP-free country or zone following the OIE pathway (Anderson and Mckay 1994), after the launch of RP eradication

Table 2.1 Different diagnostic techniques/assays/tests for diagnosis of PPR in sheep and goats

Diagnostic technique/tool	Features/advantages	References
Conventional tests are time-consuming, labour intensive, not rapid and less sensitive		
Agar gel immunodiffusion test/counter immunoelectrophoresis	Tentative diagnosis	Obi and Pattrick (1984); Obi et al. (1990)
Haemagglutination	Simple and inexpensive and provide results within a few hours	Wosu (1985) Shaila et al. (1996)
Virus isolation	*Gold standard* test – time-consuming, tedious Primary bovine and sheep cells Vero cell line Marmoset B-Lymphoblastoid – B95a	Taylor et al. (1990) Lefevre and Diallo (1990) Sreenivasa et al. (2006)
Virus or serum neutralization test	*Gold standard* test – time-consuming and tedious and requires cell culture facilities	Rossiter et al. (1985); Taylor et al. (1990); Diallo et al. (1995)
ELISAs are highly sensitive, specific and suitable for mass screening of samples		
Indirect ELISA	PPRV antigen and polyclonal antibodies	Obi and Pattrick (1984); Obi et al. (1990); Balamurugan et al. (2007)
Competitive ELISA/blocking ELISA	Neutralizing monoclonal antibodies (MAbs) – H protein MAbs – H protein	Saliki et al. (1993) Anderson and McKay (1994) Singh et al. (2004c)
Immunocapture ELISA/sandwich ELISA	Rapid differential identification and MAb – N protein	Libeau et al. (1994, 1995) Singh et al. (2004b)
Simple and aqueous phase ELISA (SNAP-ELISA)	RT-PCR/ELISA system	Forsyth et al. (2003)
Immunohistochemical detection	Detection of antigens within tissues	Saliki et al. (1994)
Dot-ELISA	Anti-M protein MAb Anti-N protein MAb	Obi and Ojeh (1989) Saravanan et al. (2006)

(continued)

Table 2.1 (continued)

Diagnostic technique/tool	Features/advantages	References
Immunofiltration		Raj et al. (2008)
Antigen-competition ELISA		Raj et al. (2008)
Recombinant antigen-based ELISA	N protein M protein F protein H protein	Ismail et al. (1995) Choi et al. (2005) Yadav et al. (2009) Liu et al. (2014a, c; Balamurugan et al. (2016) Liu et al. (2013) Wang et al. (2013a); Apsana et al. (2016) Liu et al. (2013, 2014c) Balamurugan et al. (2006b)
Luciferase immunoprecipitation system (LIPS)	Specific differentiation	Berguido et al. (2016)
Molecular nucleic acid detection or hybridization diagnostic techniques are rapid and highly sensitive		
Nucleic acid hybridization		Pandey et al. (1992) Diallo et al. (1989a, 1995)
RT-PCR	F gene	Forsyth and Barrett (1995);
	N gene	Couacy-Hymann et al. (2002); George et al. (2006)
	N and M gene	Balamurugan et al. (2006a, b) George et al. (2006) Khalafalla et al. (2010)
	H gene	Kaul (2004) Balamurugan et al. (2010b)
	Two-step RT-PCR	Shaila et al. (1989) Couacy-Hymann et al. (2002) George et al. (2006)
	One-step RT-PCR	Balamurugan et al. (2006a, b)

Diagnostic technique/tool	Features/advantages	References
PCR-ELISA	RT-PCR-ELISA	Saravanan et al. (2004)
Real-time RT-PCR	TaqMan real time – N gene Duplex real time – N gene M gene SYBR Green M gene	Bao et al. (2008); Polci et al. (2015) Balamurugan et al. (2010a); Abera and Thangavelu (2014a, b) Balamurugan et al. (2012e)
Loop-mediated isothermal amplification (LAMP)	N gene	Li et al. (2010); Dadas et al. (2012)

programme (NPRE), which stimulated the development of homologous PPR vaccine(s) by the world community. The first homologous PPR vaccine was developed using live attenuated Nigerian strain PPRV Nig 75/1 after 63 passages in Vero cells that produced a solid immunity for 3 years when tested (Diallo et al. 1989b; Diallo et al. 1995; Zahur et al. 2014). The vaccine was safe under field conditions even for pregnant animals and induced immunity in 98 % of the vaccinated animals (Diallo et al. 1995). Similarly, three other homologous PPR vaccines using Indian isolates of PPRV (goat origin, Sungri 1996 and CBE 1997; sheep origin, Arasur 1987) have been developed and evaluated (Saravanan et al. 2010a, b). It is necessary that each animal to be vaccinated should receive a minimum recommended dose (OIE), i.e. 10^3 $TCID_{50}$.

The second homologous live attenuated PPR vaccine (Sungri 1996 strain) from India has been tested extensively in in-house as well as by field trials and has been found to be safe and potent in small ruminants (Sreenivasa et al. 2000, Singh et al. 2009; Singh et al. 2010), and about 90 % of the animals have been found to have protective levels of antibodies under field conditions (Sreenivasa et al. 2000). Using this vaccine, studies have been undertaken with respect to thermostability, pathogenicity and immunogenicity at various *in vitro* passages (Sreenivasa et al. 2000; Sarkar et al. 2003; Rajak et al. 2005: Saravanan et al. 2010a, b), and the vaccine was found to induce only a transient lymphopenia, which may not induce biologically significant immunosuppression (Rajak et al. 2005). Thus, this PPR vaccine (Sungri 96) is safe for mass vaccination campaign under field conditions and is presently used throughout India to vaccinate sheep and goats with great efficacy. The third and fourth live attenuated vaccines were PPRV Arasur 87 (sheep origin) and Coimbatore 97 (goat origin), respectively, and are being used in southern states of India (Palaniswami et al. 2005). These vaccines also provide satisfactory protection against PPRV Izatnagar/94 and are also equally safe and protective as Sungri 96 in sheep and goats and suited for commercial vaccine production (Saravanan et al. 2010a, b).

Major disadvantages of these live attenuated vaccines are necessary to maintain the cold chain, and antibody responses are indistinguishable from natural infection. This makes seroepidemiological surveillance of the disease impossible in enzootic areas where a vaccination programme has been or is being implemented. Based on the earliest reports on the thermostability of the TCRP vaccine of Plowright and Ferris (1961) and Johnson (1962), developing intrinsic live attenuated thermo-adapted (Ta) PPR vaccines (goat and sheep isolates) employing thermostable Vero cell line, which is adapted to grow at 40 °C, have been successful (Balamurugan et al. 2014c). The vaccines have undergone successful in-house trials in both the sheep and goats either as single (Balamurugan et al. 2014c; Sen et al. 2009) or in combination with either sheep pox or goat pox as bivalent (unpublished data). Riyesh et al. (2011) evaluated the efficacy of different stabilizers on the thermostability of this Ta PPR vaccine. These Ta vaccines can be used as alternatives to existing vaccines in tropical countries for the control of the disease as they are considerably more stable at ambient temperatures (Riyesh et al. 2011). Similarly, promising Ta PPRV (PPRV Sungri 96) is the production of thermostable vaccine stabilized with appropriate stabilizers or heavy water (D2O) (Sen et al. 2009).

Geographic distribution of both PPR and goat or sheep pox virus infection as well as occurrence of mixed infections due to these viruses and use of a combined vaccine for control of these infections in the endemic areas would assist to a great extent in the mass immunization programs. The efficacy of the combined vaccine (PPR and goat pox) in goats showed that bivalent vaccine induced protective immune response against homologous challenge in goats (Hosamani et al. 2006). Similarly, the combined sheep pox and PPR vaccine provided safety and immunogenicity in sheep (Chaudhary et al. 2009).

The time of vaccination of young animals is the important issue in PPR control. There are few reports on duration of persistence of maternal antibodies in lambs/kids born to vaccinated dams. Maternal antibodies in young animals were detectable up to 6 months of age but fell below the protection threshold level at 3 or 4 months in lambs and kids, respectively (Ata et al. 1989; Bidjeh et al. 1999; Balamurugan et al. 2012d). Similarly, the neutralizing maternal antibodies were detectable up to 4 months compared to 3 months with competitive ELISA (Libeau et al. 1992; Ata et al. 1989). These findings lead to suggestion that lambs and kids from immunized or exposed dams should be vaccinated at 4 and 5 months of age, respectively (Libeau et al. 1992; Ata et al. 1989). Further, Cosseddu et al. (2016) evaluated humoral response and protective efficacy of an inactivated vaccine against PPRV in goats.

The H and F protein genes of several morbilliviruses have been expressed in various vector systems, and they can be used as effective subunit vaccines. A marker vaccine and a companion test that can detect infection in vaccinated animals (DIVA) are the key to facilitate the serosurveillance. Therefore, a recombinant marker vaccine (either positive or negative marker) approach may be beneficial to facilitate the serosurveillance and seromonitoring (Sen et al. 2009). The development of the reverse genetics technology for negative-strand RNA viruses has given us another means of producing marker vaccines to combat PPR. Liu et al. (2014b) systematically discussed a broad range of vaccines against PPR, including commercially available vaccines and potential vaccine candidates, and further DIVA strategies for immunization with the new generation vaccines. One of these recombinant vaccines has the RP vaccine virus genome as the backbone into which M, F and H protein genes of RPV were replaced by those of PPRV (Mahapatra et al. 2006). The resulting chimeric virus proved safe and effective vaccine, which could protect goats against virulent challenge with PPRV (Mahapatra et al. 2006). The widespread use of such vaccines, along with the diagnostic tests to identify their serological signature, would greatly improve the surveillance capabilities for disease preparedness and emergency prevention procedures. Muniraju et al. (2015) rescued a vaccine strain of PPRV and evaluated and compared in *in vivo* with standard vaccine with an aim to develop DIVA vaccine and diagnostics for control of disease. Rojas et al. (2014b) reported that vaccination with replication-defective human adenoviruses serotype 5 (Ad5) expressing the PPRV F or H proteins overcomes viral immunosuppression and induces protective immunity against PPRV challenge in sheep, and the results indicate that these could be a promising alternative to current vaccine strategies for the development of PPRV DIVA vaccines. Similarly, Herbert et al. (2014) reported recombinant adenovirus expressing the PPRV H protects goats against challenge with pathogenic virus, a DIVA vaccine for PPR. VLPs containing PPRV M protein and H or F protein are potential "differentiating infected from vaccinated animals" (DIVA) vaccine candidates for the surveillance and eradication of PPR (Liu et al. 2014c).

Prasad et al. (2004) developed an edible vaccine using HN gene of PPRV using binary vector pBI121 mobilized into *Agrobacterium tumefaciens* strain GV3 101 and subsequently expressed in pigeon pea. The goat immunized with baculovirus expressed recombinant HN glycoprotein of PPRV and produced immune response against PPRV, and antibodies generated in immunized animals could neutralize both PPRV and RPV *in vitro* (Sinnathamby et al. 2001). Recombinant *Bombyx mori* nucleopolyhedroviruses (BmNPV) expressed the immunodominant ectodomains of F glycoprotein of PPRV and the H protein of RPV, on the budded virions as well as the surface of the infected host cells, and the recombinant virus particles induced immune response in mice against PPRV or RPV (Rahman et al. 2003). Liu et al. (2013) showed expression, purification and characterization of two truncated PPRV M proteins in *Escherichia coli* and production of polyclonal antibodies against this protein in mice.

Further, Liu et al. (2014c) reported budding of PPRV-like particles from insect cell membrane based on intracellular co-expression of PPRV M, H and N proteins by recombinant baculoviruses. PPRV-like particles produced in insect cell-baculovirus system were found to be immunogenic (virus-neutralizing antibodies and promoted lymphocyte proliferation) in mice and goats (Li et al. 2014). Recombinant baculovirus co-expressed the PPRV M, H and N proteins forming virus-like particles (VLPs) and induced both complete virus-specific antibodies and virus-neutralizing antibodies in mice (Liu et al. 2015b). Liu et al. (2014a) observed formation of PPR spikeless VLPs by co-expression of M and N proteins in insect cells. Further, goats immunized with a recombinant baculovirus expressing H protein generated both humoral and cell-mediated immune responses (Sinnathamby et al. 2001) and were also RPV cross-reactive suggesting that the H protein presented by the baculovirus recombinant "resembles" the native protein present on PPRV (Sinnathamby et al. 2001). Moreover, lymphoproliferative responses were demonstrated against PPRV H and RPV H antigens (Sinnathamby et al. 2001), and mapping of N-terminal T-cell determinant and a C-terminal domain harbouring potential T-cell determinant(s) in goats was carried out (Sinnathamby et al. 2001). Though the $CD4^+$ and $CD8^+$ T cells in PBMC that responded to the recombinant protein fragments and the synthetic peptide could not be determined, this could potentially be a $CD4^+$ helper T-cell epitope, which has been shown to harbour an immunodominant H restricted epitope in mice (Sinnathamby et al. 2001). Identification of B- and T-cell epitopes on the protective antigens of PPRV would open up avenues to design novel epitope-based vaccines against PPR (Dechamma et al. 2006; Yu et al. 2015).

Rojas et al. (2014a) described two replication-defective human adenovirus serotype 5 (Ad5) containing the highly immunogenic F and H protein coding genes that inducted protective immune responses to PPRV in mice. The vaccinia virus and capripox expressing H and F glycoproteins PPRV have been shown to protect goats against PPR by cell-mediated immune responses (Jones et al. 1993; Romero et al. 1995). Apsana et al. (2015) reported immunization of sheep with DNA coding for the variable region of anti-idiotypic antibody generates humoral and cell-mediated immune responses specific for PPRV. Yin et al. (2014) reported that the recombinant PPRV expressing the FMDV VP1 protein is a potential dual live vectored vaccine against PPRV and FMDV and induced protective immune response against both PPRV and FMDV. Caufour et al. (2014) showed protective efficacy of a single immunization with *Capripoxvirus*-vectored recombinant PPR vaccines in the presence of pre-existing immunity in animals. Wang et al. (2013b) reported mice vaccinated with a suicidal DNA vaccine expressing the H glycoprotein of PPRV produce immune responses, and this suicidal vaccine could be a promising new approach for vaccine development against PPR. Similarly, Wang et al. (2015b) constructed a Semliki Forest virus (SFV) replicon-vectored DNA vaccine called as "suicidal DNA vaccine" expressing the F protein of PPRV, which induces both humoral and cell-mediated immune responses in BALB/c mice.

2.8.2 Prophylaxis Measures

The control of PPR can be ensured only through the implementation of effective prophylactic measures. All the sheep and goats of the affected flock should be under quarantine for at least one month after the last clinical case. Animal movements have to be strictly controlled in the area of the infection. Unfortunately, such sanitary and phytosanitary and control measures are difficult to maintain in a vast country with difficult terrain where PPR is endemic like in India. Therefore, the only effective way to control PPR is by mass vaccination of the animals with an effective vaccine against PPR, and quarantine measures are the only solution to control the disease effectively. Some of the other control measures (sanitary prophylaxis) that can prevent the occurrence of PPR infection are strict quarantine and control of animal movements, quarantine of newly purchased or newly arriving goats/sheep for at least 2–3 weeks before allowing them to mix with the regular flock and know the health status and the source of any new animal(s) brought into the flock, and migratory flocks are threat to local sheep and goat; therefore, contact may be avoided, effective cleaning and disinfection of contaminated areas of all premises with lipid solvent solutions of high or low pH and disinfectants including physical perimeters, equipment and clothing; dead animal/carcases should be burnt/buried deeply; monitor animals closely and frequently for any developing illness or signs of disease; isolate any sick animals from the flock, and contact the veterinarian immediately to examine sick animals in the herd/flock; use separate facilities and staff to handle isolated animals; and educate and train the employees about PPR and the signs of illness and monitoring of wild and captive animals, especially in contact with sheep and goats.

2.8.3 Control Strategies

For the control of PPR, there is a need for strong support of diagnostic methods, and timely vaccinations of the susceptible population are imperative. Control strategies may vary from country to country as per the prevalence of disease, but in developing or underdeveloped countries, the choices are limited. In a developing and underdeveloped country, stamping out by slaughter is not feasible, both for economic and sentimental reasons. In India, vaccination has become a recommended tool to support prevention and control of disease (Singh et al. 2009; Singh 2011). This way, at least the immediate loss could be prevented, and the small and marginal farmers rearing sheep and goats will be benefited. Therefore, PPR control and eradication depends mainly on rapid and accurate diagnosis or surveillance/monitoring and implementation of prompt vaccination programme. Recent success with rinderpest eradication programme (NPRE) has provided the confidence that it is required to launch a similar programme with PPR too.

Vaccination strategies for the control of PPRV would be slightly different from vaccination programmes for rinderpest. A mass vaccination campaign to cover 80 % herd or flock immunity would be needed to account for the population dynamics of sheep and goats, disparities in small ruminants' husbandry practices and the agroclimatic conditions affecting the pattern of disease (Singh 2011). The slaughtering of male goats at an early age combined with the high fecundity of the caprine species results in replacement of population (~30–40 % naïve population appears) every year. Initially, in order to reduce economic losses due to PPRV, intensive vaccination of the entire population within a specified area would need to be undertaken. Subsequent vaccinations would then be performed on younger animals at approximately 6 months of age (Singh 2011) to avoid window of susceptibility in kids to PPRV and the effort to eliminate PPR infection from susceptible populations (Balamurugan et al. 2012d). In other way, vaccinations should be focused initially on high-risk group of animals, for example, young animals (6 months to 1 year), goat population than sheep and migratory flocks (Singh 2011) in suitable period (preferably during lean periods). Alternatively, intensive vaccinations can be carried out based on populations to make disease-free areas (zone) by identifying the areas of infection and implementing vaccinations followed by screening, testing and overall revaccination if required in those areas.

2.9 Perspectives

PPR is the one of the priority animal diseases whose control is considered important for poverty alleviation in developing countries. As a transboundary animal disease (TAD), the presence of PPR can limit trade and export, import and development of intensive livestock production, which in turn diminish the consumption of animal protein in human. Now, PPR threatens the 1.7 billion-strong small ruminant population in Africa, the Middle East and the Indian subcontinent. Still PPR is a poorly recognized disease, particularly with regard to epidemiological features such as transmission dynamics under different production systems in different species, breed susceptibility, mechanism of spread of virus to other unnatural host and role of the wildlife on the epizootiology of disease. The fact that PPRV can infect cattle, buffaloes and camels gives PPR an even higher priority, particularly in the current situation where rinderpest has been eradicated. Understanding the determinants affecting virus or vaccine response, immunobiology of vaccine response in different hosts will enable us to find low- and high-responder animals and will also direct us regarding how to modulate these factors to obtain better immune protection to combat the disease.

The turnover rate of small ruminant populations is much faster than that of larger livestock, so vaccination has to be used more intensively and more frequently. Epidemics of PPR not only affect individual farmers but also the agricultural industry and, as a consequence, the national economy. Due to the occurrence of other diseases, the economic impacts of PPR are probably underestimated, but it is believed that PPR is one of the major constraints for augmenting small ruminant production. Currently, the disease has been brought under control in goats and sheep by available effective and safe live attenuated cell culture PPR vaccines in some countries. The present scenario of PPR in developing countries like India warrants the studies to be undertaken with the objective to know the effect of agroclimatic changes on the occurrence of PPR in small ruminants in different agroclimatic zones and to analyse the relationship of disease occurrence and risk factors to formulate modules for forecasting and forewarning. The epidemiology of PPR is likely to change due to vaccination as the disease occurs more severely in the naïve population. More research is needed on the host-virus interaction through cellular receptor, immunological events including protective mechanisms, development of marker vaccine to differentiate between virulent and vaccine virus antibodies and also on development of thermostable vaccine. The availability of an effective marker vaccine with thermostable nature along with its companion serological tests will greatly assist in designing effective global control programmes for eradication of disease targeted by 2030. Analytical study with statistical validity about incidence of disease and socioeconomic impact would be extremely useful and elicit widespread interest by providing sufficient additional information, which are important to support control policy decisions in the respective country.

References

Abegunde AA, Abu PO (1977) Excretion of the virus of peste des petits ruminants by goats. Bull Anim Hlth Prod Afr 25:307–311

Abera T, Thangavelu A (2014a) Development of a two-step SYBR Green I based real time RT-PCR assay for detecting and quantifying peste des petits ruminants virus in clinical samples. J Virol Methods 209:25–29

Abera T, Thangavelu A, Joy Chandran ND, Raja A (2014b) A SYBR Green I based real time RT-PCR assay for specific detection and quantitation of peste des petits ruminants virus. BMC Vet Res 10:22

Abraham G, Sintayehu A, Libeau G, Albina E, Roger F, Laekemariam Y, Abayneh D, Awoke KM (2005) Antibody seroprevalences against peste des petits ruminants (PPR) virus in camels, cattle, goats and sheep in Ethiopia. Prev Vet Med 70:51–57

Abubakar M, Mahapatra M, Muniraju M, Arshed MJ, Khan EH, Banyard AC, Ali Q, Parida S (2015) Serological detection of antibodies to peste des petits ruminants virus in large ruminants. Transbound Emerg Dis. doi: 10.1111/tbed.12392.

Abu Elzein EM, Hassanien MM, Al-Afaleq AI, Abd Elhadi MA, Housawi FM (1990) Isolation of peste des petits ruminants from goats in Saudi Arabia. Vet Rec 127(12):309–310

Abu-Elzein EME, Housawi FMT, Bashareek Y, Gameel AA, Al-Afaleq AI, Anderson E (2004) Severe PPR infection in gazelles kept under semi-free range conditions. J Vet Med B Infect Dis Vet Public Health 51:68–71

Anderson J, Mckay JA (1994) The detection of antibodies against peste des petits ruminants virus in cattle, sheep and goats and the possible implications to rinderpest control programmes. Epidemiol Infect 112:225–231

Apsana R, Isloor S, Shaila MS (2015) Immunization of sheep with DNA coding for the variable region of anti-idiotypic antibody generates humoral and cell mediated immune responses specific for peste des petits ruminants virus. Vet Res Commun 39:197–202

Apsana R, Balamurugan V, Veeragowda BM, Abraham S, raju DSN, Rathnamma D, Byregowda SM, Rahman H, Shaila MS (2016) Expresion and charcateriation of immunodominant region of fusion protein of peste des petits ruminants virus in E. coli. Small Ruminant Res 144, 75–82

Aruni AW, Latha PS, Mohan AC, Chitravelu P, Ambumani SP (1998) Histopathological study of a natural outbreak of peste des petits ruminants in goats in Tamilnadu. Small Ruminant Res 28:233–240

Asmar JA, Radwan AI, Abi Assi N, Al-Rashied A (1980) PPR-like disease in sheep of central Saudi Arabia: evidence of its immunological relation to rinderpest; prospects for a control method. In: Migahid AM (ed) 4th annual meeting of the Saudi Biological Society. University of Riyadh Press, Riyadh, pp 325–338

Ata FA, AI-Sumry HS, King GJ, Ismaili SI, Ata AA (1989) Duration of maternal immunity to peste des petits ruminants. Vet Rec 3:590–591

Baczko K, Liebert UG, Billeter M, Cattaneo R, Bridka H, ter Meulen V (1986) Expression of defective measles virus genes in brain tissues of patients with subacute sclerosing panencepahalitis. J Virol 59:472–478

Bailey D, Banyard AC, Dash P, Ozkul A, Barrett T (2005) Full genome sequence of peste des petits ruminants virus, a member of the Morbillivirus genus. Virus Res 110:119–124

Balamurugan V, Sen A, Saravanan P, Singh RP, Singh RK, Rasool TJ, Bandyopadhyay SK (2006a) One-step multiplex RT-PCR assay for the detection of Peste-des-petits-ruminants virus in clinical samples. Vet Res Commun 30:655–666

Balamurugan V, Sen A, Saravanan P, Rasool TJ, Yadav MP, Bandyopadhyay SK, Singh RK (2006b) Development and characterization of a stable vero cell line constitutively expressing peste des petits ruminants virus (PPRV) hemagglutinin protein and its potential use as antigen in enzyme-linked immunosorbent assay for serosurveillance of PPRV. Clin Vaccine Immunol 13:1367–1372

Balamurugan V, Singh RP, Saravanan P, Sen A, Sarkar J, Sahay B, Rasool TJ, Singh RK (2007) Development of an indirect ELISA for the detection of antibodies against peste des petits ruminants virus in small ruminants. Vet Res Commun 31:355–364

Balamurugan V, Sen A, Saravanan P, Bhanuprakash V, Patra RC, Swarup D, Mahesh K, Yadav MP, Singh RK (2008) Potential antiviral effect of aqueous extract of *Acacia arabica* leaves on peste des petits ruminants (PPR) virus replication. Pharm Biol 46:171–179

Balamurugan V, Sen A, Venkatesan G, Yadav V, Bhanot V, Bhanuprakash V, Singh RK (2010a) Application of semi-quantitative M gene-based hydrolysis probe (TaqMan) real-time RT-PCR assay for the detection of peste des petits ruminants virus in the clinical samples for investigation into clinical prevalence of disease. Transbound Emerg Dis 57:383–395

Balamurugan V, Sen A, Venkatesan G, Yadav V, Bhanot V, Riyesh T, Bhanuprakash V, Singh RK (2010b) Sequence and phylogenetic analyses of the structural genes of virulent isolates and vaccine strains of Peste des petites ruminants virus from India. Transbound Emerg Dis 57:352–364

Balamurugan V, Saravanan P, Sen A, Rajak KK, Bhanuprakash V, Krishnamoorthy P, Singh RK (2011) Sero-epidemiological study of peste des petits ruminants in sheep and goats in India between 2003 and 2009. Rev Sci Tech OIE 30:889–896

Balamurugan V, Krishnamoorthy P, Veeregowda BM, Sen A, Rajak KK, Bhanuprakash V, Gajendragad MR, Prabhudas K (2012a) Seroprevalence of Peste des petits ruminants in cattle and buffaloes from Southern Peninsular India. Trop Anim Health Pro 44:301–306

Balamurugan V, Saravanan P, Sen A, Rajak KK, Venkatesan G, Krishnamoorthy P, Bhanuprakash V, Singh RK (2012b) Prevalence of peste des petits ruminants among sheep and goats in India. J Vet Sci 13:279–285

Balamurugan V, Sen A, Venkatesan G, Bhanot V, Yadav V, Bhanuprakash V, Singh RK (2012c) Peste des petits ruminants virus detected in tissues from an Asiatic lion (*Panthera leo persica*) belongs to Asian lineage IV. J Vet Sci 13:203–206

Balamurugan V, Sen A, Venkatesan G, Rajak KK, Bhanuprakash V, Singh RK (2012d) Study on passive immunity: time of vaccination in kids born to goats vaccinated against Peste des petits ruminants. Virol Sin 27:228–233

Balamurugan V, Sen A, Venkatesan G, Yadav V, Bhanot V, Bhanuprakash V, Singh RK (2012e) A rapid and sensitive one step-SYBR green based semi quantitative real time RT-PCR for the detection of peste des petits ruminants virus in the clinical samples. Virol Sin 27:1–9

Balamurugan V, Hemadri D, Gajendragad MR, Singh RK, Rahman H (2014a) Diagnosis and control of peste des petits ruminants: a comprehensive review. Virusdisease 25:39–56

Balamurugan V, Krishnamoorthy P, Raju DSN, Rajak KK, Bhanuprakash V, Pandey AB, Gajendragad MR, Prabhudas K, Rahman H (2014b) Prevalence of *Peste-des-petits-ruminant virus* antibodies in cattle, buffaloes, sheep and goats in India. Virusdisease 25:85–90

Balamurugan V, Sen A, Venkatesan G, Bhanuprakash V, Singh RK (2014c) Protective immune response of live attenuated thermo-adapted peste des petits ruminants vaccine in goats. Virusdisease 25:350–357

Balamurugan V, Das S, Raju DS, Chakravarty I, Nagalingam M, Hemadri D, Govindaraj G, Ibotombi Singh N, Ltu K, Devi M, Sharma K, Gajendragad MR, Rahman H (2015a) Prevalence of peste des petits ruminants in goats in North-East India. Virusdisease 25:488–492

Balamurugan V, Rahman H, Munir M (2015b) In: Munir M (ed) Host susceptibility to PPR virus. In Peste des petitis Ruminants Virus, vol 2015. Spinger, Heidelberg, pp 39–50

Banyard AC, Parida S, Batten C, Oura C, Kwiatek O, Libeau G (2010) Global distribution of peste des petits ruminants virus and prospects for improved diagnosis and control. J Gen Virol 91:2885–2897

Balamurugan V, Roy M, Sowjanyakumari S, Abraham S, Apsana R, Nagalingam M, Hemadri D, Rahman H (2016). Development of recombinant Nucleocapsid protein based indirect ELISA for serodiagnosis of Peste des Petits ruminants in sheep and goats. Adv Anim Vet Sci 4(6): 301–310

Banyard AC, Wang Z, Parida S (2014) Peste des petits ruminants virus, eastern Asia. Emerg Infect Dis 20:2176–2178

Bao J, Li L, Wang Z, Barrett T, Suo L, Zhao W, Liu Y, Liu C, Li J (2008) Development of one-step real-time RT-PCR assay for detection and quantitation of peste des petits ruminants virus. J Virol Methods 148:232–236

Bao J, Wang Z, Li L, Wu X, Sang P, Wu G, Ding G, Suo L, Liu C, Wang J, Zhao W, Li J, Qi L (2011) Detection and genetic characterization of peste des petits ruminants virus in free-living bharals (*Pseudois nayaur*) in Tibet, China. Res Vet Sci 90:238–240

Baron J, Baron MD (2015) Development of a helper cell-dependent form of peste des petits ruminants virus: a system for making biosafe antigen. Vet Res 46:101

Baron MD, Barrett T (1995) Sequencing and analysis of the nucleocapsid (N) and polymerase (L) genes and the extragenic domains of the vaccine strain of rinderpest virus. J Gen Virol 76:593–603

Baron J, Bin-Tarif A, Herbert R, Frost L, Taylor G, Baron MD (2014a) Early changes in cytokine expression in peste des petits ruminants disease. Vet Res 45:22

Baron J, Fishbourne E, Couacy-Hyman E, Abubakar M, Jones BA, Frost L, Herbert R, Chibssa TR, Van't Klooster G, Afzal M, Ayebazibwe C, Toye P, Bashiruddin J, Baron MD (2014b) Development and testing of a field diagnostic assay for peste des petits ruminants virus. Transbound Emerg Dis 61:390–396

Barrett T, Amarel-Doel C, Kitching RP, Gusev A (1993) Use of the polymerase chain reaction in differentiating rinderpest field virus and vaccine virus in the same animals. Rev Sci Tech 12:865–872
Berguido FJ, Bodjo SC, Loitsch A, Diallo A (2016) Specific detection of peste des petits ruminants virus antibodies in sheep and goat sera by the luciferase immunoprecipitation system. J Virol Methods 227:40–46
Bhuiyan AR, Chowdhury EH, Kwiatek O, Parvin R, Rahman MM, Islam MR, Albina E, Libeau G (2014) Dried fluid spots for peste des petits ruminants virus load evaluation allowing for non-invasive diagnosis and genotyping. BMC Vet Res 10:247
Bidjeh KC, Diguimbaya P, Hendrikx V, Debet D, Tchari NS (1999) Maternal immunity in young goats or sheep whose dams were vaccinates with anti-peste des petits ruminants. Cahiers Agric 3:219–222
Birch J, Juleff N, Heaton MP, Kalbfleisch T, Kijas J, Bailey D (2013) Characterization of ovine Nectin-4, a novel peste des petits ruminants virus receptor. J Virol 87:4756–4761
Bonniwell MA (1980) The use of tissue culture rinderpest vaccine (TCRV) to protect sheep and goats against "Peste des Petits ruminants" in the Ashanti region of Ghana. Rev Sci Tech 92:1233–1238
Braide VB (1981) Peste des petits ruminants. World Anim Rev 39:25–28
Brindha K, Raj GD, Ganesan PI, Thiagarajan V, Nainar AM, Nachimuthu K (2001) Comparison of virus isolation and polymerase chain reaction for diagnosis of peste des petits ruminants. Acta Virol 45:169–172
Brown CC, Torres A (1994) Distribution of antigen in cattle infected with rinderpest virus. Vet Pathol 31:194–200
Brown CC, Mariner JC, Olander HJ (1991) An immunohistochemical study of the pneumonia caused by PPR virus. Vet Pathol 28:166–170
Bundza A, Afshan A, Dukes TW, Myus DJ, Dulac GC, Becken SAWE (1988) Experimental peste des petits ruminants (goat plague) in goats and sheep. Can J Vet Res 52:46–52
Carter MJ, Willcocks MM, TerMeulen V (1983) Defective translation of measles virus protein in a subacute sclerosing panacephalitis. Nature 305:153–155
Cattaneo R, Kaelin K, Baczko K, Billeter MA (1989) Measles virus editing provides an additional cysteine-rich protein. Cell 56:759–764
Caufour P, Rufael T, Lamien CE, Lancelot R, Kidane M, Awel D, Sertse T, Kwiatek O, Libeau G, Sahle M, Diallo A, Albina E (2014) Protective efficacy of a single immunization with capripoxvirus-vectored recombinant peste des petits ruminants vaccines in presence of pre-existing immunity. Vaccine 32:3772–3779
Chandran NDJ, Kumanan K, Venkateswan RA (1995) Differentiation of peste des petits ruminants and rinderpest viruses by neutralization indices using hyperimmune rinderpest antiserum. Trop Anim Health Prod 27:89–92
Chard LS, Bailey DS, Dash P, Banyard AC, Barrett T (2008) Full genome sequences of two virulent strains of peste-des-petits ruminants virus, the Coted'Ivoire 1989 and Nigeria 1976 strains. Virus Res 136:192–197
Chaudhary SS, Pandey KD, Singh RP, Verma PC, Gupta PK (2009) A Vero cell derived combined vaccine against sheep pox and peste des petits ruminants for sheep. Vaccine 27:2548–2553
Chaudhary K, Chaubey KK, Singh SV, Kumar N (2015) Receptor tyrosine kinase signaling regulates replication of the peste des petits ruminants virus. Acta Virol 59:78–83
Chazya R, Muma JB, Mwacalimba KK, Karimuribo E, Mkandawire E, Simuunza M (2014) A qualitative assessment of the risk of introducing peste des petits ruminants into northern zambia from Tanzania. Vet Med Int 2014:202618
Choi KS, Nah JJ, Ko YJ, Kang SY, Jo NI (2005) Rapid competitive enzyme-linked immunosorbent assay for detection of antibodies to peste des petits ruminants virus. Clin Diagn Lab Immunol 12:542–547
Chowdhury EH, Bhuiyan AR, Rahman MM, Siddique MS, Islam MR (2014) Natural peste des petits ruminants virus infection in Black Bengal goats: virological, pathological and immunohistochemical investigation. BMC Vet Res 10:263
Cosseddu GM, Polci A, Pinoni C, Capobianco Dondona A, Iapaolo F, Orsini G, Izzo F, Bortone G, Ronchi FG, Di Ventura M, El Harrak M, Monaco F (2016) Evaluation of humoral tesponse and

protective efficacy of an inactivated vaccine against peste des petits ruminants virus in goats. Transbound Emerg Dis 63(5):e447–52

Couacy-Hymann E, Roger F, Hurard C, Guillou JP, Libeau G, Diallo A (2002) A Rapid and sensitive detection of peste des petits ruminants virus by a polymerase chain reaction assay. J Virol Methods 100:17–25

Couacy-Hymann E, Bodjo C, Danho T, Libeau G, Diallo A (2007) Evaluation of the virulence of some strains of peste-des-petits-ruminants virus (PPRV) in experimentally infected West African dwarf goats. Vet J 173:178–183

Crowley JC, Dowling PC, Menonna J, Silverman JI, Schuback D, Cook SD, Blumberg BM (1988) Sequence variability and function of measles 3′ and 5′ ends and intercistronic regions. Virology 164:498–506

Curasson G (1986) Traite de pathologie exotique veterinaire et comparee. Vigot Freres, Paris, pp 28–302

Dadas RC, Muthuchelvan D, Pandey AB, Rajak KK, Sudhakar SB, Shivchandra SB, Venkatesan G (2012) Development of loop-mediated isothermal amplification (LAMP) assay for rapid detection of peste des petits ruminants virus (PPRV) genome from clinical samples. Indian J Comp Microbiol Immunol Infect Dis 33:7–13

Dardiri AH, DeBoer CJ, Hamdy EM (1976) Response of American goats and cattle to peste des petits ruminants virus. In: Proceeding of the 19th annual meeting of the American Association of Veterinary Laboratory Diagnoscians, pp 337–344.

Das SC, Baron MD, Barrett (2000) Recovery and characterization of a chimeric rinderpest virus with the glycoproteins of peste des petits ruminant virus: homologous F and H proteins are required for virus viability. J Virol 74:9039–9047

Dechamma HJ, Dighe V, Kumar CA, Singh RP, Jagadish M, Kumar S (2006) Identification of T-helper and linear B epitope in the hypervariable region of nucleocapsid protein of PPRV and its use in the development of specific antibodies to detectviral antigen. Vet Microbiol 118(3–4): 201–211

Delpeut S, Noyce RS, Richardson CD (2014) The tumor-associated marker, PVRL4 (nectin-4), is the epithelial receptor for morbilliviruses. Viruses 6:2268–2286

Devireddy LR, Raghavan R, Ramachandran S, Shaila MS (1999) The fusion protein of peste des petits ruminants virus is a hemolysin. Arch Virol 144:1241–1247

Dhanasekaran S, Biswas M, Vignesh AR, Ramya R, Raj GD, Tirumurugaan KG, Raja A, Kataria RS, Parida S, Elankumaran S (2015) Toll-like receptor responses to Peste des petits ruminants virus in goats and water buffalo. PLoS One 9:e111609

Dhar P, Sreenivasa BP, Barrett T, Corteyn M, Singh RP, Bandyopadhyay SK (2002) Recent epidemiology of peste des petits ruminants virus (PPRV). Vet Microbiol 88:153–159

Dhar P, Muthuchelvan D, Sanyal A, Kaul R, Singh RP, Singh RK, Bandyopadhyay SK (2006) Sequence analysis of the haemagglutinin and fusion protein genes of peste-des-petits ruminants vaccine virus of Indian origin. Virus Genes 32(1):71–78

Diallo A (1988) Rinderpest and peste des petits ruminanats; constant threats to animal farming in many developing countries. Impact Sci Soc 150:179–192

Diallo A (1990) Morbillivirus group: genome organization and proteins. Vet Microbiol 23:55–163

Diallo A, Barrett T, Barbras M, Shaila MS, Taylor WP (1989a) Differentiation of rinderpest and peste des petits ruminants viruses using specific cDNA clones. J Virol Methods 23:127–136

Diallo A, Taylor WP, Lefevre PC, Provost A (1989b) Attenuation d'une souche de la peste des petits ruminants candidat pour un vaccine homologue vivant. Rev Elev Med Vet Pays Trop 42:311–319

Diallo A, Barrett T, Barbron M, Meyer G, Lefevre PC (1994) Cloning of nucleocapsid protein gene of peste des petits ruminants virus: relationship to other morbilliviruses. J Gen Virol 75:233–237

Diallo A, Libeau G, Couacy-Hymann E, Barbron M (1995) Recent developments in the diagnosis of rinderpest and peste des petits ruminants. Review. Vet Microbiol 44:307–317

Diallo A, Minet C, Le Goff C, Berhe G, Albina E, Libeau G, Barrett T (2007) The threat of peste des petits ruminants: progress in vaccine development for disease control. Vaccine 25:5591–5597

Dundon WG, Adombi C, Waqas A, Otsyina HR, Arthur CT, Silber R, Loitsch A, Diallo A (2014) Full genome sequence of a peste des petits ruminants virus (PPRV) from Ghana. Virus Genes 49:497–501

El Arbi AS, El Mamy AB, Salami H, Isselmou E, Kwiatek O, Libeau G, Kane Y, Lancelot R (2014) Peste des petits ruminants virus, Mauritania. Emerg Infect Dis 20:333–336

El Hag B, Taylor TW (1984) Isolation of PPR virus from Sudan. Res Vet Sci 36:1–4

Eligulashvili R, Perl S, Stram Y, Friedgut O, Scheichat N, Samina I, Trainin Z (1999) Immunohistochemical detection of peste des petits ruminants viral antigen in formalin-fixed, paraffin-embedded tissues from cases of naturally occurring infection. J Vet Diagn Invest 11:286–288

Fentahun T, Woldie M (2012) Review on peste des petitits ruminants (PPR). Eur J Appl Sci 4:160–167

Forsyth MA, Barrett T (1995) Detection and differentiation of rinderpest and peste des petits ruminants viruses in diagnostic and experimental samples by polymerase chain reaction using P and F gene-specific primers. Virus Res 39:151–163

Forsyth MA, Parida S, Alexandersen S, Belsham GJ, Barrett T (2003) Rinderpest virus lineage differentiation using RT-PCR and SNAP-ELISA. J Virol Methods 107:29–36

Furley CW, Taylor WP, Obi TU (1987) An outbreak of *peste des petits ruminants* in a zoological collection. Vet Rec 121:443–447

Gargadennec L, Lalanne A (1942) La peste des petits ruminants. Bull Serve Zootech Epizoot Afr Occid Fr 5:16–21

George A, Dhar P, Sreenivasa BP, Singh RP, Bandyopadhyay SK (2006) The M and N genes-based simplex and multiplex PCRs are better than the F or H gene-based simplex PCR for Peste-des-petits-ruminants virus. Acta Virol 50:217–222

Gibbs EPJ, Taylor WP, Lawman MJP (1977) The isolation of adenovirus from goats affected with peste des petits ruminants in Nigeria. Res Vet Sci 23:331–335

Gibbs EP, Taylor WP, Lawman MJ, Bryant J (1979) Classification of peste des petits ruminants virus as the fourth member of the genus Morbillivirus. Intervirology 11:268–274

Gilbert Y, Monier J (1962) Adaptation des virus *peste des petits* ruminants aux cultures ruminants cellularies. Rev Elev Med Vet Pays Trop 15:321–335

Giraudon P, Wild TF (1985) Correlation between epitopes on haemagglutinin of measles virus and biological activities: passive protection by monoclonal antibodies is related to their haemagglutination inhibiting activity. Virology 144:46–58

Govindarajan R, Koteeswaran A, Venugopalan AT, Shyam G, Shaouna S, Shaila MS, Ramachandran S (1997) Isolation of pestes des petits ruminants virus from an outbreak in Indian buffalo *(Bubalus bubalis)*. Vet Rec 141:573–574

Haas L, Baron MD, Liess B, Barrett T (1995) Editing of morbillivirus P gene transcripts in infected animals. Vet Microbiol 44:299–306

Haffar A, Libeau G, Moussa A, Cècile M, Diallo A (1999) The matrix protein gene sequence analysis reveals close relationship between PPR virus and dolphin morbillivirus. Virus Res 64:69–75

Hamdy FM, Dardiri AH, Nduaka O, Breese SS, Ihemelandu EC (1976) Etiology of the stomatitis pneumoenteritis complex in Nigerian dwarf goats. Can J Comp Med 40:276–284

Haque ME, Habib S, Islam MR, Khan KA, Hannan ASMA, Nadir EUA (2004) Seromonitoring of peste des petits ruminants (PPR) antibodies in small and large animals. J Anim Vet Adv 3:453–458

Herbert R, Baron J, Batten C, Baron M, Taylor G (2014) Recombinant adenovirus expressing the haemagglutinin of peste des petits ruminants virus (PPRV) protects goats against challenge with pathogenic virus; a DIVA vaccine for PPR. Vet Res 45:24

Hinshu TV, Kher HN, Chandel BS, Jhala MK (2001) Seroprevalence of peste des petits ruminants (PPR) in Gujarat. Indian J Comp Microbiol Immunol Infect Dis 22:81

Hosamani M, Singh SK, Mondal B, Sen A, Bhanuprakash V, Bandyopadhyay SK, Yadav MP, Singh RK (2006) A bivalent vaccine against goat pox and peste des petits ruminants induces protective immune response in goats. Vaccine 24:6058–6064

Ismail TM, Yamanaka MK, Saliki JT, el-Kholy A, Mebus C, Yilma T (1995) Cloning and expression of the nucleoprotein of peste des petits ruminants virus in baculovirus for use in serological diagnosis. Virology 208:776–778

Jagtap SP, Rajak KK, Garg UK, Sen A, Bhanuprakash V, Sudhakar SB, Balamurugan V, Patel A, Ahuja A, Singh RK, Vanamayya PR (2012) Effect of immunosuppression on pathogenesis of peste des petits ruminants (PPR) virus infection in goats. Microb Pathog 52:217–226

Johnson RH (1962) Rinderpest in tissue culture. III Use of the attenuated strain as a vaccine for cattle. Brit Vet J 118:141–150

Jones L, Giavedoni L, Saliki JT, Brown C, Mebus C, Yilma T (1993) Protection of goats against peste des petits ruminants with a vaccinia virus double recombinant expressing the F and H genes of rinderpest virus. Vaccine 11:961–964

Kaul R (2004) Haemagglutinin gene based molecular epidemiology of PPR virus. PhD thesis, Indian Veterinary Research Institute, Izatnagar, Uttar Pradesh, India.

Kerur N, Jhala MK, Joshi CG (2008) Genetic characterization of Indian peste des petits ruminants virus (PPRV) by sequencing and phylogenetic analysis of fusion protein and nucleoprotein gene segments. Res Vet Sci 85:176–183

Kgotlele T, Macha ES, Kasanga CJ, Kusiluka LJ, Karimuribo ED, Van Doorsselaere J, Wensman JJ, Munir M, Misinzo G (2014) Partial genetic characterization of peste des petits ruminants virus from goats in northern and eastern Tanzania. Transbound Emerg Dis 61:56–62

Khalafallaa AI, Saeed IK, Ali YH, Abdurrahman MB, Kwiatekc O, Libeau G, Obeidaa AA, Abbas Z (2010) An outbreak of peste des petits ruminants (PPR) in camels in the Sudan. Acta Trop 116:161–165

Khan HA, Siddique M, Abubakar M, Arshad MJ, Hussain M (2008a) Prevalence and distribution of peste des petits ruminants virus infection in small ruminants. Small Ruminant Res 79:152–157

Khan HA, Siddique M, Rahman S, Abubakar M, Ashraf MJ (2008b) The detection of antibody against peste des petits ruminants virus in sheep, goats, cattle and buffaloes. Trop Anim Health Pro 40:521–527

Khandelwal N, Kaur G, Chaubey KK, Singh P, Sharma S, Tiwari A, Singh SV, Kumar N (2014) Silver nanoparticles impair peste des petits ruminants virus replication. Virus Res 190:1–7

Kihu SM, Gitao GC, Bebora LC, Njenga MJ, Wairire GG, Maingi N, Wahome RG, Oyugi JO, Lutomia E (2015) Detection of peste des petits ruminants virus in formalin-fixed tissues. Tropl Anim Health Prod 47:247–249

Kingsbury DW (1990) Paramyxoviridae and their replication. In: Fields BN, Knipe DM et al (eds) Virology, vol I, 2nd edn. Raven Press, New York, pp 945–962

Kivaria FM, Kwiatek O, Kapaga AM, Swai ES, Libeau G, Moshy W, Mbyuzi AO, Gladson J (2013) The incursion, persistence and spread of peste des petits ruminants in Tanzania: epidemiological patterns and predictions. Onderstepoort J Vet Res 80:593

Kock RA, Orynbayev MB, Sultankulova KT, Strochkov VM, Omarova ZD, Shalgynbayev EK, Rametov NM, Sansyzbay AR, Parida S (2015) Detection and genetic characterization of lineage IV peste des petits ruminant virus in Kazakhstan. Transbound Emerg Dis 62:470–479

Kumar P, Kumar R, Sharma AK, Tripathi BN (2002) Pathology of peste des petits ruminants (PPR) in goats and sheep: spontaneous study. Indian J Vet Pathol 26:15–18

Kumar P, Tripathi BN, Sharma AK, Kumar R, Sreenivasa BP, Singh RP, Dhar P, Bandyopadyaya SK (2004) Pathological and immunonohistochemical study of experimental peste des petits ruminants virus infection in goats. J Vet Med B 51:1–7

Kumar KS, Babu A, Sundarapandian G, Roy P, Thangavelu A, Kumar KS, Arumugam R, Chandran ND, Muniraju M, Mahapatra M, Banyard AC, Manohar BM, Parida S (2014a) Molecular characterisation of lineage IV peste des petits ruminants virus using multi gene sequence data. Vet Microbiol 174:39–49

Kumar N, Maherchandani S, Kashyap SK, Singh SV, Sharma S, Chaubey KK, Ly H (2014b) Peste des petits ruminants virus infection of small ruminants: a comprehensive review. Virus Res 6:2287–2327

Kwiatek O, Ali YH, Saeed IK, Khalafalla AI, Mohamed OI, Obeida AA, Rahman MBA, Osman HM, Taha KM, Abbas Z, Harrak ME, Lhor Y, Diallo A, Lancelot R, Albina E, Libeau G (2011) Asian lineage of peste des petits ruminants virus, Africa. Emerg Infect Dis 17:1223–1231

Laurent A (1968) Aspects biologiques de la multiplication du virus de la peste des petits ruminants sur les cultures cellulaire. Rev Elev Méd Vét Pays Trop 21:297–308

Lefevre PC (1980) La peste des petits ruminants (synthese bibliogrpahique). Institut d'Elevage et de Medicine Veterinaire des pays Tropicans 94700 Maisons Alfort, France, p 48
Lefevre PC, Diallo A (1990) Peste des petits ruminants virus. Rev Sci Tech 9:951–965
Li L, Bao J, Wu X, Wang Z, Wang J, Gong M, Liu C, Li J (2010) Rapid detection of peste des petits ruminants virus by a reverse transcription loop-mediated isothermal amplification assay. J Virol Methods 170:37–41
Li W, Jin H, Sui X, Zhao Z, Yang C, Wang W, Li J, Li G (2014) Self-assembly and release of peste des petits ruminants virus-like particles in an insect cell-baculovirus system and their immunogenicity in mice and goats. PLoS One 9:e104791
Libeau G, Diallo A, Calvez D, Lefevre PC (1992) A competitive ELISA using anti-N monoclonal antibodies for specific detection of RP antibodies in cattle and small ruminants. Vet Microbiol 31:147–160
Libeau G, Diallo A, Colas F, Guerre L (1994) Rapid differential diagnosis of RP and PPR using an immunocapture ELISA. Vet Rec 134:300–304
Libeau G, Prehaud C, Lancelot R, CoLas F, Guerre L, Bishop DH, Diallo A (1995) Development of a competitive ELISA for detecting antibodies to the PPR virus using a recombinant nucleoprotein. Res Vet Sci 58:50–55
Liu F, Wu X, Li L, Liu Z, Wang Z (2013) Expression, purification and characterization of two truncated peste des petits ruminants virus matrix proteins in *Escherichia coli*, and production of polyclonal antibodies against this protein. Protein Expr Purif 91:1–9
Liu F, Wu X, Li L, Liu Z, Wang Z (2014a) Formation of peste des petits ruminants spikeless virus-like particles by co-expression of M and N proteins in insect cells. Res Vet Sci 96:213–216
Liu F, Wu X, Liu W, Li L, Wang Z (2014b) Current perspectives on conventional and novel vaccines against peste des petits ruminants. Vet Res Commun 38:307–322
Liu F, Wu X, Zhao Y, Li L, Wang Z (2014c) Budding of peste des petits ruminants virus-like particles from insect cell membrane based on intracellular co-expression of peste des petits ruminants virus M, H and N proteins by recombinant baculoviruses. J Virol Methods 207:78–85
Liu W, Wu X, Wang Z, Bao J, Li L, Zhao Y, Li J (2014d) Virus excretion and antibody dynamics in goats inoculated with a field isolate of peste des petits ruminants virus. Transbound Emerg Dis 60(suppl 2):63–68
Liu F, Wu X, Zou Y, Li L, Liu S, Chi T, Wang Z (2015a) Small interfering RNAs targeting peste des petits ruminants virus M mRNA increase virus-mediated fusogenicity and inhibit viral replication *in vitro*. Antiviral Res 123:22–26
Liu F, Wu X, Zou Y, Li L, Wang Z (2015b) Peste des petits ruminants virus-like particles induce both complete virus-specific antibodies and virus neutralizing antibodies in mice. J Virol Methods 213:45–49
Lund BT, Barrett T (2000) Rinderpest virus infection in primary bovine skin fibroblasts. Arch Virol 145:1231–1237
Ma X, Yang X, Nian X, Zhang Z, Dou Y, Zhang X, Luo X, Su J, Zhu Q, Cai X (2015) Identification of amino-acid residues in the V protein of peste des petits ruminants essential for interference and suppression of STAT-mediated interferon signaling. Virology 483:54–63
Mahapatra M, Parida S, Egziabher BG, Barrett T (2003) Sequence analysis of the phosphoprotein gene of peste des petits ruminants (PPR) virus: editing of gene transcript. Virus Res 96:85–98
Mahapatra M, Parida S, Baron MD, Barrett T (2006) Matrix protein and glycoproteins F and H of Peste-des-petits-ruminants virus function better as a homologous complex. J Gen Virol 87:2021–2029
Malik YS, Singh D, Chandrashekar KM, ShuklaS SK, Vaid N, Chakravarti S (2011) Occurrence of dual infection of peste-des-petits-ruminants and goatpox in indigenous goats of central India. Transbound Emerg Dis 58:268–273
Manjunath S, Kumar GR, Mishra BP, Mishra B, Sahoo AP, Joshi CG, Tiwari AK, Rajak KK, Janga SC (2015) Genomic analysis of host – Peste des petits ruminants vaccine viral transcriptome uncovers transcription factors modulating immune regulatory pathways. Vet Res 46:15
Mbyuzi AO, Komba EV, Kimera SI, Kambarage DM (2014) Sero-prevalence and associated risk factors of peste des petits ruminants and contagious caprine pleuro-pneumonia in goats and sheep in the Southern Zone of Tanzania. Prev Vet Med 116:138–144

McIlhatton MA, Curran MD, Rima BK (1997) Nucleotide sequence analysis of the large (L) genes of phocine distemper virus and canine distemper virus (corrected sequence). J Gen Virol 78:571–576
Meng X, Dou Y, Cai X (2014) Ultrastructural features of PPRV infection in Vero cells. Virol Sin 29:311–313
Mondal AK, Chottopadhyay AP, Sarkar S, Saha GR, Bhowmik MK (1995) Report on epizootiological and clinicopathological observations on peste des petits ruminants (PPR) of goats in West Bengal. Indian J Anim Health 34:145–148
Mondal B, Sreenivasa BP, Dhar P, Singh RP, Bandyopadhyay SK (2001) Apoptosis induced by *peste des petits* ruminants virus in goat peripheral blood mononuclears cells. Virus Res 73:113–119
Mondal B, Sen A, Chand K, Biswas SK, De A, Rajak KK, Chakravarti S (2009) Evidence of mixed infection of peste des petits ruminants virus and bluetongue virus in a flock of goats as confirmed by detection of antigen, antibody and nucleic acid of both the viruses. Trop Anim Health Prod 41:1661–1667
Mornet P, Orue J, Gilbert Y, Thiery G, Sow M (1956) La Peste des Petits ruminants en Mrique occidentale francaise : ses rapports avec la Peste bovine. Rev Elev Med Vet Pays Trop 9:313–342
Munir M, Zohari S, Berg M (2013) Molecular biology and pathogenesis of peste des petits ruminants virus. Springer briefs in animal sciences. Springer, Berlin/New York, pp 23–32
Muniraju M, Mahapatra M, Ayelet G, Babu A, Olivier G, Munir M, Libeau G,Batten C, Banyard AC, Parida S (2014a) Emergence of Lineage IV Peste des Petits Ruminants Virus in Ethiopia: Complete Genome Sequence of an Ethiopian Isolate 2010. Transbound Emerg Dis 63(4):435–442
Muniraju M, Munir M, Parthiban AR, Banyard AC, Bao J, Wang Z, Ayebazibwe C, Ayelet G, El Harrak M, Mahapatra M, Libeau G, Batten C, Parida S (2014b) Molecular evolution of peste des petits ruminants virus. Emerg Infect Dis 20:2023–2033
Muniraju M, Mahapatra M, Buczkowski H, Batten C, Banyard AC, Parida S (2015) Rescue of a vaccine strain of peste des petits ruminants virus: *in vivo* evaluation and comparison with standard vaccine. Vaccine 33:465–471
Muthuchelvan D, Sanyal A, Singh RP, Hemadri D, Sen A, Sreenivasa BP, Singh RK, Bandyopadhyay SK (2005) Comparative sequence analysis of the large polymerase protein (L) gene of peste-des-petits ruminants (PPR) vaccine virus of Indian origin. Arch Virol 150:2467–2481
Muthuchelvan D, De A, Debnath B, Choudhary D, Venkatesan G, Rajak KK, Sudhakar SB, Himadri D, Pandey AB, Parida S (2014) Molecular characterization of peste-des-petits ruminants virus (PPRV) isolated from an outbreak in the Indo-Bangladesh border of Tripura state of North-East India. Vet Microbiol 175:132–138
Muthuchelvan D, Rajak KK, Ramakrishnan MA, Choudhary D, Bhadouriya S, Saravanan P, Pandey AB, Singh RK (2015) Peste-des-petits-ruminants: an Indian perspective. Adv Anim Vet Sci 3:422–429
Nanda SK, Baron MD (2006) Rinderpest virus blocks type 1 and type 2 interferon action: role of structural and non-structural proteins. J Virol 80:7555–7568
Nanda YP, Chatterjee A, Purohit AK, Diallo A, Innui K, Sharma RN, Libeau G, Thevasagayam JA, Bruning A, Kiching RP, Anderson J, Barrett T, Taylor WP (1996) The isolation of peste des petits ruminants virus from Northern India. Vet Microbiol 51:207–216
Nawthane DR, Tayler WP (1979) Experimental infection of domestic pigs with the virus of peste des petits ruminants. Trop Anim Health Prod 11:120–122
Norrby E, Utter G, Orvell C, Appel MJ (1986) Protection against canine distemper virus in dogs after immunization with isolated fusion protein. J Virol 58(2):536–541
Obi TU, Ojeh CK (1989) Dot enzyme immuno assay for visual detection of peste des petits ruminants virus antigen from infected caprine tissues. J Clin Microbiol 27:2096–2099
Obi TU, Pattrick D (1984) The detection of peste des petits ruminants (PPR) virus antigen by agar gel precipitation test and counter immunoelectrophoresis. J Hyg 93:579–586
Obi TU, Ojo MO, Durojaiye OA, Kasali OB, Akpavie S, Opasina DB (1983) Peste des petits ruminants (PPR) in goats in Nigeria: clinical, microbiological and pathological features. Zentralbl Veterinarmed B 30:751–761

Obi TU, McCullough KC, Taylor WP (1990) The production of peste des petits ruminants hyperimmune sera in rabbits and their application in virus diagnosis. J Vet Med 37:345–352
OIE -World Organisation for Animal Health (2013) Manual of diagnostic tests and vaccines for terrestrial animals. OIE, Paris, pp 1–14
OIE-World Organization for Animal Health (2014) Resolution adopted by the world assembly of delegates of the OIE. OIE, Paris
Olaleye OP, Ikede BO, Durojaiye OA (1989) Fluid, electrolyte and haematological changes in experimental peste des petits ruminants virus infection in West African dwarf goats. Cytobios 58:39–51
Ozkul A, Akca Y, Alkan F, Barrett T, Karaoglu T, Daglap SB, Anderson J, Yesilbag K, Cokcaliskan C, Genacy A, Burgu I (2002) Prevalence, distribution and host range of peste des petits ruminants virus, Turkey. Emerg Infect Dis 8:708–712
Padhi A, Ma L (2014) Genetic and epidemiological insights into the emergence of peste des petits ruminants virus (PPRV) across Asia and Africa. Sci Rep 4:7040
Palaniswami KS, Thangavelu A, Velmurugan R (2005) Development of thermostable peste des petits ruminants (PPR) virus vaccine and assessment of molecular changes in the F Gene. In: Makkar HPS, Villjoen GJ (eds) Applications of gene-based technologies for improving animal production and health in developing countries. IAEA, Springer publisher Netherlands, Dordrecht, pp 673–678
Pandey KD, Baron MD, Barrett T (1992) Differential diagnosis of rinderpest and PPR using biotinylated cDNA probes. Vet Rec 131:199–200
Parida S, Couacy-Hymann E, Pope RA, Mahapatra M, El Harrak M, Brownlie J, Banyard AC (2015a) Peste des petitis Ruminants Virus. In: Munir M (ed) Pathology of PPR. Springer, Heidelberg, pp 51–68
Parida S, Muniraju M, Mahapatra M, Muthuchelvan D, Buczkowski H, Banyard AC (2015b) Peste des petits ruminants. Vet Microbiol 181(1–2):90–106
Pawar RM, Raj GD, Kumar TM, Raja A, Balachandran C (2008) Effect of siRNA mediated suppression of signaling lymphocyte activation molecule on replication of peste des petits ruminants virus *in vitro*. Virus Res 136:118–123
Plowright W, Ferris RD (1961) Studies with rinderpest virus in tissue culture. III. The stability of cultured virus and its use in virus neutralization tests. Arch Gesamte Virusforsch 11:516–533
Polci A, Cosseddu GM, Ancora M, Pinoni C, El Harrak M, Sebhatu TT, Ghebremeskel E, Sghaier S, Lelli R, Monaco F (2015) Development and preliminary evaluation of a new real-time RT-PCR assay For detection of peste des petits ruminants virus genome. Transbound Emerg Dis 62:332–338
Prasad V, Satyavathi VV, Sanjaya KM, Valli AK, Shaila MS, Lakshmi Sita G (2004) Expression of biologically active Haemagglutinin-neuraminidase protein of peste des petits ruminants virus in transgenic pigeon pea (*Cajanus cajan* (L) Millsp). Plant Sci 166:199–205
Provost A (1970) Observation sur tes muco-antieorps nasaux des bovins. Rev Elev Med Vet Pays Trop 23:283
Provost A, Maurice Y, Borredon (1971) Protection antipestic conférée aux bovins parle virus de la rougéole II. Application aux veaux nés de mères elles-mêmes vaccines avec la souche MB 113Y. Rev Elev Méd Vét Pays Trop 24:167–172
Raghvendra L, Setty DRL, Raghavan R (2000) Clinco-biochemical changes insheep and goats experimentally infected with Vero cell adapted peste des petits ruminants (PPR) virus. Int J Anim Sci 15:171–173
Rahman MM, Shaila MS, Gopinathan KP (2003) Baculovirus display of fusion protein of Peste des petits ruminants virus and hemagglutination protein of Rinderpest virus and immunogenicity of the displayed proteins in mouse model. Virology 317:36–49
Rajak KK, Sreenivasa BP, Hosamani M, Singh RP, Singh SK, Singh RK, Bandyopadhyay SK (2005) Experimental studies on immunosuppressive effects of peste des petits ruminants (PPR) virus in goats. Comp Immunol Microbiol Infect Dis 28:287–296
Raj GD, Rajanathan TM, Kumar CS, Ramathilagam G, Hiremath G, Shaila MS (2008) Detection of peste des petits ruminants virus antigen using immunofiltration and antigen-competition ELISA methods. Vet Microbiol 129(3–4):246–251

Rey Nores JE, Anderson J, Butcher RN, Libeau G, McCullough KC (1995) Rinderpest virus infection of bovine peripheral blood monocytes. J Gen Virol 76:2779–2791
Riyesh T, Balamurugan V, Sen A, Bhanuprakash V, Venkatesan G, Yadav V, Singh RK (2011) Evaluation of efficacy of stabilizers on the thermostability of live attenuated thermo-adapted peste des petits ruminants vaccines. Virol Sin 26:324–337
Roeder PL, Abraham G, Kenfe BT (1994) PPR in Ethiopian goats. Tropl Anim Health Prod 26:69–73
Roger F, Yigezu LM, Hurard C, Libeau G, Mebratu GY, Diallo A, Faye A (2000) Investigations on a new pathological condition of camels in Ethiopia. J Camel Pract Res 7:163–165
Rojas JM, Moreno H, García A, Ramírez JC, Sevilla N, Martín V (2014a) Two replication-defective adenoviral vaccine vectors for the induction of immune responses to PPRV. Vaccine 32:393–400
Rojas JM, Moreno H, Valcárcel F, Peña L, Sevilla N, Martín V (2014b) Vaccination with recombinant adenoviruses expressing the peste des petits ruminants virus F or H proteins overcomes viral immunosuppression and induces protective immunity against PPRV challenge in sheep. PLoS One 9:e101226
Romero CH, Barrett T, Kitching RP, Bostock C, Black DN (1995) Protection of goats against peste des petits ruminants with recombinant capripoxvirus expressing the fusion and haemagglutinin protein genes of rinderpest virus. Vaccine 44:151–163
Rossiter PB, Jessett DM, Taylor WP (1985) Microneutralisation systems for use with different strains of peste des petits ruminants virus and rinderpest virus. Trop Anim Health Prod 17:75–81
Saeed IK, Ali YH, AbdulRahman MB, Mohammed ZA, Osman HM, Taha KM, Musa MZ, Khalafalla AI (2015) Mixed infection of peste des petits ruminants virus (PPRV) and other respiratory viruses in dromedary camels in Sudan, an abattoir study. Tropl Anim Health Prod 47:995–998
Saliki JT, House James A, Charles A, Mebus D, Edward J (1994) Comparison of monoclonal antibodies-based sandwich ELISA and virus isolation for detection of PPR virus in goats tissue and secretion. J Clin Microbiol 32:1349–1353
Saliki J, Libeau G, House JA, Mebus A, Dubov EJ (1993) Monoclonal antibody based blocking ELISA for specific detection and titration of PPR virus antibody in caprine and ovine sera. J Clin Microbiol 31:1075–1082
Saravanan P, Singh RP, Balamurugan V, Dhar P, Sreenivasa BP, Muthuchelvan D, Sen A, George AA, Singh RK, Bandyopadhyay SK (2004) Development of a N gene-based PCR-ELISA for detection of peste-des-petits-ruminants virus in clinical samples. Acta Virol 48:249–255
Saravanan P, Balamurugan V, Sen A, Bikash S, Singh RK (2006) Development of Dot-ELISA for diagnosis of peste des petits ruminants (PPR) in small ruminants. J Appl Anim Res 30:121–124
Saravanan P, Balamurugan V, Sen A, Sarkar J, Sahay B, Razak KK, Hosamani M, Rasool TJ, Yadav MP, Singh RK (2007) A mixed infection of peste des petits ruminants and Orf in a goat farm, Shahjhahanpur, India. Vet Rec 160:410–412
Saravanan P, Balamurugan V, Sen A, Sreenivasa BP, Singh RP, Bandyopadhyay SK, Singh RK (2010a) Long term immune response of goats to a Vero cell adapted live attenuated homologous PPR vaccine. Indian Vet J 87:1–3
Saravanan P, Sen A, Balamurugan V, Rajak KK, Bhanuprakash V, Palaniswami KS, Nachimuthu K, Thangavelu A, Dhinakarraj G, Hegde R, Singh RK (2010b) Comparative efficacy of peste des petits ruminants (PPR) vaccines. Biologicals 38:479–485
Sarkar J, Sreenivasa BP, Singh RP, Dhar P, Bandyopadhyay SK (2003) Comparative efficacy of various chemical stabilizers on the thermostability of a live-attenuated peste des petits ruminants (PPR) vaccine. Vaccine 21:4728–4735
Scheid A, Caliguiri LA, Compans RW, Choppin PW (1972) Isolation of paramyxovirus glycoproteins. Association of both haemagglutinating and neuraminidase activities with the larger SV5 glycoprotein. Virology 50:640–652
Scott GR (1990) Rinderpest and PPR. In: Virus disease of food animals. Disease monographs, vol 2. E.P.S. Gibbs Acad. Press, London

Sen A, Balamurugan V, Rajak KK, Chakravarti S, Bhanuprakash V, Singh RK (2009) Role of heavy water in biological sciences with an emphasis on thermostabilization of vaccines. Expert Rev Vaccines 8:1587–1602

Sen A, Saravanan P, Balamurugan V, Bhanuprakash V, Venkatesan G, Sarkar J, Rajak KK, Ahuja A, Yadav V, Sudhakar SB, Parida S, Singh RK (2014) Detection of subclinical peste des petits ruminants virus infection in experimental cattle. Virusdisease 25:488–492

Seth S, Shaila MS (2001) The hemagglutinin-neuraminidase protein of peste des petits ruminants virus is biologically active when transiently expressed in mammalian cells. Virus Res 75:169–177

Şevik M, Sait A (2015) Genetic characterization of peste des petits ruminants virus, Turkey, 2009–2013. Res Vet Sci 101:187–195

Sghaier S, Cosseddu GM, Ben Hassen S, Hammami S, Ammar HH, Petrini A, Monaco F (2015) Peste des petits ruminants virus, Tunisia, 2012–2013. Emerg Infect Dis 20:2184–2186

Shaila MS, Purushothaman V, Bhavasar D, Venugopal K, Venkatesan RA (1989) Peste des petits ruminants in India. Vet Rec 125:602

Shaila MS, Shamaki D, Forsyth MA, Drallo A, Groatley L, Kitching RP, Barrett T (1996) Geographic distribution and epidemiology of peste des petits ruminants viruses. Virus Res 43:149–153

Sidhu MS, Husar W, Cook SD, Dowling PC, Udem SA (1993) Canine distemper terminal and intergenic non-protein coding neucleotide sequences: completion of the entire CDV genome sequence. Virology 193:66–72

Singh RP (2011) Control strategies for peste des petits ruminants in small ruminants of India. Rev Sci Tech 30:879–887

Singh RP, Saravanan P, Sreenivasa BP, Singh RK, Bandyopadhyay SK (2004a) Prevalence and distribution of peste des petits ruminants virus infection in small ruminants in India. Rev Sci Tech 23:807–819

Singh RP, Sreenivasa BP, Dhar P, Bandyopadhyay SK (2004b) Development A sandwich-ELISA for the diagnosis of peste des petits ruminants (PPR) infection in small ruminants using anti-nucleocapsid protein monoclonal antibody. Arch Virol 149:2155–2170

Singh RP, Sreenivasa BP, Dhar P, Shah LC, Bandyopadhyay SK (2004c) Development of a monoclonal antibody based competitive-ELISA for detection and titration of antibodies to peste des petits ruminants (PPR) virus. Vet Microbiol 98:3–15

Singh RK, Balamurugan V, Bhanuprakash V, Sen A, Saravanan P, Yadav MP (2009) Control and eradication of peste des petits ruminants in sheep and goats in India: possibility. Vet Ital 45:449–462

Singh RP, De UK, Pandey KD (2010) Virological and antigenic characterization of two Peste des Petits Ruminants (PPR) vaccine viruses of Indian origin. Comp Immunol Microbiol Infect Dis 33:343–353

Sinnathamby G, Naik S, Renukaradhya GJ, Rajasekhar M, Nayak R, Shaila MS (2001) Recombinant hemagglutinin protein of rinderpest virus expressed in insect cells induces humoral and cell mediated immune responses in cattle. Vaccine 19:3870–3876

Soltan MA, Abd-Eldaim MM (2014) Emergence of peste des petits ruminants virus lineage IV in Ismailia Province, Egypt. Infect Genet Evol 28:44–47

Sreenivasa BP, Dhar P, Singh RP, Bandyopadhyay SK (2000) Evaluation of an indigenously developed homologous live-attenuated cell culture vaccine against peste des petitsruminants of small ruminants. In: Proceedings of "XX annual conference of IndianAssociation of Veterinary microbiologists, Immunologistsand specialists in infectious diseases and National symposium on trends in vaccinology for animal diseases", 14–16 Oct 2000, GBPUA&T, Pantnagar, Uttarkhand, India

Sreenivasa BP, Singh RP, Mondal B, Dhar P, Bandyopadhyay SK (2006) Marmoset B95a cells: a sensitive system for cultivation of peste des petits ruminants (PPR) virus. Vet Res Commun 30:103–108

Su W, Xing C, Wu Y, Wang Y, Ding H, He H (2015) Complete genome sequence of a novel mutant of peste des petits ruminants virus obtained from china. Genome Announc 3:e01504–e01514

Tajima M, Ushijima T (1970) The pathogenesis of rinderpest in the lymph nodes of cattle. Light and electron microscopic studies. Am J Pathol 62:221–235

Tatsuo H, Ono N, Tanaka K, Yanagi Y (2000) SLAM (CDw150) is a cellular receptor for measles virus. Nature 406:893–897

Tatsuo H, Ono N, Yanagi Y (2001) Morbilliviruses use signaling lymphocyte activation molecules (CD150) as cellular receptors. J Virol 75:5842–5850

Taylor WP (1979) Serological studies with the virus of peste des petits ruminants in Nigeria. Res Vet Sci 26:236–242

Taylor WP (1984a) The distribution and epidemiology of peste des petits ruminants in the Sultanate of Oman. Vet Microbiol 22:341–352

Taylor WP (1984b) The distribution and epidemiology of peste des petits ruminants. Prev Vet Med 2:157–166

Taylor WP, Plowright W, Pilinger R, Rampton CS, Staple REF (1965) Studies on the pathogenesis of rinderpest in experimental cattle. IV. proliferation of the virus following contact infection. J Hyg Camb 63:263

Taylor WP, Abusaidy S, Barret T (1990) The epidemiology of PPR in the sultanate of Oman. Vet Micro 22:341–352

Tober C, Seufert M, Schneider H, Billeter MA, Johnston IC, Niewiesk S, Ter Meulen V, Schneider-Schaulie S (1998) Expression of measless virus V protein is associated with pathogenicity and control of viral RNA synthesis. J Virol 72:8124–8132

Toplu N, Oguzoglu TC, Albayrak H (2012) Dual infection of fetal and neonatal small ruminants with border disease virus and peste des petits ruminants virus (PPRV): neuronal tropism of PPRV as a novel finding. J Comp Pathol 146:289–297

Tripathi BN, Chhattopadhyay SK, Parihar NS (1996a) Clinicopathological studies on an outbreak of peste des petitsruminants in goats and sheep. Indian J Vet Pathol 20:99

Tripathi BN, SK Chattopadhyay, NS Parihar (1996b) Clinical and pathological studies on peste des petits ruminants – an emerging disease in Indian goats. In: Proceeding of international conference on goats, Beijing, pp 742–744.

Truong T, Boshra H, Embury-Hyatt C, Nfon C, Gerdts V, Tikoo S, Babiuk LA, Kara P, Chetty T, Mather A, Wallace DB, Babiuk S (2014) Peste des petits ruminants virus tissue tropism and pathogenesis in sheep and goats following experimental infection. PLoS One 9:e87145

Wang Q, Dou Y, Yang X, Meng X, Zhai J, Zhu X, Luo X, Chen L, Cai X (2013a) Prokaryotic expression of F protein from PPRV and characterization of its polyclonal antibody. Monoclon Antib Immunodiagn Immunother 32:26–31

Wang Y, Liu G, Shi L, Li W, Li C, Chen Z, Jin H, Xu B, Li G (2013b) Immune responses in mice vaccinated with a suicidal DNA vaccine expressing the hemagglutinin glycoprotein from the peste des petits ruminants virus. J Virol Methods 193:525–530

Wang J, Wang M, Wang S, Liu Z, Shen N, Si W, Sun G, Drewe JA, Cai X (2015a) Peste des petits ruminants virus in Heilongjiang province, China. Emerg Infect Dis 21:677–680

Wang Y, Yue X, Jin H, Liu G, Pan L, Wang G, Guo H, Li G, Li Y (2015b) A suicidal DNA vaccine expressing the fusion protein of peste des petits ruminants virus induces both humoral and cell-mediated immune responses in mice. J Virol Methods 225:35–40

Wild TF, Malvoisin E, Buckland R (1991) Measles virus: both the haemagglutinin and fusion glycoproteins are required for fusion. J Gen Virol 72:439–442

Wohlsein P, Trautwein G, Harder TC, Liess B, Barrett T (1993) Viral antigen distribution in organs of cattle experimentally technique. Rev Elev Med Vet Pays Trop 37:11–15

Woma TY, Kalla DJ, Ekong PS, Ularamu HG, Chollom SC, Lamurde II, Bajehson DB, Tom ND, Aaron GB, Shamaki D, Bailey D, Diallo A, Quan M (2015) Serological evidence of camel exposure to peste des petits ruminants virus (PPRV) in Nigeria. Tropl Anim Health Prod 47:603–606

Woma TY, Qasim AM, Sabi AA, Abraham MN, Olaiya OD, Bailey D, Shamaki D, Quan M (2016) Co-circulation of Peste-des-Petits-Ruminants Virus Asian lineage IV with Lineage II in Nigeria. Transbound Emerg Dis 63(3):235–242

Wosu LO (1985) Agglutination of red blood cells by peste des petits ruminants (PPR) virus. Nigerian Vet J 14:54–58

Wosu LO (1991) HA test for diagnosis of PPR disease in goats with samples from live animals. J Small Anim Res 5:169–172

Wosu LO (1994) Current status of peste des petits ruminants (PPR) disease in small ruminants. A review article. Stud Res Vet Med 2:83–90

Wu X, Li L, Li J, Liu C, Wang Q, Bao JY, Zou Y, Ren W, Wang H, Zhang Y, Lv Y, Liu F, Wang S, Ma H, Wang Z (2016) Peste des Petits Ruminants Viruses Reemerging in China, 2013–2014. Transbound Emer Dis 63(5):e441–e446

Yadav V, Balamurugan V, Bhanuprakash V, Sen A, Bhanot V, Venkatesan G, Riyesh T, Singh RK (2009) Expression of peste des petits ruminants virus nucleocapsid protein in prokaryotic system and its potential use as a diagnostic antigen or immunogen. J Virol Methods 162:56–63

Yamannouchi K, Fukuda A, Kobune F, Yoshikawa Y, Chino F (1974) Pathogenesis of rinderpest virus infection in rabbits. II. Effect of rinderpest virus on the immune functions of rabbits. Infect Immun 9:206–211

Yin C, Chen W, Hu Q, Wen Z, Wang X, Ge J, Yin Q, Zhi H, Xia C, Bu Z (2014) Induction of protective immune response against both PPRV and FMDV by a novel recombinant PPRV expressing FMDV VP1. Vet Res 45:62

Yousuf RW, Sen A, Mondal B, Biswas SK, Chand K, Rajak KK, Gowane GR, Sudhakar SB, Pandey AB, Ramakrishnan MA, Muthuchelvan D (2015) Development of a single-plate combined indirect ELISA (CI-ELISA) for the detection of antibodies against peste-des-petits-ruminants and bluetongue viruses in goats. Small Ruminant Res 124:137–139

Yu R, Fan X, Xu W, Li W, Dong S, Zhu Y, He Y, Tang H, Du R, Li Z (2015) Fine mapping and conservation analysis of linear B-cell epitopes of peste des petits ruminants virus nucleoprotein. Vet Microbiol 175(1):132–138

Zahur AB, Irshad H, Ullah A, Afzal M, Latif A, Ullah RW, Farooq U, Samo MH, Jahangir M, Ferrari G, Hussain M, Ahmad MM (2014) Peste des petits ruminants vaccine (Nigerian strain 75/1) confers protection for at least 3 years in sheep and goats. JBM 2:27–23

Schmallenberg Virus

3

Virginie Doceul, Kerstin Wernike, Damien Vitour, and Eve Laloy

3.1 Timeline of Schmallenberg Virus Infection and Geographical Repartition

In summer and autumn 2011, an unidentified disease was reported in cattle in Germany, the Netherlands and Belgium (Hoffmann et al. 2012). The clinical signs were unspecific, including fever, decreased milk production and diarrhoea, but no known agent could be identified in the affected cattle. In November 2011, genomic sequences of a new virus were detected by the Friedrich-Loeffler-Institut, Germany, in a pool of blood samples taken from acutely infected cows. This newly identified virus was named Schmallenberg virus (SBV) after the place of origin of the samples, a cattle farm near the city of Schmallenberg (North Rhine-Westphalia) (Hoffmann et al. 2012). Phylogenetic analysis showed that SBV belongs to the Simbu serogroup within the *Orthobunyavirus* genus and the *Bunyaviridae* family. This was the first report of the presence of a virus from this serogroup in Europe. In December 2011, SBV was detected in samples from malformed lambs and calves in Germany, the Netherlands and Belgium (van den Brom et al. 2012). This was the result of dams being infected with SBV during summer and autumn 2011. At the beginning of 2012, other European countries including France, the United Kingdom, Luxemburg, Italy and Spain reported confirmed cases of SBV infection in

V. Doceul (✉) • D. Vitour • E. Laloy
UMR 1161 Virologie, ANSES, Laboratoire de santé animale, LabEx IBEID,
14 rue Pierre et Marie-Curie, 94701 Maisons-Alfort Cedex, France

UMR 1161 Virologie, INRA, 7 avenue du Général-de-Gaulle,
Maisons-Alfort Cedex, 94704, France

UMR 1161 Virologie, Université Paris-Est, École nationale vétérinaire d'Alfort,
7 avenue du Général de Gaulle, 94704 Maisons-Alfort Cedex, France
e-mail: virginie.doceul@vet-alfort.fr

K. Wernike
Friedrich-Loeffler-Institut, Südufer 10, 17493 Greifswald – Insel Riems, Germany

J. Bayry (ed.), *Emerging and Re-emerging Infectious Diseases of Livestock*,
DOI 10.1007/978-3-319-47426-7_3

malformed animals. Later in 2012 and in 2013, cases of SBV infection occurred in Denmark, Ireland, Switzerland, Austria, the Czech Republic, Poland, Latvia, Sweden, Finland, Norway, Estonia, Slovenia, Hungary, Croatia and Turkey (Afonso et al. 2014; Doceul et al. 2013; Yilmaz et al. 2014). Antibodies against SBV have also been found in Lithuania (Lazutka et al. 2014), Greece (Chaintoutis et al. 2014), Serbia and Romania, suggesting that the virus has been present in these countries. In the 2 years following its identification, SBV has then spread rapidly to many countries along the four cardinal directions and most likely to the whole of Europe (EFSA 2014).

3.2 Virus Classification and Origin

SBV belongs to the *Orthobunyavirus* genus within the *Bunyaviridae* family that is composed of more than 350 viruses divided into five genera: *Orthobunyavirus*, *Hantavirus*, *Nairovirus*, *Phlebovirus* and *Tospovirus*. Viruses from the latter genus infect plants, whereas viruses from the other genera infect vertebrates. The *Orthobunyavirus* genus consists of more than 170 viruses including viruses responsible for disease in humans (Oropouche virus or La Crosse virus) and in ruminants (Aino virus, Akabane virus or Cache Valley virus).The International Committee on Taxonomy of Viruses has recognised 18 serogroups and 48 species within the *Orthobunyavirus* genus (International Committee on Taxonomy of Viruses 2014). SBV belongs to the Simbu serogroup that includes more than 25 viruses such as Simbu, Oropouche, Akabane, Douglas, Sathuperi, Aino, Shamonda and Peaton viruses.

When discovered, SBV was found to share the most similar sequences with Shamonda, Aino and Akabane viruses, three viruses that can be found in cattle (Hoffmann et al. 2012). None of these viruses have been detected in continental Europe. Later, a phylogenetic study found that the M segment of SBV is derived from Sathuperi virus, while the S and L segments are derived from Shamonda virus, suggesting that SBV could be a reassortant between Sathuperi virus and Shamonda virus (Yanase et al. 2012). Another study based on phylogenetic and serologic analysis has then suggested that SBV belongs to the Sathuperi virus species and may be an ancestor of Shamonda virus (Goller et al. 2012).

In situ hybridisation and immunohistochemistry analysis did not detect the presence of SBV RNA or antigen in ruminant brain tissues archived from 1961 to 2010 in Germany (Gerhauser et al. 2014). Furthermore, SBV-specific antibodies were not found in samples from domestic or wild ruminants collected prior to summer 2011 in Germany, the Netherlands, Belgium or France (Garigliany et al. 2012a; Mouchantat et al. 2015; Veldhuis et al. 2013; Wernike et al. 2014a; Zanella et al. 2015). This suggests that the virus was not present in Northwestern Europe before 2011. However, the geographical origin of SBV introduction remains unknown. So far, closely related Simbu serogroup viruses have been detected in Africa, Asia and Australia (Yanase et al. 2005). For instance, Akabane virus has been reported in Australia, Japan, Korea, Israel, Saudi Arabia, Kenya, Sudan, Cyprus and Turkey

(Taylor and Mellor 1994); Sathuperi virus in India, Japan and Nigeria (Causey et al. 1972; Saeed et al. 2001; Yanase et al. 2004); Shamonda virus in Nigeria, Japan and Korea; and Aino virus in Japan, Korea and Australia (Lievaart-Peterson et al. 2012). It is possible that the transport by aircraft of infected midges on flowers or on exotic animals from an area where SBV is enzootic is involved in the introduction of the virus in Europe (Gale et al. 2015). It is also interesting to notice that SBV has emerged in the same area as bluetongue virus serotype 8 (BTV-8) 5 years earlier (Wilson and Mellor 2009). One could hypothesise that both viruses have been introduced in Northern Europe via a similar route.

3.3 Virus Structure and Genome

Members of the *Bunyaviridae* family are enveloped viruses with a genome composed of three single-sense negative-stranded RNA segments designed as L (large), M (medium) and S (small) (Fig. 3.1a). These segments encode four structural proteins and two nonstructural proteins (Fig. 3.1b). The S segment encodes the nucleoprotein N and the nonstructural protein NSs. The M segment encodes a polyprotein, which is cleaved into two surface glycoproteins (Gn and Gc) and the nonstructural protein NSm. The L segment encodes the RNA-dependent RNA polymerase (RdRp or L protein). Electron microscopic analysis has revealed that SBV virions are about 100 nm in diameter and confirmed that they contain three genome segments (Fig. 3.1a).

Knowledge on SBV structure and life cycle remains poor and is mostly deduced from data accumulated on other bunyaviruses. Their genome is encapsidated as a ribonucleoprotein (RNP) within the virion where each segment is associated to many copies of the nucleoprotein N and a few copies of the polymerase (Briese et al. 2013). The nucleoprotein N is the most abundant protein in the virion and in infected cells. Its major role is encapsidation of the genome, but it may be involved in transcription and replication of the viral RNA (Dong et al. 2013; Elliott 2014). The surface glycoproteins Gc and Gn of bunyaviruses are type I integral membrane proteins that are embedded in the envelope. Their C-terminal cytoplasmic tails are towards the intraviral space, while the N-terminal parts are in contact with the outer environment. Their functions have not been fully elucidated but they are necessary for the budding of orthobunyaviruses. They may also be involved in virus fusion and entry into the cells (Strandin et al. 2013). Similarly to Akabane virus (Kobayashi et al. 2007), the M segment of SBV is the most variable one as it possesses a hypervariable region within the sequence coding for the glycoprotein Gc (Coupeau et al. 2013; Fischer et al. 2013a; Hofmann et al. 2015; Rosseel et al. 2012). This hypervariable region was predominantly identified in malformed offspring or after several passages of the virus in cell culture. In viraemic animals, the virus seems to be more stable (Wernike et al. 2015a). The L protein of bunyaviruses has a double role. It acts both as an RNA polymerase and an endonuclease. As an endonuclease, the L protein cleaves cellular messenger RNAs, leading to capped primers that initiate transcription of viral messenger RNAs; this process is known as 'cap-snatching' (Briese et al. 2013). The NSs protein of orthobunyaviruses is not essential for virus

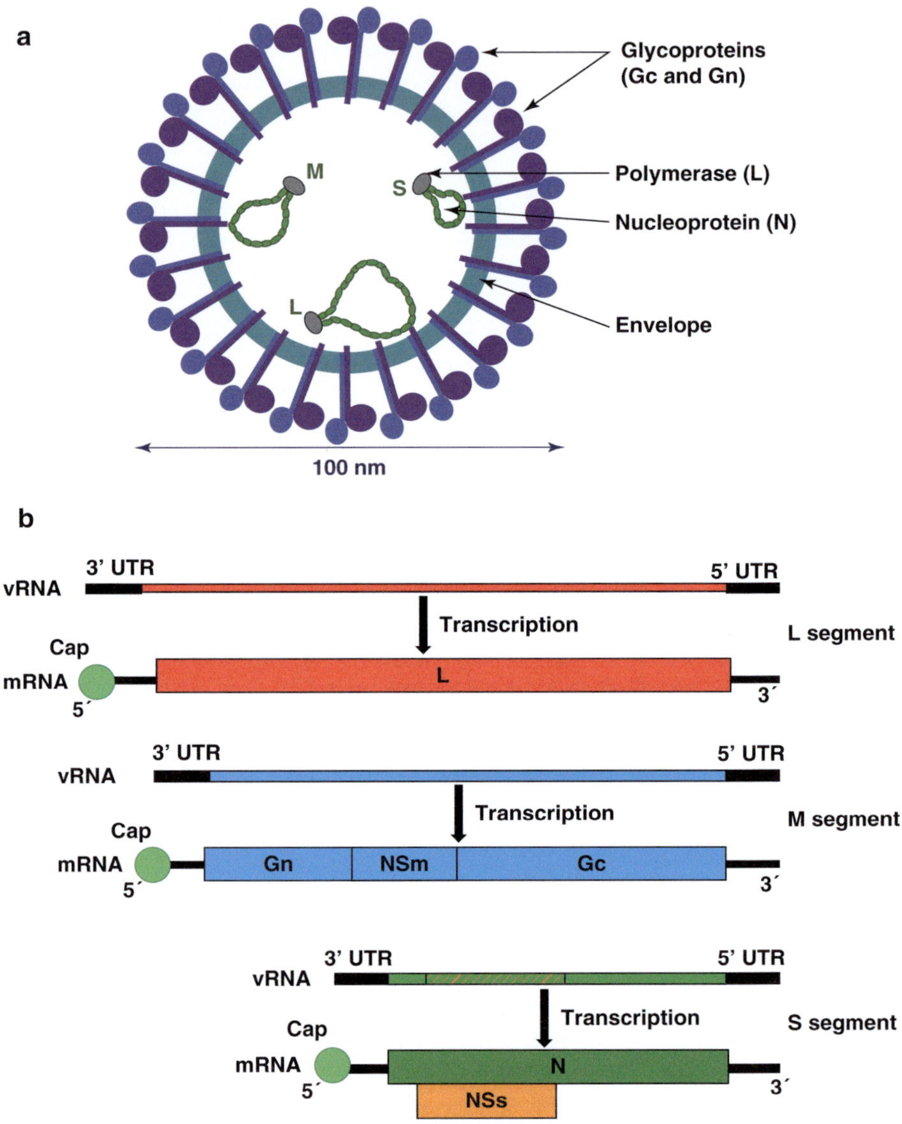

Fig. 3.1 Schematic representation of the SBV virion (**a**) and genome organisation (**b**). (**a**) The genome is composed of three negative-sense single-stranded RNA segments (large (L), medium (M) and small (S)) that form complexes with the nucleoprotein and the viral polymerase to constitute viral RNPs. (**b**) Each viral segment includes a coding region flanked by 3′ and 5′ non-coding sequences (UTR). The nucleotides at the terminal 3′ and 5′ ends are complementary, leading to the formation of stable panhandle structures used as a promoter for transcription and replication. The L polymerase transcribes genomic RNAs (vRNA) into mRNAs, which contain host-derived sequences at the 5′ end (cap). L segment mRNA is translated into the L protein. Translation of the M segment leads to the synthesis of a polyprotein that is proteolytically cleaved to generate the viral glycoproteins Gn and Gc and the nonstructural protein NSm. S segment mRNA is translated into the N protein and a second nonstructural protein NSs, which is translated from an overlapping reading frame by leaky ribosomal scanning

growth but is involved in viral pathogenesis as it inhibits host cell transcription and interferon (IFN) production (Barry et al. 2014; Varela et al. 2013). The function of the NSm protein is still unknown, but it may participate in viral assembly, probably in the Golgi apparatus (Eifan et al. 2013).

3.4 Host Range

3.4.1 Ruminant Species

SBV infects predominantly ruminants. Direct and indirect detection of SBV associated with clinical signs has been identified in adult cattle, sheep and goats or in their offspring (van den Brom et al. 2012; Garigliany et al. 2012b; Herder et al. 2012; Wernike et al. 2014b). SBV infection has also been detected in several species among wild and exotic ruminants (alpacas, water buffalo, elk, European bison, red deer, fallow deer, roe deer, sika deer, llama, reindeer, mouflon and chamois) (EFSA 2014; Laloy et al. 2014). After experimental inoculation, SBV RNA was detected in alpacas and llamas up to 7 days after infection without clinical signs (Schulz et al. 2015). SBV RNA was also found in the serum of a 6-month-old elk found in a national park in Poland (Larska et al. 2013a). In addition, antibodies against SBV have been detected in several ruminant species from two zoological parks in the United Kingdom (bongo, banteng, Congo buffalo, European bison, gaur, gemsbok, greater kudu, moose, Nile lechwe, Pere David's deer, reindeer, roan antelope, sitatunga, scimitar-horned oryx, sitatunga, giraffe and yak) (EFSA 2014; Molenaar et al. 2015). However, clinical signs related to SBV infection have not been described in these wild and exotic species.

3.4.2 Non-ruminant Species

SBV has been detected indirectly in several non-ruminant species. Antibodies against SBV have been found in free-ranging wild boars in Belgium and Germany (Desmecht et al. 2013; Mouchantat et al. 2015). After experimental SBV infection, some domestic pigs transiently seroconverted but did not show clinical signs, and no SBV RNA was detected (Poskin et al. 2014), suggesting they are receptive to SBV but do not develop the disease. SBV infection can also occur in dogs, albeit few cases have been reported. In Sweden, antibodies against SBV have been found in one dog (Wensman et al. 2013). In France, SBV RNA was found in the brain of a puppy showing torticollis and degenerative encephalopathy, while antibodies against SBV were detected in the mother (Sailleau et al. 2013a). However, serologic surveys performed on samples from 132 dogs in Belgium (Garigliany et al. 2013) and from wild carnivores sampled between 2011 and 2013 in Germany (Mouchantat et al. 2015) did not detect any antibodies against SBV. Moreover, antibodies against SBV have been detected in onager, babirusa, Grevy's zebra and Asian elephant in two zoological parks in the United Kingdom (EFSA 2014; Molenaar et al. 2015). As several viruses from the genus *Orthobunyavirus* can cause disease in humans, the possibility of SBV transmission from animals to humans was one of the most important questions to answer at the beginning of the epizootics. Molecular and serological testing has been performed on exposed populations in Germany and in the Netherlands. SBV RNA and antibodies against SBV were not detected (Ducomble et al. 2012; Reusken et al. 2012). It was concluded that the public health risk for SBV infection is 'absent or extremely low' (Reusken et al. 2012).

3.5 Clinical Signs and Lesions

In domestic ruminants, SBV is responsible for a range of clinical signs that are similar to those found in association with Akabane disease. In non-pregnant dairy cows, SBV causes short-lasting non-specific clinical signs including fever, decreased milk production and diarrhoea (Hoffmann et al. 2012). One report suggests that goats infected by SBV may also experience diarrhoea and reduction in milk yield (Helmer et al. 2013). Fever, diarrhoea and decreased milk production have been sometimes reported in sheep from the field, without an obvious causal link with SBV infection (Luttikholt et al. 2014). Under experimental conditions, the viraemic period is short, lasting for 2–6 days (Hoffmann et al. 2012; Laloy et al. 2015; Wernike et al. 2013a).

In contrast to the mild symptoms observed in adult animals, SBV can cause severe congenital malformations summarised as 'arthrogryposis-hydranencephaly syndrome' (AHS), stillbirth or premature birth or the birth of mummified foetuses, when dams are infected during a critical phase of gestation (Fig. 3.2a). After the discovery of SBV in non-pregnant adult cows in Germany in 2011, epizootics of congenital malformations were reported in sheep in the Netherlands. They were associated in about half of the cases with detection of SBV RNA in the brains from the affected newborns (van den Brom et al. 2012), showing the ability of SBV to cross the placenta in pregnant ewes and to cause lesions in the growing embryo or foetus. During pregnancy, SBV might be associated with embryonic or foetal death and abortion. In flocks affected by SBV, increased numbers of repeated oestrus and increased rate of abortions suggest early embryonic or foetal loss and abortions in ewes (Dominguez et al. 2014; Lievaart-Peterson et al. 2012; Luttikholt et al. 2014; Saegerman et al. 2014) and goats (Dominguez et al. 2014; Helmer et al. 2013). In cattle flocks affected by SBV, increased numbers of repeated oestrus and of early embryonic deaths have also been reported (Dominguez et al. 2014). Embryonic or foetal death can result either from anomalies in the environment (the mother, the placenta) or from anomalies in the embryo or foetus itself. The birth of one malformed and one healthy offspring within the same litter has been described in SBV-infected cattle (Wernike et al. 2014b) and sheep (van den Brom et al. 2012). In affected sheep flocks, lambs were either malformed, «dummy» with inability to suckle or normal (van den Brom et al. 2012). Neurological signs have also been described in a newborn calf infected in utero by SBV: hypertonicity, hyperreflexia, depression, blindness, ventrolateral strabismus and inability to stand (Garigliany

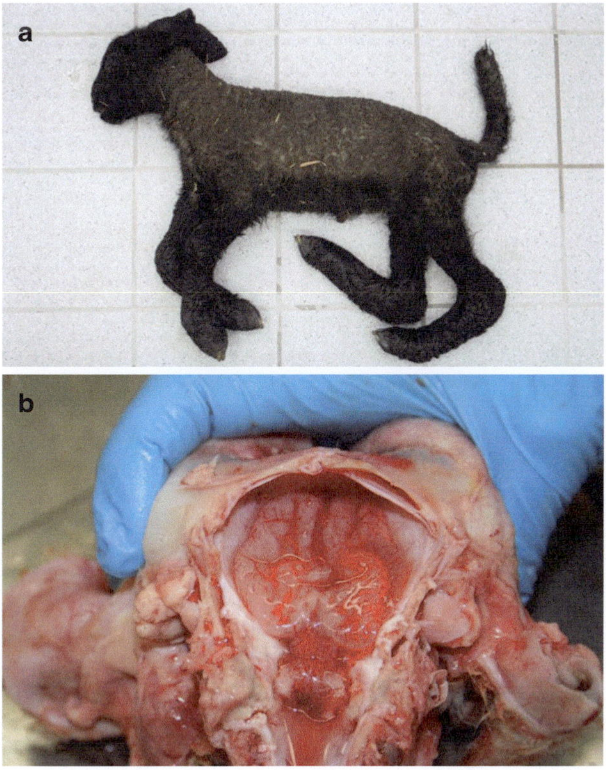

Fig. 3.2 SBV-associated arthrogryposis in a lamb (**a**) and virus-induced hydranencephaly and hypoplasia of the cerebellum (**b**) ((**a**) was kindly provided by M. Peters (Staatliches Veterinäruntersuchungsamt Arnsberg, Germany))

et al. 2012b). These neurological signs are suggestive of SBV-induced lesions in the central nervous system (CNS).

No lesions have been described in association with SBV infection in non-pregnant adult ruminants in the field. In contrast, a wide range of lesions has been described in foetuses and newborns with proven or presumptive SBV infection occurring during gestation (Table 3.1). The most common lesions affect the skeletal muscle, the CNS and the axial skeleton; they can occur in combination (Herder et al. 2012; Peperkamp et al. 2014; Seehusen et al. 2014). The animals may show arthrogryposis associated with histological evidence of muscular hypoplasia, characterised by a reduction in number and diameter of the myofibrils, with or without loss of cross-striation in myofibrils and fatty replacement (Herder et al. 2012; Seehusen et al. 2014). In the CNS, the most common lesions are hydranencephaly, porencephaly, hydrocephalus, cerebellar hypoplasia and micromyelia; they are sometimes associated with nonsuppurative inflammation and neuronal degeneration and necrosis (Fig. 3.2b). The animals often display vertebral malformations, including lordosis, kyphosis, scoliosis and torticollis, and may show brachygnathia inferior (Herder et al. 2012). The nature of the musculoskeletal and nervous lesions associated with SBV is similar in cattle and sheep. However, the severity of the CNS lesions may be higher in lambs than in calves (Peperkamp et al. 2014).

Table 3.1 Lesions in foetuses and newborns associated with SBV infection of naïve domestic ruminants during a critical phase of gestation

	Gross lesions	Histological lesions	References
Bovine	Porencephaly, hydrocephalus, hydranencephaly, brain stem hypoplasia, cerebellar hypoplasia, cerebellar dysplasia, micromyelia, arthrogryposis, torticollis, lordosis, scoliosis, kyphosis, cranial malformations, brachygnathism inferior, prognathism, ectopia cordis, lung hypoplasia, ventricular septal defect	Nonsuppurative meningoencephalitis, nonsuppurative poliomyelitis, skeletal muscle hypoplasia, lymphoid depletion in thymus and lymph node, chronic hepatitis	Bayrou et al. (2014), Garigliany et al. (2012b), Herder et al. (2012), Peperkamp et al. (2012, 2014), Seehusen et al. (2014)
Ovine	Arthrogryposis, torticollis, lordosis, scoliosis, kyphosis, brachygnathism inferior, domed skull, flattened skull, hydranencephaly, hydrocephalus, micrencephaly, macrocephaly, brainstem hypoplasia, cerebral hypoplasia, cerebellar hypoplasia, cerebellar dysplasia, micromyelia, cardiac ventricular septal defect, unilateral hydronephrosis, colonic atresia	Nonsuppurative meningoencephalitis, skeletal muscle hypoplasia, lymphoid depletion in spleen or lymph node, cataract, decreased haematopoietic cellularity in bone marrow	van den Brom et al. (2012), Herder et al. (2012), Peperkamp et al. (2014)
Caprine	Arthrogryposis, vertebral deformities, brachygnathia inferior, hydrocephalus, porencephaly, cerebellar hypoplasia, lung hypoplasia	Nonsuppurative meningoencephalitis, nonsuppurative poliomyelitis	Herder et al. (2012), Wagner et al. (2014)

3.6 Pathogenesis

Owing to its recent discovery, studies on the pathogenesis of SBV infection are limited. Given that SBV is phylogenetically closely related to Akabane virus and causes very similar clinical signs and lesions, it is likely that they are involved in common mechanisms of disease induction. As described for Akabane virus (Kurogi et al. 1975, 1977), only a small proportion of SBV infections leads to the birth of malformed and/or dead animals (Martinelle et al. 2015b; Veldhuis et al. 2014a; Wernike et al. 2014b). The susceptibility of the growing embryo or foetus to Akabane virus infection may depend on the maturity of the placentomes. If the infection of the pregnant female takes place before maturity, the embryo may be protected from viral invasion (Kirkland et al. 1988; Parsonson et al. 1988). In one

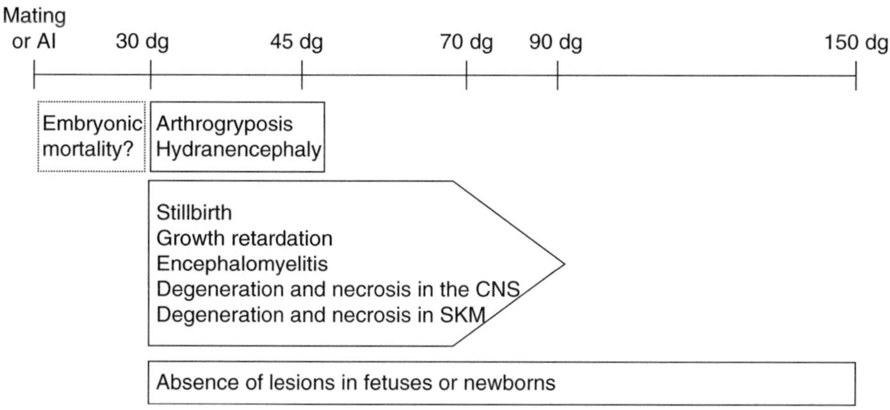

Fig. 3.3 Hypothetic consequences of SBV infection in pregnant goats and ewes depending on the stage of gestation. Seroconversion might occur in the foetus, provided the infection takes place after 30 days of gestation. By analogy with data from experimental infection with Akabane virus (Hashiguchi et al. 1979; Konno and Nakagawa 1982; Narita et al. 1979). *AI* artificial insemination, *CNS* central nervous system, *dg* day of gestation, *SKM* skeletal muscle

study investigating gestating sheep experimentally infected with SBV, it was suggested that placentomes at 45 days of pregnancy are not sufficiently developed to sustain intensive viral replication in contrast to placentomes at 60 days of gestation (Martinelle et al. 2015b). Susceptibility to Akabane virus infection also depends on the stage of maturity of the target organs (McClure et al. 1988). Neurons in the cerebral cortex, brainstem and spinal cord are target cells of SBV in infected calves and lambs (Peperkamp et al. 2012; Varela et al. 2013). In young mice, Akabane virus infects neuronal stem cells before spreading to differentiated neurons (Murata et al. 2015). Foetuses might be more susceptible to SBV infection before the blood-brain barrier is developed (Varela et al. 2013) and before the end of neurogenesis (Martinelle et al. 2015b). Some studies have suggested that muscle cells are also target cells of SBV infection (Balseiro et al. 2015; Martinelle et al. 2015b) and that direct infection of skeletal muscle cells is involved in the pathogenesis of arthrogryposis, as reported for Akabane virus (Konno and Nakagawa 1982). The effects of an infection caused by Simbu serogroup viruses on the growing embryo or foetus also depend on the stage of development of the foetal immune system. Transplacentally infected ovine foetuses can produce neutralising antibodies against Akabane virus as soon as 64 days of gestation (Hashiguchi et al. 1979). Pre-colostral antibodies may be detectable in newborn calves when the dam is infected with SBV from the fourth month of gestation onwards (Wernike et al. 2014b).

By analogy with Akabane virus, hypotheses can be drawn on the effects of SBV infection in pregnant females depending on the stage of gestation. Figure 3.3 shows the putative consequences of SBV infection in pregnant ewes and goats. In cattle, the critical phase for Akabane virus infection seems to be between the third and sixth month of the 9-month gestation (Kirkland et al. 1988; Konno and Nakagawa 1982; Kurogi et al. 1977). In one study, the birth of malformed or dead calves was recorded after infection of the mothers between days 60 and 144, which is within this critical timeframe (Wernike et al. 2014b).

To study the pathogenesis of SBV in the brain, a mouse model of SBV infection has been developed in NIH Swiss mice. Intracerebral inoculation resulted in death and severe brain lesions with malacia and haemorrhage in the cerebral cortex, multifocal vacuolation in the white matter of the cerebrum as well as lymphocytic perivascular cuffing in the grey matter. These lesions were associated with SBV antigen in neurons (Varela et al. 2013), which is in favour of a direct role for SBV in inducing this range of lesions. In addition, activated caspase-3 staining has been associated with SBV in the brain of NIH Swiss mice infected intracerebrally (Barry et al. 2014).

To explore the role of SBV NSs as a virulence factor, a NSs deletion mutant (SBVΔNSs) was produced by reverse genetics. Its virulence was then tested in NIH Swiss mice infected by intracerebral inoculation. SBVΔNSs showed an attenuated phenotype, characterised by a delay in the time of death in comparison to wild-type SBV. This defined SBV NSs as a virulence factor (Varela et al. 2013). In vitro, SBVΔNSs was able to induce the production of IFN in several cell lines, while wild-type SBV was not, showing that SBV NSs inhibits the IFN response of the host (Kraatz et al. 2015; Varela et al. 2013). SBV NSs has also the ability to induce the degradation of the RPB1 subunit of RNA polymerase II *in vitro* and subsequently to inhibit transcription and protein synthesis. The inhibition of the IFN response by NSs may be a consequence of this global inhibition of transcription (Barry et al. 2014). Besides, a transcriptomic study has shown that in vitro, SBV NSs causes a shutdown in the expression of genes involved in innate immunity. Nevertheless, this shutdown is incomplete as a few antiviral genes were still induced during SBV infection (Blomström et al. 2015). SBV NSs is also able to enhance the rate of apoptotic cell death in vitro (Barry et al. 2014).

3.7 Transmission

3.7.1 Vectors

Similarly to other Simbu viruses, SBV is an arbovirus transmitted by small blood-sucking biting midges from the *Culicoides* genus. SBV RNA has been found in Europe in *C. obsoletus* complex, *C. dewulfi* and *C. chiopterus* (Elbers et al. 2013, 2015; Goffredo et al. 2013; Rasmussen et al. 2012, 2014; De Regge et al. 2012a). *C. punctatus* and *C. nubeculosus* may also be involved (Balenghien et al. 2014; Larska et al. 2013b). Vector competence has only been established under laboratory conditions for *C. scoticus* (Balenghien et al. 2014; Veronesi et al. 2013). Nevertheless, field data have confirmed the vector competence of midges of the *C. obsoletus* complex, *C. dewulfi* and *C. chiopterus* (De Regge et al. 2012a, 2014). A few studies have also been carried out to elucidate the role of mosquitoes in SBV transmission. No SBV RNA was found in 50,000 mosquitoes trapped in Germany in 2011 (Wernike et al. 2014a). Experimental oral infection of *Aedes albopictus* and *Culex pipiens* mosquitoes did not result in SBV replication to transmissible levels, suggesting these two species are not vectors of SBV (Balenghien et al. 2014).

3.7.2 Vertical Transmission in Ruminants

As described above, SBV can be transmitted vertically from the pregnant female to its offspring. Congenital malformations associated with the presence of SBV RNA have been found in newborns, stillborn or aborted animals in sheep, cattle and goats (van den Brom et al. 2012; Garigliany et al. 2012b; Herder et al. 2012). Vertical transmission in other species has not been reported.

3.7.3 Horizontal Transmission in Ruminants

SBV RNA has been detected in faecal, oral and nasal swabs in subcutaneously inoculated cows (Wernike et al. 2013b). However, oral inoculation of cattle and nasal inoculation of sheep failed to produce RNA-emia or seroconversion (Martinelle et al. 2015a; Wernike et al. 2013b), and naïve cattle or sheep in contact with viraemic animals did not show RNA-emia and remained SBV seronegative (Wernike et al. 2012, 2013a, b). Direct transmission of SBV from an infected ruminant to a naïve one by contact or by the oral or nasal route seems therefore unlikely. Whether SBV can be sexually transmitted is still unknown. Bulls can excrete SBV in their semen as shown by the detection of infectious SBV in bovine semen samples from the field (Hoffmann et al. 2013; Ponsart et al. 2014; Schulz et al. 2014). Cows have been successfully infected with Akabane virus via the uterine route at the time of artificial insemination (Parsonson et al. 1988), but it is not known whether SBV is also able to infect females via this route, and if so, the epidemiological relevance of this route remains to be evaluated.

3.7.4 Overwintering

In May 2012, 8 months after the likely introduction date of SBV into France (Zanella et al. 2013), evidence of acute infection was found in cows, suggesting that SBV could overwinter or was reintroduced (Sailleau et al. 2013b). SBV infection recurrence was also recorded in adult ruminants in Germany during summer and fall 2012 (Conraths et al. 2013). SBV has then overwintered in 2013 and 2014, with new cases being reported each vector season (Wernike et al. 2015a). Several mechanisms could account for SBV overwintering. SBV may persist in its host. This hypothesis seems unlikely in domestic ruminants as viremia is short-lasting (Hoffmann et al. 2012) and SBV RNA is not often detected in malformed newborns (De Regge et al. 2013). Transovarial transmission of the virus in the vector might be involved as one study reported the presence of SBV RNA in nulliparous females of *C. obsoletus* complex and *C. punctatus* (Larska et al. 2013b). It is also possible that SBV persists in adult midges during winter. Adult midges of the *C. obsoletus* complex are able to survive, without blood meal, for 10 days at 4°C and up to 92 days at temperatures between 17°C and 25°C (Goffredo et al. 2004). Moreover, *C. obsoletus* midges have the ability to live indoors (Koenraadt et al. 2014). Infected midges could then survive in the coldest months in barns and infect vertebrates once the temperature rises. This hypothesis is supported by evidence of transmission of SBV to sheep in winter 2013 in Germany, after a rise in temperature above 5°C for a few days (Wernike et al. 2013c). Finally, vectors that have not been identified so far could play a role in SBV transmission and overwintering.

3.8 Risk Factors

The risk of SBV infection seems to differ among ruminant species. Regarding domestic ruminants, within-herd prevalence was found to be lower in sheep than in cattle in Germany (Helmer et al. 2015; Wernike et al. 2014a). In addition, goats have been shown to have a lower risk of SBV infection than sheep (Helmer et al. 2015). This could be due to intrinsic differences in the susceptibility of these species to SBV or, alternatively, to differences in the exposure of these species to midges. Housing and breeding conditions are also a risk factor for SBV infection. Dairy cow herds kept indoors showed lower SBV seroprevalence than outdoor herds, probably due to lower midge exposure inside stables (Tarlinton and Daly, 2013; Veldhuis et al. 2014a). Housing conditions were also found to influence seroprevalence in goats, with outdoor herds tending to have a higher within-herd seroprevalence in France and in Germany (Helmer et al. 2013; Valas et al. 2015).

3.9 Impact on Livestock Farming

SBV impact in infected herds originated mainly from stillbirth or malformations in foetuses and neonates (Dominguez et al. 2014; Poskin et al. 2015; Raboisson et al. 2014). Affected offspring could not be bred and sold, leading to money loss. Besides, emotional impact has been reported in sheep farmers, due to the sight of malformations and to the stress raised by this new disease (Harris et al. 2014). The impact on adults was evaluated as limited, resulting mainly from dystocia in sheep and cattle (Poskin et al. 2015) and from milk loss in dairy cattle (Veldhuis et al. 2014a). In dairy cattle, some fertility parameters were slightly reduced in SBV-affected herds in Germany and the Netherlands in 2011 (Veldhuis et al. 2014b). In 2012, in SBV-affected farms, a twofold reduction in prolificacy was observed in ewes in Belgium (Saegerman et al. 2014), while goat and sheep farmers reported more frequent repeated oestrus or early embryonic deaths in France (Dominguez et al. 2014). Between species, morbidity and malformation rates associated with SBV infection differed. In SBV-affected herds in France in 2012–2013, the median frequency of offspring morbidity (stillbirth or malformation) was significantly higher in lambs than in calves or kids (Dominguez et al. 2014). In addition to its impact on livestock and farming, SBV had an economic impact on Europe due to trade restrictions. Several countries outside Europe restricted cattle semen trade, leading to an overall drop of 11–26 % in the number of semen doses exported outside Europe in 2012 (EFSA 2014). The impact on wild and exotic ruminants remains unknown.

3.10 Diagnostics

Due to the combination of high sensitivity, low time requirement and the possibility of high-throughput screening, direct SBV detection is based primarily on the detection of the viral genome by real-time reverse transcription polymerase chain reaction (RT-qPCR). To detect all members of the Simbu serogroup, an RT-qPCR assay has been established which allows the possibility of subsequent species classification via sequencing (Fischer et al. 2013b). For the specific and direct detection of SBV, different PCR systems that target either the S, M or L segment have been developed (Bilk et al. 2012; Fischer et al. 2013b; Hoffmann et al. 2012). Of those, the S segment-based assay is the most suitable system for a specific and sensitive detection of the SBV genome (Fischer et al. 2013b). In addition to samples from affected ruminants, RT-qPCR assays may also be applied to screen insect vectors for the presence of viral RNA (Balenghien et al. 2014; Veronesi et al. 2013). As an additional tool, the virus can be isolated on either insect or mammalian cell lines such as KC (*Culicoides variipennis* larvae), BHK (baby hamster kidney) or Vero (African green monkey) cells (Hoffmann et al. 2012; Loeffen et al. 2012; Wernike et al. 2013a). However, due to the short-lived viraemia of only up to 6 days in acutely infected adult animals (Hoffmann et al. 2012; Wernike et al. 2013a, b, 2015a), the detection of SBV-specific antibodies is more promising than direct virus detection. In foetuses or newborns, the detection of antibodies is also a valuable tool to confirm congenital SBV infection when serum samples are taken before colostrum intake or when foetal heart blood is used as sample matrix (van Maanen et al. 2012; De Regge et al. 2013; Wernike et al. 2014b). Systems for antibody detection include various commercially available or in-house ELISAs, micro-neutralisation and indirect immunofluorescence tests (Bréard et al. 2013; van der Heijden et al. 2013; Humphries and Burr, 2012; Loeffen et al. 2012; Wernike et al. 2013b). Although protocols were carried out slightly differently between laboratories, the neutralisation test was more sensitive than several ELISAs assessed in a limited European-wide ring trial (van der Poel et al. 2014). Furthermore, the neutralisation test is highly specific and allows the differentiation between different members of the Simbu serogroup (Goller et al. 2012; Hechinger et al. 2013; Kinney and Calisher 1981), while the S segment-based ELISAs may lead to cross-reaction between SBV and other Simbu viruses.

The choice of the sample material is critical for both virus and antibody detection. In adult ruminants, the preferred material for detecting acute infections and antibodies is serum. Suitable sample materials for SBV detection by RT-qPCR in malformed foetuses or newborns include different parts of the brain, placenta, meconium and hair swabs (Bilk et al. 2012; De Regge et al. 2012b, 2013). As mentioned above, the detection of specific neutralising antibodies in pre-colostral serum or body fluids may be used as an additional tool to confirm congenital SBV exposure since the viral genome is not detectable in all lambs or calves suspected of being infected (van Maanen et al. 2012; De Regge et al. 2013). A critical sample material for the detection of viral genome is frozen diluted or undiluted bull semen due to potential inhibitory effects of the material itself. SBV-infected bulls may excrete the virus in their semen over an extended period or in some cases intermittently (Van Der Poel et al. 2014; Hoffmann et al. 2013; Ponsart et al. 2014; Schulz et al. 2014). The most reliable system to detect SBV genome is based on lysis of the sample using TRIzol® LS Reagent in combination or not with a magnetic bead-based commercial extraction kit (Hoffmann et al. 2013). To exclude an inhibitory effect of the sample matrix, the simultaneous amplification of an internal control based on either housekeeping genes or heterologous RNA is highly recommended.

3.11 Control Measures

Since SBV is an insect-transmitted virus, the use of insecticides or repellents could be taken into consideration to prevent potentially infected vectors from biting susceptible animals. However, a case-control study provided no evidence for protective effects of those treatments (Wernike et al. 2014c). An intelligent management system could be used to prevent transplacental transmission of the virus to the developing foetus and thereby the induction of malformation. The mating period could be adjusted in order to avoid that susceptible animals are in the critical phase of gestation during the season of the highest activity of the insect vectors responsible for virus transmission. A combination of an intelligent breeding system with further protective measures like housing and insecticide treatment might potentially result in a reduction of clinical cases (Helmer et al. 2013). Furthermore, a grazing management could be adapted and young stock kept outside during the major vector season. Thereby, young animals would be more likely exposed to the vector and potentially SBV infection before heifers conceive for the first time. Since anti-SBV antibodies are detectable for at least 2–3 years after infection (Elbers et al. 2014; Wernike et al. 2015b), the immunity acquired as calf/lamb may prevent foetal infection during a later pregnancy. However, a more reliable way of prophylaxis than exposure to potentially infected vectors is vaccination. A classical approach for vaccine development is the use of chemically inactivated whole viruses; such inactivated vaccine formulations are able to prevent clinical diseases caused by closely related viruses (Kim et al. 2011). A Japanese multivalent vaccine against Akabane virus, Aino virus and the likewise teratogenic reovirus Chuzan virus has already been developed and prevents reproductive disorders in immunised ruminants. However, this vaccine was not able to protect against SBV infection in an experimental setting (Hechinger et al. 2013). In contrast to this heterologous vaccine, SBV-specific inactivated candidate vaccines effectively protected cattle against an SBV infection after two immunisations (Wernike et al. 2013d) and sheep even after a single vaccination (Hechinger et al. 2014). Provisional marketing authorisations have been granted for inactivated whole virus vaccines against SBV for the British and French market in 2013, and in May 2015 such vaccine was licenced for the complete European Union. Apart from inactivated vaccine formulations, which require high virus titre pre-inactivation, the use of adjuvants and in most cases repeated inoculations, modified live vaccines may be used. However, live attenuated vaccines might potentially revert to a virulent form when attenuation is achieved by mutations of only a few nucleotides within the viral genome (Luongo et al. 2012; Shams 2005). This safety risk can be circumvented by complete or partial deletion of genes that are non-essential for replication and/or induction of immunity. For Rift Valley fever virus (RVFV), a mosquito-transmitted phlebovirus from the *Bunyaviridae* family, mutant viruses lacking both nonstructural proteins, NSs and NSm, were shown to replicate efficiently in cell culture and confer protection from viraemia after experimental infection of rats and sheep. In addition, the differentiation between infected and vaccinated animals was possible making the NSs/NSm double deletion RVFV mutants ideal modified live vaccine candidates (Bird et al. 2011). In regard to these observations, SBV double deletion mutants as well as viruses lacking either NSs or NSm have been produced and tested in cattle, one of the major target species of SBV (Kraatz et al. 2015). As for other members of the bunyavirus family, SBV NSs and NSm alone or the combination of both nonstructural proteins is not essential for viral growth in vitro. The same holds true for NSm in vivo. SBV lacking this protein induced viraemia and seroconversion similarly to the wild-type virus in cattle. However, the NSs and the NSs/NSm deletion mutants caused no clinical disease and no virus replication was detected after vaccination. Moreover, three out of four animals immunised with the virus lacking NSs and all cattle that received the double deletion mutant were fully protected against a virulent SBV challenge infection (Kraatz et al. 2015).

References

Afonso A, Abrahantes JC, Conraths F, Veldhuis A, Elbers A, Roberts H, Van der Stede Y, Méroc E, Gache K, Richardson J (2014) The Schmallenberg virus epidemic in Europe-2011-2013. Prev Vet Med 116:391–403

Balenghien T, Pagès N, Goffredo M, Carpenter S, Augot D, Jacquier E, Talavera S, Monaco F, Depaquit J et al (2014) The emergence of Schmallenberg virus across Culicoides communities and ecosystems in Europe. Prev Vet Med. 116(4):360–9

Balseiro A, Royo LJ, Gómez Antona A, García Marín JF (2015) First confirmation of Schmallenberg virus in cattle in Spain: tissue distribution and pathology. Transbound Emerg Dis 62:e62–e65

Barry G, Varela M, Ratinier M, Blomström A-L, Caporale M, Seehusen F, Hahn K, Schnettler E, Baumgärtner W et al (2014) NSs protein of Schmallenberg virus counteracts the antiviral response of the cell by inhibiting its transcriptional machinery. J Gen Virol 95:1640–1646

Bayrou C, Garigliany M-M, Sarlet M, Sartelet A, Cassart D, Desmecht D (2014) Natural intrauterine infection with schmallenberg virus in malformed newborn calves. Emerg Infect Dis 20:1327–1330

Bilk S, Schulze C, Fischer M, Beer M, Hlinak A, Hoffmann B (2012) Organ distribution of Schmallenberg virus RNA in malformed newborns. Vet Microbiol 159:236–238

Bird BH, Maartens LH, Campbell S, Erasmus BJ, Erickson BR, Dodd KA, Spiropoulou CF, Cannon D, Drew CP et al (2011) Rift Valley fever virus vaccine lacking the NSs and NSm genes is safe, nonteratogenic, and confers protection from viremia, pyrexia, and abortion following challenge in adult and pregnant sheep. J Virol 85:12901–12909

Blomström A-L, Gu Q, Barry G, Wilkie G, Skelton JK, Baird M, McFarlane M, Schnettler E, Elliott RM et al (2015) Transcriptome analysis reveals the host response to Schmallenberg virus in bovine cells and antagonistic effects of the NSs protein. BMC Genomics 16:324

Bréard E, Lara E, Comtet L, Viarouge C, Doceul V, Desprat A, Vitour D, Pozzi N, Cay AB et al (2013) Validation of a commercially available indirect ELISA using a nucleocapside recombinant protein for detection of Schmallenberg virus antibodies. PLoS One 8:e53446

Briese T, Calisher CH, Higgs S (2013) Viruses of the family Bunyaviridae: are all available isolates reassortants? Virology 446:207–216

Van den Brom R, Luttikholt SJM, Lievaart-Peterson K, Peperkamp NHMT, Mars MH, van der Poel WHM, Vellema P (2012) Epizootic of ovine congenital malformations associated with Schmallenberg virus infection. Tijdschr Diergeneeskd 137:106–111

Causey OR, Kemp GE, Causey CE, Lee VH (1972) Isolations of Simbu-group viruses in Ibadan, Nigeria 1964–69, including the new types Sango, Shamonda, Sabo and Shuni. Ann Trop Med Parasitol 66:357–362

Chaintoutis SC, Kiossis E, Giadinis ND, Brozos CN, Sailleau C, Viarouge C, Bréard E, Papanastassopoulou M, Zientara S et al (2014) Evidence of Schmallenberg virus circulation in ruminants in Greece. Trop Anim Health Prod 46:251–255

Conraths FJ, Peters M, Beer M (2013) Schmallenberg virus, a novel orthobunyavirus infection in ruminants in Europe: potential global impact and preventive measures. N Z Vet J 61:63–67

Coupeau D, Claine F, Wiggers L, Kirschvink N, Muylkens B (2013) In vivo and in vitro identification of a hypervariable region in Schmallenberg virus. J Gen Virol 94:1168–1174

Van Der Poel WHM, Parlevliet JM, Verstraten ER, Kooi EA, Hakze-Van Der Honing R, Stockhofe N (2014a) Schmallenberg virus detection in bovine semen after experimental infection of bulls. Epidemiol Infect 142:1495–1500

Desmecht D, Garigliany M-M, Beer M, Schirrmeier H, Paternostre J, Volpe R, Linden A (2013) Detection of antibodies against Schmallenberg virus in wild boars, Belgium, 2010–2012. Proc 31th Congr Int Union Game Biol Bruss Belg

Doceul V, Lara E, Sailleau C, Belbis G, Richardson J, Bréard E, Viarouge C, Dominguez M, Hendrikx P et al (2013) Epidemiology, molecular virology and diagnostics of Schmallenberg virus, an emerging orthobunyavirus in Europe. Vet Res 44:31

Dominguez M, Gache K, Touratier A, Perrin J-B, Fediaevsky A, Collin E, Bréard E, Sailleau C, Viarouge C et al (2014) Spread and impact of the Schmallenberg virus epidemic in France in 2012–2013. BMC Vet Res 10:248

Dong H, Li P, Elliott RM, Dong C (2013) Structure of Schmallenberg orthobunyavirus nucleoprotein suggests a novel mechanism of genome encapsidation. J Virol 87:5593–5601

Ducomble T, Wilking H, Stark K, Takla A, Askar M, Schaade L, Nitsche A, Kurth A (2012) Lack of evidence for Schmallenberg virus infection in highly exposed persons, Germany, 2012. Emerg Infect Dis 18:1333–1335

EFSA (2014) Schmallenberg virus: state of art. EFSA J 12(5):3681

Eifan S, Schnettler E, Dietrich I, Kohl A, Blomström A-L (2013) Non-structural proteins of arthropod-borne bunyaviruses: roles and functions. Viruses 5:2447–2468

Elbers ARW, Meiswinkel R, van Weezep E, Sloet van Oldruitenborgh-Oosterbaan MM, Kooi EA (2013) Schmallenberg virus in Culicoides spp. biting midges, the Netherlands, 2011. Emerg Infect Dis 19:106–109

Elbers ARW, Stockhofe-Zurwieden N, van der Poel WHM (2014) Schmallenberg virus antibody persistence in adult cattle after natural infection and decay of maternal antibodies in calves. BMC Vet Res 10:103

Elbers ARW, Meiswinkel R, van Weezep E, Kooi EA, van der Poel WHM (2015) Schmallenberg virus in culicoides biting midges in the Netherlands in 2012. Transbound Emerg Dis 62:339–342

Elliott RM (2014) Orthobunyaviruses: recent genetic and structural insights. Nat Rev Microbiol 12:673–685

Fischer M, Hoffmann B, Goller KV, Höper D, Wernike K, Beer M (2013a) A mutation 'hot spot' in the Schmallenberg virus M segment. J Gen Virol 94:1161–1167

Fischer M, Schirrmeier H, Wernike K, Wegelt A, Beer M, Hoffmann B (2013b) Development of a pan-Simbu real-time reverse transcriptase PCR for the detection of Simbu serogroup viruses and comparison with SBV diagnostic PCR systems. Virol J 10:327

Gale P, Kelly L, Snary EL (2015) Pathways for entry of livestock arboviruses into Great Britain: assessing the strength of evidence. Transbound Emerg Dis 62:115–123

Garigliany M-M, Desmecht D, Bayrou C, Peeters D (2013) No serologic evidence for emerging Schmallenberg virus infection in dogs (Canis domesticus). Vector Borne Zoonotic Dis Larchmt N 13:830–833

Garigliany M-M, Bayrou C, Kleijnen D, Cassart D, Desmecht D (2012a) Schmallenberg virus in domestic cattle, Belgium, 2012. Emerg Infect Dis 18:1512–1514

Garigliany M-M, Hoffmann B, Dive M, Sartelet A, Bayrou C, Cassart D, Beer M, Desmecht D (2012b) Schmallenberg virus in calf born at term with porencephaly, Belgium. Emerg Infect Dis 18:1005–1006

Gerhauser I, Weigand M, Hahn K, Herder V, Wohlsein P, Habierski A, Varela M, Palmarini M, Baumgärtner W (2014) Lack of schmallenberg virus in ruminant brain tissues archived from 1961 to 2010 in Germany. J Comp Pathol 150:151–154

Goffredo M, Romeo G, Monaco F, Di Gennaro A, Savini G (2004) Laboratory survival and blood feeding response of wild-caught *Culicoides obsoletus* Complex (Diptera: Ceratopogonidae) through natural and artificial membranes. Vet Ital 40:282–285

Goffredo M, Monaco F, Capelli G, Quaglia M, Federici V, Catalani M, Montarsi F, Polci A, Pinoni C et al (2013) Schmallenberg virus in Italy: a retrospective survey in Culicoides stored during the bluetongue Italian surveillance program. Prev Vet Med 111:230–236

Goller KV, Höper D, Schirrmeier H, Mettenleiter TC, Beer M (2012) Schmallenberg virus as possible ancestor of Shamonda virus. Emerg Infect Dis 18:1644–1646

Harris KA, Eglin RD, Hayward S, Milnes A, Davies I, Cook AJC Downs SH (2014) The impact of Schmallenberg virus on British sheep farms during the 2011/2012 lambing season. Vet Rec. 175(7):172

Hashiguchi Y, Nanba K, Kumagai T (1979) Congenital abnormalities in newborn lambs following Akabane virus infection in pregnant ewes. Natl Inst Anim Health Q (Tokyo) 19:1–11

Hechinger S, Wernike K, Beer M (2013) Evaluating the protective efficacy of a trivalent vaccine containing Akabane virus, Aino virus and Chuzan virus against Schmallenberg virus infection. Vet Res 44:114

Hechinger S, Wernike K, Beer M (2014) Single immunization with an inactivated vaccine protects sheep from Schmallenberg virus infection. Vet Res 45:79

Van der Heijden HMJF, Bouwstra RJ, Mars MH, van der Poel WHM, Wellenberg GJ, van Maanen C (2013) Development and validation of an indirect Enzyme-linked immunosorbent assay for the detection of antibodies against Schmallenberg virus in blood samples from ruminants. Res Vet Sci 95:731–735

Helmer C, Eibach R, Tegtmeyer PC, Humann-Ziehank E, Ganter M (2013) Survey of Schmallenberg virus (SBV) infection in German goat flocks. Epidemiol Infect 141:2335–2345

Helmer C, Eibach R, Tegtmeyer PC, Humann-Ziehank E, Runge M, Ganter M (2015) Serosurvey of Schmallenberg virus infections in sheep and goat flocks in Lower Saxony, Germany. Transbound Emerg Dis 62:425–436

Herder V, Wohlsein P, Peters M, Hansmann F, Baumgärtner W (2012) Salient lesions in domestic ruminants infected with the emerging so-called Schmallenberg virus in Germany. Vet Pathol 49:588–591

Hoffmann B, Scheuch M, Höper D, Jungblut R, Holsteg M, Schirrmeier H, Eschbaumer M, Goller KV, Wernike K et al (2012) Novel orthobunyavirus in Cattle, Europe, 2011. Emerg Infect Dis 18:469–472

Hoffmann B, Schulz C, Beer M (2013) First detection of Schmallenberg virus RNA in bovine semen, Germany, 2012. Vet Microbiol 167:289–295

Hofmann MA, Mader M, Flückiger F, Renzullo S (2015) Genetic stability of Schmallenberg virus in vivo during an epidemic, and in vitro, when passaged in the highly susceptible porcine SK-6 cell line. Vet Microbiol 176:97–108

Humphries D, Burr P (2012) Schmallenberg virus milk antibody ELISA. Vet Rec 171:511–512

International Committee on Taxonomy of Viruses (2014) ICTV virus taxonomy 2014
Kim Y-H, Kweon C-H, Tark D-S, Lim SI, Yang D-K, Hyun B-H, Song J-Y, Hur W, Park SC (2011) Development of inactivated trivalent vaccine for the teratogenic Aino, Akabane and Chuzan viruses. Biol J Int Assoc Biol Stand 39:152–157
Kinney RM, Calisher CH (1981) Antigenic relationships among Simbu serogroup (Bunyaviridae) viruses. Am J Trop Med Hyg 30:1307–1318
Kirkland PD, Barry RD, Harper PA, Zelski RZ (1988) The development of Akabane virus-induced congenital abnormalities in cattle. Vet Rec 122:582–586
Kobayashi T, Yanase T, Yamakawa M, Kato T, Yoshida K, Tsuda T (2007) Genetic diversity and reassortments among Akabane virus field isolates. Virus Res 130:162–171
Koenraadt CJ, Balenghien T, Carpenter S, Ducheyne E, Elbers AR, Fife M, Garros C, Ibáñez-Justicia A, Kampen H et al (2014) Bluetongue, Schmallenberg – what is next? Culicoides-borne viral diseases in the 21st Century. BMC Vet Res 10:77
Konno S, Nakagawa M (1982) Akabane disease in cattle: congenital abnormalities caused by viral infection. Experimental disease. Vet Pathol 19:267–279
Kraatz F, Wernike K, Hechinger S, König P, Granzow H, Reimann I, Beer M (2015) Deletion mutants of Schmallenberg virus are avirulent and protect from virus challenge. J Virol 89:1825–1837
Kurogi H, Inaba Y, Goto Y, Miura Y, Takahashi H (1975) Serologic evidence for etiologic role of Akabane virus in epizootic abortion-arthrogryposis-hydranencephaly in cattle in Japan, 1972–1974. Arch Virol 47:71–83
Kurogi H, Inaba Y, Takahashi E, Sato K, Satoda K (1977) Congenital abnormalities in newborn calves after inoculation of pregnant cows with Akabane virus. Infect Immun 17:338–343
Laloy E, Riou M, Barc C, Belbis G, Bréard E, Breton S, Cordonnier N, Crochet D, Delaunay R et al (2015) Schmallenberg virus: experimental infection in goats and bucks. BMC Vet Res 11:221
Laloy E, Bréard E, Sailleau C, Viarouge C, Desprat A, Zientara S, Klein F, Hars J, Rossi S (2014) Schmallenberg virus infection among red deer, France, 2010–2012. Emerg Infect Dis 20:131–134
Larska M, Krzysiak M, Smreczak M, Polak MP, Zmudziński JF (2013a) First detection of Schmallenberg virus in elk (*Alces alces*) indicating infection of wildlife in Białowieża National Park in Poland. Vet J 198:279–281
Larska M, Lechowski L, Grochowska M, Żmudziński JF (2013b) Detection of the Schmallenberg virus in nulliparous *Culicoides obsoletus*/scoticus complex and *C. punctatus* – the possibility of transovarial virus transmission in the midge population and of a new vector. Vet Microbiol 166:467–473
Lazutka J, Zvirbliene A, Dalgediene I, Petraityte-Burneikiene R, Spakova A, Sereika V, Lelesius R, Wernike K, Beer M, Sasnauskas K (2014) Generation of recombinant schmallenberg virus nucleocapsid protein in yeast and development of virus-specific monoclonal antibodies. J Immunol Res 2014:160316
Lievaart-Peterson K, Luttikholt SJM, van den Brom R, Vellema P (2012) Schmallenberg virus infection in small ruminants – First review of the situation and prospects in Northern Europe. Small Rumin Res 106:71–76
Loeffen W, Quak S, de Boer-Luijtze E, Hulst M, van der Poel W, Bouwstra R, Maas R (2012) Development of a virus neutralisation test to detect antibodies against Schmallenberg virus and serological results in suspect and infected herds. Acta Vet Scand 54:44
Luongo C, Winter CC, Collins PL, Buchholz UJ (2012) Increased genetic and phenotypic stability of a promising live-attenuated respiratory syncytial virus vaccine candidate by reverse genetics. J Virol 86:10792–10804
Luttikholt S, Veldhuis A, van den Brom R, Moll L, Lievaart-Peterson K, Peperkamp K, van Schaik G, Vellema P (2014) Risk factors for malformations and impact on reproductive performance and mortality rates of schmallenberg virus in sheep flocks in the Netherlands. PLoS One 9:e100135

Van Maanen C, van der Heijden H, Wellenberg GJ, Witteveen G, Luttikholt S, Bouwstra R, Kooi B, Vellema P, Peperkamp K, Mars J (2012) Schmallenberg virus antibodies in bovine and ovine fetuses. Vet Rec 171:299

Martinelle L, Poskin A, Dal Pozzo F, Mostin L, Van Campe W, Cay AB, De Regge N, Saegerman C (2015a) Three different routes of inoculation for experimental infection with Schmallenberg virus in sheep. Transbound Emerg Dis. DOI: 10.1111/tbed.12356

Martinelle L, Poskin A, Dal Pozzo F, De Regge N, Cay B, Saegerman C (2015b) Experimental infection of sheep at 45 and 60 days of gestation with Schmallenberg virus readily led to placental colonization without causing congenital malformations. PLoS One 10:e0139375

McClure S, McCullagh P, Parsonson IM, McPhee DA, Della-Porta AJ, Orsini A (1988) Maturation of immunological reactivity in the fetal lamb infected with Akabane virus. J Comp Pathol 99:133–143

Molenaar FM, La Rocca SA, Khatri M, Lopez J, Steinbach F, Dastjerdi A (2015) Exposure of Asian elephants and other exotic ungulates to Schmallenberg virus. PLoS One 10:e0135532

Mouchantat S, Wernike K, Lutz W, Hoffmann B, Ulrich RG, Börner K, Wittstatt U, Beer M (2015) A broad spectrum screening of Schmallenberg virus antibodies in wildlife animals in Germany. Vet Res 46:99

Murata Y, Uchida K, Shioda C, Uema A, Bangphoomi N, Chambers JK, Akashi H, Nakayama H (2015) Histopathological Studies on the Neuropathogenicity of the Iriki and OBE-1 Strains of Akabane Virus in BALB/cAJcl Mice. J Comp Pathol 153:140–149

Narita M, Inui S, Hashiguchi Y (1979) The pathogenesis of congenital encephalopathies in sheep experimentally induced by Akabane virus. J Comp Pathol 89:229–240

Parsonson IM, McPhee DA, Della-Porta AJ, McClure S, McCullagh P (1988) Transmission of Akabane virus from the ewe to the early fetus (32 to 53 days). J Comp Pathol 99:215–227

Peperkamp K, Dijkman R, van Maanen C, Vos J, Wouda W, Holzhauer M, van Wuijckhuise L, Junker K, Greijdanus S, Roumen M (2012) Polioencephalo-myelitis in a calf due to infection with Schmallenberg virus. Vet Rec 170:570

Peperkamp NH, Luttikholt SJ, Dijkman R, Vos JH, Junker K, Greijdanus S, Roumen MP, van Garderen E, Meertens, N. et al (2014) Ovine and bovine congenital abnormalities associated with intrauterine infection with Schmallenberg virus. Vet Pathol. 52:1057–1066

Van der Poel WHM, Cay B, Zientara S, Steinbach F, Valarcher JF, Bøtner A, Mars MH, Hakze-van der Honing R, Schirrmeier H, Beer M (2014b) Limited interlaboratory comparison of Schmallenberg virus antibody detection in serum samples. Vet Rec 174:380

Ponsart C, Pozzi N, Bréard E, Catinot V, Viard G, Sailleau C, Viarouge C, Gouzil J, Beer M et al (2014) Evidence of excretion of Schmallenberg virus in bull semen. Vet Res 45:37

Poskin A, Méroc E, Behaeghel I, Riocreux F, Couche M, Van Loo H, Bertels G, Delooz L, Quinet C et al (2015) Schmallenberg virus in Belgium: estimation of impact in cattle and sheep herds. Transbound Emerg Dis. doi:10.1111/tbed.12367

Poskin A, Van Campe W, Mostin L, Cay B, De Regge N (2014) Experimental Schmallenberg virus infection of pigs. Vet Microbiol 170:398–402

Raboisson D, Waret-Szkuta A, Rushton J, Häsler B, Alarcon P (2014) Application of integrated production and economic models to estimate the impact of Schmallenberg virus for various beef suckler production systems in France and the United Kingdom. BMC Vet Res 10:254

Rasmussen LD, Kristensen B, Kirkeby C, Rasmussen TB, Belsham GJ, Bødker R, Bøtner A (2012) Culicoids as vectors of Schmallenberg virus. Emerg Infect Dis 18:1204–1206

Rasmussen LD, Kirkeby C, Bødker R, Kristensen B, Rasmussen TB, Belsham GJ, Bøtner A (2014) Rapid spread of Schmallenberg virus-infected biting midges (Culicoides spp.) across Denmark in 2012. Transbound Emerg Dis 61:12–16

De Regge N, Deblauwe I, De Deken R, Vantieghem P, Madder M, Geysen D, Smeets F, Losson B, van den Berg T, Cay AB (2012a) Detection of Schmallenberg virus in different Culicoides spp. by real-time RT-PCR. Transbound Emerg Dis 59:471–475

De Regge N, Van loo H, Bertels G, Saulmont M, Vandenbussche F, Van den Berg T, Cay AB (2012b) Comparative analysis to determine the most appropriate tissue for Schmallenberg virus detection by RT-PCR in aborted lambs and calves. Presented at the proceedings of the EPIZONE 6th annual meeting, Brighton

De Regge N, van den Berg T, Georges L, Cay B (2013) Diagnosis of Schmallenberg virus infection in malformed lambs and calves and first indications for virus clearance in the fetus. Vet Microbiol 162:595–600

De Regge N, Madder M, Deblauwe I, Losson B, Fassotte C, Demeulemeester J, Smeets F, Tomme M, Cay AB (2014) Schmallenberg virus circulation in culicoides in Belgium in 2012: field validation of a real time RT-PCR approach to assess virus replication and dissemination in midges. PloS One 9:e87005

Reusken C, van den Wijngaard C, van Beek P, Beer M, Bouwstra R, Godeke G-J, Isken L, van den Kerkhof H, van Pelt W et al (2012) Lack of evidence for zoonotic transmission of Schmallenberg virus. Emerg Infect Dis 18:1746–1754

Rosseel T, Scheuch M, Höper D, De Regge N, Caij AB, Vandenbussche F, Van Borm S (2012) DNase SISPA-next generation sequencing confirms Schmallenberg virus in Belgian field samples and identifies genetic variation in Europe. PLoS One 7:e41967

Saeed MF, Li L, Wang H, Weaver SC, Barrett AD (2001) Phylogeny of the Simbu serogroup of the genus Bunyavirus. J Gen Virol 82:2173–2181

Saegerman C, Martinelle L, Dal Pozzo F, Kirschvink N (2014) Preliminary survey on the impact of Schmallenberg virus on sheep flocks in South of Belgium. Transbound Emerg Dis 61:469–472

Sailleau C, Boogaerts C, Meyrueix A, Laloy E, Bréard E, Viarouge C, Desprat A, Vitour D, Doceul V et al (2013a) Schmallenberg virus infection in dogs, France, 2012. Emerg Infect Dis 19:1896–1898

Sailleau C, Bréard E, Viarouge C, Desprat A, Doceul V, Lara E, Languille J, Vitour D, Attoui H, Zientara S (2013b) Acute Schmallenberg virus infections, France, 2012. Emerg Infect Dis 19:321–322

Schulz C, Wernike K, Beer M, Hoffmann B (2014) Infectious Schmallenberg virus from bovine semen, Germany. Emerg Infect Dis 20:338–340

Schulz C, Beer M, Hoffmann B (2015) Schmallenberg virus infection in South American camelids: field and experimental investigations. Vet Microbiol 180:171–179

Seehusen F, Hahn K, Herder V, Weigand M, Habierski A, Gerhauser I, Wohlsein P, Peters M, Varela M et al (2014) Skeletal muscle hypoplasia represents the only significant lesion in peripheral organs of ruminants infected with Schmallenberg virus during gestation. J Comp Pathol 151:148–152

Shams H (2005) Recent developments in veterinary vaccinology. Vet J Lond Engl 170:289–299

Strandin T, Hepojoki J, Vaheri A (2013) Cytoplasmic tails of bunyavirus Gn glycoproteins-could they act as matrix protein surrogates? Virology 437:73–80

Tarlinton R, Daly J (2013) Testing for Schmallenberg virus. Vet Rec 172:190

Taylor WP, Mellor PS (1994) The distribution of Akabane virus in the Middle East. Epidemiol Infect 113:175–185

Valas S, Baudry C, Ehrhardt N, LeVen A, Thirion M, Aubert C, Vialard J (2015) Serosurvey of Schmallenberg virus infection in the highest goat-specialized region of France. Transbound Emerg Dis 62:e85–e88

Varela M, Schnettler E, Caporale M, Murgia C, Barry G, McFarlane M, McGregor E, Piras IM, Shaw A et al (2013) Schmallenberg virus pathogenesis, tropism and interaction with the innate immune system of the host. PLoS Pathog 9:e1003133

Veldhuis AMB, van Schaik G, Vellema P, Elbers ARW, Bouwstra R, van der Heijden HMJF, Mars MH (2013) Schmallenberg virus epidemic in the Netherlands: spatiotemporal introduction in 2011 and seroprevalence in ruminants. Prev Vet Med 112:35–47

Veldhuis AMB, Carp-van Dijken S, van Wuijckhuise L, Witteveen G, van Schaik G (2014a) Schmallenberg virus in Dutch dairy herds: potential risk factors for high within-herd seroprevalence and malformations in calves, and its impact on productivity. Vet Microbiol 168:281–293

Veldhuis AMB, Santman-Berends IMGA, Gethmann JM, Mars MH, van Wuyckhuise L, Vellema P, Holsteg M, Höreth-Böntgen D, Conraths FJ, van Schaik G (2014b) Schmallenberg virus epidemic: impact on milk production, reproductive performance and mortality in dairy cattle in the Netherlands and Kleve district Germany. Prev Vet Med. 116:412–422

Veronesi E, Henstock M, Gubbins S, Batten C, Manley R, Barber J, Hoffmann B, Beer M, Attoui H et al (2013) Implicating Culicoides biting midges as vectors of Schmallenberg virus using semi-quantitative RT-PCR. PLoS One 8:e57747

Wagner H, Eskens U, Nesseler A, Riesse K, Kaim U, Volmer R, Hamann H-P, Sauerwald C, Wehrend A (2014) Pathologic-anatomical changes in newborn goats caused by an intrauterine Schmallenberg virus infection. Berl Münch Tierärztl Wochenschr 127:115–119

Wensman JJ, Blomqvist G, Hjort M, Holst BS (2013) Presence of antibodies to Schmallenberg virus in a dog in Sweden. J Clin Microbiol 51:2802–2803

Wernike K, Eschbaumer M, Breithaupt A, Hoffmann B, Beer M (2012) Schmallenberg virus challenge models in cattle: infectious serum or culture-grown virus? Vet Res 43:84

Wernike K, Hoffmann B, Bréard E, Bøtner A, Ponsart C, Zientara S, Lohse L, Pozzi N, Viarouge C et al (2013a) Schmallenberg virus experimental infection of sheep. Vet Microbiol 166:461–466

Wernike K, Eschbaumer M, Schirrmeier H, Blohm U, Breithaupt A, Hoffmann B, Beer M (2013b) Oral exposure, reinfection and cellular immunity to Schmallenberg virus in cattle. Vet Microbiol 165:155–159

Wernike K, Kohn M, Conraths FJ, Werner D, Kameke D, Hechinger S, Kampen H, Beer M (2013c) Transmission of Schmallenberg virus during winter, Germany. Emerg Infect Dis 19:1701–1703

Wernike K, Nikolin VM, Hechinger S, Hoffmann B, Beer M (2013d) Inactivated Schmallenberg virus prototype vaccines. Vaccine 31:3558–3563

Wernike K, Jöst H, Becker N, Schmidt-Chanasit J, Beer M (2014a) Lack of evidence for the presence of Schmallenberg virus in mosquitoes in Germany, 2011. Parasit Vectors 7:402

Wernike K, Holsteg M, Schirrmeier H, Hoffmann B, Beer M (2014b) Natural infection of pregnant cows with Schmallenberg virus – a follow-up study. PLoS One 9:e98223

Wernike K, Conraths F, Zanella G, Granzow H, Gache K, Schirrmeier H, Valas S, Staubach C, Marianneau P et al. (2014c) Schmallenberg virus—two years of experiences. Prev Vet Med. 116:423–34

Wernike K, Hoffmann B, Conraths FJ, Beer M (2015a) Schmallenberg Virus Recurrence, Germany, 2014. Emerg Infect Dis 21:1202–1204

Wernike K, Holsteg M, Saßerath M, Beer M (2015b) Schmallenberg virus antibody development and decline in a naturally infected dairy cattle herd in Germany, 2011–2014. Vet Microbiol. 181:294–297

Wilson AJ, Mellor PS (2009) Bluetongue in Europe: past, present and future. Philos Trans R Soc Lond Ser B Biol Sci 364:2669–2681

Yanase T, Fukutomi T, Yoshida K, Kato T, Ohashi S, Yamakawa M, Tsuda T (2004) The emergence in Japan of Sathuperi virus, a tropical Simbu serogroup virus of the genus Orthobunyavirus. Arch Virol 149:1007–1013

Yanase T, Maeda K, Kato T, Nyuta S, Kamata H, Yamakawa M, Tsuda T (2005) The resurgence of Shamonda virus, an African Simbu group virus of the genus Orthobunyavirus, in Japan. Arch Virol 150:361–369

Yanase T, Kato T, Aizawa M, Shuto Y, Shirafuji H, Yamakawa M, Tsuda T (2012) Genetic reassortment between Sathuperi and Shamonda viruses of the genus Orthobunyavirus in nature: implications for their genetic relationship to Schmallenberg virus. Arch Virol 157:1611–1616

Yilmaz H, Hoffmann B, Turan N, Cizmecigil UY, Richt JA, Van der Poel WHM (2014) Detection and partial sequencing of Schmallenberg virus in cattle and sheep in Turkey. Vector Borne Zoonotic Dis Larchmt N 14:223–225

Zanella G, Raballand C, Durand B, Sailleau C, Pelzer S, Benoit F, Doceul V, Zientara S, Bréard E (2015) Likely introduction date of Schmallenberg virus into France according to monthly serological surveys in cattle. Transbound Emerg Dis 62:e76–e79

Equine Coronavirus Infection

4

Nicola Pusterla, Ron Vin, Christian Leutenegger,
Linda D. Mittel, and Thomas J. Divers

4.1 Etiology

Coronaviruses are members of the *Coronaviridae* family, all of which are single-stranded, positive-sense, non-segmented, enveloped RNA viruses responsible for enteric, respiratory, hepatic, or neurologic disease in a variety of mammalian and avian species (Wege et al. 1982). The family *Coronaviridae* is subdivided in two subfamilies, *Torovirinae* and *Coronavirinae*, with the latter subfamily containing four genera defined on the basis of serological cross-reactivity and genetic differences: *Alphacoronavirus*, *Betacoronavirus*, *Deltacoronavirus*, and *Gammacoronavirus* (Woo et al. 2012). Equine coronavirus (ECoV) is classified within the *Betacoronavirus* genus, along with human OC43 and HKU1 coronavirus, bovine coronavirus, porcine hemagglutinating encephalomyelitis virus, canine respiratory coronavirus, mouse hepatitis virus, and sialodacryoadenitis virus (Zhang et al. 2007).

The nucleocapsid protein of coronaviruses complexes with the genomic RNA to form a helical capsid structure found within the viral envelope. Trimers of the spike protein form peplomers embedded in the envelope giving the virion its corona or

N. Pusterla (✉)
Department of Medicine and Epidemiology, School of Veterinary Medicine, University of California, Davis, CA 95616, USA
e-mail: npusterla@ucdavis.edu

R. Vin • C. Leutenegger
IDEXX Laboratories, Inc., West Sacramento, CA 95605, USA

L.D. Mittel
The Animal Health Diagnostic Center, College of Veterinary Medicine, Cornell University, Ithaca, NY 14853, USA

T.J. Divers
College of Veterinary Medicine, Cornell University, Ithaca, NY 14853, USA

J. Bayry (ed.), *Emerging and Re-emerging Infectious Diseases of Livestock*,
DOI 10.1007/978-3-319-47426-7_4

crown-like morphology. In ECoV, the hemagglutinin-esterase protein forms smaller spikes on the membrane. There are two fully transmembraneous structural proteins referred to as the membrane and small membrane proteins in the coronavirus virion (Weiss and Leibowitz 2011).

4.2 Epidemiology

Epidemiological information for ECoV is sparse at this time and either extrapolated from close-related coronaviruses such as BCoV or largely based on non-published preliminary observations (Table 4.1).

Clinical cases are either sporadic or epizootic. Collective data from three veterinary diagnostic laboratories in the USA (IDEXX Laboratories, Inc., West Sacramento; Animal Health Diagnostic Center (AHDC), Cornell University, Ithaca; Real-time PCR Research and Diagnostics Core Facility, University of California, Davis) shows that the overall number of ECoV qPCR-positive cases is steadily

Table 4.1 Documented or observed epidemiological factors for BCoV and ECoV (Pusterla et al. 2015a)

Epidemiological factors	BCoV (winter dysentery)	ECoV
Case presentation	Epizootic	Sporadic and epizootic
Age of affected animals	Predominantly cattle	Predominantly adult horses
Seasonality	Usually during winter month (November to April)	Year around with increased reports during cold months (October to April)
Transmission route	Feco-oral	Suspected feco-oral for natural disease, naso-esophageal for experimental disease
Geographic distribution	More common in northern states	No geographic predilection
Morbidity rate	High (30–100 %)	Variable (10–80 %)
Mortality rate	Rare	Low
Incubation time	2–8 days	2–3 days
Outbreak duration	Less than 2 weeks	2–3 weeks
Period of illness	Few days to 1 week	Few days to 1 week
Clinical signs	Explosive diarrhea, anorexia, depression, reduced milk production	Anorexia, depression, fever, less frequently diarrhea, colic, neurological signs
Diagnosis	Antigen-capture ELISA, fluorescent antibody, qPCR, electron microscopy	qPCR
Recovery	Fast, generally self-limiting disease	Fast, generally self-limiting disease
Prevention	Vaccination and biosecurity	Biosecurity

increasing since 2010 (approximate time when ECoV testing by qPCR was introduced by molecular diagnostic laboratories) and that similar to BCoV, the case number is higher during the colder months of the year (October to April). Clinical ECoV infection has predominantly been reported in adult horses (Oue et al. 2011, 2013; Pusterla et al. 2013; Fielding et al. 2015). Outbreaks have been reported in riding, racing, and show horses and less frequently in breeding animals. One hypothesis, behind the lack of large numbers of documented clinical cases at larger breeding farms, is the frequent circulation of ECoV between asymptomatic young animals and adult resident horses, conferring protection against clinical disease in adult horses.

Preliminary epidemiological observations support a feco-oral route of transmission. This is supported by a recent experimental study in which three 9-to-10-month old Japanese draft horses were successfully infected via naso-esophageal intubation using fecal material from a previously characterized ECoV-infected horse (Nemoto et al. 2014). Two out of the three draft horses developed clinical disease, and all of them shed large amounts of ECoV in feces. Bovine coronavirus (BCoV) is closely related to ECoV and considered a pneumoenteric virus, causing not only enteric disease but also mild upper respiratory signs (Saif et al. 1986; Tsunemitsu et al. 1999; Tråvén et al. 2001; O'Neill et al. 2014). Recently, a French group screened 395 feces and 200 respiratory specimens submitted to a veterinary diagnostic laboratory for the presence of ECoV (Miszczak et al. 2014). The samples had been collected from foals and adult horses suffering from mild respiratory or enteric disease. In that study, the researchers found ECoV by qPCR in a total of 12 samples (11 fecal samples and one respiratory specimen). In a study testing nasal secretions from 2437 horses with signs of fever and/or acute onset of upper respiratory tract infection submitted to a commercial diagnostic laboratory in the USA from January 2013 to December 2014 for the detection of common respiratory pathogens, ECoV was detected by qPCR in only 17 (0.7 %) horses (Pusterla et al. 2015b). Collectively, both these studies show that ECoV is infrequently detected in nasal secretions from horses with infectious upper respiratory tract disease, suggesting lack of tropism for ECoV to the respiratory epithelium. In the three experimentally infected Japanese draft horses (Nemoto et al. 2014), nasal secretions were ECoV PCR positive during the time of peak fecal shedding, but it could not be determined if this was due to nasal replication and shedding of the virus or from environmental contamination from the positive feces or both.

Morbidity of ECoV in affected herds is variable and has been reported to range between 10 and 83 % (Oue et al. 2011, 2013; Pusterla et al. 2013; Fielding et al. 2015). Mortality rates have been reported to be low (Oue et al. 2011, 2013; Pusterla et al. 2013), although a mortality rate of 27 % was recently reported by Fielding et al. (2015) in American miniature horses. Mortality has been associated with endotoxemia, septicemia, and hyperammonemia-associated encephalopathy (Pusterla et al. 2013; Fielding et al. 2015; Giannitti et al. 2015). A large number of documented clinical cases will be needed to determine the effect of various host, viral, treatment, and environmental factors on outcome, as well as the identification of risk factors associated with risk of morbidity.

Similar to BCoV the incubation period is short, and clinical disease develops between 48 and 72 h after either natural exposure or experimental infection (Nemoto et al. 2014; Fielding et al. 2015). Clinical signs persist for a few days to 1 week and generally resolve with minimal supportive care. Occasionally, clinical signs can last longer as recently identified in two ECoV cases from Florida that experienced colic and systemic illness for 14 days (Mittel personal communication 2016). Shedding post-experimental infection as documented by qPCR ranged between 10 and 12 days (Nemoto et al. 2014). Fecal shedding of ECoV under natural conditions has been reported to range between three and 25 days (Pusterla et al. 2013; Fielding et al. 2015; Mittel personal communication 2016). It is not clear how long ECoV will persist in the environment and potentially act as the source of infection. It is reasonable to draw comparisons from other coronaviral pathogens such as severe acute respiratory syndrome-associated coronavirus (SARS-CoV) which has been shown in vitro to persist for 2 days in hospital wastewater, domestic sewage, and dechlorinated tap water while 3 days in feces, 14 days in PBS, and 17 days in urine at 20 °C. However, at 4 °C, the SARS-CoV could persist for 14 days in wastewater and at least 17 days in feces or urine (Wang et al. 2005). Enhanced replication of the virus in feces in the environment during colder weather is one possible explanation for the apparent higher prevalence of test-positive fecal samples and clinical disease during cooler weather. ECoV disease can occur in one or more horses on a farm with no recent exposure to new horses (divers, personal communication 2016). Hypothetically this might be explained by indirect ECoV transmission by non-equine animal species or that colder weather might allow increased fecal replication of the virus from a chronically infected horse.

4.3 Clinical Presentation

Clinical information collected from 16 outbreaks during the time period of November 2011 to December 2014 on a total of 406 horses showed that 122 horses (30 %) showed clinical signs (Pusterla et al. 2013; Pusterla, personal communication 2016). The main clinical signs reported were anorexia (98 %), lethargy (89 %), and fever (84 %). The rectal temperature of febrile horses ranged from 38.6 to 41.0 °C (median 39.9 °C). Changes in fecal character, ranging from soft formed to watery consistency (Fig. 4.1), and colic were observed in 25 % and 18 % of horses with any clinical signs, respectively. Gastrointestinal signs are generally preceded by systemic signs of anorexia and fever (Fig. 4.2). These results are in agreement with recent outbreaks from Japan in adult draft horses reporting that anorexia and pyrexia were the main clinical signs, while specific gastrointestinal signs were observed in about 10 % of affected horses (Oue et al. 2011, 2013). In 3 % of infected horses, signs of encephalopathy including circling, head pressing, ataxia, proprioceptive deficits, nystagmus, recumbency, and seizure have also been reported (Fig. 4.3; Pusterla et al. 2013; Fielding et al. 2015).

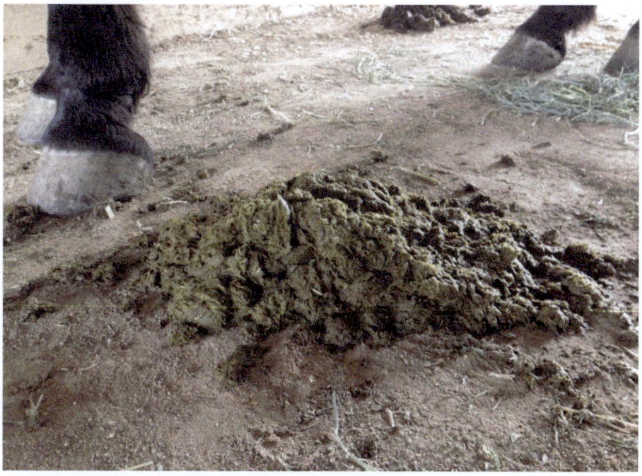

Fig. 4.1 Soft-formed feces from a horse with clinical ECoV infection and concurrent depression, anorexia, and fever

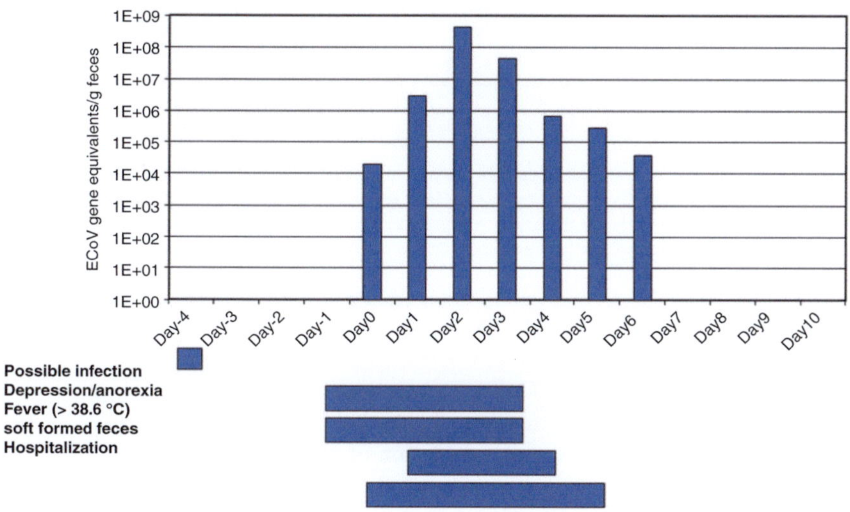

Fig. 4.2 Diagram showing temporal clinical signs and fecal shedding of ECoV in an adult horse presented to a referring hospital (day 0 was the first day of hospitalization) because of anorexia, depression, and fever

Although clinical disease is apparent in 10–83 % of ECoV-infected horses, one needs to take into account that some horses remain asymptomatic after infection. Asymptomatic infection is defined as lack of clinical disease in a horse from which ECoV is detected in feces by qPCR. The percentage of asymptomatic horses during an outbreak of ECoV has been observed to range between 11 and 83 % (Pusterla et al. 2013).

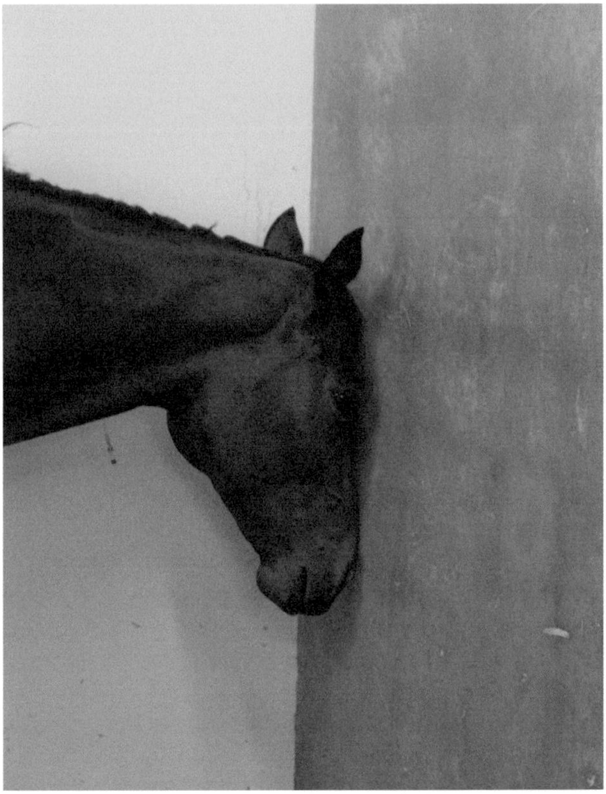

Fig. 4.3 Head pressing in an adult thoroughbred gelding with clinical ECoV infection

4.4 Diagnostic Evaluation

The antemortem diagnosis of ECoV relies on the presence of clinical signs compatible with ECoV infection, neutropenia, and/or lymphopenia, the exclusion of other infectious causes, and the molecular detection of ECoV in feces.

The consistently observed hematological abnormalities with ECoV infection are leukopenia due to neutropenia and/or lymphopenia. The blood work from 73 clinical cases with suspected ECoV infection showed that the total nucleated cell count ranged from 520 to 7800/μl (median 3580/μl; reference interval 5000–11,600/μl; Fig. 4.4). The white cell blood count in the 73 diseased horses showed leukopenia in 25 %, neutrophilia in 66 %, and lymphopenia in 72 % of the horses. The neutrophil and lymphocyte counts in these horses ranged from 180 to 5070/μl (median 1650/μl; reference interval 2600–6800/μl) and 80–3580/μl (median 1090/μl; reference interval 1600–5800/μl), respectively. Both the complete blood count (CBC) and white cell differential were unremarkable in only 11 % of the horses. Additionally, less consistent hematological abnormalities included the presence of band neutrophils and shifts in monocyte counts (most typically shifting from a low or a normal count to a high normal or rebound monocytosis). Occasional rebound

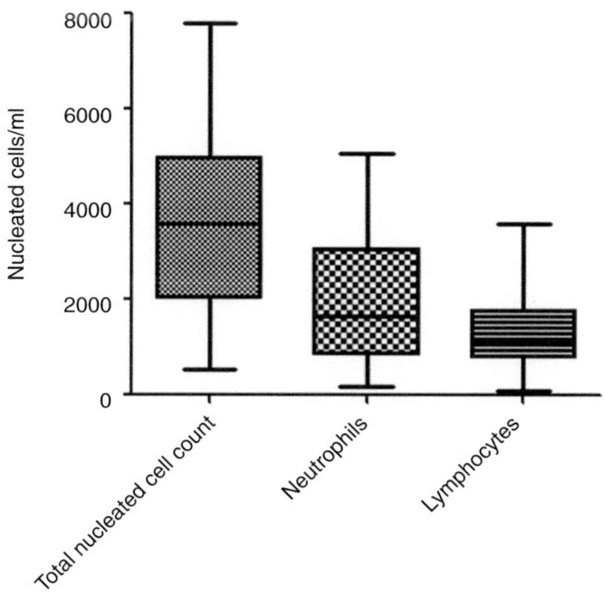

Fig. 4.4 Hematological findings in 73 adult horses with laboratory confirmation of ECoV infection. The results are expressed as box and whiskers (Reproduced from Pusterla et al. 2015a)

leukocytosis due to neutrophilia and monocytosis during the disease course and recovery was observed as well. CBC abnormalities are expected to resolve within 5–7 days as long as no complications associated with the disruption of the gastrointestinal barrier occur. Biochemical parameters may be unremarkable, but elevation of total and indirect bilirubin due to partial or complete anorexia, electrolyte changes consistent with enterocolitis, transient elevation of liver enzymes, and renal parameters suggestive of prerenal azotemia have been observed in some of the cases. It is judicious to measure blood ammonia in horses with suspected ECoV infection and concurrent signs of encephalopathy. Fielding et al. (2015) reported on a case of severe hyperammonemia (677 μmol/L; reference interval ≤60 μmol/L) with encephalopathic signs that subsequently died. Hyperammonemia associated with ECoV infection is likely due to increased ammonia production within or absorption from the gastrointestinal tract due to gastrointestinal barrier breakdown. An increase in enteric ammonia production could also be the result of bacterial microbiome changes associated with EcoV infection.

Isolation of ECoV in human rectal adenocarcinoma cells, although difficult, has been previously reported (Guy et al. 2000; Oue et al. 2011). The need for cost-effective and timely detection methods has restricted the diagnosis of ECoV infection to direct demonstration of coronavirus antigens or specific nucleic acids in biological samples. Historically, coronavirus detection in feces had been based on negative-stain EM and antigen-capture ELISA (Davis et al. 2000; Guy et al. 2000). However, the sensitivity and specificity of these diagnostic modalities have not been evaluated, and detection may be unsuccessful if viral particles are not present in sufficient numbers. Sensitive laboratory diagnosis of ECoV is through fecal quantitative PCR (qPCR). A recent study evaluated the overall accuracy of qPCR and determined 90 % accuracy between clinical status and PCR detection of ECoV in various outbreak populations (Pusterla et al. 2013). The authors have documented a few cases of ECoV infection, which tested qPCR negative during early disease. These few horses ended testing qPCR positive on a 24–48 h recheck fecal sample. It is hypothesized that during peracute stages of infection or when diseased horses experience gastrointestinal stasis due to colic, there are not enough viral particles in the feces to be detected. The recommendation is to retest fecal matter in a suspected index case at a later time point or collect multiple samples for pooled testing. Viral kinetics of ECoV in feces from experimentally infected horses showed that horses began to shed ECoV RNA in their feces at 3 or 4 days postinfection and continued shedding virus until 12 or 14 days postinfection (Nemoto et al. 2014). Peak shedding is generally seen on day 3 to 4 following the development of clinical signs (Fig. 4.2), and qPCR detection of ECoV in naturally infected horses can last for 3–25 days (Pusterla et al. 2013; Fielding et al. 2015).

Interestingly, the fecal viral load measured by qPCR in foals (up to 6 months) appears to be lower compared to horses older than 12 months of age, although the difference is not statistically significant ($P = 0.307$). This observation suggests that viral replication in the gut is comparable (Leutenegger, personal communication 2016) between foals and older horses. In human coronavirus infections, viral load is a strong prognostic indicator for clinical outcome and mortality (Hung et al. 2009). A similar observation has recently been reported for ECoV (Fielding et al. 2015). The viral load that is shed in humans infected with coronavirus is associated with polymorphisms of genes involved in innate immunity, an individual's genetic makeup, and immunologic host response to the virus (Chen et al. 2006). In addition, coronavirus strain variations can influence the replication ability and viral load as shown with feline coronavirus (FCoV) which exists in two genotypes: a benign feline enteric coronavirus genotype (FECV) with low replication competency in enterocytes and a highly virulent feline infectious peritonitis virus (FIPV) with an enhanced ability to replicate in different cell populations (Chang et al. 2012). Single-point mutations on the spike gene are responsible for dramatic changes in cell tropism, replication competency, and clinical manifestations.

Necropsy cases of suspected enteritis should have feces or gastrointestinal content tested by qPCR for ECoV and other gastrointestinal infectious agents. Further, formalin-fixed intestinal tissue samples can also be tested by immunochemistry and direct fluorescent antibody testing using BCoV reagents (Giannitti et al. 2015).

4.5 Pathogenesis

Enteritis caused by ECoV has been suspected in foals for many years, but the direct pathogenicity of ECoV has only been described in one neonate (Davis et al. 2000). Further, the high and similar frequency of ECoV shedding detected in healthy foals and foals with gastrointestinal diseases from Central Kentucky suggests that ECoV commonly circulates among young horses with subclinical disease (Slovis et al. 2014).

An interesting observation is that all ECoV infections in foals with gastrointestinal disease were associated with coinfections, while most healthy foals infected with ECoV displayed a mono-infection (Slovis et al. 2014). This observation may indicate that in foals, pre-existing ECoV infection may predispose to opportunistic secondary viral, bacterial, or protozoal infections as shown for enteric and respiratory coronavirus in other species (Pakpinyo et al. 2003; Srikumaran et al. 2007; Brockmeier et al. 2008). Other hypothetical reasons for the lack of clinical signs in foals infected solely with ECoV are that host factors such as the absence of specific receptor binding sites or presence of ECoV-specific colostrally derived antibodies prevent the development of enteritis. However, in the vast majority of clinical adult horses, ECoV is a mono-infection, demonstrated by enteric panel qPCR testing. This suggests a unique pathogenicity, coinfection with still unknown pathogens, or a distinct difference in immunological reaction between foals and adult horses toward the ECoV infection similar to FCoV. The development of a humoral, cell-mediated, or mixed immune response against the highly virulent FIPV version of FCoV distinctly influences the clinical outcome of the infection. Interestingly, the clinical outcome of FIPV has been correlated with a severe suppression of NK cells and T regulatory cells confirming earlier reports that a weakened cell-mediated immune response is associated with fatal FIPV infection (Vermeulen et al. 2013).

In adult horses, the pathology of ECoV has recently been described in three equids (Giannitti et al. 2015). The ECoV-infected equids displayed severe diffuse necrotizing enteritis with marked villus attenuation, epithelial cell necrosis in the tips of the villi, neutrophilic and fibrin extravasation into the small intestinal lumen (pseudomembrane formation), as well as crypt necrosis, microthrombosis, and hemorrhage (Fig. 4.5). ECoV was detected by qPCR in small intestinal tissue, gastrointestinal content, and/or feces, and coronavirus antigen was detected by immunohistochemistry and/or direct fluorescent antibody testing in the small intestine of all cases (Fig. 4.6).

4.6 Treatment and Prevention

Most adult horses with clinical ECoV infection recover spontaneously in a few days without specific treatment. Horses with persistent elevated rectal temperature, anorexia, and depression are routinely treated with nonsteroidal anti-inflammatory drugs such as flunixin meglumine (0.5–1.1 mg/kg BWT q12–24 h IV or PO) or phenylbutazone (2–4 mg/kg BWT q12–24 h IV or PO) for 24–48 h, as long as their hydration status is believed normal. Horses with colic, persistent depression, and anorexia and/or diarrhea have been treated more intensively with fluid and electrolyte per nasogastric intubation or intravenous administration of polyionic fluids until clinical signs have resolved. Additionally, antimicrobials and gastrointestinal protectants should be considered in horses developing signs of endotoxemia and/or septicemia secondary to disruption of the gastrointestinal barrier. While hyperammonemia-associated encephalopathy only occurs in a small percentage of horses with ECoV infection, early recognition and treatment is associated with a

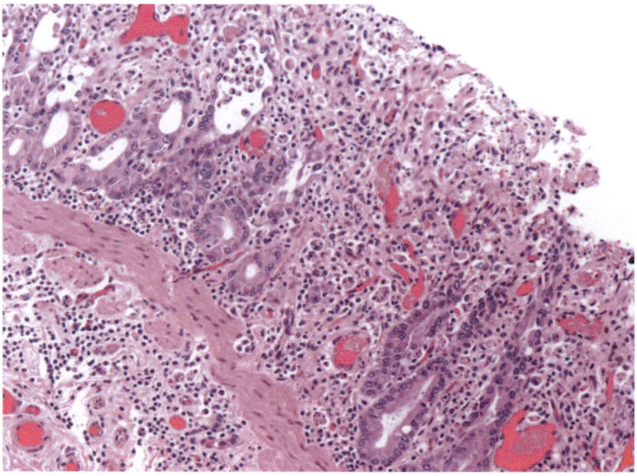

Fig. 4.5 H&E stain of jejunum from an adult horse with ECoV infection. There is loss of crypts and few remaining crypts are dilated, lined by attenuated epithelium, and contain sloughed necrotic enterocytes. The lamina propria and superficial submucosa are expanded by inflammatory infiltrates. Capillaries and venules in the mucosa and submucosa are occluded by fibrin thrombi (Reproduced from Pusterla et al. 2015a)

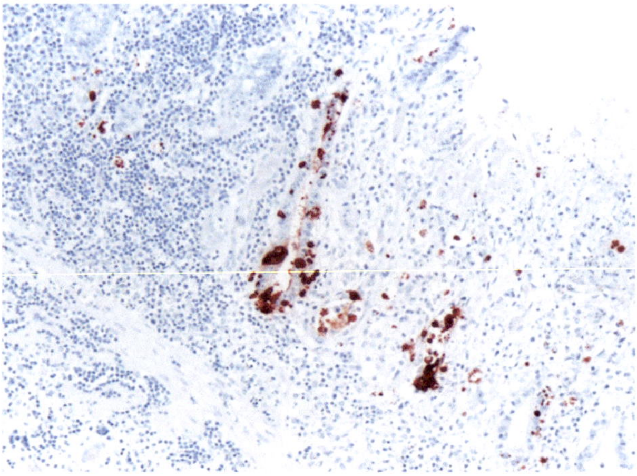

Fig. 4.6 Bovine coronavirus immunohistochemistry and hematoxylin counterstain of jejunum from an adult horse with ECoV infection. Strong granular/globular immunoreactivity is seen in the cytoplasm of deep gland enterocytes (Reproduced from Pusterla et al. 2015a)

positive outcome. Horses with suspected or documented hyperammonemia are treated with oral lactulose (0.1–0.2 ml/kg BWT q6-q12 h PO), neomycin sulfate (4–8 mg/kg BWT q8 h PO), or fecal transfaunation and crystalloid fluids.

Immunization strategies have been best described in cattle for the prevention of winter dysentery infection using a modified-live commercially available BCoV vaccine (Welter 1998). The use of BCoV vaccine in horses for the prevention of ECoV has to the author's knowledge not been investigated and cannot be recommended at this time due to the lack of safety and efficacy data and sporadic incidence of disease. Cattle that recovered from winter dysentery after experimental infection with

BCoV maintained a very long-lasting BCoV-specific serum (IgA and IgG) and local (IgA) antibody response (Tråvén et al. 2001).

4.7 Biosecurity Recommendations

The prevention of ECoV infection should focus on the implementation of routine management practices aimed at reducing the likelihood of introducing and disseminating ECoV at any horse-based premise (boarding facility, showground, veterinary hospital). Due to the highly contagious nature of ECoV, any horse developing or presenting with significant fever, anorexia, and depression with or without enteric signs (colic, diarrhea) should be strictly isolated until a diagnosis is secured. Such an approach can prevent the quarantine of an entire horse population should ECoV infection be diagnosed subsequently. Once ECoV infection is confirmed, strict isolation procedures and secondary quarantine of the source stable of the particular horse should be employed. Postinfection testing of clinical cases should be done to prevent viral spread to other horses. At stables and farms, all newly arrived horses should ideally be isolated for at least 3 weeks. Once an ECoV infection is suspected or confirmed, strict biosecurity measures including footbaths and the use of personal protective equipment should be provided and adequately maintained for sanitary purposes. Separate equipment, tack, bedding, and feedstuffs should be used in the care of these animals. Grooms and other personnel should be instructed to work with these animals last in the course of their daily routine. Exercise periods should be confined to a time when other horses are not present in the training areas, and riders should wear protective clothing and clean and thoroughly disinfect their boots, tack, and hands after contact with such animals. Horses returning from shows or extended traveling events should be isolated according to their particular circumstances. All horse vans and trailers should be thoroughly cleaned and disinfected after use. Equine coronavirus is susceptible to common disinfectants including sodium hypochlorite, povidone-iodine, chlorhexidine gluconate, phenols, quaternary ammonium compounds, accelerated hydrogen peroxide, and peroxygen compounds. The examination of at-risk horses for clinical signs of disease, including twice daily assessment of rectal temperature, remains the most effective tool in determining possible sources of virus introduction.

References

Brockmeier SL, Loving CL, Nicholson TL et al (2008) Coinfection of pigs with porcine respiratory coronavirus and *Bordetella bronchiseptica*. Vet Microbiol 128:36–47

Chang H, Egberink HF, Halpin R et al (2012) Spike protein fusion peptide and feline coronavirus virulence. Emerg Infect Dis 18:1089–1095

Chen W, Yang J, Lin J et al (2006) Nasopharyngeal shedding of severe acute respiratory syndrome–associated coronavirus is associated with genetic polymorphisms. Clin Infect Dis 42:1561–1569

Davis E, Rush BR, Cox J et al (2000) Neonatal enterocolitis associated with coronavirus infection in a foal: a case report. J Vet Diagn Investig 12:153–156

Fielding CL, Higgins JK, Higgins JC et al (2015) Disease associated with equine coronavirus infection and high case fatality rate. J Vet Intern Med 29:307–310

Giannitti F, Diab S, Mete A et al (2015) Necrotizing enteritisi and hyperammonemic encephalopathy associated with equine coronavirus infection in equids. Vet Pathol 52:1148–1156
Guy JS, Breslin JJ, Breuhaus B et al (2000) Characterization of a coronavirus isolated from a diarrheic foal. J Clin Microbiol 38:4523–4526
Hung IF, Lau SK, Woo PC et al (2009) Viral loads in clinical specimens and SARS manifestations. Hong Kong Med J 15:20–22
Miszczak F, Tesson V, Kin N et al (2014) First detection of equine coronavirus (ECoV) in Europe. Vet Microbiol 171:206–209
Nemoto M, Oue Y, Morita Y et al (2014) Experimental inoculation of equine coronavirus into Japanese draft horses. Arch Virol 159:3329–3334
O'Neill R, Mooney J, Connaghan E (2014) Patterns of detection of respiratory viruses in nasal swabs from calves in Ireland: a retrospective study. Vet Rec 175:351
Oue Y, Ishihara R, Edamatsu H et al (2011) Isolation of an equine coronavirus from adult horses with pyrogenic and enteric disease and its antigenic and genomic characterization in comparison with the NC99 strain. Vet Microbiol 150:41–48
Oue Y, Morita Y, Kondo T et al (2013) Epidemic of equine coronavirus at Obihiro Racecourse, Hokkaido, Japan in 2012. J Vet Med Sci 75:1261–1265
Pakpinyo S, Ley DH, Barnes HJ et al (2003) Enhancement of enteropathogenic *Escherichia coli* pathogenicity in young turkeys by concurrent turkey coronavirus infection. Avian Dis 47:396–405
Pusterla N, Mapes S, Wademan C et al (2013) Emerging outbreaks associated with equine coronavirus in adult horses. Vet Microbiol 162:228–231
Pusterla N, Vin R, Leutenegger C et al (2015a) Equine coronavirus: an emerging enteric virus of adult horses. Equine Vet Educ. doi:10.1111/eve.12453
Pusterla N, Holzenkaempfer N, Mapes S et al (2015b) Prevalence of equine coronavirus in nasal secretions from horses with fever and upper respiratory tract infection. Vet Rec. doi:10.1136/vr.103263
Saif LJ, Redman DR, Moorhead PD et al (1986) Experimentally induced coronavirus infections in calves: viral replication in the respiratory and intestinal tracts. Am J Vet Res 47:1426–1432
Slovis NM, Elam J, Estrada M et al (2014) Infectious agents associated with diarrhoea in neonatal foals in central Kentucky: a comprehensive molecular study. Equine Vet J 46:311–316
Srikumaran S, Kelling CL, Ambagala A (2007) Immune evasion by pathogens of bovine respiratory disease complex. Anim Health Res Rev 8:215–229
Tråvén M, Näslund K, Linde N et al (2001) Experimental reproduction of winter dysentery in lactating cows using BCV – comparison with BCV infection in milk-fed calves. Vet Microbiol 81:127–151
Tsunemitsu H, Smith DR, Saif LJ (1999) Experimental inoculation of adult dairy cows with bovine coronavirus and detection of coronavirus in feces by RT-PCR. Arch Virol 144:167–175
Vermeulen BL, Devriendt B, Olyslaegers DA et al (2013) Suppression of NK cells and regulatory T lymphocytes in cats naturally infected with feline infectious peritonitis virus. Vet Microbiol 164:46–59
Wang XW, Li JS, Jin M et al (2005) Study on the resistance of severe acute respiratory syndrome-associated coronavirus. J Virol Methods 126:171–177
Wege H, Siddell S, Ter Meulen V (1982) The biology and pathogenesis of coronavirus. Curr Top Microbiol Immunol 99:165–200
Weiss SR, Leibowitz JL (2011) Coronavirus pathogenesis. Adv Virus Res 81:85–164
Woo PC, Lau SK, Lam CS et al (2012) Discovery of seven novel mammalian and avian coronaviruses in the genus Deltacoronavirus supports bat coronaviruses as the gene source of Alphacoronavirus and Betacoronavirus and avian coronaviruses as the gene source of Gammacoronavirus and Deltacoronavirus. J Virol 86:3995–4008
Welter MW (1998) Adaptation and serial passage of bovine coronavirus in an established diploid swine testicular cell line and subsequent development of a modified live vaccine. Adv Exp Med Biol 440:707–711
Zhang J, Guy JS, Snijder EJ et al (2007) Genomic characterization of equine coronavirus. Virology 369:92–104

Coronaviridae: Infectious Bronchitis Virus

5

Ahmed S. Abdel-Moneim

5.1 History

Infectious bronchitis was first reported in 1931 who had observed the disease in North Dakota in the spring of 1930 (Schalk and Hawn 1931), and in 1936, the virus etiology was established (Beach and Schalm 1936). Initially, IBV was recognized as primarily a disease of young chickens; however it was later recorded to be common in semi-mature and laying flocks. Other manifestations of IBV include decline in egg production in laying flocks noted following the typical respiratory disease in the 1940s, kidney lesions observed in the 1960s (Cavanagh and Gelb 2008), enteric lesions observed in 1985, and more recently proventriculus affection in 1998.

5.2 Classification

IBV is a large, enveloped, positive-stranded RNA gammacoronavirus that is related to the family *Coronaviridae*, subfamily *Coronavirinae*, and within the order *Nidovirales* (Table 5.1). The coronaviruses possess the largest RNA genome of all RNA viruses and replicate by a unique mechanism associated with a multiple subgenomic nested set of mRNAs and high frequency of recombination. The subfamily *Coronavirinae* contains four distinct genera: *Alphacoronavirus*, *Betacoronavirus*, *Deltacoronavirus*, and *Gammacoronavirus*. To date, viruses of *Alphacoronavirus* and *Betacoronavirus* have been isolated from mammals, while deltacoronaviruses have been isolated from birds and pigs (Table 5.1) (Woo et al. 2012). Meanwhile, gammacoronaviruses are

A.S. Abdel-Moneim
Department of Microbiology (Virology Division), College of Medicine, Taif University, Al-Taif 21944, Saudi Arabia

Department of Virology, Faculty of Veterinary Medicine, Beni-Suef University, Beni-Suef 62511, Egypt
e-mail: asa@bsu.edu.eg

J. Bayry (ed.), *Emerging and Re-emerging Infectious Diseases of Livestock*,
DOI 10.1007/978-3-319-47426-7_5

Table 5.1 Taxonomy of coronaviruses

Order: *Nidovirales* Family: *Coronaviridae* Subfamily: *Coronavirinae*	
Genus: *Alphacoronavirus*	*Alphacoronavirus* 1a Canine coronavirus (CCoV) Feline coronavirus (FCoV)
	Alphacoronavirus 1b Human coronavirus 229E (HCoV-229E) Human coronavirus NL63 (HCoV-NL63)
Genus: *Betacoronavirus*	*Betacoronavirus* A Human coronavirus OC43 (HCoV-OC43) Human coronavirus HKU1 (HCoV-HKU1) Bovine coronavirus (BCoV) Murine hepatitis coronavirus (MHV) Canine respiratory coronavirus Dromedary camel coronavirus HKU23 Equine coronavirus Porcine hemagglutinating encephalomyelitis virus
	Betacoronavirus B Severe acute respiratory syndrome (SARS)-related coronavirus
	Betacoronavirus C Pipistrellus bat coronavirus HKU5 Tylonycteris bat coronavirus HKU4 Middle East Respiratory Syndrome (MERS-CoV) *Betacoronavirus* D Rousettus bat coronavirus HKU9 (BtCoV-HKU9)
Genus: *Deltacoronavirus*	Munia coronavirus HKU13 Porcine coronavirus HKU15 Sparrow coronavirus HKU17
Genus: *Gammacoronavirus*	Infectious bronchitis virus (IBV) Turkey coronavirus (TCoV) Duck coronavirus Goose coronavirus Pigeon coronavirus Pheasant coronavirus Beluga whale coronavirus SW1

found in birds, except for the coronaviruses detected in beluga whale and bottlenose dolphin (Mihindukulasuriya et al. 2008; Woo et al. 2010).

5.3 Morphology and Structure

The virus possesses a round structure that is often 100 to 160 nm in diameter and with long, petal-shaped spikes on the virus surface (Gonzalez et al. 2003). Inside the virion is a single-stranded, positive-sense linear RNA genome. The helical nucleocapsid, unusual for positive-stranded RNA animal viruses, is enclosed by a lipoprotein envelope that contains long petal-shaped spike glycoprotein (S); an integral membrane glycoprotein (M) which spans the lipid bilayer three times; and an

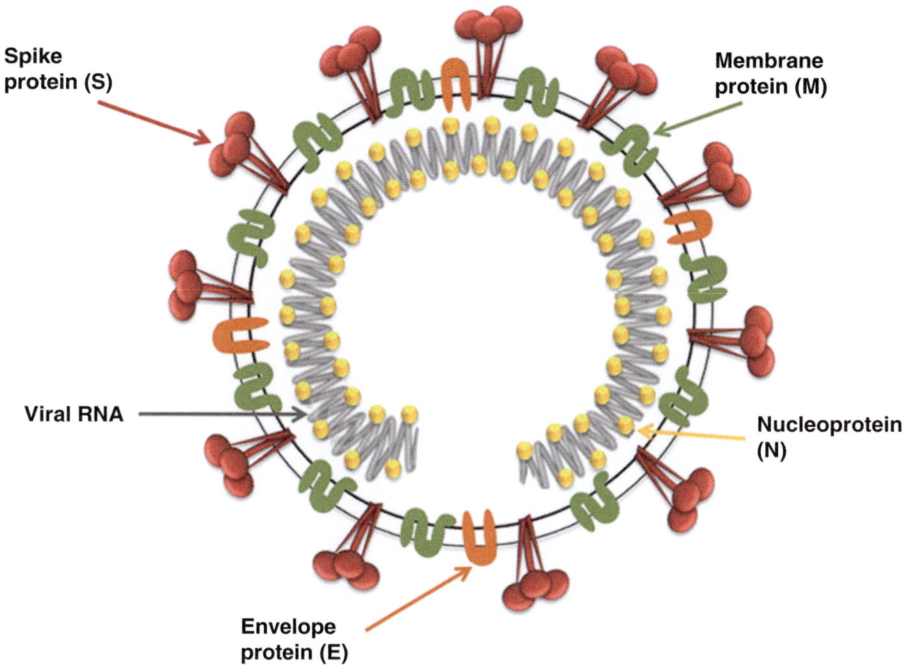

Fig. 5.1 Schematic diagram of the infectious bronchitis virus

envelope or small membrane (E) protein which is present in much smaller amounts than the other viral envelope proteins (Fig. 5.1).

5.4 Genome

IBV genome is a single-stranded, positive-sense linear genome with a cap at the 5′ end and poly(A) tail at the 3′ end (Boursnell et al. 1987). The viral genome is 27,620–27,661 nucleotides (nts) in length excluding the polyadenylated tail. At the 5′ end of the genome, there is a leader sequence (64 nt), which is followed by 5′ untranslated region (5′-UTR) of 528 nts (Ammayappan et al. 2008; Zhang et al. 2010; Abro et al. 2012). At the 3′ end of the RNA genome, there is 507–528 nts UTR, followed by a poly(A) sequence of variable length. At least ten open reading frames (ORFs) were detected (Zhang et al. 2010; Ammayappan et al. 2008): ORF1ab nonstructural protein (nsp) (529–20,360), ORF2 spike S glycoprotein (20,311–23,820, 3489 nts and 1162 amino acids [aa]), ORF3abc [3a, (23,820–23,993, 174 nts, 57 aa), 3b (23,993–24,187, 195 nts, 64aa), 3c small envelope protein (E) (24,168–24,491, 330 nts, 109 aa), ORF4, membrane glycoprotein (M) (24,469–25,140, 678 nts; 225 aa), ORF5ab [5a 198 nts (25,500–25,697), 5b 294 nts (25,694–25,942)], and ORF6 nucleoprotein N (25,885–27,114, 1230 nt, 409 aa). The genome organization of classical IBV is 5'UTR-ORF1a-ORF1b-S-3a-3b-E-M-5a-5b-N- UTR 3′ (Fig. 5.2); however, different genetic organizations were recorded 5′ UTR-Pol-S-X1-E-M-N-UTR-3′ or 5′ UTR -Pol-S-X1-E-M-5b-N-UTR3′ (Mardani et al. 2008).

Fig. 5.2 Schematic diagram of IBV genome organization

5.5 Structural Proteins

5.5.1 Spike Protein

The spike protein is petal-shaped protrusions of about 20 nm in length that emerge from the virion envelope. A cleaved N-terminal signal peptide (Binns et al. 1985) directs the S protein toward the endoplasmic reticulum (ER) where it undergoes terminal N-linked glycosylation (Cavanagh 1983a, b). After glycosylation, the monomers oligomerize to form dimers or trimers (Cavanagh 1983a, b; Delmas and Laude 1990; Lewicki and Gallagher 2002). The S protein of IBV is cleaved by a furin-like host cell protease at the highly basic motif RRFRR generating S1 (90 kDa) and S2 (84 kDa) subunits of about 500 and 600 amino acids in size, respectively (Cavanagh 1983a, b). The N-terminal part of S1 forms an ectodomain, while the C-terminal S2 subunit comprises a narrow stalk ectodomain, short transmembrane, and endodomain. All the receptor-binding domains (RBD) of IBV are located in S1 domain (Masters and Perlman 2013; Promkuntod et al. 2014). After endocytosis, conformational changes in the S protein are triggered by exposure to acidic pH in endosomes (Chu et al. 2006), resulting in fusion of the viral envelope with the cellular membrane. The nucleotide sequence of the S1 subunit is used for genotyping IBV isolates (OIE 2013). S protein contains epitopes for neutralization (Cavanagh 1983a, b; Kant et al. 1992; Koch et al. 1990; Mockett et al. 1984; Niesters et al. 1987b). In the S1 subunit, three hypervariable regions (HVRs) are located within amino acids 38–67, 91–141, and 274–387 (Kant et al. 1992; Koch et al. 1991). Neutralizing-serotype-specific epitopes are associated within the defined serotypes (Cavanagh et al. 1988; Niesters et al. 1987a; Jia et al. 1996). N38S, H43Q, P63S, and T69I amino acid substitutions lead to loss of the ability of M41 strain to bind to the trachea (Promkuntod et al. 2014).

5.5.2 Matrix Protein

Small domain of the M glycoprotein (25–33 kDa) is exposed to the exterior of the viral envelope. There is a triple membrane and a large carboxyl-terminal domain inside the viral envelope (Lai and Cavanagh 1997). M protein is glycosylated by N linkage (Lai and Cavanagh 1997). The M proteins are targeted to the pre-Golgi region. The M protein plays a key role in virus assembly and interacts with both N and S proteins (Kuo and Masters 2002; de Haan et al. 2002). The M protein may also be critical for packaging viral RNA into nucleocapsids, by specifically interacting with the viral RNA packaging signal (Narayanan et al. 2003).

5.5.3 Nucleocapsid Protein

The N protein is a phosphoprotein of 50 to 60 kDa that binds to the genomic RNA to form a helical ribonucleoprotein complex (Jayaram et al. 2005). The N protein interacts with M, leading to the incorporation of nucleocapsid into virus particles (Kuo and Masters 2002). It plays a role in the induction of cytotoxic T lymphocytes (Seo et al. 1997; Collisson et al. 2000). In addition, novel linear B-cell epitope peptides were found in N-terminal domain of N protein (Yu et al. 2010).

5.5.4 Envelope Protein

It is 9 to 12 kDa protein associated with the viral envelope (Godet et al. 1992). The E protein transverses the lipid bilayer twice, with both termini of the protein present in the virus lumen (Maeda et al. 2001). Both the M and E proteins are required for budding from infected cells (Vennema et al. 1996). The expression of E alone is sufficient for vesicle release from transfected cells (Maeda et al. 1999). This protein is associated with viral envelope formation, assembly, budding, ion channel activity, and apoptosis (Corse and Machamer 2003; Wilson et al. 2006).

5.6 IBV Genotypes

It is suggested that the emergence of IBV appears to be a regular influx, and up to date, more than 65 different types do exist worldwide (Table 5.2). Different serotypes generally have large differences (20–50 %) in the deduced amino acid sequences of the S1 subunit (Kusters et al. 1989). IBV serotypes that share more than 95 % amino acid identity in S1 should have cross protection, whereas IBV strains of other serotypes share less than 85 % amino acid identity did not cross protect each other (Cavanagh and Gelb 2008). Poor cross protection was found in viruses that are clearly distinguishable in only 2–3% differences in amino acid sequences (Cavanagh 1991; Abdel-Moneim et al. 2006). This diversity in S1 probably results from mutation, recombination, and strong positive selection in vivo (Cavanagh et al. 1988, 1990). The widespread use of live attenuated vaccine strains and the subsequent selective pressure induced by neutralizing antibodies against the spike may force the adaptation of the virus to escape immunity and hence result in faster evolutionary rates (Jackwood 2012). Error prone during replication is not expected to constitute a major role in the evolution of IBV, since RdRp possesses exoribonuclease (ExoN) activity that provides some proofreading errors during coronavirus replication (Minskaia et al. 2006). During the replication of the IBV, both full genomic minus-strand template and the subgenomic minus-strand templates are generated by continuous and discontinuous unique mechanisms, respectively; the latter allows recombination between RNA viruses (Sawicki and Sawicki 1995). Although recombination was found throughout the whole IBV genome, hot spots of recombination have been found in the upstream of S glycoprotein gene in

Table 5.2 IBV genotypes in different countries

Mass	Worldwide
IBV types distributed worldwide or in multiple countries	
793B(CR88/ 4-91vaccine) (Cavanagh et al. 2005)	UK (Gough et al. 1992)/Brazil (De Wit et al. 2015)/France (Cavanagh et al. 2005)/India (Sumi et al. 2012)/Egypt (Sultan et al. 2004)/Israel (Gelb Jr et al. 2005)/India (Elankumaran et al. 1999)/Spain (Worthington et al. 2008)/Ukrania (Ovchinnikova et al. 2011)/Nigeria (Ducatez et al. 2009)/ Mexico (Jackwood et al. 2005; Cook et al. 1996)/Thailand (Promkuntod et al. 2015)/China (Han et al. 2011)/Japan (Ariyoshi et al. 2010)/Thailand (Cook et al. 1996)/Canada (Martin et al. 2014)/Russia (Bochkov et al. 2006)/Morocco (Fellahi et al. 2015)
China-type I (LX4-type)/QX	China (Han et al. 2011)/Russia (Bochkov et al. 2006)/Europe (Worthington et al. 2008)/Korea(K-II) (Lim et al. 2012)/Japan (Ariyoshi et al. 2010)/South Africa (Sigrist et al. 2012) (Knoetze et al. 2014)/Thailand (Promkuntod et al. 2015)
China-type IV(LDL/Q1)	China (Han et al. 2011)/Taiwan (Chen et al. 2009)/Colombia (Jackwood 2012)/Chile (Jackwood 2012)/Italy (Toffan et al. 2013a)/Canada (Martin et al. 2014)/Saudi Arabia (Ababneh et al. 2012), Jordan (Ababneh et al. 2012), Iraq (Ababneh et al. 2012)
D207 (D274)	Europe (Davelaar et al. 1984; Worthington et al. 2008)/ Nigeria (Ducatez et al. 2009)/Egypt (Madbouly et al. 2002)/ Russia (Bochkov et al. 2006)
Arkansas [Gray/JMK]	Kazakhstan (Ovchinnikova et al. 2011)/Mexico (Quiroz et al. 1993)/Japan (Ariyoshi et al. 2010)/Brazil (De Wit et al. 2015)
USA/Connecticut	USA/Canada (Martin et al. 2014)/Mexico (Jackwood et al. 2005)/Argentina (Rimondi et al. 2009)/Colombia (Alvarado et al. 2005)
Italy-02	Europe (Jones et al. 2005)/Morocco (Fellahi et al. 2015)/ Ukraine148]/Slovania (Ovchinnikova et al. 2011) Russia (Bochkov et al. 2006)
Eg-Var-I/IS-Var II	Egypt (Abdel-Moneim et al. 2002;Abdel-Moneim et al. 2012)/Israel (Gelb Jr et al. 2005)/Turkey[HM802259.1]/Iraq (Mahmood et al. 2011)/Libya (Awad et al. 2014)/Oman (Al-Shekaili et al. 2015)
Eg-Var-II	Egypt (Abdel-Moneim et al. 2012)/Libya (Awad et al. 2014)/ Oman (Al-Shekaili et al. 2015)
B1648	Russia (Bochkov et al. 2006)/Belgium (Reddy et al. 2015)/ Nigeria (Ducatez et al. 2009)/Cuba (Acevedo et al. 2013)
Australia/Group I (Vic.S, N1/62, N3/62, N9/74)	Australia (Ignjatovic et al. 2006) New Zealand (McFarlane and Verma 2008)/China (Han et al. 2011;Jackwood 2012)
IBV types restricted to certain region or country	
USA/California/CA 99 USA (Mondal and Cardona 2007)/ Canada (Martin et al. 2014)	Netherlands/D3128(Davelaar et al. 1984) Egypt (El-Kady 1989)
USA/California / CA/557/03(Jackwood et al. 2007)	Italy/624/I (Capua et al. 1994)

Table 5.2 (continued)

Mass	Worldwide
USA/California CA/1737/04 USA (Jackwood et al. 2007)/ Canada (Martin et al. 2014)/ Cuba (Acevedo et al. 2013)	Turkey/IBV/Turkey/BB012/VIR9657/2012 [C404845]
USA/Delaware 072 USA (Gelb et al. 1997)/Canada (Martin et al. 2014)	Russia/RF1(Bochkov et al. 2006)
USA/Georgia/GA98 (Lee et al. 2001)	Russia/RF1(Bochkov et al. 2006)
USA/Georgia/GA11 (Jackwood 2012)	Russia/RF2 (Bochkov et al. 2006)
USA/Georgia/GA08 (Jackwood et al. 2010b)	Russia/RF3(Bochkov et al. 2006)
USA/Georgia/GA07 (Jackwood 2012)	Russia/RF4(Bochkov et al. 2006)
USA/PA/Wolgemuth/98 USA (Ziegler et al. 2002)/Canada (Martin et al. 2014)	Russia/RF5(Bochkov et al. 2006)
USA/PA/1220/98 USA (Ziegler et al. 2002)/Canada (Martin et al. 2014)	Russia/RF6(Bochkov et al. 2006)
Canada/Qu_mv (Martin et al. 2014)	China-type II (CK/CH/LSC/99I–type) (Han et al. 2011)
Mexico/47/UNAM/01 (Jackwood 2012)	China-type III (KM-91-like)(Korea/K-II) (Han et al. 2011) (Lim et al. 2012)
Mexico/7277/99 (Gelb et al. 2001)	China/BJ (Han et al. 2011)
Mexico/07,484/98 (Callison et al. 2001)	China/CK/CH/LHLJ/95I–type (Han et al. 2011)
Mexico/UNAM-97/97 (Escorcia et al. 2000)	Japan/JP-I (Ariyoshi et al. 2010)
Mexico/2001/47/UNAM [EU526405.1]	Japan/JP-II (Ariyoshi et al. 2010)
Argentina/Clus A (Rimondi et al. 2009)	Korea/K-I (Lim et al. 2012)
Argentina/Clus B (Rimondi et al. 2009)	Korea/New cluster 1 (Lim et al. 2012)
Argentina/Clus C (Rimondi et al. 2009)	Korea/New cluster 2 (Lim et al. 2012)
Brazil/01 (De Wit et al. 2015)	Taiwan/Group I (Ma et al. 2012)
Brazil/02(De Wit et al. 2015)	Taiwan /Group II (Taiwan/China)(Ma et al. 2012)
Brazil/03(De Wit et al. 2015)	Thailand/THA001(Promkuntod et al. 2015)
Brazil/04(De Wit et al. 2015)	Malaysia/MH5365/95 (Zulperi et al. 2009)

(continued)

Table 5.2 (continued)

Mass	Worldwide
Australia Group II (N1/88, Q3/88 / V18/91) (Ignjatovic et al. 2006)	India/PDRC/Pune/Ind/1/00 (Bayry et al. 2005)
Australia/subgroup 3/ (N1/03, N4/02, N5/03, N4/03) (Ignjatovic et al. 2006)	Tunisia/TN20/00 (Bourogaa et al. 2009)
Netherlands/D212 (D1466 vaccine) (Davelaar et al. 1984)	Morocco/Moroccan type (Fellahi et al. 2015)

the nonstructural proteins 2, 3, and 16, in the E and M genes as well as the area near the 3′ UTR (Thor et al. 2011). Recombination in different genes of IBV could affect the pathogenicity and virus virulence, but recombination of the S gene may result in the emergence of new strains, new serotypes, or even new viruses infecting other hosts (Jackwood et al. 2010a). Natural intergenic and intertypic recombination occurs naturally in an extensive manner (Cavanagh et al. 1992b; Wang et al. 1993; Jia et al. 1995; Lee and Jackwood 2000; Brooks et al. 2004; Bochkov et al. 2007; Ammayappan et al. 2008; Kuo et al. 2010; Mardani et al. 2010; Pohuang et al. 2011; Ovchinnikova et al. 2011; Thor et al. 2011; Liu et al. 2013; Song et al. 2013; Zhao et al. 2013; Hewson et al. 2014; Zhang et al. 2015). Interestingly, mosaic S1-containing recombinants from three different genotypes (H120, QX, D274) were reported in Russia (Ovchinnikova et al. 2011). In addition, recombination of distant unrecognized gammacoronavirus with a known IBV strain resulted in the evolution of gammacoronavirus able to infect turkeys (Jackwood et al. 2010a).

5.7 Replication

5.7.1 Attachment

The first step in the viral replication cycle is the binding of virions to the plasma membranes of the target cells. The cell receptor for IBV has yet to be elucidated. Only α-2, 3-linked sialic acid has shown to be essential for spike attachment (Wickramasinghe et al. 2011; Winter et al. 2008; Abd El Rahman et al. 2009; Promkuntod et al. 2014). After the virus binds to a specific receptor, it enters the cell, a step that involves fusion of the viral envelope with plasma membrane.

5.7.2 Penetration and Uncoating

The binding of virus with the receptor induces a conformational change of the S protein that activates the membrane fusion activity. After virus-membrane fusion, the viral nucleocapsid is released into the cytoplasm, and the RNA is uncoated to become available for translation and transcription.

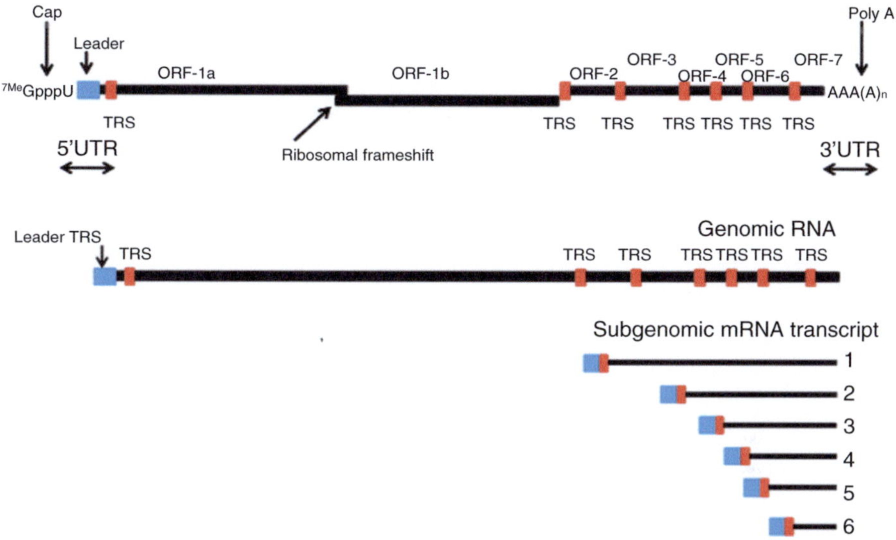

Fig. 5.3 Schematic diagram of the IBV genomic RNA and subgenomic mRNA transcripts. The nested set of seven IBV mRNAs (genome and sg mRNAs 2–6) is depicted below. The *blue box* is leader TRS, while *red boxes* indicate the position of the body TRSs

5.7.3 Transcription and Translation of Viral RNA

After the release of the viral RNA into the cytoplasm, the ORFs 1a and 1b are translated into functional nonstructural proteins, which comprise the RNA replicase-transcriptase complex. This replicase-transcriptase complex synthesizes a full-length negative-sense RNA copy, which is used as a template for the transcription of full-length and six subgenomic mRNAs that possess identical 3′ ends but different lengths (Fig. 5.3) (Sawicki and Sawicki 1990; Sethna et al. 1989). The initiation point of each mRNA corresponds to a stretch of consensus sequences, called intergenic sequences or transcription-regulatory sequences (TRSs, 5′ CT(T/G) AACAA(A/T)3′) that are found at the 3′ end of the leader sequence and at different positions upstream of genes in the genomic 3′-proximal domain of IBV. The 5′ two-thirds of the genome, 1a and 1b, encoding polyprotein precursor that is translated into a large polyprotein, 1ab, through a ribosomal frameshift mechanism (Brierley et al. 1989) and processed into 15 nonstructural proteins (nsp2–16) involved in virus replication. Papain-like proteinase (PL^{pro}), main protease (M^{pro}) or $3CL^{pro}$ (because it has some similarities to the 3C proteases of picornaviruses), adenosine diphosphate-ribose 1-phosphatase (nsp3), RNA-dependent RNA polymerase (nsp12, RdRp), and RNA helicase (nsp13), exonuclease (nsp14), endoribonuclease (nsp15), and 2-O-methyltransferase (nsp16) (Snijder et al. 2003; Fang et al. 2010) are among the important replication enzymes encoded by the replicase gene. Exonuclease and endoribonuclease are involved in processing RNA (Ivanov et al. 2004; Fang et al. 2010). The remaining 3′ third of the genome encoding the structural genes in addition to accessory genes interspersed within the structural gene region. Each viral subgenomic mRNA is used for translation of a single viral protein. The four structural proteins, spike (S), nucleocapsid (N), membrane (M), and envelope (E) proteins, are translated from separate mRNA. The accessory genes encode gene products although not essential for virus replication, but their deletion often causes viral attenuation (de Haan et al. 2002).

5.7.4 Replication of Viral Genomic RNA

IBV genome replication occurs through continuous transcription, while the subgenomic RNA synthesis occurs through discontinuous transcription (Fig. 5.3) (Masters 2006; Pasternak et al. 2006; Sawicki and Sawicki 2005; Tan et al. 2012). In addition to the replicase gene, the 5′ and 3′ end UTR sequences, with some specific secondary structures, are required for genomic RNA replication. The nucleocapsid (N) is also required for efficient viral RNA synthesis (Verheije et al. 2010; Zuniga et al. 2010). The genome-size transcripts are packaged into progeny virions.

5.7.5 Assembly and Release

IBV assembles and buds intracellularly into the lumen of a smooth-walled, tubulovesicular compartment located intermediately between the rough endoplasmic reticulum and Golgi (Klumperman et al. 1994). After budding, virus particles are transported through a functional Golgi stack and are released out of the host cells by the exocytic pathway. A strong interaction between IBV E and M occurs where E protein provides a temporary anchor to relocate M in the pre-Golgi compartments, as it “prepares” the membranes for budding (Raamsman et al. 2000). The spike (S) protein contains a canonical dilysine endoplasmic reticulum retrieval signal (–KKXX-COOH) in its cytoplasmic tail that plays an important role in protein accumulation near the budding sites (Ujike and Taguchi 2015). The virus nucleocapsid is enclosed by a lipoprotein envelope during virus budding from intracellular membranes. The envelope contains S, M, and E proteins.

5.8 Epizootiology

5.8.1 Hosts

All ages of chicken are susceptible to infection with IBV. The virus induces more severe disease in baby chicks, and the severity decreases as the age increases. IBV infection was also recorded in peafowl and also in non-galliform birds, e.g., the teal (Liu et al. 2005).

5.8.2 Transmission

IBV is a highly contagious airborne infection (Cumming 1970; OIE 2013) that can be easily transmitted directly by chicken to chicken through aerosols and indirectly contact via contamination of personnel or equipment, egg packing materials, litter, and farm visits (OIE 2013; Cavanagh and Gelb 2008). IBV can establish persistent infections when it affects the genital system of birds during early days of life; virus shedding is detected approximately when the egg production started. Reports of extended and intermittent shedding through nasal and fecal discharge are evident and could constitute a potential risk of flock-to-flock transmission (Jones and Ambali 1987; Adzhar et al. 1996; Alexander and Gough 1978; Cook 1968; Alexander and Gough 1977).

5.8.3 Incubation Period

The incubation period of IBV is very short 18–36 h and it depends on the infecting dose of the virus, and the clinical signs appear within 24–48 h of virus exposure (Hofstad and Yoder 1966).

5.8.4 Clinical Signs

The clinical picture includes decreased in the general bird vitality, huddling under a heat source, and decrease in both food and water consumption. The respiratory clinical form of IBV infection in chicks includes: nasal discharge, sneezing, coughing, and gasping. Some chicks may develop wet eyes and swollen sinuses. In chickens more than 6 weeks of age and older, the signs are similar to those in chicks, and the respiratory clinical form occurs but in a milder form (Cavanagh and Gelb 2008). Nephropathogenic viruses induce respiratory distress in addition to signs of ruffled feathers, wet droppings, increased water intake, and mortality (Winterfield and Hitchner 1962). In laying flocks, declines in egg production and quality are seen in addition to respiratory signs. About 6 to 8 weeks may elapse before production returns to the pre-infection level, but in some cases, this is never attained. The severity of the production declines may vary with the period of lay (van Eck 1983). In addition to production declines, the number of eggs unacceptable for setting is increased, hatchability is reduced, and soft-shelled, misshapen, and rough-shelled eggs are produced (Crinion 1972). The albumen may be thin and watery without definite demarcation between the thick and thin albumen of the normal fresh egg. Infectious bronchitis virus infection of 1-day-old chicks can produce permanent damage to oviducts leading to reduced egg production and inferior quality eggs when the chickens come into lay. The severity of oviduct lesions is likely to be less in infections of older chickens, and some serotypes may fail to produce any pathologic change even in infections of 1-day-old chicks. The presence of specific maternal antibody was also shown to protect the oviduct from damage due to IBV infection in early life (Chew et al. 1997).

5.8.5 Gross Lesions

Infected chicken showed petechial lesions in the larynx and tracheal exudate, which can be serous or caseous. Cloudy air sacs may be noticed in some birds. Caseous plug in the tracheal bifurcation could also be seen in some birds. Small areas of pneumonia may be observed in the lungs (Cavanagh and Gelb 2008). In nephropathogenic strains, the kidneys are swollen and the ureters are distended with urates (Ziegler et al. 2002; Abdel-Moneim et al. 2005). Some IBV strains are associated with thickening of the proventricular wall with congestion at the point of emergence of the glandular ducts (Toffan et al. 2013b). Cystic oviducts were observed in layer birds infected very early during the first days of life. Birds infected at the time of lay have reduced size and weight of the oviduct and regression of the ovaries. The fluid yok material may be observed in the abdominal cavity.

5.8.6 Histopathology

Loss of cilia of the tracheal mucosa and minor infiltration of heterophils and lymphocytes are detected 18–24 h after infection. Hyperplasia is followed by massive lymphocytic infiltration of the lamina propria may be present after 7 days. In nephrogenic strains, interstitial nephritis, infiltration of heterophils in the interstitium, and (Cavanagh and Gelb 2008; Abdel-Moneim et al. 2006) sometime renal hemorrhages are observed (Abdel-Moneim et al. 2005; Abdel-Moneim et al. 2006) (Fig. 5.4). The oviduct of mature hens showed decreased height and loss of cilia, infiltration by lymphocytes, and edema as well as fibroplasia of the mucosa of all regions of the oviduct (Sevoian and Levine 1957). Multifocal erosion and necrosis of the tunica mucosa and glandular epithelium of the proventriculus are associated with lymphocytic infiltration and fibroplasia in the lamina propria (Toffan et al. 2013b).

5.9 Pathogenicity

IBV initially infects ciliated and mucus-secreting cells of the upper respiratory tract (Raj and Jones 1997). Maximum virus shedding occurs 3–5 days after infection in the nose and trachea (Cavanagh 2003; Hofstad and Yoder 1966; Ambali and Jones 1990). High virus titers occur also in the lungs and air sacs (Raj and Jones 1997).

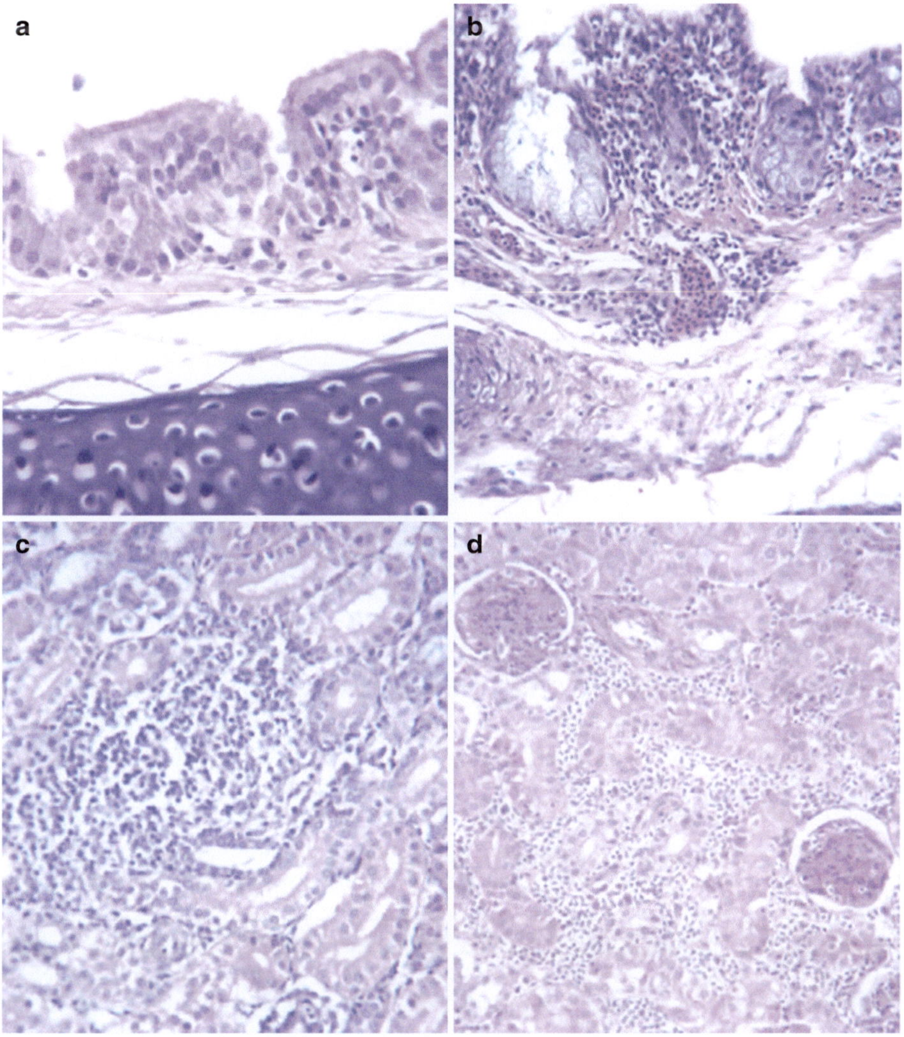

Fig. 5.4 Trachea and kidney histopathology following experimental infection of 1-day-old chicken with Egypt/F/03 (Abdel-Moneim et al. 2006). Trachea and kidney stained with H&E. (**a**) Trachea of chickens 5 days postinfection with Egypt/F/03 showing hyperplasia, lymphocytic infiltration, and edema (40 ×). (**b**) Trachea of chickens 7 days postinfection with Egypt/F/03 showing diffuse lymphocytic aggregation, degeneration of the epithelium mucus, and hemorrhages (20 ×). (**c**) Kidney of chickens 5 days postinfection with Egypt/F/03 showing focal lymphocytic aggregation in the interstitium and glomeruli, as well as degenerative changes in tubular epithelium (40 ×). (**d**) Kidney of chickens 7 days postinfection with Egypt/F/03 showing massive renal hemorrhages and degeneration of renal tubular epithelium (20 ×)

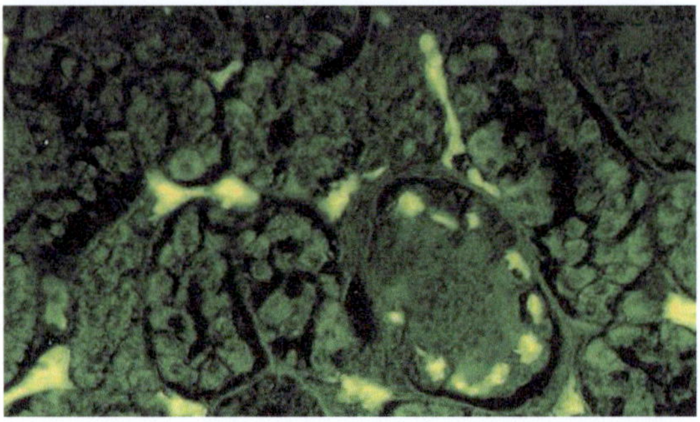

Fig. 5.5 Immunofluorescent staining of kidney paraffin section of kidney 5 days postinfection with Egypt/Beni-Suef/01 (Abdel-Moneim et al. 2005). Intracytoplasmic fluorescence in glomerular tuft and endothelial lining of renal blood vessels in the intertubular areas (40 ×)

IBV grows also in the epithelial lining of the kidney, oviduct, testes, esophagus, proventriculus, duodenum, jejunum, spleen, bursa of Fabricius, cecal tonsils, Harderian gland, rectum, and cloaca (Cavanagh 2003; Raj and Jones 1997; Abdel-Moneim et al. 2005; Ambali and Jones 1990; Seo et al. 1997) with minimal pathological effect. The virus commonly persists in the alimentary tract in young chickens (Ambali and Jones 1990; Alvarado et al. 2006) and in layers in the absence of clinical disease (Jones and Ambali 1987). Proventricular-type IBV (QX) reported in 1996 in China induces hemorrhagic ulceration of proventriculi and diarrhea followed by obvious signs of respiratory disease and high mortality (Zhou et al. 1998; GenCheng et al. 1998). Nephropathogenic strains result in considerable mortalities in meat-type birds (Cook et al. 2001; Lambrechts et al. 1993; Li and Yang 2001; Pensaert and Lambrechts 1994). The virus replicates in renal tubules and ducts, distal convoluted tubules, and Henle's loops (Chen and Itakura 1996) but may also replicate in the renal glomeruli (Fig. 5.5) (Abdel-Moneim et al. 2005). Modest to high titers of IBV in the kidney do not necessarily correlate with overt kidney disease, and there may be no gross kidney lesion (Ambali and Jones 1990). IBV infection of the chicken reproductive system leads to decreased egg production and quality due to the infection of the oviduct. In roosters, the virus results in epididymal stones, decreases sperm production, and decreases serum testosterone concentrations (Boltz et al. 2004). Infection is commonly followed by secondary bacterial infections, which may increase the mortality and complicate the clinical situation (Vandekerchove et al. 2004). Infection of enteric tissues usually does not manifest itself clinically.

5.10 Immunity

5.10.1 Innate Immunity

Hyperplasia of the goblet cells and alveolar mucous glands with subsequent increase in seromucous nasal discharge and catarrhal exudates in the trachea are the first response of the innate immunity against IBV infection (Nakamura et al. 1991). Toll-like receptor (TLR) 21 is stimulated by the presence of deoxyoligonucleotides containing CpG motifs, and it induces NF-κB production, leading to enhanced transcription of a number of cytokines (Brownlie et al. 2009). A rapid influx of macrophages to the infected tissue, detected within hours postinfection, plays an important role in limiting the replication of IBV within respiratory tissues. Heterophils are responsible for the destruction of IBV-infected cells during initial infection by phagocytosis and oxidative lysosomal enzyme release (Fulton et al. 1997). However, at the tracheal epithelium, heterophils do not reduce virus replication but worsen the severity of lesions (Raj et al. 1997). Interferon production in the plasma and all over body tissues (Otsuki et al. 1987), with simultaneous upregulation of mRNA levels of pro-inflammatory cytokines (IL-6 and IL-1β) and lipopolysaccharide-induced tumor necrosis factor (TNF)-α factor, is produced during IBV infection. This coincides with the highest viral loads and microscopic lesions, indicating a potential role of these cytokines with high virus loads and the development of tracheal and kidney lesions (Okino et al. 2014; Jang et al. 2013; Chhabra et al. 2015). In contrast, *il6* gene expression and upregulation of IFN-γ, IL-8 (CXCLi2), and MIP-1β genes together with mannose-binding lectin (MBL), which activates complement, inhibit the propagation of the virus (Juul-Madsen et al. 2007). Apoptosis is another nonspecific defense mechanism against IBV infection by premature lysis of infected cells, thereby aborting viral multiplication (Cong et al. 2013).

5.10.2 Role of Antibodies in Protection

Circulating antibody titers do not highly correlate with protection from IBV infection (Raggi and Lee 1965; Gough and Alexander 1979). In contrast, it has also been reported that high titers of humoral antibodies correlate well with the absence of virus re-isolation from the kidneys and genital tract (Gough et al. 1977; Macdonald et al. 1981; Yachida et al. 1985) and protection against a drop in egg production (Box et al. 1988). IBV-specific antibodies were suggested to be involved in limiting IBV spread by viremia from the trachea to other susceptible organs, including the kidneys and oviduct (Raj and Jones 1997). In general, serum antibody levels do not closely correlate with tissue protection, but local antibodies may contribute to the protection of the respiratory tract (Ignjatovic and McWaters 1991; Raggi and Lee 1965). Furthermore, IBV-specific IgA antibodies were first detected in tears and later in serum, which suggests that IgA is important in neutralizing IBV at mucosal surfaces and is thought to play a role in the control of IBV locally (Davelaar et al. 1982; Gelb et al. 1998). However, IgA might not be important in protection against IBV infection of the upper respiratory tract, whereas locally produced IgY, after a secondary immunization, provided effective protection against IBV by neutralizing this virus (Guo et al. 2008; Orr-Burks et al. 2014).

5.10.3 Cellular Immunity

IBV-specific cytotoxic T cell lymphocyte (CTL) activity is dependent on the S and N proteins of IBV (Collisson et al. 2000), while of $CD4^+$ T cells do not appear to be important in initially containing IBV infection in chickens (Seo et al. 2000); however, $CD4^+$ T cells and B cells could be more critical for long-term virus control (Chhabra et al. 2015). S1 and N but not the M protein proteins of IBV generated cytotoxic T cell responses. The whole N protein and its carboxy terminal region but not its amino terminal region were reported to induce a CTL response (Seo et al. 1997; Guo et al. 2010).

5.10.4 Maternally Derived Antibodies

Chicks hatched with high levels of maternally derived antibodies are protected against IBV challenge at 1 day of age but not at 7 days (>30 %) (Mondal and Naqi 2001). Protection is correlated with levels of local antibody but not humoral antibody (Mondal and Naqi 2001).

5.11 Diagnosis

5.11.1 Virus Isolation

5.11.1.1 Sampling

Samples should be obtained as soon as possible after the appearance of the clinical signs. Laryngotracheal swabs from live birds or tracheal and lung tissues from fresh carcasses can be used for laboratory diagnosis of IBV. Kidney, oviduct, or proventriculus samples are collected from birds with nephritis, egg production, or proventriculitis, respectively. All samples should be placed in virus transport medium containing penicillin (10,000 International Units [IU]/ml) and streptomycin (10 mg/ml) and kept in ice and then frozen (OIE 2013).

5.11.1.2 Virus Isolation in Embryonated Chicken Eggs

Specific pathogen-free embryonated chicken egg (SPF-ECE) is recommended for primary isolation of IBV. Processed samples (10–20 % w/v) in phosphate-buffered saline (PBS) are used for egg inoculation, after being clarified by low-speed centrifugation and filtration through bacteriological filters. 100–200 μl of the processed sample is inoculated into the allantoic cavity of 9–11-day-old embryos (Delaplane 1947). Embryo mortalities within the first 24 h is considered nonspecific death. The allantoic fluids of inoculated eggs (36–48h post-inoculation) are harvested and pooled (Cunningham 1973; Cunningham and El Dardiry 1948). Blind passage into another set of eggs for up to a total of three to four passages is conducted. The last passage is left for 7 days to screen the presence of pathognomonic embryonic changes: stunted and curled embryos (Fig. 5.6) with feather dystrophy and urate

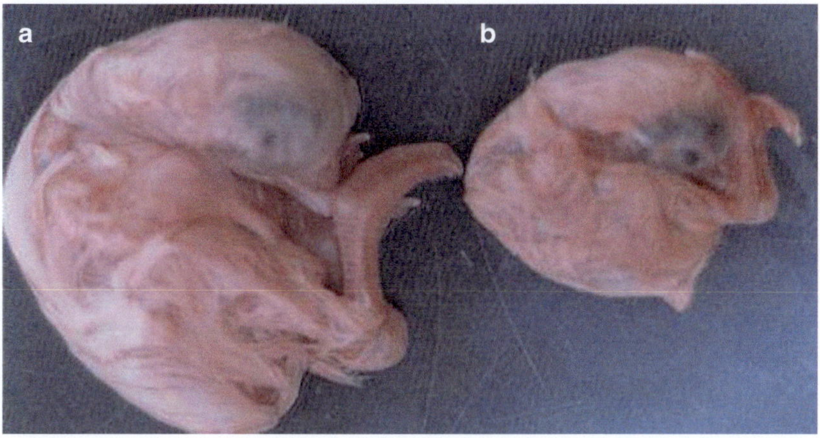

Fig. 5.6 Normal embryo (**a**) and stunted and dwarfed embryo following inoculation of specific pathogen-free embryonated chicken eggs with IBV (**b**)

deposits in the mesonephros. These lesions could also appear as early as the second passage (Delaplane 1947). The embryo-adapted strains induce more embryo mortalities. Isolation of IBV must be confirmed by serum neutralization or reverse transcription polymerase chain reaction (RT-PCR).

5.11.1.3 Tracheal Ring Culture

Tracheal ring culture (0.5–1.0 mm thick) from 19- to 20-day-old embryos can be used for primary isolation of IBV directly from field samples (Cook et al. 1976). The rings are maintained in Eagle's N-2-hydroxyethylpiperazine-N′-2- ethanesulfonic acid (HEPES) in roller drums (15 rev/hour) (OIE 2013). Ciliostasis within 24–48 h is an indication for virus multiplication; however, other viruses could produce similar lesions, so subsequent virus identification is needed.

5.11.2 Biological and Immunological Identification

IBV exerts hemagglutination (HA) activity only after phospholipase C treatment of concentrated virus infected allanto-aminiotic fluids (Bingham et al. 1975). A rapid plate HA test to detect neuraminidase-treated IBV in the allantoic fluid of ECE was introduced into the routine procedure of IBV identification and was found to correlate with the RT-PCR during the early stages of IBV detection and identification and isolation in ECE (Ruano et al. 2000). Such technique depends on the principle that IBV acquires its HA activity after removal of α 2, 3-linked N-acetyl neuraminic acid from the virion surface (Schultze et al. 1992). IBV can also be detected using immunofluorescence or immunoperoxidase on the tracheal or kidney section from the field isolates or on the chorioallantoic membrane or TOC from the inoculated embryos (Handberg et al. 1999; Abdel-Moneim et al. 2009; Bhattacharjee et al. 1994). However, nonspecific reactions or lower sensitivity especially in field samples may occur (Braune and Gentry 1965; Yagyu and Ohta 1990; Benyeda et al. 2010). The specificity of IFA may possibly be improved by using monoclonal antibodies (MAbs) (Naqi 1990; Yagyu and Ohta 1990; De Wit et al. 1995). Agar gel precipitation can be used for IBV identification, however, it possesses lower sensitivity in comparison to other assays (De Wit et al. 1992). Enzyme immunoassays are quick, inexpensive, and sensitive assays, which are suitable for screening large number of samples, IBV diagnosis, and serotype identification as well (Naqi 1990; Ignjatovic and McWaters 1991; Cavanagh et al. 1992a; Karaca and Syed 1993).

5.11.3 Molecular Identification

In situ hybridization can be used to detect viral nucleic acid (Collisson et al. 1990). RT-PCR and restriction fragment length polymorphism (RFLP) are used to genetically identify IBV (Kwon et al. 1993).

5.11.4 Serotyping and Genotyping of IBV Strains

Serotyping of IBV isolates has been conducted using hemagglutination inhibition (HI) (Alexander et al. 1983; King and Hopkins 1984) and virus neutralization (VN) tests in chick embryos (Dawson and Gough 1971), TOCs (Darbyshire et al. 1979), and cell cultures (Hopkins 1974). Enzyme-linked immunosorbent assays (ELISA) using MAbs are successfully used in serotyping IBV strains (Ignjatovic and McWaters 1991). The limitations of MAb analysis for IBV serotype definition are the lack of availability of MAbs or hybridomas and the need to produce new MAbs with appropriate specificity to keep pace with the ever-growing number of emerging IBV-variant serotypes (Karaca et al. 1992). There is a good correlation between the S1 sequence results and the VN serotyping (OIE 2013). The emergence of vast majority of the strains circulated worldwide (Jackwood 2012) renders serotyping impossible in many cases, and hence genotyping methods replaced HI and VN typing of IBV strains. Restriction fragment length polymorphism (RFLP) analysis of the S1 gene following RT-PCR amplification has been used to identify IBV serotypes (Lin et al. 1991; Kwon et al. 1993). Identification of IBV serotype is also conducted using serotype-specific S1 gene primer. Despite the success of both RFLP and serotype-specific RT-PCR, RFLP-derived restriction patterns of some IBV serotypes may be difficult to distinguish from others. Furthermore, samples containing mixture of more than one serotype may be difficult to be differentiated (Keeler et al. 1998). On the other hand, a mutation at a specific primer site or at an endonuclease recognition site may result in false negative in both RT-PCR and RFLP techniques. Direct sequencing of the S1 gene provides the ability to rapidly identify field strains including unrecognized variant virus serotypes (Kingham et al. 2000; Kusters et al. 1989).

5.11.5 Determination of IBV Protectotypes

Antigenic and genetic variations among IBV alone are not adequate to define cross protection between strains (Cavanagh et al. 1997; Raggi and Lee 1965); hence, the term "protectotype" was suggested (Lohr 1988) to determine the cross protection afforded by the existing vaccines against the emerged serotypes/genotypes. Cross immunity tests (CIT) in experimental birds have been performed (Lambrechts et al. 1993; Darbyshire 1985, 1980); the use of tracheal organ cultures (TOCs) from IBV-immunized birds was also suggested (Darbyshire 1980) and used successfully (Hinze et al. 1991). Since IBV has a tropism for epithelial cells of the respiratory tract, kidney, oviduct, and gut of chickens, IBV vaccines are evaluated on the basis of protection afforded at the level of the trachea (McMartin 1993), the kidneys for nephropathogenic IBV (Lambrechts et al. 1993), and the oviduct level (Dhinakar Raj and Jones 1996).

5.11.6 Serodiagnosis

VN test may be performed in ECE, CKC, or tracheal organ culture (TOC). The test may be conducted using the constant serum-diluted virus or diluted virus content serum method (Gelb 1989). VN is highly specific and highly sensitive; it is rarely used because it is too expensive and time-consuming. HI test detects antibody earlier than NV and could be used for serology (Kaufhold et al. 1988; Gough and Alexander 1979, 1977). AGPT is proved to be specific but with poor sensitivity (De Wit et al. 1997). ELISA is used on a more frequent basis to measure IBV antibody (Garcia and Bankowski 1981; Marquardt et al. 1981; Soula and Moreau 1981; Snyder et al. 1985). Among the advantages of ELISA are the increased sensitivity and specificity (Garcia and Bankowski 1981; Marquardt et al. 1981; De Wit et al. 1997) and the automation of the ELISA steps and calculations (Snyder et al. 1983a, b).

5.12 Treatment and Vaccination

5.12.1 Treatment

No specific antiviral therapy is available to control IBV field infection. On the other hand, antimicrobial therapy may reduce the effect of the complicating bacterial infections. Increasing the ambient temperature may reduce mortalities in cold weather. Reduced mortalities in nephrogenic strains can be achieved by reducing the protein concentrations in ration, providing electrolytes in drinking water, and using diuretics.

5.12.2 Vaccination

5.12.2.1 Live Attenuated Vaccines

Live attenuated IB vaccines are used to control IBV infection. Live vaccines are frequently attenuated by serial passage in embryonated chicken eggs (Klieve and Cumming 1988); however, extensive passage should be avoided to prevent the reduction in immunogenicity. There is an evidence that some attenuated vaccines showed increased in virulence after back passage in chickens (Hopkins and Yoder 1986). Vaccination is conducted by drinking water or coarse spray at 1 day or within the first week of age. Live vaccination of 1-day-old chicks induced a rapid decline in maternally derived antibodies due to binding and partial neutralization of vaccine viruses (Mondal and Naqi 2001). Booster vaccination is carried out 2–3 weeks after the first vaccination (Cavanagh 2003). The vaccine is administered individually by eyedrop, intratracheal, or intranasal route. Mass application by coarse spray, aerosol, and drinking water is used. In case of drinking water, removal of sanitizers and the incorporation of 1:4000 skim milk help to stabilize the vaccine titer during vaccination (Gentry and Braune 1972). Live attenuated IBV with NDV is used frequently; however, if excess IBV component is present, IBV may interfere with the NDV response (Thornton and Muskett 1975). Most of the commercially available live attenuated vaccines are derived from Massachusetts-based M41 serotype and the Dutch H52 and H120 strains, although some strains with regional impact have been introduced in different parts of the world in addition to Mass serotype (Lee et al. 2010; Bande et al. 2015). In the USA, strains belong to Connecticut, and Arkansas serotypes are used, whereas other serotypes like DE072 are used regionally. In some parts of Europe, D274, D1466, 4/91, and QX are used. In Australia, strains B and C subtypes are used (Klieve and Cumming 1988). In Egypt, MASS + CONN and 4/91 live attenuated vaccines and D274/M41 inactivated are used (Abdel-Moneim et al. 2006). In China, LDT3-A and QX live vaccines are used (Feng et al. 2015). Limitations of live attenuated vaccines include reversion to virulence, tissue damage, and interference by MDA. H52 and H120 IBV vaccines have been found to induce considerable pathology in the trachea (Bijlenga et al. 2004; Zhang et al. 2010). Potential recombination between vaccine strains and virulent field strains may lead to the emergence of new IBV serotypes (Lee et al. 2010; McKinley et al. 2008).

5.12.2.2 Inactivated or Killed Vaccines

Inactivated IBV vaccines are administered by injection to layers and breeders at point of lay (13–18 weeks of age). The inactivated vaccine may contain two IBV types and in association with other virus vaccines including NDV, egg drop syndrome, and others. Of course, inactivated vaccines require priming with live attenuated vaccines. In addition, inactivated autogenous vaccines prepared from specific local isolates can be used to immunize commercial layers and breeder chickens.

5.12.2.3 Recombinant Vaccines

Recombinant IBV Beaudette with S proteins of virulent M41 (Hodgson et al. 2004; Hodgson et al., 2004) or 4/91 (S) (Armesto et al. 2011) or replacing the S1 ectodomain of the Beaudette with that of H120 (Wei et al. 2014) kept the viruses attenuated and provided homologous protection. Fowl pox virus vaccine expressing IBV-S1-gene and chicken interferon-γ gene [rFPV-IFNγS1] and fowl adenovirus vectors (Shi et al. 2011; Johnson et al. 2003) as well as BacMam (baculovirus with mammalian expression system) expressing S and N genes (Abdel-Moneim et al. 2014) or S1 gene(Zhang et al. 2014) could be good candidates for IBV vaccines, since vectors replicate well in the bird's respiratory tract (Cavanagh 2007).

References

Ababneh M, Dalab AE, Alsaad S, Al-Zghoul M (2012) Presence of infectious bronchitis virus strain CK/CH/LDL/97I in the middle east. ISRN Vet Sci 2012:201721. doi:10.5402/2012/201721

Abd El Rahman S, El-Kenawy AA, Neumann U, Herrler G, Winter C (2009) Comparative analysis of the sialic acid binding activity and the tropism for the respiratory epithelium of four different strains of avian infectious bronchitis virus. Avian Pathol 38(1):41–45. doi:10.1080/03079450802632049

Abdel-Moneim A, Madbouly H, Gelb J, Ladman B (2002) Isolation and identification of Egypt/Beni-Suef/01 a novel genotype of infectious bronchitis virus. Vet Med 50(4):1065–1078

Abdel-Moneim A, Madbouly H, El-Kady M (2005) In vitro characterization and pathogenesis of Egypt/Beni-Suef/01; a novel genotype of infectious bronchitis virus. Beni Suef Vet Med J 15(2):127–133

Abdel-Moneim AS, El-Kady MF, Ladman BS, Gelb J Jr (2006) S1 gene sequence analysis of a nephropathogenic strain of avian infectious bronchitis virus in Egypt. Virol J 3:78. doi:10.1186/1743-422x-3-78

Abdel-Moneim AS, Zlotowski P, Veits J, Keil GM, Teifke JP (2009) Immunohistochemistry for detection of avian infectious bronchitis virus strain M41 in the proventriculus and nervous system of experimentally infected chicken embryos. Virol J 6:15. doi:10.1186/1743-422x-6-15

Abdel-Moneim AS, Afifi MA, El-Kady MF (2012) Emergence of a novel genotype of avian infectious bronchitis virus in Egypt. Arch Virol 157(12):2453–2457. doi:10.1007/s00705-012-1445-1

Abdel-Moneim AS, Giesow K, Keil GM (2014) High-level protein expression following single and dual gene cloning of infectious bronchitis virus N and S genes using baculovirus systems. Viral Immunol 27(2):75–81. doi:10.1089/vim.2013.0114

Abro SH, Renstrom LH, Ullman K, Belak S, Baule C (2012) Characterization and analysis of the full-length genome of a strain of the European QX-like genotype of infectious bronchitis virus. Arch Virol 157(6):1211–1215. doi:10.1007/s00705-012-1284-0

Acevedo AM, Martínez N, Brandão P, Perera CL, Frías MT, Barrera M, Pérez LJ (2013) Phylogenetic and molecular characterization of coronavirus affecting species of bovine and birds in Cuba. Biotecnol Apl 30:228–231

Adzhar A, Shaw K, Britton P, Cavanagh D (1996) Universal oligonucleotides for the detection of infectious bronchitis virus by the polymerase chain reaction. Avian Pathol 25(4):817–836. doi:10.1080/03079459608419184

Alexander DJ, Gough RE (1977) Isolation of avian infectious bronchitis virus from experimentally infected chickens. Res Vet Sci 23(3):344–347

Alexander DJ, Gough RE (1978) A long-term study of the pathogenesis of infection of fowls with three strains of avian infectious bronchitis virus. Res Vet Sci 24(2):228–233

Alexander DJ, Allan WH, Biggs PM, Bracewell CD, Darbyshire JH, Dawson PS, Harris AH, Jordan FT, MacPherson I, McFerran JB, Randall CJ, Stuart JC, Swarbrick O, Wilding GP (1983) A standard technique for haemagglutination inhibition tests for antibodies to avian infectious bronchitis virus. Vet Rec 113(3):64

Al-Shekaili T, Baylis M, Ganapathy K (2015) Molecular detection of infectious bronchitis and avian metapneumoviruses in Oman backyard poultry. Res Vet Sci 99:46–52. doi:10.1016/j.rvsc.2014.12.018

Alvarado IR, Villegas P, Mossos N, Jackwood MW (2005) Molecular characterization of avian infectious bronchitis virus strains isolated in Colombia during 2003. Avian Dis 49(4):494–499. doi:10.1637/7202-050304r.1

Alvarado I, Villegas P, El-Attrache J, Jackwood M (2006) Detection of Massachusetts and Arkansas serotypes of infectious bronchitis virus in broilers. Avian Dis 50(2):292–297

Ambali A, Jones R (1990) Early pathogenesis in chicks of infection with an enterotropic strain of infectious bronchitis virus. Avian Dis 34:809–817

Ammayappan A, Upadhyay C, Gelb J Jr, Vakharia VN (2008) Complete genomic sequence analysis of infectious bronchitis virus Ark DPI strain and its evolution by recombination. Virol J 5:157. doi:10.1186/1743-422x-5-157

Ariyoshi R, Kawai T, Honda T, Tokiyoshi S (2010) Classification of IBV S1 genotypes by direct reverse transcriptase-polymerase chain reaction (RT-PCR) and relationship between serotypes and genotypes of strains isolated between 1998 and 2008 in Japan. J Vet Med Sci Jpn Soc Vet Sci 72(6):687–692

Armesto M, Evans S, Cavanagh D, Abu-Median AB, Keep S, Britton P (2011) A recombinant avian infectious bronchitis virus expressing a heterologous spike gene belonging to the 4/91 serotype. PLoS One 6(8):e24352. doi:10.1371/journal.pone.0024352

Awad F, Baylis M, Ganapathy K (2014) Detection of variant infectious bronchitis viruses in broiler flocks in Libya. Int J Vet Sci Med 2(1):78–82

Bande F, Arshad SS, Bejo MH, Moeini H, Omar AR (2015) Progress and challenges toward the development of vaccines against avian infectious bronchitis. J Immunol Res 2015:424860. doi:10.1155/2015/424860

Bayry J, Goudar MS, Nighot PK, Kshirsagar SG, Ladman BS, Gelb J Jr, Ghalsasi GR, Kolte GN (2005) Emergence of a nephropathogenic avian infectious bronchitis virus with a novel genotype in India. J Clin Microbiol 43(2):916–918. doi:10.1128/jcm.43.2.916-918.2005

Beach J, Schalm O (1936) A filterable virus, distinct from that of laryngotracheitis, the cause of a respiratory disease of chicks. Poult Sci 15(3):199–206

Benyeda Z, Szeredi L, Mató T, Süveges T, Balka G, Abonyi-Tóth Z, Rusvai M, Palya V (2010) Comparative histopathology and immunohistochemistry of QX-like, Massachusetts and 793/B serotypes of infectious bronchitis virus infection in chickens. J Comp Pathol 143(4):276–283

Bhattacharjee P, Naylor C, Jones R (1994) A simple method for immunofluorescence staining of tracheal organ cultures for the rapid identification of infectious bronchitis virus. Avian Pathol 23(3):471–480

Bijlenga G, Cook JK, Gelb J Jr, de Wit JJ (2004) Development and use of the H strain of avian infectious bronchitis virus from the Netherlands as a vaccine: a review. Avian Pathol 33(6):550–557. doi:10.1080/03079450400013154

Bingham RW, Madge MH, Tyrrell DA (1975) Haemagglutination by avian infectious bronchitis virus-a coronavirus. J Gen Virol 28(3):381–390. doi:10.1099/0022-1317-28-3-381

Binns MM, Boursnell ME, Cavanagh D, Pappin DJ, Brown TD (1985) Cloning and sequencing of the gene encoding the spike protein of the coronavirus IBV. J Gen Virol 66(Pt 4):719–726. doi:10.1099/0022-1317-66-4-719

Bochkov YA, Batchenko GV, Shcherbakova LO, Borisov AV, Drygin VV (2006) Molecular epizootiology of avian infectious bronchitis in Russia. Avian Pathol 35(5):379–393. doi:10.1080/03079450600921008

Bochkov YA, Tosi G, Massi P, Drygin VV (2007) Phylogenetic analysis of partial S1 and N gene sequences of infectious bronchitis virus isolates from Italy revealed genetic diversity and recombination. Virus Genes 35(1):65–71. doi:10.1007/s11262-006-0037-0

Boltz DA, Nakai M, Bahra JM (2004) Avian infectious bronchitis virus: a possible cause of reduced fertility in the rooster. Avian Dis 48(4):909–915. doi:10.1637/7192-040808r1

Bourogaa H, Miled K, Gribaa L, El Behi I, Ghram A (2009) Characterization of new variants of avian infectious bronchitis virus in Tunisia. Avian Dis 53(3):426–433. doi:10.1637/8666-022609-Reg.1

Boursnell ME, Brown TD, Foulds IJ, Green PF, Tomley FM, Binns MM (1987) Completion of the sequence of the genome of the coronavirus avian infectious bronchitis virus. J Gen Virol 68(Pt 1):57–77. doi:10.1099/0022-1317-68-1-57

Box PG, Holmes HC, Finney PM, Froymann R (1988) Infectious bronchitis in laying hens: the relationship between haemagglutination inhibition antibody levels and resistance to experimental challenge. Avian Pathol 17(2):349–361. doi:10.1080/03079458808436453

Braune MO, Gentry RF (1965) Standardization of the fluorescent antibody technique for the detection of avian respiratory viruses. Avian Dis 9(4):535–545

Brierley I, Digard P, Inglis SC (1989) Characterization of an efficient coronavirus ribosomal frameshifting signal: requirement for an RNA pseudoknot. Cell 57(4):537–547

Brooks JE, Rainer AC, Parr RL, Woolcock P, Hoerr F, Collisson EW (2004) Comparisons of envelope through 5B sequences of infectious bronchitis coronaviruses indicates recombination occurs in the envelope and membrane genes. Virus Res 100(2):191–198. doi:10.1016/j.virusres.2003.11.016

Brownlie R, Zhu J, Allan B, Mutwiri GK, Babiuk LA, Potter A, Griebel P (2009) Chicken TLR21 acts as a functional homologue to mammalian TLR9 in the recognition of CpG oligodeoxynucleotides. Mol Immunol 46(15):3163–3170. doi:10.1016/j.molimm.2009.06.002

Callison S, Jackwood M, Hilt D (2001) Molecular characterization of infectious bronchitis virus isolates foreign to the United States and comparison with United States isolates. Avian Dis 45:492–499

Capua I, Gough RE, Mancini M, Casaccia C, Weiss C (1994) A 'novel' infectious bronchitis strain infecting broiler chickens in Italy. Zentralbl Veterinarmed B 41(2):83–89

Cavanagh D (1983a) Coronavirus IBV glycopolypeptides: size of their polypeptide moieties and nature of their oligosaccharides. J Gen Virol 64(Pt 5):1187–1191. doi:10.1099/0022-1317-64-5-1187

Cavanagh D (1983b) Coronavirus IBV: structural characterization of the spike protein. J Gen Virol 64(Pt 12):2577–2583. doi:10.1099/0022-1317-64-12-2577

Cavanagh D (1991) Sequencing approach to IBV antigenicity and epizootiology. In: Proceedings of the second international symposium on infectious bronchitis, Rauischholzhausen, pp 147–160

Cavanagh D (2003) Severe acute respiratory syndrome vaccine development: experiences of vaccination against avian infectious bronchitis coronavirus. Avian Pathol 32(6):567–582. doi:10.1080/03079450310001621198

Cavanagh D (2007) Coronavirus avian infectious bronchitis virus. Vet Res 38(2):281–297. doi:10.1051/vetres:2006055

Cavanagh D, Gelb JJ (2008) Infectious bronchitis. In: Saif YM, Fadly AM, Glisson JR, McDougald LR, Nolan MK, Swayne DE (eds) Diseases of poultry, 12th edn. Ames/Iowa, BlackWell, p 117–136

Cavanagh D, Davis PJ, Mockett AP (1988) Amino acids within hypervariable region 1 of avian coronavirus IBV (Massachusetts serotype) spike glycoprotein are associated with neutralization epitopes. Virus Res 11(2):141–150

Cavanagh D, Davis P, Cook J, Li D (1990) Molecular basis of the variation exhibited by avian infectious bronchitis coronavirus (IBV). Adv Exp Med Biol 276:369–372

Cavanagh D, Davis P, Cook JK, Li D, Kant A, Koch G (1992a) Location of the amino acid differences in the S1 spike glycoprotein subunit of closely related serotypes of infectious bronchitis virus. Avian Pathol 21(1):33–43

Cavanagh D, Davis PJ, Cook JK (1992b) Infectious bronchitis virus: evidence for recombination within the Massachusetts serotype. Avian Pathol 21(3):401–408. doi:10.1080/03079459208418858
Cavanagh D, Elus MM, Cook JK (1997) Relationship between sequence variation in the S1 spike protein of infectious bronchitis virus and the extent of cross-protection in vivo. Avian Pathol 26(1):63–74. doi:10.1080/03079459708419194
Cavanagh D, Picault JP, Gough R, Hess M, Mawditt K, Britton P (2005) Variation in the spike protein of the 793/B type of infectious bronchitis virus, in the field and during alternate passage in chickens and embryonated eggs. Avian Pathol 34(1):20–25. doi:10.1080/03079450400025414
Chen BY, Itakura C (1996) Cytopathology of chick renal epithelial cells experimentally infected with avian infectious bronchitis virus. Avian Pathol 25(4):675–690. doi:10.1080/03079459608419174
Chen HW, Huang YP, Wang CH (2009) Identification of Taiwan and China-like recombinant avian infectious bronchitis viruses in Taiwan. Virus Res 140(1–2):121–129. doi:10.1016/j.virusres.2008.11.012
Chew PH, Wakenell PS, Farver TB (1997) Pathogenicity of attenuated infectious bronchitis viruses for oviducts of chickens exposed in ovo. Avian Dis 41(3):598–603
Chhabra R, Chantrey J, Ganapathy K (2015) Immune responses to virulent and vaccine strains of infectious bronchitis viruses in chickens. Viral Immunol. doi:10.1089/vim.2015.0027
Chu VC, McElroy LJ, Chu V, Bauman BE, Whittaker GR (2006) The avian coronavirus infectious bronchitis virus undergoes direct low-pH-dependent fusion activation during entry into host cells. J Virol 80(7):3180–3188. doi:10.1128/jvi.80.7.3180-3188.2006
Collisson EW, Li JZ, Sneed LW, Peters ML, Wang L (1990) Detection of avian infectious bronchitis viral infection using in situ hybridization and recombinant DNA. Vet Microbiol 24(3–4):261–271
Collisson EW, Pei J, Dzielawa J, Seo SH (2000) Cytotoxic T lymphocytes are critical in the control of infectious bronchitis virus in poultry. Dev Comp Immunol 24(2–3):187–200
Cong F, Liu X, Han Z, Shao Y, Kong X, Liu S (2013) Transcriptome analysis of chicken kidney tissues following coronavirus avian infectious bronchitis virus infection. BMC Genomics 14:743. doi:10.1186/1471-2164-14-743
Cook JK (1968) Duration of experimental infectious bronchitis in chickens. Res Vet Sci 9(6):506–514
Cook JK, Darbyshire JH, Peters RW (1976) The use of chicken tracheal organ cultures for the isolation and assay of avian infectious bronchitis virus. Arch Virol 50(1–2):109–118
Cook JK, Orbell SJ, Woods MA, Huggins MB (1996) A survey of the presence of a new infectious bronchitis virus designated 4/91 (793B). Vet Rec 138(8):178–180
Cook JK, Chesher AJ, Baxendale W, Greenwood N, Huggins MB, Orbell SJ (2001) Protection of chickens against renal damage caused by a nephropathogenic infectious bronchitis virus. Avian Pathol 30:423–426
Corse E, Machamer CE (2003) The cytoplasmic tails of infectious bronchitis virus E and M proteins mediate their interaction. Virology 312(1):25–34
Crinion RA (1972) Egg quality and production following infectious bronchitis virus exposure at one day old. Poult Sci 51(2):582–585
Cumming RB (1970) Studies on Australian infectious bronchitis virus. IV. Apparent farm-to-farm airborne transmission of infectious bronchitis virus. Avian Dis 14(1):191–195
Cunningham C (1973) A laboratory guide in virology. 7th edn. Burgess, Minneapolis, p 67–80
Cunningham CH, El Dardiry A (1948) Distribution of the virus of infectious bronchitis of chickens in embryonated chicken eggs. Cornell Vet 38:381–388
Darbyshire JH (1980) Assessment of cross-immunity dm chickens to strains of avian infectious bronchitis virus using tracheal organ cultures. Avian Pathol 9(2):179–184. doi:10.1080/03079458008418401
Darbyshire JH (1985) A clearance test to assess protection in chickens vaccinated against avian infectious bronchitis virus. Avian Pathol 14(4):497–508. doi:10.1080/03079458508436252
Darbyshire JH, Rowell JG, Cook JK, Peters RW (1979) Taxonomic studies on strains of avian infectious bronchitis virus using neutralisation tests in tracheal organ cultures. Arch Virol 61(3):227–238

Davelaar FG, Noordzij A, Vanderdonk JA (1982) A study on the synthesis and secretion of immunoglobulins by the Jarderian gland of the fowl after eyedrop vaccination against infectious bronchitis at 1-day-old. Avian Pathol 11(1):63–79. doi:10.1080/03079458208436082

Davelaar FG, Kouwenhoven B, Burger AG (1984) Occurrence and significance of infectious bronchitis virus variant strains in egg and broiler production in the Netherlands. Vet Q 6(3):114–120. doi:10.1080/01652176.1984.9693924

Dawson PS, Gough RE (1971) Antigenic variation in strains of avian infectious bronchitis virus. Arch Gesamte Virusforsch 34(1):32–39

de Haan CA, Masters PS, Shen X, Weiss S, Rottier PJ (2002) The group-specific murine coronavirus genes are not essential, but their deletion, by reverse genetics, is attenuating in the natural host. Virology 296(1):177–189. doi:10.1006/viro.2002.1412

De Wit J, Davelaar F, Braunius W (1992) Comparison of the enzyme linked immunosorbent assay, the haemagglutination inhibition test and the agar gel precipitation test for the detection of antibodies against infectious bronchitis and Newcastle disease in commercial broilers. Avian Pathol 21(4):651–658

De Wit JJ, Koch G, Kant A, Van Roozelaar DJ (1995) Detection by immunofluorescent assay of serotype-specific and group-specific antigens of infectious bronchitis virus in tracheas of broilers with respiratory problems. Avian Pathol 24(3):465–474. doi:10.1080/03079459508419086

De Wit J, Mekkes D, Kouwenhoven B, Verheijden J (1997) Sensitivity and specificity of serological tests for infectious bronchitis virus antibodies in broilers. Avian Pathol 26(1):105–118

De Wit JJ, Brandao P, Torres CA, Koopman R, Villarreal LY (2015) Increased level of protection of respiratory tract and kidney by combining different infectious bronchitis virus vaccines against challenge with nephropathogenic Brazilian genotype subcluster 4 strains. Avian Pathol 44(5):352–357. doi:10.1080/03079457.2015.1058916

Delaplane J (1947) Technique for the isolation of infec tious bronchitis or Newcastle virus including observa tions on the use of streptomycin in overcoming bacterial contaminants. In: NE. Conf. Lab. Workers Pullorum Dis. Control Proc. 19, p 11–13

Delmas B, Laude H (1990) Assembly of coronavirus spike protein into trimers and its role in epitope expression. J Virol 64(11):5367–5375

Dhinakar Raj G, Jones RC (1996) Protectotypic differentiation of avian infectious bronchitis viruses using an in vitro challenge model. Vet Microbiol 53(3–4):239–252

Ducatez MF, Martin AM, Owoade AA, Olatoye IO, Alkali BR, Maikano I, Snoeck CJ, Sausy A, Cordioli P, Muller CP (2009) Characterization of a new genotype and serotype of infectious bronchitis virus in Western Africa. J Gen Virol 90(Pt 11):2679–2685. doi:10.1099/vir.0.012476-0

Elankumaran S, Balachandran C, Chandran ND, Roy P, Albert A, Manickam R (1999) Serological evidence for a 793/B related avian infectious bronchitis virus in India. Vet Rec 144(11):299–300

El-Kady MF (1989) Studies on the epidemiology and means of control of infectious bronchitis disease in chickens in Egypt. Faculty of Veterinary Medicine, Egypt

Escorcia M, Jackwood MW, Lucio B, Petrone VM, Lopez C, Fehervari T, Tellez G (2000) Characterization of Mexican strains of avian infectious bronchitis isolated during 1997. Avian Dis 44(4):944–947

Fang S, Shen H, Wang J, Tay FP, Liu DX (2010) Functional and genetic studies of the substrate specificity of coronavirus infectious bronchitis virus 3C-like proteinase. J Virol 84(14):7325–7336. doi:10.1128/jvi.02490-09

Fellahi S, El Harrak M, Ducatez M, Loutfi C, Koraichi SI, Kuhn JH, Khayi S, El Houadfi M, Ennaji MM (2015) Phylogenetic analysis of avian infectious bronchitis virus S1 glycoprotein regions reveals emergence of a new genotype in Moroccan broiler chicken flocks. Virol J 12:116. doi:10.1186/s12985-015-0347-8

Feng K, Xue Y, Wang J, Chen W, Chen F, Bi Y, Xie Q (2015) Development and efficacy of a novel live-attenuated QX-like nephropathogenic infectious bronchitis virus vaccine in China. Vaccine 33(9):1113–1120. doi:10.1016/j.vaccine.2015.01.036

Fulton RM, Thacker HL, Reed WM, DeNicola DB (1997) Effect of cytoxan-induced heteropenia on the response of specific-pathogen-free chickens to infectious bronchitis. Avian Dis 41(3):511–518

Garcia Z, Bankowski R (1981) Comparison of a tissue-culture virus-neutralization test and the enzyme-linked immunosorbent assay for measurement of antibodies to infectious bronchitis. Avian Dis 25:121–130

Gelb J Jr (1989) Infectious bronchitis. A laboratory manual for the isolation and identification of avian pathogens, 3rd edn, Dubuque Iowa, Kendall/Hunt Publishing Co, 124–127

Gelb J Jr, Keeler CL Jr, Nix WA, Rosenberger JK, Cloud SS (1997) Antigenic and S-1 genomic characterization of the Delaware variant serotype of infectious bronchitis virus. Avian Dis 41(3):661–669

Gelb J Jr, Nix WA, Gellman SD (1998) Infectious bronchitis virus antibodies in tears and their relationship to immunity. Avian Dis 42(2):364–374

Gelb J Jr, Ladman BS, Tamayo M, Gonzalez M, Sivanandan V (2001) Novel infectious bronchitis virus S1 genotypes in Mexico 1998-1999. Avian Dis 45(4):1060–1063

Gelb J Jr, Weisman Y, Ladman BS, Meir R (2005) S1 gene characteristics and efficacy of vaccination against infectious bronchitis virus field isolates from the United States and Israel (1996 to 2000). Avian Pathol 34(3):194–203. doi:10.1080/03079450500096539

GenCheng P, YiHai J, YongLing W, YuDong W, ZiChun Z, Jiang D, XiangE L, YinLin Y, YuXiang O, FuYong W (1998) Isolation and identification of glandular stomach-pathogenic type infectious bronchitis virus in Beijing area. Chin J Vet Sci Technol 28(10):21–22

Gentry RF, Braune MO (1972) Prevention of virus inactivation during drinking water vaccination of poultry. Poult Sci 51(4):1450–1456

Godet M, L'Haridon R, Vautherot JF, Laude H (1992) TGEV corona virus ORF4 encodes a membrane protein that is incorporated into virions. Virology 188(2):666–675

Gonzalez JM, Gomez-Puertas P, Cavanagh D, Gorbalenya AE, Enjuanes L (2003) A comparative sequence analysis to revise the current taxonomy of the family Coronaviridae. Arch Virol 148(11):2207–2235. doi:10.1007/s00705-003-0162-1

Gough R, Alexander D (1977) Comparison of serological tests for the measurement of the primary immune response to avian infectious bronchitis virus vaccines. Vet Microbiol 2(4):289–301

Gough RE, Alexander DJ (1979) Comparison of duration of immunity in chickens infected with a live infectious bronchitis vaccine by three different routes. Res Vet Sci 26(3):329–332

Gough RE, Allan WH, Nedelciu D (1977) Immune response to monovalent and bivalent Newcastle disease and infectious bronchitis inactivated vaccines. Avian Pathol 6(2):131–142. doi:10.1080/03079457708418221

Gough RE, Randall CJ, Dagless M, Alexander DJ, Cox WJ, Pearson D (1992) A 'new' strain of infectious bronchitis virus infecting domestic fowl in Great Britain. Vet Rec 130(22):493–494

Guo X, Rosa AJ, Chen DG, Wang X (2008) Molecular mechanisms of primary and secondary mucosal immunity using avian infectious bronchitis virus as a model system. Vet Immunol Immunopathol 121(3–4):332–343. doi:10.1016/j.vetimm.2007.09.016

Guo Z, Wang H, Yang T, Wang X, Lu D, Li Y, Zhang Y (2010) Priming with a DNA vaccine and boosting with an inactivated vaccine enhance the immune response against infectious bronchitis virus. J Virol Methods 167(1):84–89. doi:10.1016/j.jviromet.2010.03.016

Han Z, Sun C, Yan B, Zhang X, Wang Y, Li C, Zhang Q, Ma Y, Shao Y, Liu Q, Kong X, Liu S (2011) A 15-year analysis of molecular epidemiology of avian infectious bronchitis coronavirus in China. Infect Genet Evol J Mol Epidemiol Evol Genet Infect Dis 11(1):190–200. doi:10.1016/j.meegid.2010.09.002

Handberg K, Nielsen O, Pedersen M, J⊘rgensen PH (1999) Detection and strain differentiation of infectious bronchitis virus in tracheal tissues from experimentally infected chickens by reverse transcription-polymerase chain reaction. Comparison with an immunohistochemical technique. Avian Pathol 28 (4):327–335

Hewson KA, Noormohammadi AH, Devlin JM, Browning GF, Schultz BK, Ignjatovic J (2014) Evaluation of a novel strain of infectious bronchitis virus emerged as a result of spike gene recombination between two highly diverged parent strains. Avian Pathol 43(3):249–257. doi:10.1080/03079457.2014.914624

Hinze V, Lohr J, Kaleta E (1991) Infectious bronchitis virus strain differentiation attempts by cross-immunity studies in tracheal organ cultures derived from immunised chickens. In: Proc.

of the 2nd Int. Symp. on Infectious Bronchitis. World's Poultry Science Association, European group no. 7 Rauischholzhausen, pp 73–86

Hodgson T, Casais R, Dove B, Britton P, Cavanagh D (2004) Recombinant infectious bronchitis coronavirus Beaudette with the spike protein gene of the pathogenic M41 strain remains attenuated but induces protective immunity. J Virol 78(24):13804–13811. doi:10.1128/jvi.78.24.13804-13811.2004

Hofstad M, Yoder HW (1966) Avian infectious bronchitis: virus distribution in tissues of chicks. Avian Dis 10(2):230–239

Hopkins SR (1974) Serological comparisons of strains of infectious bronchitis virus using plaque-purified isolants. Avian Dis 18(2):231–239

Hopkins SR, Yoder HW Jr (1986) Reversion to virulence of chicken-passaged infectious bronchitis vaccine virus. Avian Dis 30(1):221–223

Ignjatovic J, McWaters PG (1991) Monoclonal antibodies to three structural proteins of avian infectious bronchitis virus: characterization of epitopes and antigenic differentiation of Australian strains. J Gen Virol 72(Pt 12):2915–2922. doi:10.1099/0022-1317-72-12-2915

Ignjatovic J, Gould G, Sapats S (2006) Isolation of a variant infectious bronchitis virus in Australia that further illustrates diversity among emerging strains. Arch Virol 151(8):1567–1585. doi:10.1007/s00705-006-0726-y

Ivanov KA, Hertzig T, Rozanov M, Bayer S, Thiel V, Gorbalenya AE, Ziebuhr J (2004) Major genetic marker of nidoviruses encodes a replicative endoribonuclease. Proc Natl Acad Sci U S A 101(34):12694–12699. doi:10.1073/pnas.0403127101

Jackwood MW (2012) Review of infectious bronchitis virus around the world. Avian Dis 56(4):634–641. doi:10.1637/10227-043012-Review.1

Jackwood MW, Hilt DA, Lee CW, Kwon HM, Callison SA, Moore KM, Moscoso H, Sellers H, Thayer S (2005) Data from 11 years of molecular typing infectious bronchitis virus field isolates. Avian Dis 49(4):614–618. doi:10.1637/7389-052905r.1

Jackwood MW, Hilt DA, Williams SM, Woolcock P, Cardona C, O'Connor R (2007) Molecular and serologic characterization, pathogenicity, and protection studies with infectious bronchitis virus field isolates from California. Avian Dis 51(2):527–533. doi:10.1637/0005-2086(2007)51[527:mascpa]2.0.co;2

Jackwood MW, Boynton TO, Hilt DA, McKinley ET, Kissinger JC, Paterson AH, Robertson J, Lemke C, McCall AW, Williams SM, Jackwood JW, Byrd LA (2010a) Emergence of a group 3 coronavirus through recombination. Virology 398(1):98–108. doi:10.1016/j.virol.2009.11.044

Jackwood MW, Hilt DA, Sellers HS, Williams SM, Lasher HN (2010b) Rapid heat-treatment attenuation of infectious bronchitis virus. Avian Pathol 39(3):227–233. doi:10.1080/03079451003801516

Jang H, Koo BS, Jeon EO, Lee HR, Lee SM, Mo IP (2013) Altered pro-inflammatory cytokine mRNA levels in chickens infected with infectious bronchitis virus. Poult Sci 92(9):2290–2298. doi:10.3382/ps.2013-03116

Jayaram J, Youn S, Collisson EW (2005) The virion N protein of infectious bronchitis virus is more phosphorylated than the N protein from infected cell lysates. Virology 339(1):127–135. doi:10.1016/j.virol.2005.04.029

Jia W, Karaca K, Parrish CR, Naqi SA (1995) A novel variant of avian infectious bronchitis virus resulting from recombination among three different strains. Arch Virol 140(2):259–271

Jia W, Wang X, Parrish CR, Naqi SA (1996) Analysis of the serotype-specific epitopes of avian infectious bronchitis virus strains Ark99 and Mass41. J Virol 70(10):7255–7259

Johnson MA, Pooley C, Ignjatovic J, Tyack SG (2003) A recombinant fowl adenovirus expressing the S1 gene of infectious bronchitis virus protects against challenge with infectious bronchitis virus. Vaccine 21(21–22):2730–2736

Jones R, Ambali A (1987) Re-excretion of an enterotropic infectious bronchitis virus by hens at point of lay after experimental infection at day old. Vet Rec 120(26):617–618

Jones RC, Worthington KJ, Gough RE (2005) Detection of the Italy O2 strain of infectious bronchitis virus in the UK. Vet Rec 156(8):260

Juul-Madsen HR, Norup LR, Handberg KJ, Jorgensen PH (2007) Mannan-binding lectin (MBL) serum concentration in relation to propagation of infectious bronchitis virus (IBV) in chickens. Viral Immunol 20(4):562–570. doi:10.1089/vim.2007.0036

Kant A, Koch G, van Roozelaar DJ, Kusters JG, Poelwijk FA, van der Zeijst BA (1992) Location of antigenic sites defined by neutralizing monoclonal antibodies on the S1 avian infectious bronchitis virus glycopolypeptide. J Gen Virol 73(Pt 3):591–596. doi:10.1099/0022-1317-73-3-591

Karaca K, Syed N (1993) A monoclonal antibody blocking ELISA to detect serotype-specific infectious bronchitis virus antibodies. Vet Microbiol 34(3):249–257

Karaca K, Naqi S, Gelb J Jr (1992) Production and characterization of monoclonal antibodies to three infectious bronchitis virus serotypes. Avian Dis 36(4):903–915

Kaufhold C, Lohr JE, Kosters J (1988) Comparison of serum neutralization and haemagglutination inhibition tests for various serotypes of avian infectious bronchitis virus. Tierarztl Umsch 45:595–600

Keeler CL Jr, Reed KL, Nix WA, Gelb J Jr (1998) Serotype identification of avian infectious bronchitis virus by RT-PCR of the peplomer (S-1) gene. Avian Dis 42(2):275–284

King DJ, Hopkins SR (1984) Rapid serotyping of infectious bronchitis virus isolates with the hemagglutination-inhibition test. Avian Dis 28(3):727–733

Kingham BF, Keeler CL Jr, Nix WA, Ladman BS, Gelb J Jr (2000) Identification of avian infectious bronchitis virus by direct automated cycle sequencing of the S-1 gene. Avian Dis 44(2):325–335

Klieve AV, Cumming RB (1988) Immunity and cross-protection to nephritis produced by Australian infectious bronchitis viruses used as vaccines. Avian Pathol 17(4):829–839. doi:10.1080/03079458808436505

Klumperman J, Locker JK, Meijer A, Horzinek MC, Geuze HJ, Rottier PJ (1994) Coronavirus M proteins accumulate in the Golgi complex beyond the site of virion budding. J Virol 68(10):6523–6534

Knoetze AD, Moodley N, Abolnik C (2014) Two genotypes of infectious bronchitis virus are responsible for serological variation in KwaZulu-Natal poultry flocks prior to 2012. Onderstepoort J Vet Res 81(1). doi:10.4102/ojvr.v81i1.769

Koch G, Hartog L, Kant A, van Roozelaar DJ (1990) Antigenic domains on the peplomer protein of avian infectious bronchitis virus: correlation with biological functions. J Gen Virol 71(Pt 9):1929–1935. doi:10.1099/0022-1317-71-9-1929

Koch G, Kant A, Cook J, Cavanagh D (1991) Epitopes of neutralizing antibodies are localized within three regions of the S1 spike protein of infectious bronchitis virus. In: International Symposium of Infectious Bronchitis, World Veterinary Poultry Association, Rauischholzhausen, pp 154–160

Kuo L, Masters PS (2002) Genetic evidence for a structural interaction between the carboxy termini of the membrane and nucleocapsid proteins of mouse hepatitis virus. J Virol 76(10):4987–4999

Kuo SM, Wang CH, Hou MH, Huang YP, Kao HW, Su HL (2010) Evolution of infectious bronchitis virus in Taiwan: characterisation of RNA recombination in the nucleocapsid gene. Vet Microbiol 144(3–4):293–302. doi:10.1016/j.vetmic.2010.02.027

Kusters JG, Niesters HG, Lenstra JA, Horzinek MC, van der Zeijst BA (1989) Phylogeny of antigenic variants of avian coronavirus IBV. Virology 169(1):217–221

Kwon HM, Jackwood MW, Gelb J Jr (1993) Differentiation of infectious bronchitis virus serotypes using polymerase chain reaction and restriction fragment length polymorphism analysis. Avian Dis 37:194–202

Lai MM, Cavanagh D (1997) The molecular biology of coronaviruses. Adv Virus Res 48:1–100

Lambrechts C, Pensaert M, Ducatelle R (1993) Challenge experiments to evaluate cross-protection induced at the trachea and kidney level by vaccine strains and Belgian nephropathogenic isolates of avian infectious bronchitis virus. Avian Pathol 22(3):577–590. doi:10.1080/03079459308418945

Lee CW, Jackwood MW (2000) Evidence of genetic diversity generated by recombination among avian coronavirus IBV. Arch Virol 145(10):2135–2148

Lee CW, Hilt DA, Jackwood MW (2001) Identification and analysis of the Georgia 98 serotype, a new serotype of infectious bronchitis virus. Avian Dis 45(1):164–172

Lee HJ, Youn HN, Kwon JS, Lee YJ, Kim JH, Lee JB, Park SY, Choi IS, Song CS (2010) Characterization of a novel live attenuated infectious bronchitis virus vaccine candidate derived from a Korean nephropathogenic strain. Vaccine 28(16):2887–2894. doi:10.1016/j.vaccine.2010.01.062

Lewicki DN, Gallagher TM (2002) Quaternary structure of coronavirus spikes in complex with carcinoembryonic antigen-related cell adhesion molecule cellular receptors. J Biol Chem 277(22):19727–19734. doi:10.1074/jbc.M201837200

Li H, Yang H (2001) Sequence analysis of nephropathogenic infectious bronchitis virus strains of the Massachusetts genotype in Beijing. Avian Pathol 30(5):535–541. doi:10.1080/03079450120078734

Lim TH, Kim MS, Jang JH, Lee DH, Park JK, Youn HN, Lee JB, Park SY, Choi IS, Song CS (2012) Live attenuated nephropathogenic infectious bronchitis virus vaccine provides broad cross protection against new variant strains. Poult Sci 91(1):89–94. doi:10.3382/ps.2011-01739

Lin Z, Kato A, Kudou Y, Ueda S (1991) A new typing method for the avian infectious bronchitis virus using polymerase chain reaction and restriction enzyme fragment length polymorphism. Arch Virol 116(1–4):19–31

Liu S, Chen J, Chen J, Kong X, Shao Y, Han Z, Feng L, Cai X, Gu S, Liu M (2005) Isolation of avian infectious bronchitis coronavirus from domestic peafowl (*Pavo cristatus*) and teal (Anas). J Gen Virol 86(3):719–725

Liu X, Shao Y, Ma H, Sun C, Zhang X, Li C, Han Z, Yan B, Kong X, Liu S (2013) Comparative analysis of four Massachusetts type infectious bronchitis coronavirus genomes reveals a novel Massachusetts type strain and evidence of natural recombination in the genome. Infect Genet Evol J Mol Epidemiol Evol Genet Infect Dis 14:29–38. doi:10.1016/j.meegid.2012.09.016

Lohr J (1988) Differentiation of IBV strains. In: Proceedings of the 1st International Symposium on Infectious Bronchitis. Deutsche Veterinarmedizinische Gesellschaft eV, Giessen, pp 199–207

Ma H, Shao Y, Sun C, Han Z, Liu X, Guo H, Liu X, Kong X, Liu S (2012) Genetic diversity of avian infectious bronchitis coronavirus in recent years in China. Avian Dis 56(1):15–28. doi:10.1637/9804-052011-Reg.1

Macdonald JW, Randall CJ, McMartin DA, Dagless MD (1981) Immunity following vaccination with the H120 strain of infectious bronchitis virus via the drinking water. Avian Pathol 10(3):295–301. doi:10.1080/03079458108418478

Madbouly HM, Abdel-Moneim AS, Gelb JJ, Ladman BS, Rofael NB (2002) Molecular characterization of three Egyptian infectious bronchitis virus isolates. Vet Med 50(4):1053–1064

Maeda J, Maeda A, Makino S (1999) Release of coronavirus E protein in membrane vesicles from virus-infected cells and E protein-expressing cells. Virology 263(2):265–272. doi:10.1006/viro.1999.9955

Maeda J, Repass JF, Maeda A, Makino S (2001) Membrane topology of coronavirus E protein. Virology 281(2):163–169. doi:10.1006/viro.2001.0818

Mahmood ZH, Sleman RR, Uthman AU (2011) Isolation and molecular characterization of Sul/01/09 avian infectious bronchitis virus, indicates the emergence of a new genotype in the Middle East. Vet Microbiol 150(1–2):21–27. doi:10.1016/j.vetmic.2010.12.015

Mardani K, Noormohammadi AH, Hooper P, Ignjatovic J, Browning GF (2008) Infectious bronchitis viruses with a novel genomic organization. J Virol 82(4):2013–2024. doi:10.1128/jvi.01694-07

Mardani K, Noormohammadi AH, Ignjatovic J, Browning GF (2010) Naturally occurring recombination between distant strains of infectious bronchitis virus. Arch Virol 155(10):1581–1586. doi:10.1007/s00705-010-0731-z

Marquardt W, Snyder D, Schlotthober B (1981) Detection and quantification of antibodies to infectious bronchitis virus by enzyme-linked immunosorbent assay. Avian Dis 25:713–722

Martin EA, Brash ML, Hoyland SK, Coventry JM, Sandrock C, Guerin MT, Ojkic D (2014) Genotyping of infectious bronchitis viruses identified in Canada between 2000 and 2013. Avian Pathol 43(3):264–268. doi:10.1080/03079457.2014.916395

Masters P, Perlman S (2013) Coronaviridae. In: Howley P, Knipe DM (eds) Fields of virology. Kluwer, Wolters

Masters PS (2006) The molecular biology of coronaviruses. Adv Virus Res 66:193–292. doi:10.1016/s0065-3527(06)66005-3

McFarlane R, Verma R (2008) Sequence analysis of the gene coding for the S1 glycoprotein of infectious bronchitis virus (IBV) strains from New Zealand. Virus Genes 37(3):351–357. doi:10.1007/s11262-008-0273-6

McKinley ET, Hilt DA, Jackwood MW (2008) Avian coronavirus infectious bronchitis attenuated live vaccines undergo selection of subpopulations and mutations following vaccination. Vaccine 26(10):1274–1284. doi:10.1016/j.vaccine.2008.01.006

McMartin DA (1993) Infectious bronchitis. Virus infections of vertebrates virus infections of birds, vol 4. Elsevier Science Publishers, Amsterdam/New York, pp 249–275

Mihindukulasuriya KA, Wu G, St Leger J, Nordhausen RW, Wang D (2008) Identification of a novel coronavirus from a beluga whale by using a panviral microarray. J Virol 82(10):5084–5088. doi:10.1128/jvi.02722-07

Minskaia E, Hertzig T, Gorbalenya AE, Campanacci V, Cambillau C, Canard B, Ziebuhr J (2006) Discovery of an RNA virus 3'- > 5' exoribonuclease that is critically involved in coronavirus RNA synthesis. Proc Natl Acad Sci U S A 103(13):5108–5113. doi:10.1073/pnas.0508200103

Mockett AP, Cavanagh D, Brown TD (1984) Monoclonal antibodies to the S1 spike and membrane proteins of avian infectious bronchitis coronavirus strain Massachusetts M41. J Gen Virol 65(Pt 12):2281–2286. doi:10.1099/0022-1317-65-12-2281

Mondal SP, Cardona CJ (2007) Genotypic and phenotypic characterization of the California 99 (Cal99) variant of infectious bronchitis virus. Virus Genes 34(3):327–341. doi:10.1007/s11262-006-0014-7

Mondal SP, Naqi SA (2001) Maternal antibody to infectious bronchitis virus: its role in protection against infection and development of active immunity to vaccine. Vet Immunol Immunopathol 79(1–2):31–40

Nakamura K, Cook JK, Otsuki K, Huggins MB, Frazier JA (1991) Comparative study of respiratory lesions in two chicken lines of different susceptibility infected with infectious bronchitis virus: histology, ultrastructure and immunohistochemistry. Avian Pathol 20(2):241–257. doi:10.1080/03079459108418761

Naqi SA (1990) A monoclonal antibody-based immunoperoxidase procedure for rapid detection of infectious bronchitis virus in infected tissues. Avian Dis 34(4):893–898

Narayanan K, Chen CJ, Maeda J, Makino S (2003) Nucleocapsid-independent specific viral RNA packaging via viral envelope protein and viral RNA signal. J Virol 77(5):2922–2927

Niesters HG, Bleumink-Pluym NM, Osterhaus AD, Horzinek MC, van der Zeijst BA (1987a) Epitopes on the peplomer protein of infectious bronchitis virus strain M41 as defined by monoclonal antibodies. Virology 161(2):511–519

Niesters HG, Kusters JG, Lenstra JA, Spaan WJ, Horzined MC, van der Zeijst BA (1987b) The neutralization epitopes on the spike protein of infectious bronchitis virus and their antigenic variation. Adv Exp Med Biol 218:483–492

OIE (2013) Avian infectious bronchitis virus. Terrestrial Manual. Chapter 2.3.2., pp. 1–15.

Okino CH, dos Santos IL, Fernando FS, Alessi AC, Wang X, Montassier HJ (2014) Inflammatory and cell-mediated immune responses in the respiratory tract of chickens to infection with avian infectious bronchitis virus. Viral Immunol 27(8):383–391. doi:10.1089/vim.2014.0054

Orr-Burks N, Gulley SL, Toro H, van Ginkel FW (2014) Immunoglobulin A as an early humoral responder after mucosal avian coronavirus vaccination. Avian Dis 58(2):279–286. doi:10.1637/10740-120313-Reg.1

Otsuki K, Nakamura T, Kubota N, Kawaoka Y, Tsubokura M (1987) Comparison of two strains of avian infectious bronchitis virus for their interferon induction, viral growth and development of virus-neutralizing antibody in experimentally-infected chickens. Vet Microbiol 15(1–2):31–40

Ovchinnikova EV, Bochkov YA, Shcherbakova LO, Nikonova ZB, Zinyakov NG, Elatkin NP, Mudrak NS, Borisov AV, Drygin VV (2011) Molecular characterization of infectious bronchitis virus isolates from Russia and neighbouring countries: identification of intertypic recombination in the S1 gene. Avian Pathol 40(5):507–514. doi:10.1080/03079457.2011.605782

Pasternak AO, Spaan WJ, Snijder EJ (2006) Nidovirus transcription: how to make sense...? J Gen Virol 87(Pt 6):1403–1421. doi:10.1099/vir.0.81611-0

Pensaert M, Lambrechts C (1994) Vaccination of chickens against a Belgian nephropathogenic strain of infectious bronchitis virus B1648 using attenuated homologous and heterologous strains. Avian Pathol 23(4):631–641. doi:10.1080/03079459408419033

Pohuang T, Chansiripornchai N, Tawatsin A, Sasipreeyajan J (2011) Sequence analysis of S1 genes of infectious bronchitis virus isolated in Thailand during 2008-2009: identification of natural recombination in the field isolates. Virus Genes 43(2):254–260. doi:10.1007/s11262-011-0635-3

Promkuntod N, Thongmee S, Yoidam S (2015) Analysis of the S1 gene of the avian infectious bronchitis virus (IBV) reveals changes in the IBV genetic groups circulating in southern Thailand. Res Vet Sci 100:299–302. doi: 10.1016/j.rvsc.2015.05.002

Promkuntod N, van Eijndhoven RE, de Vrieze G, Grone A, Verheije MH (2014) Mapping of the receptor-binding domain and amino acids critical for attachment in the spike protein of avian coronavirus infectious bronchitis virus. Virology 448:26–32. doi:10.1016/j.virol.2013.09.018

Quiroz M, Retana A, Tamayo M (1993) Determinacion de la presencia del serotipe Arkansas a partir de aislamintos del virus de bronquitos infecciosa aviar en Mexico. Jornada Medico Avicola Coyoacan Mexico 4:191–198

Raamsman MJ, Locker JK, de Hooge A, de Vries AA, Griffiths G, Vennema H, Rottier PJ (2000) Characterization of the coronavirus mouse hepatitis virus strain A59 small membrane protein E. J Virol 74(5):2333–2342

Raggi LG, Lee GG (1965) Lack of correlation between infectivity, serologic response and challenge results in immunization with an avian infectious bronchitis vaccine. J Immunol (Baltimore, Md : 1950) 94:538–543

Raj GD, Jones RC (1997) Infectious bronchitis virus: Immunopathogenesis of infection in the chicken. Avian Pathol 26(4):677–706. doi:10.1080/03079459708419246

Raj GD, Savage CE, Jones RC (1997) Effect of heterophil depletion by 5-fluorouracil on infectious bronchitis virus infection in chickens. Avian Pathol 26(2):427–432. doi:10.1080/03079459708419224

Reddy VR, Theuns S, Roukaerts ID, Zeller M, Matthijnssens J, Nauwynck HJ (2015) Genetic Characterization of the Belgian Nephropathogenic Infectious Bronchitis Virus (NIBV) Reference Strain B1648. Viruses 7(8):4488–4506. doi:10.3390/v7082827

Rimondi A, Craig MI, Vagnozzi A, Konig G, Delamer M, Pereda A (2009) Molecular characterization of avian infectious bronchitis virus strains from outbreaks in Argentina (2001-2008). Avian Pathol 38(2):149–153. doi:10.1080/03079450902737821

Ruano M, El-Attrache J, Villegas P (2000) A rapid-plate hemagglutination assay for the detection of infectious bronchitis virus. Avian Dis 44(1):99–104

Sawicki SG, Sawicki DL (1990) Coronavirus transcription: subgenomic mouse hepatitis virus replicative intermediates function in RNA synthesis. J Virol 64(3):1050–1056

Sawicki SG, Sawicki DL (1995) Coronaviruses use discontinuous extension for synthesis of subgenome-length negative strands. Adv Exp Med Biol 380:499–506

Sawicki SG, Sawicki DL (2005) Coronavirus transcription: a perspective. Curr Top Microbiol Immunol 287:31–55

Schalk A, Hawn M (1931) An apparently new respiratory disease of baby chicks. J Am Vet Med Assoc 78(413–422):19

Schultze B, Cavanagh D, Herrler G (1992) Neuraminidase treatment of avian infectious bronchitis coronavirus reveals a hemagglutinating activity that is dependent on sialic acid-containing receptors on erythrocytes. Virology 189(2):792–794

Seo SH, Pei J, Briles WE, Dzielawa J, Collisson EW (2000) Adoptive transfer of infectious bronchitis virus primed alphabeta T cells bearing CD8 antigen protects chicks from acute infection. Virology 269(1):183–189. doi:10.1006/viro.2000.0211

Seo SH, Wang L, Smith R, Collisson EW (1997) The carboxyl-terminal 120-residue polypeptide of infectious bronchitis virus nucleocapsid induces cytotoxic T lymphocytes and protects chickens from acute infection. J Virol 71(10):7889–7894

Sethna PB, Hung SL, Brian DA (1989) Coronavirus subgenomic minus-strand RNAs and the potential for mRNA replicons. Proc Natl Acad Sci U S A 86(14):5626–5630

Sevoian M, Levine P (1957) Effects of infectious bronchitis on the reproductive tracts, egg production, and egg quality of laying chickens. Avian Dis 1(2):136–164
Shi XM, Zhao Y, Gao HB, Jing Z, Wang M, Cui HY, Tong GZ, Wang YF (2011) Evaluation of recombinant fowlpox virus expressing infectious bronchitis virus S1 gene and chicken interferon-gamma gene for immune protection against heterologous strains. Vaccine 29(8):1576–1582. doi:10.1016/j.vaccine.2010.12.102
Sigrist B, Tobler K, Schybli M, Konrad L, Stockli R, Cattoli G, Luschow D, Hafez HM, Britton P, Hoop RK, Vogtlin A (2012) Detection of Avian coronavirus infectious bronchitis virus type QX infection in Switzerland. J Vet Diagn Invest 24(6):1180–1183. doi:10.1177/1040638712463692
Snijder EJ, Bredenbeek PJ, Dobbe JC, Thiel V, Ziebuhr J, Poon LL, Guan Y, Rozanov M, Spaan WJ, Gorbalenya AE (2003) Unique and conserved features of genome and proteome of SARS-coronavirus, an early split-off from the coronavirus group 2 lineage. J Mol Biol 331(5):991–1004
Snyder D, Marquardt W, Mallinson E, Allen D, Savage P (1985) An enzyme-linked immunosorbent assay method for the simultaneous measurement of antibody titer to multiple viral, bacterial or protein antigens. Vet Immunol Immunopathol 9(4):303–317
Snyder D, Marquardt W, Mallinson E, Russek E (1983a) Rapid serological profiling by enzyme-linked immunosorbent assay. I. Measurement of antibody activity titer against Newcastle disease virus in a single serum dilution. Avian Dis 27:161–170
Snyder D, Marquardt W, Russek E (1983b) Rapid serological profiling by enzyme-linked immunosorbent assay. II. Comparison of computational methods for measuring antibody titer in a single serum dilution. Avian Dis 27:474–484
Song JE, Jeong WG, Sung HW, Kwon HM (2013) Sequencing, phylogenetic analysis, and potential recombination events of infectious bronchitis viruses isolated in Korea. Virus Genes 46(2):371–374. doi:10.1007/s11262-012-0856-0
Soula A, Moreau Y (1981) Antigen requirements and specificity of a microplate enzyme-linked immunosorbent assay (ELISA) for detecting infectious bronchitis viral antibodies in chicken serum. Arch Virol 67(4):283–295
Sultan HA, Tantawi L, Youseif AI, Ahmed AAS (2004) Urolithiathis in white commercial egg laying chickens associated with an infectious bronchitis virus. Paper presented at the 6th Sci Conf Egyp Vet Poult Assoc, Giza, Egypt
Sumi V, Singh SD, Dhama K, Gowthaman V, Barathidasan R, Sukumar K (2012) Isolation and molecular characterization of infectious bronchitis virus from recent outbreaks in broiler flocks reveals emergence of novel strain in India. Tropl Anim Health Prod 44(7):1791–1795. doi:10.1007/s11250-012-0140-2
Tan YW, Hong W, Liu DX (2012) Binding of the 5'-untranslated region of coronavirus RNA to zinc finger CCHC-type and RNA-binding motif 1 enhances viral replication and transcription. Nucleic Acids Res 40(11):5065–5077. doi:10.1093/nar/gks165
Thor SW, Hilt DA, Kissinger JC, Paterson AH, Jackwood MW (2011) Recombination in avian gamma-coronavirus infectious bronchitis virus. Viruses 3(9):1777–1799. doi:10.3390/v3091777
Thornton DH, Muskett JC (1975) Effect of infectious bronchitis vaccination on the performance of live Newcastle disease vaccine. Vet Rec 96(21):467–468
Toffan A, Bonci M, Bano L, Bano L, Valastro V, Vascellari M, Capua I, Terregino C (2013a) Diagnostic and clinical observation on the infectious bronchitis virus strain Q1 in Italy. Vet Ital 49(4):347–355. doi:10.12834/VetIt.1303.01
Toffan A, Bonci M, Bano L, Valastro V, Vascellari M, Capua I, Terregino C (2013b) Diagnostic and clinical observation on the infectious bronchitis virus strain Q1 in Italy. Vet Ital 49(4):347–355
Ujike M, Taguchi F (2015) Incorporation of spike and membrane glycoproteins into coronavirus virions. Viruses 7(4):1700–1725. doi:10.3390/v7041700
van Eck JH (1983) Effects of experimental infection of fowl with EDS'76 virus, infectious bronchitis virus and/or fowl adenovirus on laying performance. Vet Q 5(1):11–25. doi:10.1080/01652176.1983.9693868
Vandekerchove D, Herdt PD, Laevens H, Butaye P, Meulemans G, Pasmans F (2004) Significance of interactions between *Escherichia coli* and respiratory pathogens in layer hen flocks suffering from colibacillosis-associated mortality. Avian Pathol 33(3):298–302

Vennema H, Godeke GJ, Rossen JW, Voorhout WF, Horzinek MC, Opstelten DJ, Rottier PJ (1996) Nucleocapsid-independent assembly of coronavirus-like particles by co-expression of viral envelope protein genes. EMBO J 15(8):2020–2028

Verheije MH, Hagemeijer MC, Ulasli M, Reggiori F, Rottier PJ, Masters PS, de Haan CA (2010) The coronavirus nucleocapsid protein is dynamically associated with the replication-transcription complexes. J Virol 84(21):11575–11579. doi:10.1128/jvi.00569-10

Wang L, Junker D, Collisson EW (1993) Evidence of natural recombination within the S1 gene of infectious bronchitis virus. Virology 192(2):710–716

Wei YQ, Guo HC, Dong H, Wang HM, Xu J, Sun DH, Fang SG, Cai XP, Liu DX, Sun SQ (2014) Development and characterization of a recombinant infectious bronchitis virus expressing the ectodomain region of S1 gene of H120 strain. Appl Microbiol Biotechnol 98(4):1727–1735. doi:10.1007/s00253-013-5352-5

Wickramasinghe IN, de Vries RP, Grone A, de Haan CA, Verheije MH (2011) Binding of avian coronavirus spike proteins to host factors reflects virus tropism and pathogenicity. J Virol 85(17):8903–8912. doi:10.1128/jvi.05112-11

Wilson L, Gage P, Ewart G (2006) Hexamethylene amiloride blocks E protein ion channels and inhibits coronavirus replication. Virology 353(2):294–306. doi:10.1016/j.virol.2006.05.028

Winter C, Herrler G, Neumann U (2008) Infection of the tracheal epithelium by infectious bronchitis virus is sialic acid dependent. Microbes Infect 10(4):367–373. doi:10.1016/j.micinf.2007.12.009

Winterfield RW, Hitchner SB (1962) Etiology of an infectious nephritis-nephrosis syndrome of chickens. Am J Vet Res 23:1273–1279

Woo PC, Huang Y, Lau SK, Yuen KY (2010) Coronavirus genomics and bioinformatics analysis. Viruses 2(8):1804–1820. doi:10.3390/v2081803

Woo PC, Lau SK, Lam CS, Lau CC, Tsang AK, Lau JH, Bai R, Teng JL, Tsang CC, Wang M, Zheng BJ, Chan KH, Yuen KY (2012) Discovery of seven novel Mammalian and avian coronaviruses in the genus deltacoronavirus supports bat coronaviruses as the gene source of alphacoronavirus and betacoronavirus and avian coronaviruses as the gene source of gammacoronavirus and deltacoronavirus. J Virol 86(7):3995–4008. doi:10.1128/jvi.06540-11

Worthington KJ, Currie RJ, Jones RC (2008) A reverse transcriptase-polymerase chain reaction survey of infectious bronchitis virus genotypes in Western Europe from 2002 to 2006. Avian Pathol 37(3):247–257. doi:10.1080/03079450801986529

Yachida S, Aoyama S, Sawaguchi K, Takahashi N, Iritani Y, Hayashi Y (1985) Relationship between several criteria of challenge-immunity and humoral immunity in chickens vaccinated with avian infectious bronchitis vaccines. Avian Pathol 14(2):199–211. doi:10.1080/03079458508436222

Yagyu K, Ohta S (1990) Detection of infectious bronchitis virus antigen from experimentally infected chickens by indirect immunofluorescent assay with monoclonal antibody. Avian Dis 34(2):246–252

Yu D, Han Z, Xu J, Shao Y, Li H, Kong X, Liu S (2010) A novel B-cell epitope of avian infectious bronchitis virus N protein. Viral Immunol 23(2):189–199. doi:10.1089/vim.2009.0094

Zhang J, Chen XW, Tong TZ, Ye Y, Liao M, Fan HY (2014) BacMam virus-based surface display of the infectious bronchitis virus (IBV) S1 glycoprotein confers strong protection against virulent IBV challenge in chickens. Vaccine 32(6):664–670. doi:10.1016/j.vaccine.2013.12.006

Zhang T, Han Z, Xu Q, Wang Q, Gao M, Wu W, Shao Y, Li H, Kong X, Liu S (2015) Serotype shift of a 793/B genotype infectious bronchitis coronavirus by natural recombination. Infect Genet Evol J Mol Epidemiol Evol Genet Infect Dis 32:377–387. doi:10.1016/j.meegid.2015.03.034

Zhang Y, Wang HN, Wang T, Fan WQ, Zhang AY, Wei K, Tian GB, Yang X (2010) Complete genome sequence and recombination analysis of infectious bronchitis virus attenuated vaccine strain H120. Virus Genes 41(3):377–388. doi:10.1007/s11262-010-0517-0

Zhao F, Zou N, Wang F, Guo M, Liu P, Wen X, Cao S, Huang Y (2013) Analysis of a QX-like avian infectious bronchitis virus genome identified recombination in the region containing the ORF 5a, ORF 5b, and nucleocapsid protein gene sequences. Virus Genes 46(3):454–464. doi:10.1007/s11262-013-0884-4

Zhou J, Yu L, Hong J (1998) Isolation, identification and pathogenicity of virus causing proventricular-type infectious bronchitis. Chi J Anim Poult Infect Dis 20:62–65

Ziegler AF, Ladman BS, Dunn PA, Schneider A, Davison S, Miller PG, Lu H, Weinstock D, Salem M, Eckroade RJ, Gelb J Jr (2002) Nephropathogenic infectious bronchitis in Pennsylvania chickens 1997-2000. Avian Dis 46(4):847–858. doi:10.1637/0005-2086(2002)046[0847:nibipc]2.0.co;2

Zulperi ZM, Omar AR, Arshad SS (2009) Sequence and phylogenetic analysis of S1, S2, M, and N genes of infectious bronchitis virus isolates from Malaysia. Virus Genes 38(3):383–391. doi:10.1007/s11262-009-0337-2

Zuniga S, Cruz JL, Sola I, Mateos-Gomez PA, Palacio L, Enjuanes L (2010) Coronavirus nucleocapsid protein facilitates template switching and is required for efficient transcription. J Virol 84(4):2169–2175. doi:10.1128/jvi.02011-09

6 Norovirus Infection

Amauri Alcindo Alfieri, Raquel Arruda Leme,
and Alice Fernandes Alfieri

Norovirus (NoV) was first reported as causative agent of gastroenteritis in 1972, when students and staff of an elementary school located in Norwalk, Ohio, USA, presented vomitus and diarrhoea (Atmar and Estes 2001). After this, a number of studies definitively associated the NoV infection with outbreaks of acute gastroenteritis (Karst et al. 2015). Nowadays, NoV is considered the leading cause of non-bacterial gastroenteritis outbreaks and severe childhood diarrhoea worldwide, including water- and food-borne outbreaks (Patel et al. 2009; Karst et al. 2014).

In animals NoV infections have been reported in swine, bovine, ovine, canine, feline, and murine. The pathogenic role of NoV infection and its impact in animal health are not completely clear. Most of the epidemiological studies detected NoV in asymptomatic animal hosts worldwide (van der Poel et al. 2000; Keum et al. 2009; L'Homme et al. 2009). However, there are studies that associated the NoV infection with disease, especially enteritis, in canine (Martella et al. 2008), feline (Martella et al. 2007), swine (Shen et al. 2012), and bovine (Otto et al. 2011).

6.1 The *Norovirus*

The *Caliciviridae* family comprises the genera *Norovirus*, *Lagovirus*, *Nebovirus*, *Sapovirus*, and *Vesivirus* (Fig. 6.1). *Norwalk virus*, previously named *Norovirus*, is the single representative species of *Norovirus* genus (ICTV 2014).

A.A. Alfieri (✉) • R.A. Leme • A.F. Alfieri
Laboratory of Animal Virology, Department of Veterinary Preventive Medicine, Universidade Estadual de Londrina, Campus Universitário, PO Box 10011, 86057-970 Londrina, Paraná, Brazil

Multi-User Animal Health Laboratory, Molecular Biology Unit, Department of Veterinary Preventive Medicine, Universidade Estadual de Londrina, Campus Universitário, PO Box 10011, 86057-970 Londrina, Paraná, Brazil
e-mail: alfieri@uel.br

J. Bayry (ed.), *Emerging and Re-emerging Infectious Diseases of Livestock*,
DOI 10.1007/978-3-319-47426-7_6

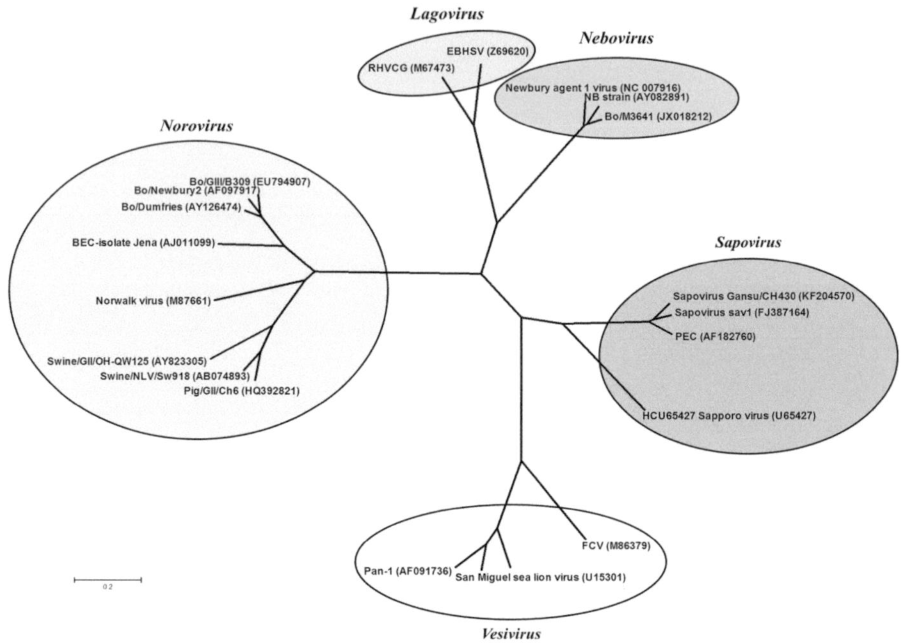

Fig. 6.1 Genetic relationship of virus members of *Caliciviridae* family, including the representative strains of each of the five genera. The GenBank accession numbers of the sequences are indicated between parentheses. *BEC* bovine enteric calicivirus, *RHVCG* rabbit haemorrhagic disease virus – calicivirus genome, *EBHSV* European brown hare syndrome virus, *PEC* porcine enteric *calicivirus*, *FCV* feline *calicivirus*, *VESV* vesicular exanthema of swine virus

Members of *Norovirus* genus are small, non-enveloped, with an icosahedral capsid, and present 27–40 nm in diameter (Fig. 6.2). Virus particle's buoyant density is 1.33–1.41 g/cm^3 in caesium chloride gradient (Kapikian et al. 1973). The NoV has a linear, single-stranded, positive-sense RNA genome of approximately 7.5 kb in size. The 3′ end of the RNA genome is polyadenylated, and the 5′ end has a covalently linked protein, named VPg. This protein acts primarily in the replication process by binding initiation factors for the translation of the virus RNA.

The genome of NoV is organised into three open reading frames (ORFs). The ORF1 encodes a polyprotein with approximately 200 kDa that is cleaved by the cysteine proteinase, encoded by the virus genome. The cleaved polyprotein gives origin to six nonstructural proteins, including the 2C helicase, 3C protease, and 3D RNA-dependent RNA polymerase (RdRp). The ORF2 and ORF3 encode two structural proteins that are, respectively, the VP1, which is the major capsid protein, and the VP2, a minor structural protein that is responsible for the virus genome packaging. The VP1 is functionally divided into shell (S) and protrusion (P) domains. The last is divided into P1, which is formed by two noncontiguous regions, and the hypervariable P2 that is inserted between the P1 subregions (Fig. 6.3) (Jiang et al. 1993; Ettayebi and Hardy 2003). A fourth ORF, tentatively named virulence factor

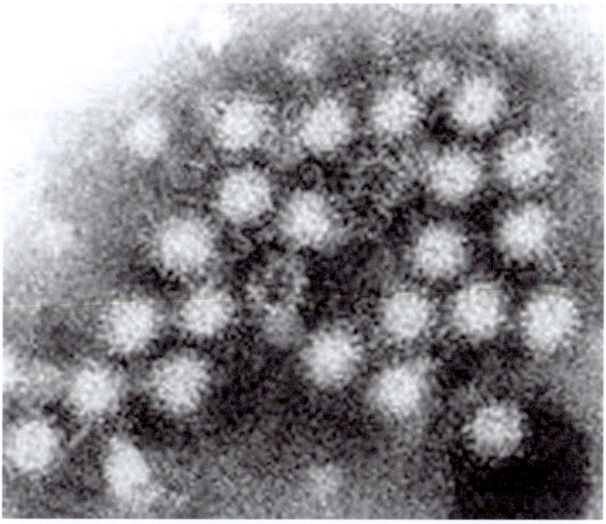

Fig. 6.2 Electron microscopy of norovirus particles present in human faecal samples stained with 2 % potassium phosphotungstate (Courtesy of Dr. J. Vinje, National Calicivirus Laboratory, Centers of Disease Control and Prevention (CDC), Atlanta, GA, USA)

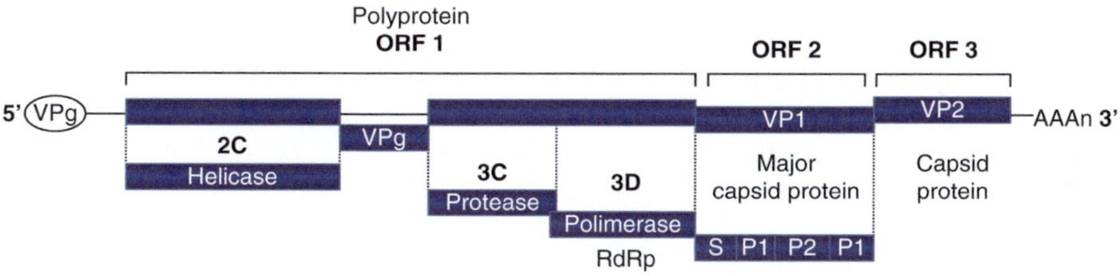

Fig. 6.3 Schematic genome organization of *Norovirus*. Open reading frames (ORFs) and the respective virus proteins for each genomic region are shown. *ORF1* encodes a polyprotein that is cleaved into nonstructural proteins. *ORF2* and *ORF3* encode the structural proteins *VP1* and *VP2*, respectively, both composing the virus capsid. The VP1 is divided into shell (*S*) and protrusion (*P*) domains. *P* domain is externally exposed

1 (VF1), which encodes a protein involved in the regulation of innate immune response has been described in the murine NoV genome (McFadden et al. 2011).

Based on the complete deduced amino acid (aa) sequence of the VP1 capsid protein, NoVs are classified into six genogroups (GI-GVI) with the proposed seventh genogroup (GVII) (Vinje 2015). Genogroups have been further subdivided into at least 31 genotypes (Fig. 6.4). Human NoV strains are organised into GI, GII, and GIV. The human NoV GII is the most commonly detected and is classified into other distinct genotypes, of which the GII.4 is the most frequently detected in the majority of the recent large outbreaks (Karst et al. 2015).

Differently from the human NoV, animal NoVs are less genetically variable. The canine NoV is classified as GIV and GVI, while the feline NoV is classified as GIV

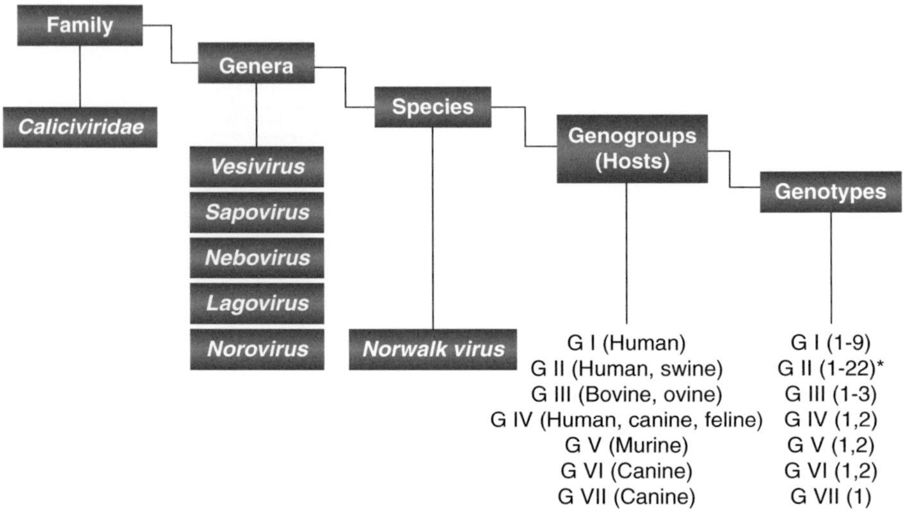

Fig. 6.4 *Norovirus* (NoV) is classified into seven genogroups and their genotypes based on the VP1 capsid protein. GI.1 is the NoV prototype strain. *The GII.11, GII.18, and GII.19 NoV strains infect swine; GIII.1 and GIII.2 bovine; GIII.3 ovine; GIV.2 canine and feline; GV.1 mice; and GV.2 rats. The GVI and GVII NoV strains were reported in canine

(Martella et al. 2008; Pinto et al. 2012). The murine NoV strains are in the GV (Zheng et al. 2006; McFadden et al. 2011).

The bovine and ovine NoV isolates are in the genogroup GIII. The molecular characterisation of bovine NoV genome showed that there are two different bovine NoV genotypes within the GIII genogroup. The prototypes were named Jena agent (Bo/Jena/80/DE) for the genotype 1 and Newbury agent-2 (Bo/Newbury2/76/UK) for the genotype 2 (Di Martino et al. 2014).

Porcine NoV strains are classified into GII and are closest to the most prevalent human NoV isolates. The porcine NoV GII strains detected to date are distributed into the three genotypes GII.11, .18, and .19, while the human NoV GII is classified into other distinct genotypes (Zheng et al. 2006). Since the porcine NoV strains have grouped only into these three genotypes, it was suggested that these strains have been adapted and its occurrence is restricted to pigs (L'Homme et al. 2009). However, the porcine NoV GII.18 was shown to be genetically and antigenically most closely related to human GII NoV, raising questions of whether pigs may be reservoirs for emergence of new human NoV strains (Wang et al. 2005). Figure 6.5 presents a dendrogram showing the NoV classification into genotypes and their genetic relationship.

6.2 Pathogenesis of the Disease

NoV infection can be symptomatic or asymptomatic; however, the pathogenesis of norovirus in humans and animals is not fully clarified. The incubation period after exposure to NoV is short, varying between 24 and 48 h. Symptoms of infection include

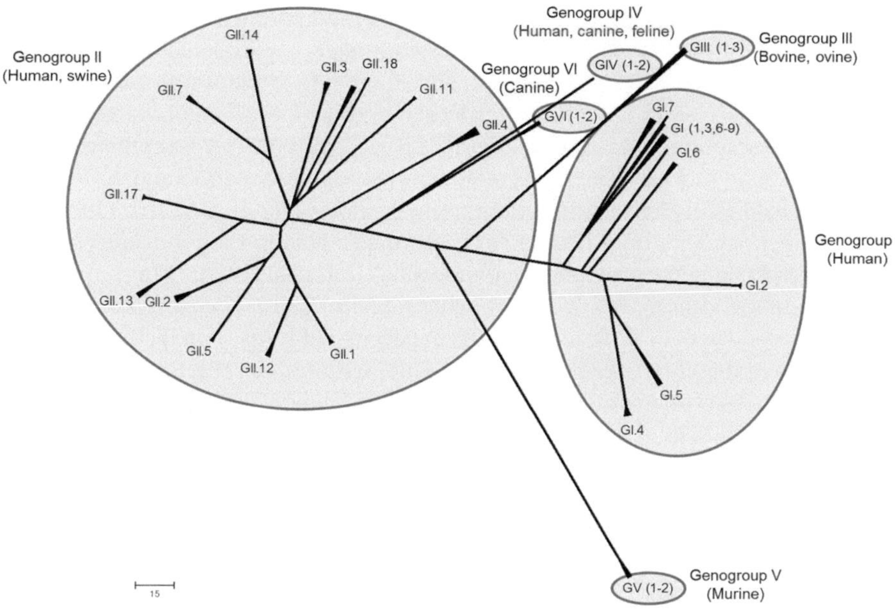

Fig. 6.5 *Norovirus* (NoV) classification into genotypes based on the VP1 capsid protein (From Vinje (2015), Copyright© American Society for Microbiology (2015), with permission)

acute enteritis with non-haemorrhagic diarrhoea, vomitus (characteristic sign in outbreaks), nausea, anorexia, abdominal pain, and mild fever. However, asymptomatic infections occur in one third of individuals experimentally infected. The disease duration is short (12–60 h) and self-limiting; however, immunosuppressed patients may have chronic diarrhoea and excrete viruses for months or years (Karst et al. 2015).

Previous studies with NoV performed in volunteers showed that the target cells for virus replication are primarily enterocytes of the proximal portion of the intestine, leading to malabsorption diarrhoea. Although the intestinal epithelium appears to remain intact, there are specific histopathological lesions by infection of human NoV in the jejunum, including atrophy in intestinal villi, breakdown of intestinal epithelial cells, hyperplasia crypt cells, and vacuolated and mononuclear inflammatory infiltrate in the lamina propria of villi. The malabsorption is related to the shortened microvilli and decreased brush border enzyme activity, both observed in acute infection (Karst et al. 2015).

Bovine NoV has been detected in diarrhoeic and non-diarrhoeic cattle faecal samples. Newborn calves that were inoculated with the bovine NoV GIII.1 (Jena agent) by the oral route presented severe diarrhoea in a very short incubation period (Otto et al. 2011). Experimental infection with the bovine NoV GIII.2 (Newbury agent-2) showed calves presenting diarrhoea 3–4 days postinoculation (dpi), with short duration (1 day), and no diarrhoea was observed after the re-challenge of the calves (Jor et al. 2010). This and other bovine NoV experimental-based studies showed that infected calves presented reduced appetite at the fourth and fifth dpi, non-haemorrhagic enteritis, mild to moderate diarrhoea, transient anorexia, and/or xylose malabsorption; discrete or no diarrhoea was observed in conventionally kept calves at 1–8 days of age. The rectal temperatures were between 37 and 40 °C, with pulse and respiratory rates kept within normal ranges (Jor et al. 2010; Otto et al. 2011). Infections with both bovine NoV genotypes lead to the villus atrophy and crypt hyperplasia in the proximal small intestine (Hall et al. 1984; Otto et al. 2011).

Other enteric virus agents, such as bovine rotavirus and *Torovirus*, infect primarily the tips and bases of villi; however, an experimental study with the bovine NoV GIII.1 showed that this virus infects all the enteroabsorptive cells. Since the bovine NoV GIII.1 causes severe villus atrophy and loss of mature enterocytes, it was suggested that these facts may limit the infection duration due to the reduction in the number of susceptible cells to the virus infection (Otto et al. 2011). On the other hand, the bovine NoV GIII.2 experimental studies showed that calves shed the NoV in faeces for at least 30 days after inoculation, regardless of the faecal consistency (diarrhoeic or not) and the duration of the clinical signs (Jor et al. 2010; Jung et al. 2014). Diarrhoea and prolonged faecal shedding of bovine NoV GIII.2 were observed even in calves that were not presented with major histological changes in the intestine, including no necrosis of intestinal epithelium, villous atrophy, or inflammatory lesions (Jung et al. 2014).

The importance of the porcine NoV as diarrhoea-causative agent in pigs is not yet fully understood. In an experimental challenge of piglets, the porcine NoV incubation period was of only 1 dpi, and the diarrhoea persisted for 2–6 days. Piglets presented mild to moderate villous atrophy and mild to moderate and multifocal villous fusion in the small intestine (Shen et al. 2012).

Previously, an experimental study inoculated piglets with the human NoV strain GII.4. The incubation period varied from 24 to 48 h; the diarrhoea was mild and self-limiting, persisting for 1–3 days. As well, the virus shedding was shown to be short, from 1 to 4 days. The virus antigen was detected in the cytoplasm of the small intestine cells. The histopathological lesions that were multifocal atrophy of the intestinal villi, enterocytes infected with low columnar morphology, and oedema of the lamina propria duodenal occurred at low frequency and were considered to be of low intensity. Another finding of this study was the increase in the number of apoptotic enterocytes (Cheetham et al. 2006).

Replication of NoV may not be restricted to enterocytes. Of all the potential experimental models studied to better understand the pathogenesis of noroviruses, the only norovirus which replicates in vitro is the murine NoV. This agent replicates in macrophages and dendritic cells derived from cultures of bone marrow cells and in mouse macrophage cell lines (RAW 264.7) (Wobus et al. 2004). The murine NoV-1 infection in knockout mice for recombination-activating gene 2 (RAG2) and signal transducer and activator of transcription 1 (STAT-1) genes, RAG2/STAT-1, showed tropism for haematopoietic cell (macrophages and dendritic cells) and development of systemic disease. Clinical signs include pneumonia, hepatitis, encephalitis, and vasculitis in brain capillaries and can be observed even in inoculation in serial passages (Karst et al. 2003; Wobus et al. 2006; Scipioni et al. 2008b). It was also demonstrated that the murine NoV can naturally infect wild and immunodeficient mice. The infection also occurs following oral or intranasal inoculation. However, although other strains of murine NoV have already been isolated from faecal samples of infected mice, it is not yet clear whether this virus is an effective enteric pathogen in this animal species (Hsu et al. 2005).

6.3 Epidemiology

The transmission of NoV occurs predominantly by the faecal-oral route for both human and animals. Differently of other viruses that depend on high virus concentrations for causing disease, NoV requires a low infectious dose (<10–100 virions) to establish the infection (Atmar and Estes 2006).

NoV showed a long-term survival in suspensions at environmental temperature, indicating that transmission by routes involving surface or drinking water, moisture fomites, or workplace surfaces is possible (Duizer et al. 2004). Additionally, the respiratory tract has been considered as another natural route of NoV transmission by the inhalation of aerosolised particles in vomitus (Atmar and Estes 2006).

NoV is considered a waterborne virus of primary concern, together with other virus agents, such as hepatitis A virus, hepatitis E virus, adenovirus, astrovirus, enterovirus, and rotavirus (Gibson 2014). Outbreaks of human NoV infection were associated with contaminated drinking water in different countries (Duizer et al. 2004). Studies revealed a widespread occurrence of human enteric viruses in both individual and municipal wells, showing that groundwater can be pathogen contaminated, including with NoV, and that groundwater-sourced public water systems producing water without disinfection can represent a risk of waterborne illness (Gibson 2014), including for animals.

The faecal contamination in water, food, and fomites and the direct individual-to-individual contacts are responsible for the major occurrence of gastroenteritis outbreaks determined by NoV. Since the surface/drinking water and groundwater quality can be affected by multiple sources of pathogens, vegetables also can be contaminated with NoV by irrigation with contaminated water. The food may be contaminated with virus particles since its production or crop, as in the case of oysters and fresh produce, or is contaminated on site preparation by means of handling by infected people, as in the case of cold food, sandwiches, and salads (Atmar and Estes 2006).

High rates of secondary attacks (≥30 %) among people who had contact with infected individuals lead the outbreak amplification in places where there is overcrowding, such as hospital wards, cruise ships, and shelters (Atmar and Estes 2006). In addition to the low infectious dose required for NoV transmission, continuous NoV infection is a result of the difficult elimination of the virus due to its resistance to disinfectants and many chemical products, the facilitated survival by organic debris of the clinical specimens (faeces/vomitus) in which the virus is shed, and the NoV aggregate formation that protects the virus from the environmental conditions (Kotwal and Cannon 2014).

In symptomatic animals, the virus shedding appears shortly before or during the first clinical signs and is prolonged, even after resolving of the symptoms (Scipioni et al. 2008b); the individuals with asymptomatic NoV infection also shed the virus. The period of virus shedding may range between 5 and 60 days, with a medium of 30 days. In human NoV infection, the virus is excreted in high amounts; the peak of virus RNA titres may vary from 10^9 to 10^{12} genomic copies per gramme stool and may be 1–2 log lower in symptomatic and asymptomatic individuals, respectively (Newman and Leon 2015). These facts and the NoV infectious stability for weeks or months in the environment may facilitate the NoV persistence and the virus transmission among infected and susceptible hosts (Mathijs et al. 2012).

The murine NoV is one of the most prevalent pathogens of murine, being a causative agent of systemic infection and lethal disease in immunodeficient laboratory mice (Karst et al. 2003). However, murine NoV strains were also identified from immunocompetent laboratory mice with silent infection (Hsu et al. 2005; Wobus et al. 2006). As the murine NoV is the only norovirus that replicates in cell culture, this virus is considered an excellent model to comprehend the basic mechanisms of norovirus replication in vitro and in vivo (Wobus et al. 2006).

Dogs were first evidenced to be susceptible to NoV infection in 2007 (Martella et al. 2008). Later, serological and molecular studies showed that canine NoV is spread in dogs from Europe (Ntafis et al. 2010; Caddy et al. 2013; Mesquita et al. 2014) and Asia (Tse et al. 2012; Soma et al. 2015). Norovirus also was detected from enteric disease-affected feline of Italy (Martella et al. 2007), Japan (Soma et al. 2015; Takano et al. 2015), the United States (USA (Pinto et al. 2012), and Brazil (Castro et al. 2015). Although canine and feline noroviruses are likely worldwide disseminated, their pathogenic importance has to be further studied.

Epidemiological studies have demonstrated that noroviruses are spread in livestock animals worldwide. Enteric caliciviruses morphologically similar to the human NoV were identified from diarrhoeic calves in the United Kingdom and Germany in 1978 and 1980, respectively (Woode and Bridger 1978; Gunther and Otto 1987), and were molecularly classified as bovine NoV in 1999 (Liu et al. 1999). Since then, bovine NoV has been detected in diarrhoeic and non-diarrhoeic faecal samples of beef and dairy cattle with young animals being more frequently described with the infection (Table 6.1). High seroprevalence of bovine NoV has been reported from cattle herds in Europe and North America (Table 6.2). The bovine NoV GIII.2 was prevalent in most of the studies.

The porcine NoV was first reported in Japan, where the virus RNA was recovered from caecum content of asymptomatic pigs (Sugieda et al. 1998). Subsequently, porcine NoV was reported from faecal samples of diarrhoeic and

Table 6.1 Frequency of bovine norovirus detection from cattle faecal samples of distinct countries by using molecular assays

Continent	Country	Animal data: Age	Animal data: Health status	No. of samples evaluated	Results: No. positive	Results: Prevalence (%)	Diagnostic technique	Reference
Eurasia	South Korea	2–90 days	Diarrhoeic	645	60	9.3	Nested PCR	Park et al. (2007)
	Turkey	1–60 days	Diarrhoeic	70	6	8.6	qPCR	Shen et al. (2009)
	The Netherlands	1–52 weeks 4–6 years	N.I.	75 herds 43 herds	33 herds 0 herds	44 0	RT-PCR; hybridisation	van der Poel et al. (2000)[a]
	The Netherlands	1 week–> 2 year	Asymptomatic/diarrhoeic	555	90	16.2	RT-PCR; hybridisation	van der Poel et al. (2003)
	Belgium	1 week–6 months	Diarrhoeic	133	10	7.5	RT-PCR	Mauroy et al. (2009)
	Slovenia	4–5 months Adult	Asymptomatic	108 11	2 0	1.9 0	RT-PCR	Mijovski et al. (2010)
	Norway	0–356 days	Asymptomatic/ diarrhoeic	419	208	49.6	RT-qPCR	Jor et al. (2010)
	Italy	7–20 days	Diarrhoeic	101	21	20.8	RT-PCR	Di Bartolo et al. (2011)
	Italy	0–6 weeks	Asymptomatic	104	11	10.6	RT-PCR	Di Martino et al. (2014)
Africa	Tunisia	3–90 days	Diarrhoeic	169	28	16.6	RT-PCR	Hassine-Zaafrane et al. (2012)

(continued)

Table 6.1 (continued)

		Animal data		No. of samples evaluated	Results			
Continent	Country	Age	Health status		No. positive	Prevalence (%)	Diagnostic technique	Reference
America	Canada	N.I.	N.I.	179	3	1.7	RT-PCR	Mattison et al. (2007)
	Venezuela	2 months	Asymptomatic	129	1	0.8	RT-PCR	Alcala et al. (2003)

RT-PCR reverse transcription-polymerase chain reaction assay, *RT-qPCR* reverse transcription-quantitative PCR, *N.I.* not informed

[a]This study informed the number of evaluated and positive herds

Table 6.2 Prevalence of bovine NoV antibodies from cattle herds of distinct countries by using ELISA technique

		Samples		Results		
Country	Age of the animals evaluated	Specimen	No. tested	No. positive	Prevalence (%)	Reference
Germany	10 weeks–9 years	Serum/ plasma	824	817	99.1	Deng et al. (2003)
Germany	6 months Adult	Serum/ plasma	200	137	68.5	Oliver et al. (2007)
United Kingdom	6 months Adult	Serum/ plasma	200	175	87.5	
Belgium	1 week–>9 years	Serum	439	409	93.2	Mauroy et al. (2009)
United States	6–7 months	Serum	343	326	95	Thomas et al. (2014)
	7–10 days	Serum	47	44	93.6	

Enzyme-linked immunosorbent assay

non-diarrhoeic pigs at different ages in European, Asian, Oceania, and American countries. Table 6.3 shows the frequency of porcine NoV detection by molecular assays, primarily the reverse transcription-polymerase chain reaction (RT-PCR), in different countries, and the age group studied, the number of samples evaluated, the number of positive animals, and the prevalence found. Older pigs have been more frequently detected with NoV infection; however, the virus also has been reported from suckling piglets and nursing pigs. A seroprevalence-based study showed that NoV infection is common in domestic pigs, with 71 % (78/110) and 36 % (95/226) of pig serum samples from the United States and Japan, respectively, presenting antibodies against porcine NoV (Farkas et al. 2005).

In both bovine and swine species, NoV infections with other enteric virus agents may occur. Co-infections with rotavirus, coronavirus, bovine viral diarrhoea virus (Park et al. 2007), circovirus (Shen et al. 2012), torque teno sus virus (Leme et al. 2014, unpublished data), and even with other caliciviruses (Hassine-Zaafrane et al. 2012) were reported.

To date, there are limited epidemiological data regarding NoV infection in ovine. In 2007, a study conducted in New Zealand screened sheep for the virus. In that case, animals that were positive for NoV were not presenting clinical signs suggestive of the infection. Another study was conducted in Belgium in the same year; however, sheep were not detected with the virus (Mathijs et al. 2012).

Interestingly, a seasonality pattern of NoV infection has been reported. Although the infection can occur in different seasons, wintertime seasonality is observed for different host species, including livestock animals (Hassine-Zaafrane et al. 2012; Ahmed et al. 2013; Silva et al. 2015).

Table 6.3 Frequency of porcine norovirus detection from pig faecal samples of distinct countries by using molecular assays

		Animal data			Results			
Continent	Country	Age	Health status	No. of samples evaluated	No. of positive	Prevalence (%)	Diagnostic technique	Reference
Asia	Japan	N.I.	Asymptomatic	1117	4	0.4	RT-PCR	Sugieda et al. (1998)
	South Korea	Nursing finisher	Asymptomatic/diarrhoeic	537	10	1.9	Semi-nested PCR	Keum et al. (2009)
	Korea	Post-weaning nursing grower sows	N.I.	567	3	0.5	RT-PCR	Song et al. (2011)
	China	<4 weeks 8–12 weeks >12 weeks	N.I.	904	2	0.2	RT-PCR	Shen et al. (2009)
	China	Piglets	Diarrhoeic	12	2	16.7	RT-PCR	Shen et al. (2012)
Oceania	New Zealand	<20 weeks	N.I.	23	2	8.7	RT-qPCR	Wolf et al. (2009)
Europe	The Netherlands	12–36 weeks	N.I.	100 herds	2 herds	2	RT-PCR	van der Poel et al. (2000)[a]
	Hungary	1–12 days 2–6 months Adult	Asymptomatic/diarrhoeic	17	1	5.9	RT-PCR	Reuter et al. (2007)
	Italy	12 days 1–3 months 12 months	Asymptomatic/diarrhoeic	290	1	0.3	RT-PCR	Di Bartolo et al. (2014)
	Belgium	Young adult	Diarrhoeic	43	2	4.6	RT-PCR	Mauroy et al. (2008)

Continent	Country	Animal data		No. of samples evaluated	Results		Diagnostic technique	Reference
		Age	Health status		No. of positive	Prevalence (%)		
	Slovenia	<3 weeks 4–10 weeks >11 weeks	Asymptomatic	406	5	1.2	RT-PCR	Mijovski et al. (2010)
	Germany	Finishing	Asymptomatic	120	17	14.2	RT-qPCR	Machnowska et al. (2014)
America	Canada	N.I.	N.I.	120	30	25	RT-PCR	Mattison et al. (2007)
	United States	10–24 weeks		275	6	2.2	RT-PCR	Wang et al. (2005)
	United States	1 week to >1 year		621	124	20.0	RT-PCR; microwell hybridisation assay	Wang et al. (2006)
	United States	Finishing	Asymptomatic	413	78	18.9	RT-PCR; hybridisation assay	Scheuer et al. (2013)
	Brazil	>1 week 4–10 weeks 10–26 weeks Adult	Asymptomatic/diarrhoeic	96	1	1.04	RT-PCR	Cunha et al. (2010)
	Brazil	9–24 weeks	Asymptomatic	112	58	51.8	RT-PCT	Silva et al. (2015)

RT-PCR reverse transcription-polymerase chain reaction assay, *RT-qPCR* reverse transcription-quantitative PCR, *N.I.* not informed
[a]This study informed the number of evaluated and positive herds

6.3.1 Zoonotic Transmission

A concern regarding NoV infection is the possibility of cross infection among human and animal NoV strains, which raises questions about the zoonotic transmission potential of noroviruses. Studies based on the molecular epidemiology of noroviruses have shown a closer genetic relationship of human and animal noroviruses and the emergence of recombinant NoV strains in different hosts (Koopmans 2008).

Serological surveys have revealed antibodies against animal NoV in humans, including general population and veterinarians (Widdowson et al. 2005; Menon et al. 2013), and antibodies against human NoV in swine (Farkas et al. 2005). Molecular studies also revealed the presence of human NoV strains, including the most prevalent in the major recent large epidemics (GII.4), in cattle, in pig faeces, and in raw pork meat (Mattison et al. 2007). Additionally, the experimental challenge of gnotobiotic calves and piglets with human NoV showed virus replication and seroconversion (Cheetham et al. 2006; Souza et al. 2008).

Although the zoonotic transmission is likely, this hypothesis has not been proved. A possible explanation for the detection of antibodies against animal NoVs in human may be the existence of cross-reactive epitopes between the different NoV strains, as shown between human and bovine NoVs (Scipioni et al. 2008b). Noroviruses are thought to be a species-specific pathogen (Karst et al. 2015), and further studies are needed to fully comprehend the role of animals as reservoirs for human NoV infection.

6.4 Progress in Diagnosis and Management of NoV Infection

6.4.1 Diagnostic Tests

The techniques used for the laboratory diagnosis of both human and animal NoVs include the direct detection of intact virus particles by electron microscopy (EM) and of the virus RNA by RT-PCR. Currently, the development of other techniques for detecting and quantifying the NoV RNA by RT-quantitative PCR (RT-qPCR) has been reported for the diagnosis of human (Vinje 2015), swine (Machnowska et al. 2014), and bovine (Jor et al. 2010; Yilmaz et al. 2011) NoV strains. Additionally, the next-generation sequencing (NGS) has been a useful tool for the identification of evolutionary changes in relation to NoV epidemiological data (Bavelaar et al. 2015). Immunoassays, such as enzyme-linked immunosorbent assay (ELISA), are able to detect viral antigens or antibodies, both techniques being associated with the production of recombinant virus-like particles (VLPs) (Wang et al. 2007; Mauroy et al. 2009). Other techniques, such as microarray hybridisation (Wang et al. 2006; Scheuer et al. 2013) and immunohistochemistry (Otto et al. 2011), may also be used for the virus diagnosis. Table 6.4 presents the advantages and disadvantages of techniques used for the laboratory diagnosis of NoV infection.

Samples In general, faecal samples of clinically affected animals present higher concentrations of NoV particles, and, therefore, faeces are considered the best samples for diagnosis of the infection, although the virus can be detected from rectal swabs and vomitus (Vinje 2015). The first reports of NoV infections were based on EM, since other diagnostic methods were not yet available for the virus detection

Table 6.4 Advantages and disadvantages of techniques used for the laboratory diagnosis of norovirus infection

Diagnostic assay	Advantages	Disadvantages
Electron microscopy	Ability to detect multiple virus pathogens	Low sensitive technique (detection limit of 10^6 enteric virus particle per mL of stool sample) Unable to differentiate NoV from other small round enteric viruses Requires highly skilled experts Expensive equipment (not widely available in diagnostic microbiology laboratories) Time-consuming process
Conventional RT-PCR Nested PCR	High sensitive High specificity Allow the virus classification into genogroups/genotypes	RT-PCR products have to be analysed in agarose gel Results have to be confirmed by sequencing or hybridisation The sensitivity/specificity may be affected by the sample quality, nucleic acid extraction and purification methods, oligonucleotide primers, and/or assay conditions Nested PCR increases the risk of cross-contamination
RT-qPCR one-step RT-qPCR	Increased sensitivity and specificity, even compared with conventional RT-PCR assays; Do not require agarose gel or hybridisation analyses Less sample handling (decrease the cross-contamination risk) Allow the determination of virus nucleic acid amount in a sample Rapid test	A RT-qPCR equipment is required The supplies are expensive The technique does not allow the virus genotyping
NGS	High sensitive High specificity Detection and characterization of the virus in a single assay Identification of the microbial contents of clinical specimens in a single test May elucidate the infection causative agent of unexplained cases of suspect viral gastroenteritis	Expensive equipment and supplies (not widely available in veterinary diagnostic laboratories) Requires skilled staff

(continued)

Table 6.4 (continued)

Diagnostic assay	Advantages	Disadvantages
Enzyme immunoassays	Detect both viral particles and soluble antigens Direct ELISA was shown to be a sensitive technique (detection limit of 0.025 ng of capsid protein and 1:10,000 dilutions of viral antigen in the stools) Useful for rapid screening of multiple faecal samples	High specificity in VLP-based EIAs (the circulation of antigenically distinct NoV strains may lead to underestimated result interpretations) Cross-reaction of porcine NoV VLP-based antibody ELISA with human NoV antibodies

Source: Wang et al. (2007), Scipioni et al. (2008a, b), Vinje (2015), Bavelaar et al. (2015)
NoV Norovirus, *RT-PCR* reverse transcription-polymerase chain reaction assay, *RT-qPCR* reverse transcription-quantitative PCR, *NGS* next-generation sequencing, *ELISA* enzyme-linked immunosorbent assay, *VLPs* virus-like particles

at that time. However, NoV particles are morphologically similar to other small round viruses, such as sapovirus, rotavirus, astrovirus, and enteric picornavirus, and the differentiation among the viruses is harder (Wang et al. 2007). The immune EM (IEM) and solid-phase IEM are variations of the technique that are based on antigen-antibody reaction and visualised by negative staining; both can be used for NoV detection. Nowadays, the EM is generally used for research purposes and not for routine diagnosis of NoV infection.

RT-PCR and RT-qPCR Since the first reports of NoV from different hosts, the development of diagnostic methods has been hampered by the lack of a cell culture system for NoVs, other than murine NoV (Scipioni et al. 2008b). With the development of the molecular tools for NoV diagnosis, such as conventional or endpoint RT-PCR in the mid-1990s and later the RT-qPCR and their variations, an increased number of NoV sequences became available, and the infection diagnosis became faster and trusty (Vinje 2015).

Among the factors that can affect the sensibility and/or specificity of the RT-PCR is the nucleic acid extraction method, which has to efficiently recover the virus RNA and remove the RT-PCR inhibitors. The conserved RdRp gene is frequently targeted in the RT-PCR assays for the virus diagnosis, while the VP1 region is used for the virus genotyping. The highly variable NoV strains become the oligonucleotide primer selection a hard work, since a single primer pair likely will not be able to detect all the circulating NoV strains (Wang et al. 2007). Although several primer sets can be used for the detection of genetically diverse virus strains, the design of oligonucleotide primers based on the regional NoV strains in circulation is reasonable and acceptable. For this, molecular epidemiological surveillance-based studies should be conducted.

A second round of the RT-PCR, a semi-nested or a nested PCR, can be performed to increase the sensitivity of the reaction. The increased sensitivity with the use of a nested PCR can range from 10 to 1000 times; however, this technique also increases the risk of cross-contamination (Wang et al. 2007) and its use should be under closer attention.

The RT-qPCR has major advantages in comparison with the conventional RT-PCR assays, including higher sensibility and specificity, no need of agarose gel and sequencing analyses, and the ability to determine the amount of virus nucleic acid in the clinical sample. The one-step RT-qPCR is another variation of the technique, in which the reverse transcription and cDNA amplification are performed in a single reaction. With this, less sample handling is required, which decreases the risk of cross-contamination (Vinje 2015). However, this technique does not allow depth phylogenetic studies for NoV genotyping, and the conventional RT-PCR assay is required to complement the genomic analysis.

The RT-qPCR assays have been considered the gold standard for the rapid and sensitive detection of NoV in faeces, vomitus, and serum samples of clinically affected persons, as well as in food, water, and environmental samples (Vinje 2015). Nevertheless, these techniques are not largely used by the veterinary diagnostic laboratories. For the detection of NoV in bovine, both SYBR Green- (Park et al. 2009) and TaqMan (Scipioni et al. 2008a)-based RT-qPCR assays are available; however, the SYBR Green RT-qPCR has been more frequently used for this animal species host. For swine, although a TaqMan RT-qPCR has been validated for the NoV detection (Wolf et al. 2009), this technique has not currently being used. The most likely cause for the limited use of the RT-qPCR assays for animal NoV diagnosis is their high cost, even though the prices of equipment and reagents have decreased in the last years (Wang et al. 2007).

Since the recombination is common in NoV, the surveillance of emerging strains is important for understanding the NoV evolution and global epidemiology. For this, the NGS is a powerful tool in the detection and characterisation of different types of norovirus from clinical specimens (Bavelaar et al. 2015). However, most of the routine veterinary diagnostic laboratories are not equipped with the necessary tools or staff to perform the NGS (Bavelaar et al. 2015).

Enzyme Immunoassays (EIA) NoV capsid proteins are expressed in baculovirus system to provide VLPs that are used as antigen in the EIA. The ELISA is the most frequently EIA used for animal NoV infection diagnosis. Hyperimmune polyclonal serum obtained after animal (mice, rabbits, guinea pigs) immunisation with VLPs is used as capture antibody for the ELISA-based detection of viral antigen (Scipioni et al. 2008b). The large amounts of viral soluble proteins in stool are likely responsible for the sensitivity of antigen ELISA similar to that of conventional RT-PCR. However, the hyperimmune anti-VLP serum is often highly specific with the homologous strains or viruses within the same genotype, which make the technique highly specific (Wang et al. 2007). This fact limits the use of the technique in diagnostic laboratories, since there are highly diverse NoV strains in circulation (Wang et al. 2007; Scipioni et al. 2008b). Despite this, antigen ELISA is useful for screening large number of samples since it is a rapid and simple test (Scipioni et al. 2008b) and can be used for epidemiological-based studies from host species in which larger antigenic and/or molecular NoV genomic variations were not identified.

Antibody ELISA It is more broadly reactive than the antigen detection and is more suitable to identify heterotypic NoV infection (Scipioni et al. 2008b). Since this assay presents a good sensitivity and specificity for NoV-specific antibody detection, it has been largely used in human epidemiological studies. However, the cross-reaction of the baculovirus-expressed VLP antibodies against GI and GII human NoV was reported (Wang et al. 2007).

Antigen and Antibody ELISA They are available for bovine (Deng et al. 2003; Mauroy et al. 2009) and swine (Farkas et al. 2005; Cheetham et al. 2006) NoV infection diagnosis. Nevertheless, it is important to mention that three common epitopes are shared among NoVs, of which one is in the same genogroup GI (human), the second is between GII (human, swine) and GIII (bovine), and the third between GI and GIII (Scipioni et al. 2008b). Therefore, as for human ELISA, cross-reaction among different genogroups can occur. Studies reported that VLPs of porcine NoV cross reacted with antibodies against human NoV (Farkas et al. 2005) and the presence of antibodies against bovine NoV GIII.2 in veterinarians in the Netherlands (Widdowson et al. 2005).

RT-PCR-based Multiplex Molecular Diagnosis Tests In the recent years, commercial RT-PCR-based multiplex molecular diagnosis tests for the detection of several different multi-gastrointestinal pathogens have been developed. These diagnostic platforms are able to simultaneously detect enteric viruses, including NoV GI and GII, bacteria, and parasites. However, the interpretation of data generated by these multi-pathogen systems may be a challenge, since high number of mixed infections can be detected, making it difficult to determine which pathogen is responsible for gastroenteritis (Vinje 2015). Regardless of the advantages and disadvantages of the commercial tests, these multi-gastrointestinal diagnostic platforms are not yet available for animals.

6.4.2 Management of NoV Infection

Livestock animals affected with NoV infection usually present mild to moderate non-haemorrhagic diarrhoea. The NoV faecal shedding can be prolonged, even after the diarrhoea recovery and/or in symptomatic animals. In this context, closer attention should be given to the livestock and environmental managements in order to provide clinical support to the affected animals and to avoid the virus exposure and transmission to susceptible animals.

Livestock animals suspected or confirmed with NoV infection should be isolated from the other herd animals. Adequate fluid and electrolyte replacement and maintenance are key points to manage NoV-induced diarrhoea that is generally a self-limiting infection, typically resolved within days after exposure. The fluid and electrolyte therapies are easier to be implemented for calves relative to piglets. For piglets, oral products intended to re-establish the hydroelectrolytic balance can be used. In cases in which severe diarrhoea is present, the use of broad-spectrum antibiotic therapy is recommended to avoid/control secondary infections. It is likely that maternal antibodies in colostrum and milk limit infection and damage in the gut of nursing piglets. Even not preventing infection, the colostrum intake may have some impact in the bovine NoV infection by limiting the infection dissemination and reducing the severity and duration of diarrhoea (Otto et al. 2011). However, further pathogenic studies under field conditions should be performed to evaluate the effects of the colostrum on the course of the bovine and swine NoV infections.

6.5 Development of Vaccines and Prophylaxis Measures

There are no vaccines available against bovine and swine NoV infections. Therefore, the prophylaxis measures are fundamental to prevent the disease. Faeces removal, clean facilities, disinfection, and other sanitary precautions should control and/or prevent NoV infection and persistence in the environment. Since NoV is a waterborne pathogen, adequate water management practices should be adopted, including for the drinking water that is offered to the animals, which has to be of good quality, clean, and adequately chlorine treated. As NoVs are considered potentially zoonotic, management focused on public health also should be addressed.

In general, NoV is stable in the environment and may be resistant to inactivation with certain chemicals and heat, depending on the conditions in which both are used. NoV appears to be resistant to the effects of freeze/thaw (Nims and Plavsic 2013). Inactivation of NoV by heat is time and temperature dependent, and it should exceed 56 °C to have some reduction of NoV infectivity, with extensive inactivation requiring exposure times of 30 min or higher (Duizer et al. 2004; Nims and Plavsic 2013). More consistent and extensive inactivation is observed at temperatures in excess of 60 °C, regardless of calicivirus species or strain (Nims and Plavsic 2013).

Ultraviolet Radiation Caliciviruses are susceptible to inactivation by ultraviolet radiation in the C range (UV-C, 254 nm) at fluence of 22–40 mJ/cm^2. The UV-C radiation for the inactivation of caliciviruses appears to be less effective compared to parvovirus and circovirus that are inactivated by exposure to 8–13 mJ/cm^2 and 5–11 mJ/cm^2, respectively. On the other hand, the calicivirus UV-C inactivation is generally similar to those displayed by other small, non-enveloped viruses. The UV radiation in the B range (UV-B, 280–320 nm) at fluence of ~60 mJ/cm^2 would be required to achieve calicivirus inactivation (Nims and Plavsic 2013).

Inactivation by Chemicals Different chemical products can be used for the calicivirus inactivation, in compliance with specific conditions for each one (Table 6.5). The NoV particles display resistance to inactivation by low pH, with incomplete inactivation achieved at pH 2.7 for 3 h at room temperature (Duizer et al. 2004). Ethanol or hypochlorite solution contacts for a short period of time (1–10 min

Table 6.5 Chemical products, concentrations, and exposure time for substantial inactivation of caliciviruses at environmental temperature

Chemical product	Concentration/condition	Time
Formaldehyde	≥7000 ppm	≥30 min
Free chlorine	>2500 ppm	≥30 s
Glutaraldehyde	≥2500 ppm	≥5 min
Hypochlorous acid	Fogging	≥5 min
Hydrogen peroxide	Vapour	≥15 min
Ozone	20–25 ppm	≥20 min

Source: Nims and Plavsic (2013)
ppm parts per million, *min* minutes, *s* seconds

depending on the concentrations) also were reported to incomplete inactive caliciviruses (Duizer et al. 2004). Differences to the susceptibility to inactivation with these chemical agents may not afford adequate NoV inactivation.

Treatment of Animal Wastes To improve the agriculture, livestock manure applications to field land are practised worldwide because of its potential as biofertilizer. Another common practice in rural areas is the artificial subsurface drainage that facilitates the crop production. However, it is an efficient means by which agricultural pollutants from field systems, including faecal pollution, can enter the surface water environment (Wilkes et al. 2014). The land application of manures should be based on safety parameters, and agricultural beneficial management practices, including waste treatment in cattle and pig farm units, should be implemented to reduce the persistence of NoV in the animal waste and to mitigate the manure-borne faecal contamination of the environment, including water resources. Although it is not clear if enteric viruses are efficiently inactivated by anaerobic biodigestion, the anaerobic biodigester system is a tool that can be used to treat livestock manure in order to provide safe biofertilizers.

References

Ahmed SM, Lopman BA, Levy K (2013) A systematic review and meta-analysis of the global seasonality of norovirus. PLoS One 8(10):e75922

Alcala AC, Hidalgo MA, Obando C, Vizzi E, Liprandi F, Ludert JE (2003) Molecular identification of bovine enteric calciviruses in Venezuela (In Spanish, with English abstract). Acta Cient Venez 54(2):148–152

Atmar RL, Estes MK (2001) Diagnosis of noncultivatable gastroenteritis viruses, the human caliciviruses. Clin Microbiol Rev 14(1):15–37

Atmar RL, Estes MK (2006) The epidemiologic and clinical importance of norovirus infection. Gastroenterol Clin N Am 35(2):275–290

Bavelaar HH, Rahamat-Langendoen J, Niesters HG, Zoll J, Melchers WJ (2015) Whole genome sequencing of fecal samples as a tool for the diagnosis and genetic characterization of norovirus. J Clin Virol 72:122–125

Caddy S, Emmott E, El-Attar L, Mitchell J, de Rougemont A, Brownlie J, Goodfellow I (2013) Serological evidence for multiple strains of canine norovirus in the UK dog population. PLoS One 8(12):e81596

Castro TX, Garcia RCNC, Fumian TM, Costa EM, Mello R, White PA, Leite JPG (2015) Detection and molecular characterization of caliciviruses (vesivirus and norovirus) in an outbreak of acute diarrhea in kittens from Brazil. Vet J 206(1):115–117

Cheetham S, Souza M, Meulia T, Grimes S, Han MG, Saif LJ (2006) Pathogenesis of a genogroup II human norovirus in gnotobiotic pigs. J Virol 80(21):10372–10381

Cunha JB, de Mendonca MC, Miagostovich MP, Leite JP (2010) First detection of porcine norovirus GII.18 in Latin America. Res Vet Sci 89(1):126–129

Deng Y, Batten CA, Liu BL, Lambden PR, Elschner M, Gunther H, Otto P, Schnurch P, Eichhorn W, Herbst W, Clarke IN (2003) Studies of epidemiology and seroprevalence of bovine noroviruses in Germany. J Clin Microbiol 41(6):2300–2305

Di Bartolo I, Ponterio E, Monini M, Ruggeri FM (2011) A pilot survey of bovine norovirus in northern Italy. Vet Rec 169(3):73

Di Bartolo I, Tofani S, Angeloni G, Ponterio E, Ostanello F, Ruggeri FM (2014) Detection and characterization of porcine caliciviruses in Italy. Arch Virol 159(9):2479–2484

Di Martino B, Di Profio F, Di Felice E, Melegari I, Ceci C, Mauroy A, Thiry E, Martella V, Marsilio F (2014) Genetic heterogeneity of bovine noroviruses in Italy. Arch Virol 159(10):2717–2722

Duizer E, Bijkerk P, Rockx B, De Groot A, Twisk F, Koopmans M (2004) Inactivation of caliciviruses. Appl Environ Microbiol 70(8):4538–4543

Ettayebi K, Hardy ME (2003) Norwalk virus nonstructural protein p48 forms a complex with the SNARE regulator VAP-A and prevents cell surface expression of vesicular stomatitis virus G protein. J Virol 77(21):11790–11797

Farkas T, Nakajima S, Sugieda M, Deng X, Zhong W, Jiang X (2005) Seroprevalence of noroviruses in swine. J Clin Microbiol 43(2):657–661

Gibson KE (2014) Viral pathogens in water: occurrence, public health impact, and available control strategies. Curr Opin Virol 4:50–57

Gunther H, Otto P (1987) Diarrhea in young calves. 7. "Zackenvirus" (Jena agent 117/80)-a new diarrhea pathogen in calves [In Germany]. Arch Exp Veterinarmed 41(6):934–938

Hall GA, Bridger JC, Brooker BE, Parsons KR, Ormerod E (1984) Lesions of gnotobiotic calves experimentally infected with a calicivirus-like (Newbury) agent. Vet Pathol 21(2):208–215

Hassine-Zaafrane M, Kaplon J, Sdiri-Loulizi K, Aouni Z, Pothier P, Aouni M, Ambert-Balay K (2012) Molecular prevalence of bovine noroviruses and neboviruses detected in central-eastern Tunisia. Arch Virol 157(8):1599–1604

Hsu CC, Wobus CE, Steffen EK, Riley LK, Livingston RS (2005) Development of a microsphere-based serologic multiplexed fluorescent immunoassay and a reverse transcriptase PCR assay to detect murine norovirus 1 infection in mice. Clin Diagn Lab Immunol 12(10):1145–1151

ICTV (2014) Virus taxonomy: 2014 release. International Committee on Taxonomy of Viruses. http://www.ictvonline.org/virustaxonomy.asp. Accessed 3 Jan 2016

Jiang X, Wang M, Wang K, Estes MK (1993) Sequence and genomic organization of Norwalk virus. Virology 195(1):51–61

Jor E, Myrmel M, Jonassen CM (2010) SYBR Green based real-time RT-PCR assay for detection and genotype prediction of bovine noroviruses and assessment of clinical significance in Norway. J Virol Methods 169(1):1–7

Jung K, Scheuer KA, Zhang Z, Wang Q, Saif LJ (2014) Pathogenesis of GIII.2 bovine norovirus, CV186-OH/00/US strain in gnotobiotic calves. Vet Microbiol 168(1):202–207

Kapikian AZ, Gerin JL, Wyatt RG, Thornhill TS, Chanock RM (1973) Density in cesium chloride of the 27 nm "8FIIa" particle associated with acute infectious nonbacterial gastroenteritis: determination by ultra-centrifugation and immune electron microscopy. Proc Soc Exp Biol Med 142(3):874–877

Karst SM, Wobus CE, Lay M, Davidson J, Virgin HW (2003) STAT1-dependent innate immunity to a Norwalk-like virus. Science 299(5612):1575–1578

Karst SM, Wobus CE, Goodfellow IG, Green KY, Virgin HW (2014) Advances in norovirus biology. Cell Host Microbe 15(6):668–680

Karst SM, Zhu S, Goodfellow IG (2015) The molecular pathology of noroviruses. J Pathol 235(2):206–216

Keum HO, Moon HJ, Park SJ, Kim HK, Rho SM, Park BK (2009) Porcine noroviruses and sapoviruses on Korean swine farms. Arch Virol 154(11):1765–1774

Koopmans M (2008) Progress in understanding norovirus epidemiology. Curr Opin Infect Dis 21(5):544–552

Kotwal G, Cannon JL (2014) Environmental persistence and transfer of enteric viruses. Curr Opin Virol 4:37–43

L'Homme Y, Sansregret R, Plante-Fortier E, Lamontagne AM, Lacroix G, Ouardani M, Deschamps J, Simard G, Simard C (2009) Genetic diversity of porcine Norovirus and Sapovirus: Canada, 2005-2007. Arch Virol 154(4):581–593

Liu BL, Lambden PR, Gunther H, Otto P, Elschner M, Clarke IN (1999) Molecular characterization of a bovine enteric calicivirus: relationship to the Norwalk-like viruses. J Virol 73(1):819–825

Machnowska P, Ellerbroek L, Johne R (2014) Detection and characterization of potentially zoonotic viruses in faeces of pigs at slaughter in Germany. Vet Microbiol 168(1):60–68

Martella V, Campolo M, Lorusso E, Cavicchio P, Camero M, Bellacicco AL, Decaro N, Elia G, Greco G, Corrente M, Desario C, Arista S, Banyai K, Koopmans M, Buonavoglia C (2007) Norovirus in captive lion cub (*Panthera leo*). Emerg Infect Dis 13(7):1071–1073

Martella V, Lorusso E, Decaro N, Elia G, Radogna A, D'Abramo M, Desario C, Cavalli A, Corrente M, Camero M, Germinario CA, Banyai K, Di Martino B, Marsilio F, Carmichael LE, Buonavoglia C (2008) Detection and molecular characterization of a canine norovirus. Emerg Infect Dis 14(8):1306–1308

Mathijs E, Stals A, Baert L, Botteldoorn N, Denayer S, Mauroy A, Scipioni A, Daube G, Dierick K, Herman L, Van Coillie E, Uyttendaele M, Thiry E (2012) A review of known and hypothetical transmission routes for noroviruses. Food Environ Virol 4(4):131–152

Mattison K, Shukla A, Cook A, Pollari F, Friendship R, Kelton D, Bidawid S, Farber JM (2007) Human noroviruses in swine and cattle. Emerg Infect Dis 13(8):1184–1188

Mauroy A, Scipioni A, Mathijs E, Miry C, Ziant D, Thys C, Thiry E (2008) Noroviruses and sapoviruses in pigs in Belgium. Arch Virol 153(10):1927–1931

Mauroy A, Scipioni A, Mathijs E, Saegerman C, Mast J, Bridger JC, Ziant D, Thys C, Thiry E (2009) Epidemiological study of bovine norovirus infection by RT-PCR and a VLP-based antibody ELISA. Vet Microbiol 137(3-4):243–251

McFadden N, Bailey D, Carrara G, Benson A, Chaudhry Y, Shortland A, Heeney J, Yarovinsky F, Simmonds P, Macdonald A, Goodfellow I (2011) Norovirus regulation of the innate immune response and apoptosis occurs via the product of the alternative open reading frame 4. PLoS Pathog 7(12):e1002413

Menon VK, George S, Shanti AA, Saravanabavan A, Samuel P, Ramani S, Estes MK, Kang G (2013) Exposure to human and bovine noroviruses in a birth cohort in southern India from 2002 to 2006. J Clin Microbiol 51(7):2391–2395

Mesquita JR, Delgado I, Costantini V, Heenemann K, Vahlenkamp TW, Vinje J, Nascimento MS (2014) Seroprevalence of canine norovirus in 14 European countries. Clin Vaccine Immunol 21(6):898–900

Mijovski JZ, Poljsak-Prijatelj M, Steyer A, Barlic-Maganja D, Koren S (2010) Detection and molecular characterisation of noroviruses and sapoviruses in asymptomatic swine and cattle in Slovenian farms. Infect Genet Evol 10(3):413–420

Newman KL, Leon JS (2015) Norovirus immunology: of mice and mechanisms. Eur J Immunol 45(10):2742–2757

Nims R, Plavsic M (2013) Inactivation of caliciviruses. Pharmaceuticals 3:358–392

Ntafis V, Xylouri E, Radogna A, Buonavoglia C, Martella V (2010) Outbreak of canine norovirus infection in young dogs. J Clin Microbiol 48(7):2605–2608

Oliver SL, Wood E, Asobayire E, Wathes DC, Brickell JS, Elschner M, Otto P, Lambden PR, Clarke IN, Bridger JC (2007) Serotype 1 and 2 bovine noroviruses are endemic in cattle in the United kingdom and Germany. J Clin Microbiol 45(9):3050–3052

Otto PH, Clarke IN, Lambden PR, Salim O, Reetz J, Liebler-Tenorio EM (2011) Infection of calves with bovine norovirus GIII.1 strain Jena virus: an experimental model to study the pathogenesis of norovirus infection. J Virol 85(22):12013–12021

Park SI, Jeong C, Kim HH, Park SH, Park SJ, Hyun BH, Yang DK, Kim SK, Kang MI, Cho KO (2007) Molecular epidemiology of bovine noroviruses in South Korea. Vet Microbiol 124(1-2):125–133

Park SI, Park DH, Saif LJ, Jeong YJ, Shin DJ, Chun YH, Park SJ, Kim HJ, Hosmillo M, Kwon HJ, Kang MI, Cho KO (2009) Development of SYBR Green real-time RT-PCR for rapid detection, quantitation and diagnosis of unclassified bovine enteric calicivirus. J Virol Methods 159(1):64–68

Patel MM, Hall AJ, Vinje J, Parashar UD (2009) Noroviruses: a comprehensive review. J Clin Virol 44(1):1–8

Pinto P, Wang Q, Chen N, Dubovi EJ, Daniels JB, Millward LM, Buonavoglia C, Martella V, Saif LJ (2012) Discovery and genomic characterization of noroviruses from a gastroenteritis outbreak in domestic cats in the US. PLoS One 7(2):e32739

Reuter G, Biro H, Szucs G (2007) Enteric caliciviruses in domestic pigs in Hungary. Arch Virol 152(3):611–614

Scheuer KA, Oka T, Hoet AE, Gebreyes WA, Molla BZ, Saif LJ, Wang Q (2013) Prevalence of porcine noroviruses, molecular characterization of emerging porcine sapoviruses from finisher swine in the United States, and unified classification scheme for sapoviruses. J Clin Microbiol 51(7):2344–2353

Scipioni A, Bourgot I, Mauroy A, Ziant D, Saegerman C, Daube G, Thiry E (2008a) Detection and quantification of human and bovine noroviruses by a TaqMan RT-PCR assay with a control for inhibition. Mol Cell Probes 22(4):215–222

Scipioni A, Mauroy A, Vinje J, Thiry E (2008b) Animal noroviruses. Vet J 178(1):32–45

Shen Q, Zhang W, Yang S, Chen Y, Ning H, Shan T, Liu J, Yang Z, Cui L, Zhu J, Hua X (2009) Molecular detection and prevalence of porcine caliciviruses in eastern China from 2008 to 2009. Arch Virol 154(10):1625–1630

Shen Q, Zhang W, Yang S, Yang Z, Chen Y, Cui L, Zhu J, Hua X (2012) Recombinant porcine norovirus identified from piglet with diarrhea. BMC Vet Res 8:155

Silva PF, Alfieri AF, Barry AF, de Arruda LR, Gardinali NR, van der Poel WH, Alfieri AA (2015) High frequency of porcine norovirus infection in finisher units of Brazilian pig-production systems. Trop Anim Health Prod 47(1):237–241

Soma T, Nakagomi O, Nakagomi T, Mochizuki M (2015) Detection of Norovirus and Sapovirus from diarrheic dogs and cats in Japan. Microbiol Immunol 59(3):123–128

Song YJ, Yu JN, Nam HM, Bak HR, Lee JB, Park SY, Song CS, Seo KH, Choi IS (2011) Identification of genetic diversity of porcine Norovirus and Sapovirus in Korea. Virus Genes 42(3):394–401

Souza M, Azevedo MS, Jung K, Cheetham S, Saif LJ (2008) Pathogenesis and immune responses in gnotobiotic calves after infection with the genogroup II.4-HS66 strain of human norovirus. J Virol 82(4):1777–1786

Sugieda M, Nagaoka H, Kakishima Y, Ohshita T, Nakamura S, Nakajima S (1998) Detection of Norwalk-like virus genes in the caecum contents of pigs. Arch Virol 143(6):1215–1221

Takano T, Kusuhara H, Kuroishi A, Takashina M, Doki T, Nishinaka T, Hohdatsu T (2015) Molecular characterization and pathogenicity of a genogroup GVI feline norovirus. Vet Microbiol 178(3-4):201–207

Thomas C, Jung K, Han MG, Hoet A, Scheuer K, Wang Q, Saif LJ (2014) Retrospective serosurveillance of bovine norovirus (GIII.2) and nebovirus in cattle from selected feedlots and a veal calf farm in 1999 to 2001 in the United States. Arch Virol 159(1):83–90

Tse H, Lau SK, Chan WM, Choi GK, Woo PC, Yuen KY (2012) Complete genome sequences of novel canine noroviruses in Hong Kong. J Virol 86(17):9531–9532

van der Poel WH, Vinje J, van der Heide R, Herrera MI, Vivo A, Koopmans MP (2000) Norwalk-like calicivirus genes in farm animals. Emerg Infect Dis 6(1):36–41 van der Poel WH, van der Heide R, Verschoor F, Gelderblom H, Vinje J, Koopmans MP (2003) Epidemiology of Norwalk-like virus infections in cattle in The Netherlands. Vet Microbiol 92(4):297–309

Vinje J (2015) Advances in laboratory methods for detection and typing of norovirus. J Clin Microbiol 53(2):373–381

Wang QH, Han MG, Cheetham S, Souza M, Funk JA, Saif LJ (2005) Porcine noroviruses related to human noroviruses. Emerg Infect Dis 11(12):1874–1881

Wang QH, Souza M, Funk JA, Zhang W, Saif LJ (2006) Prevalence of noroviruses and sapoviruses in swine of various ages determined by reverse transcription-PCR and microwell hybridization assays. J Clin Microbiol 44(6):2057–2062

Wang QH, Costantini V, Saif LJ (2007) Porcine enteric caliciviruses: genetic and antigenic relatedness to human caliciviruses, diagnosis and epidemiology. Vaccine 25(30):5453–5466

Widdowson MA, Rockx B, Schepp R, van der Poel WH, Vinje J, van Duynhoven YT, Koopmans MP (2005) Detection of serum antibodies to bovine norovirus in veterinarians and the general population in the Netherlands. J Med Virol 76(1):119–128

Wilkes G, Brassard J, Edge TA, Gannon V, Gottschall N, Jokinen CC, Jones TH, Khan IU, Marti R, Sunohara MD, Topp E, Lapen DR (2014) Long-term monitoring of waterborne pathogens and microbial source tracking markers in paired agricultural watersheds under controlled and conventional tile drainage management. Appl Environ Microbiol 80(12):3708–3720

Wobus CE, Karst SM, Thackray LB, Chang KO, Sosnovtsev SV, Belliot G, Krug A, Mackenzie JM, Green KY, Virgin HW (2004) Replication of Norovirus in cell culture reveals a tropism for dendritic cells and macrophages. PLoS Biol 2(12):e432

Wobus CE, Thackray LB, Virgin HW (2006) Murine norovirus: a model system to study norovirus biology and pathogenesis. J Virol 80(11):5104–5112

Wolf S, Williamson W, Hewitt J, Lin S, Rivera-Aban M, Ball A, Scholes P, Savill M, Greening GE (2009) Molecular detection of norovirus in sheep and pigs in New Zealand farms. Vet Microbiol 133(1-2):184–189

Woode GN, Bridger JC (1978) Isolation of small viruses resembling astroviruses and caliciviruses from acute enteritis of calves. J Med Microbiol 11(4):441–452

Yilmaz H, Turan N, Altan E, Bostan K, Yilmaz A, Helps CR, Cho KO (2011) First report on the phylogeny of bovine norovirus in Turkey. Arch Virol 156(1):143–147

Zheng DP, Ando T, Fankhauser RL, Beard RS, Glass RI, Monroe SS (2006) Norovirus classification and proposed strain nomenclature. Virology 346(2):312–323

Caprine Arthritis-Encephalitis

7

Michelle Macugay Balbin and Claro Niegos Mingala

7.1 Introduction

Caprine arthritis-encephalitis (CAE) is a persistent disease in goats caused by caprine arthritis-encephalitis virus (CAEV), a small ruminant lentivirus (SRLV), of the *Retroviridae* family. The *Retroviridae* consist of non-oncogenic viruses that produce multi-organ disease characterized by long incubation period and persistent infection. In general, the affected animal does not show clinical signs of infection due to very low seroconversion, which may take several months to years. The common clinical signs of CAEV infection are polyarthritis and mastitis in adult goats and leukoencephalitis or progressive paresis among kids. Pneumonia can also be observed in infected animals (Clements and Zink 1996; Smith and Sherman 2009). This virus is known to be transmitted mainly through ingestion of virus-infected colostrum or milk and horizontal transmission through direct contact with the infected animal. Currently, there is no known treatment or vaccine for this virus. The spread of CAEV is usually controlled by eradicating the animals infected with the virus.

The occurrence of CAEV infection has been reported worldwide. Importation of goats from countries where it is reported is believed to be the main reason for its widespread. The CAEV causes substantial loss especially in the dairy industry due to the occurrence of mastitis. It is known that mastitis decreases the production performance of the infected animal in terms of milk quantity and quality.

Currently, different diagnostic techniques and methods are used in the detection of the virus among goats. The Office International des Epizooties (OIE)-approved tests for CAEV in goats for international trade are agar gel immunodiffusion test (AGID) and enzyme-linked immunosorbent assay (ELISA) (OIE 2008; Hermann

M.M. Balbin (✉) • C.N. Mingala (✉)
Animal Health and Biosafety Unit, Philippine Carabao Center, National Headquarters and Genepool, Science City of Muñoz, Nueva Ecija 3120, Philippines
e-mail: michellebalbin218@gmail.com; cnmingala@hotmail.com

J. Bayry (ed.), *Emerging and Re-emerging Infectious Diseases of Livestock*,
DOI 10.1007/978-3-319-47426-7_7

and Hoesing 2010; OIE 2012). The abovementioned tests have been proven to be useful in the detection and control of the spread of the virus. Virus isolation on the other hand is also useful but may take several days to weeks before obtaining a result. Molecular tests such as polymerase chain reaction (PCR) are also becoming common as diagnostic method in most of the laboratories due to its sensitivity as well as specificity.

7.2 Caprine Arthritis-Encephalitis Virus: General Characteristics

7.2.1 Taxonomical Classification

CAEV is a SRLV, which belongs to the genus *Lentivirus*, subfamily *Orthoretrovirinae*, of the family *Retroviridae*. Other viruses that are classified under *Lentivirus* are human immunodeficiency virus (HIV), simian immunodeficiency virus (SIV), feline immunodeficiency virus (FIV), bovine leukemia virus (BLV), equine infectious anemia virus (EIAV), and maedi-visna virus (MVV) (Minguijon et al. 2015). The MVV is the closest relative of the CAEV since both these viruses are SRLV (Gjerset et al. 2006; Santry et al. 2013). Like other viruses that belong to the *Retroviridae* family, the CAEV is also characterized by the ability to reverse transcribe viral RNA to double-stranded DNA (dsDNA) through the action of reverse transcriptase (RT) (Minguijon et al. 2015). This virus has a high degree of genetic variability, resulting from factors such as low reverse transcriptase fidelity, lack of 3′ exonuclease proofreading capabilities, and repair mechanisms during replication (Zanoni 1998).

7.2.2 Properties of Caprine Arthritis-Encephalitis Virus

In general, the SRLVs have a diameter of 80–100 nm (Minguijon et al. 2015). It is comprised of nucleocapsid (NC), capsid (CA), matrix (MA), and envelope (ENV). The NC contains genome of the virus (two positive single-strand RNA chains of 8.4–9.2 kb), 30 molecules of RT, and the enzymes protease (PR) and integrase (IN). The CA surrounds the NC, which has icosahedral morphology about 60 nm in diameter and is surrounded by MA. The virion has an outer envelope which is formed by a lipid bilayer of cellular origin and the ENV glycoproteins: transmembrane (TM) and surface (SU) (Petursson et al. 1992; Murphy et al. 1999; Munguijon et al. 2015). According to Petursson et al. (1992), the lentiviruses are composed of approximately 60 % protein, 35 % lipids, 3 % carbohydrates, and 1 % RNA.

The CAEV encodes three structural genes, *gag* (group-specific antigen), *pol* (polymerase), and *env* (envelope), and accessory genes, *tat*, *rev*, and *vif* (Abelson and Schoborg 2003; Gjerset et al. 2006; Reina et al. 2009; Santry et al. 2013; Munguijon et al. 2015). These structural genes are being flanked by the noncoding long terminal repeat (LTR) at both ends (Santry et al. 2013; Munguijon et al. 2015) (Fig. 7.1).

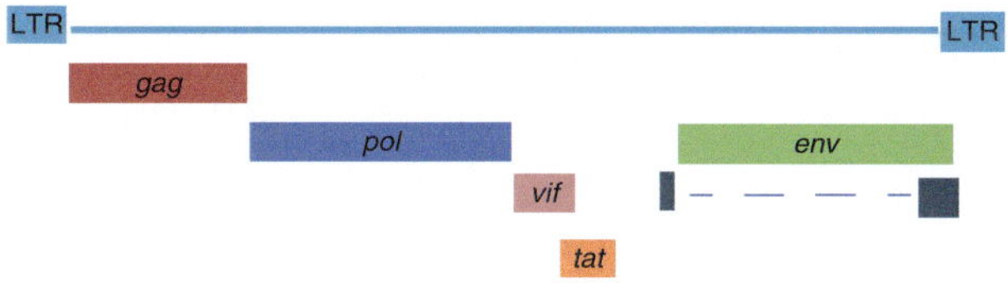

Fig. 7.1 The genome structure of CAEV

The *gag* gene is expressed as Gag polyprotein (p55^{gag} kDa). This gene is the precursor for virion capsid proteins (CA, p28 kDa), the matrix protein (MA, p19 kDa), and the nucleocapsid protein (NP or NC, p16 kDa) (Abelson and Schoborg 2003; Munguijon et al. 2015). The *pol* gene encodes the virion enzymes PR, RT, RNase H, IN, and dUTPase. The dUTPase subunit has been found to be significant in transcription and protein synthesis during viral replication (Clements and Zink 1996; Reina et al. 2009). The *env* on the other hand encodes for the envelope glycoproteins gp42 and gp135 or SU (Abelson and Schoborg 2003). Antigenic variation in the SU protein occurs as a result of *env* mutation and is responsible for the biological and serological variation of CAEV isolates (Concha-Bermejillo and Dela 2003).The accessory genes *tat* and *rev* are encoded by overlapping reading frames (ORFs) of the CAEV. The *tat* gene or the *vpr*-like gene increases the viral load, tissue distribution, and inflammatory lesion in the infected animal (Reina et al. 2009). The *rev* gene is known to encode a protein necessary for transport of unspliced and singly spliced viral RNA (*gag*, *pol*, and e*nv*) from the nucleus to the cytoplasm (Abelson and Schoborg 2003). The *vif* plays a significant role in the virus replication as it protects the viral genome from attack of the host's cellular defenses. The LTR is important in establishing the provirus in the host cell DNA and contains signals that are involved in expressing the viral genes. Deletions and mutation in the LTR have shown to be associated with variation in virulence or pathogenicity of the virus (Reina et al. 2009).

The proviral DNA of the CAEV is transcribed and produced in the cell nucleus of the infected cell. According to previous reports, the infected cell contains few proviruses that are integrated in the host cell DNA. The reverse transcription is prone to errors (Munguijon et al. 2015), which contributes to mutation and deletions and explains why there is a great divergence that forms different strains and quasispecies in the CAEV (Santry et al. 2013). Despite this high rate of mutation, the structural genes *gag* and *pol* are conserved among CAEV strains (L'Homme et al. 2011; Munguijon et al. 2015). This is the reason why these genes are often used in investigation of the virus phylogeny and evolution (Reina et al. 2009) and in development of molecular and serological-based diagnostic techniques (L'homme et al. 2011).

7.2.3 Virus Tropism

The CAEV is SRLV that mainly infects monocytes and dendritic cells in goats (Gendelman et al. 1985). CAEV is typically transmitted by direct contact specifically through ingestion of virus-infected colostrum (De Boer et al. 1979; Greenwood et al. 1995; Abelson and Schoborg 2003; Blacklaws et al. 2004; Peterhans et al. 2004;

Shah et al. 2004; de Andre's et al. 2005; APHIS 2007; CFSPH 2007; Martinez-Navlon et al. 2007; Brajon et al. 2012; Elfahal et al. 2013) and horizontal transmission through respiratory exudates (Blacklaws et al. 2004). CAEV is associated with slow and persistent inflammatory diseases like encephalitis, pneumonitis, arthritis, and mastitis (Zanoni et al. 1998; Minguijon et al. 2015).

Elfahal et al. (2013) reported that lymph nodes are important reservoirs for the virus. They show moderate virus replication relative to other lentiviruses affecting primates. However, the mammary gland and milk cells are still the preferred organ for replication of CAEV.

Though CAEV has a tropism to monocyte-macrophage cell lineage (Narayan et al. 1982, 1983; Reddy et al. 1993; de Andre's et al. 2005; Kaba et al. 2011; Ali Al Ahmad et al. 2012; Brajon et al. 2012; Munguijon et al. 2015), the bone marrow serves as a reservoir of infection (Gendelman et al. 1985). This lentivirus replicates in macrophages but does not infect the monocytes (Munguijon et al. 2015). These viruses replicate slowly (Martinez-Navlon et al. 2007; Munguijon et al. 2015), and animals may be latently infected for many years before developing clinical signs of infection (de Andre's et al. 2005; Martinez-Navlon et al. 2007; Brajon et al. 2012).

7.3 Epidemiology of Caprine Arthritis-Encephalitis

7.3.1 Disease Distribution

The neurologic form of CAEV was first described in the United States by Cork et al. in 1974 while the causative virus was first isolated from a case of arthritis in 1980 (Crawford et al. 1980). The disease caused by this virus was initially termed as leukoencephalomyelitis of goats, but through continuous investigation and research, it was changed to caprine arthritis-encephalitis (Smith and Sherman 2009).

The transmission of CAEV has been reported in different regions. The movement of CAEV-infected animals from one country to another resulted to spread of this virus throughout the world. Similarity between South African sheep lentivirus and European strains such as EV1 provides phylogenetic evidence for the introduction of SRLV through animal trade. Introduction of disease was reported when Iceland had imported live animals in their country. SRLV transmission was also reported due to importation of live animals from Denmark to Norway, Scotland to Canada, England to Hungary, Holland to France, and Sweden to Finland (Peterhans et al. 2004).

In earlier studies, evidence of SRLV infection was reported in Kenya (Wander 1970), Morroco (Mahin et al. 1984), Nigeria (Belino and Ezeifeka 1984), Mozambique (Pereira et al. 1989) and Algeria (De Boer et al. 1979). The infection of SRLV in Africa was due to importation of goats from Europe for genetic improvement program. Similarly, import of French Alpine goats from France was reported to be an underlying reason for the severe outbreak of CAEV in Spain during 1984 (APHIS 2007).

It is also reported that Japan (Konishi et al. 2004) and Mexico (Torres-Acosta et al. 2003) imported their goats prior to first report of disease in those countries. Investigation of CAEV outbreak in Mexico points out to importation of goats from the United States. Poland has also imported goats from France during the 1990s. Investigation of CAEV in one Polish herd containing imported goats showed 75 % prevalence compared to 15 % prevalence in herds native to Poland (APHIS 2007).

Live animal trade is a major risk factor in the dispersion of CAEV from one herd to another (APHIS 2007). This supports the claim or report that the transmission of the virus was due to importation of breeds or stocks from one place to another without prior testing (De Boer et al. 1979; Adams et al. 1980).

The prevalence of the CAEV varies in each country (APHIS 2007), but it was observed that the infection is highest where there is intensive dairy management (Grewal 1986; Cutlip et al. 1992). This virus is rarely found in the indigenous breeds unless they were exposed or had a direct contact with the imported goats (Waseem et al. 2015; CFSPH 2007).

According to Smith and Sherman (2009), countries reported to have high prevalence of CAEV infection include Australia (82 %) (Grewal 1986), United States (31–81 %) (Crawford and Adams 1981; Cutlip et al. 1992; Sanchez et al. 2001), Norway (49.5 %) (Nord et al. 1998), Brazil (14.1–36.5 %) (Lilenbaum et al. 2007; Bandeira et al. 2009; Martins et al. 2012), Spain (20.6 %) (Sanchez et al. 2001), Japan (63.3 %) (Konoshi et al. 2004), and Poland (12.1 %) (Kuzmak et al. 2007). Lower to high prevalence was also reported using cELISA in Italy (6.58 %) (Gufler et al. 2007), the UK (4.3 % according to Dawson and Wilesmith 1985 and 54.5 % as per Synge and Ritchie 2010), Switzerland (2 % as per Krieg and Peterhans 1990 and 1 % according to CFSPH 2007), Turkey (1.9 %) (Burgu et al. 1994), Mexico (0.4 %) (Torres-Acosta et al. 2003), and China (0.2–30 %) (Qu et al. 2005). According to Peterhans et al. (2004), data on the seroprevalence studies show that no European country can be considered free from SRLV infection following the definition of the OIE (<1 % of herds infected, probability 99 %).

Investigation in five states of Sudan revealed 5.8 % seroprevalence of CAEV (Elfahal et al. 2013). In Somalia and Jordan, the seroprevalence of CAEV was reported to be 6 % and 23.2 % (Ghanem et al. 2009; Al-Qudah et al. 2006).

Recent studies conducted in the eastern provinces of Philippines showed 3.03 % seroprevalence (Lluz et al. 2011), while Gonzales et al. (2013) reported 5.73 % prevalence in samples collected in Luzon island, Philippines. In Thailand, the reported seroprevalence was 31 % (Thant et al. 2011), while 2.73 % was reported in Korea (Oem et al. 2012).

CAEV is more commonly reported among dairy goat breeds. This is attributed to genetics, management, and farming practices. The feeding of milk from a single dam to a large number of kids is one of the practices that is considered as a risk. The prevalence is also affected by age but not by sex. Al-Qudah et al. (2006) reported that CAEV seroprevalence in Jordan was observed highest in goats older than 3 years and younger than 6 years of age.

7.3.2 Genetic Diversity of Caprine Arthritis-Encephalitis Virus

The CAEV infection has been reported worldwide. The spread of this virus is believed to be caused by continuous importation of CAEV-infected goats. The cross-species infection between goats and sheep with CAEV and MVV, respectively, has been reported due to their close relatedness.

Sequence analysis of the gag-pol regions of the virus revealed five principal sequence groups, A–E, which differs 25–37 % in the genetic sequence (Shah et al. 2004; Grego et al. 2007). Group A is related to the prototype MVV (Shah et al. 2004; Grego et al. 2007). This is composed of a large heterogenous group and is further divided into ten subtypes, A1–A10 (Shah et al. 2004; Grego et al. 2007; Pisoni et al. 2010). Group B is related to the CAEV prototype and is further divided into 3 subtypes, B1–B3 (Shah et al. 2004; Bertolotti et al. 2011). Group C is comprised of a sequence of CAEV isolated in Norway (Gjerset et al. 2006). The CAEV sequence from Switzerland and Spain belongs to group D (Shah et al. 2004; Reina et al. 2006), while those from the northern part of Italy is clustered in Group E (Reina et al. 2006; Grego et al. 2007). According to Muz et al. (2012), ancestors of SRLVs have been located in Turkey, and establishment of different strains was attributed to the migration of sheep from Middle East to Europe (Munguijon et al. 2015).

The wide diversity is known to be a characteristic of lentiviruses. Knowledge on the antigenic variation of CAEV is essential as it may affect the reliability of diagnostic tests (Santry et al. 2013; Munguijon et al. 2015), as well as molecular-based epidemiological studies (Shah et al. 2004; Gjerset et al. 2006; Munguijon et al. 2015). This diversity is generated through continuous mutation and selection pressure by the host's immune system.

7.4 Pathogenesis of Caprine Arthritis-Encephalitis

The pathogenesis of CAEV is not fully understood. However, SRLVs such as CAEV mainly infect the leukocytes (Kaba et al. 2011). The infected monocytes and macrophages in the colostrum and milk are absorbed intact through the mucosa (Smith and Sherman 1994). The virus is disseminated throughout the body via infected mononuclear cells in the bloodstream (de Andre's et al. 2005). The differentiation of the monocyte into macrophage triggers the viral transcription (Minguijon et al. 2015).

Periodic viral replication and macrophage maturation induce characteristic lymphoproliferative lesions in target tissues such as lungs, synovium, choroid plexus, and udder (Ravazollo et al. 2006; Minguijon et al. 2015). The replicative capacity and cytopathic effects of SRLVs are highly variable (Minguijon et al. 2015). CAEV is a slow growing virus as it replicates very slow and yields low titer (Peterhans et al. 2004). However, high virus production can be achieved in vitro in macrophages and choroid plexus (Glaria et al. 2009).

Persistence of CAEV in the host is facilitated by its ability to become sequestered as provirus in host cells. Only small amount of infected goats develop or show clinical signs of the infection. Clinical signs of infection include progressive arthritis, which would result to lameness and eventually recumbency, indurative mastitis with hypogalactia, chronic interstitial pneumonia, and wasting syndrome (Kaba et al. 2011).

7.4.1 Transmission of the Disease

CAE can be described as infectious, multisystemic disease that affects different ages of the caprine independent of gender, breed, or production status (Brajon et al. 2012; Lara et al. 2005). Understanding the route of transmission is significant enough in establishing the prevalence as well as management and control programs within a herd (Peterhans et al. 2004).

In most cases, the virus is transmitted during postnatal period when kid ingests CAEV-contaminated colostrum and milk (Abelson and Schoborg 2003; de Andre's et al. 2005; APHIS 2007; CFSPH 2007; Shah et al. 2004; Blacklaws et al. 2004; Martinez-Navlon et al. 2007; Peterhans et al. 2004; Brajon et al. 2012; Elfahal et al. 2013). The presence of virus in the mammary gland and lacteal secretions increases the probability of CAEV to be transferred from dam to kid (Bolea et al. 2006; Pisoni et al. 2010).

7.4.1.1 Horizontal Transmission of Caprine Arthritis-Encephalitis

Horizontal transmission has been reported as a result of close contact with infected animals (Adams et al. 1983; Narayan and Cork 1985; Rowe and East 1997; Gufler 2004; de Andre's et al. 2005; Ali Al Ahmad et al. 2008b; Brajon et al. 2012). Exposure to respiratory tract secretions of the infected animals was also reported as a mean to transfer the virus (Blacklaws et al. 2004; Martinez-Navlon et al. 2007), as the lung is one of the major target organs of this virus (Peterhans et al. 2004). Aerosol transmission between animals in close contact or even up to few meters appears to be significant in transfer of the virus (Rowe et al. 1997; Peterhans et al. 2004). Recent studies suggest that this virus can be isolated and possibly be transmitted through other body fluids (Ali Al Ahmad et al. 2008b, 2012). Transmission by contaminated milking machines, use of contaminated needles, tattoo and surgical equipment, and even humans as a fomite, has been documented (CFSPH 2007).

CAEV is noted to have a tropism for the monocyte and macrophage cell lineage and causes chronic inflammatory disease (Narayan et al. 1983). All body secretions and excretions therefore can be a potential source of infection (Ali Al Ahmad et al. 2012), and so vertical transmission to kid from dam or horizontally by prolonged exposure or contact between infected and healthy animals can occur (Adams et al. 1983; Narayan and Cork 1985; Rowe and East 1997; Gufler 2004; de Andre's et al. 2005; Ali Al Ahmad et al. 2008b; Brajon et al. 2012).

A recent study conducted by Villoria et al. (2013) reported detection of SRLV in environmental samples of air and water. The source of samples in this research was drinking water, pen air, and air exhaled from the infected animals. According to this study, the frequency of detection was highest in samples from bowls, which could be due to smaller volume of liquid, therefore higher concentrations of saliva and nasal fluid. CAEV was also detected in the air samples from animals that are due for culling (Villoria et al. 2013). The result of the study implicates that there is indeed a high probability of transmitting the virus horizontally or among animals within the herd.

7.4.1.2 Vertical Transmission of Caprine Arthritis-Encephalitis

Vertical transmission, on the other hand, is still unclear (Fieni et al. 2003). Some researchers reported the possibility of vertical transmission of CAEV during pregnancy of the infected goats. A study conducted by Martinez-Navlon et al. (2007) revealed that transplacental infection is of little significance in the transmission of CAEV from dam to kid. This finding is in agreement to the result of the study conducted by Cork and Narayan (1980).

The risk of transmission of the virus is also seen among selected goat breed. This breed of goats has been selected for genetic improvement programs. However, pathogens can also be transmitted using these animals. Less efficient way of transmission includes sexual transmission, which is dependent on the presence of virus-infected cells within the genital tracts of the breeding animals (Ali Al Ahmad et al. 2012).

The presence of caprine arthritis-encephalitis was detected from experimentally infected male goats. CAEV proviral DNA was identified in the preputial cells, seminal plasma, and in the non-spermatic cells of the ejaculate (Travassos et al. 1999). This was confirmed when Ali Al Ahmad et al. (2008a, b) detected CAEV in the semen and genital tract tissues of male goats. These findings therefore support the first step in transmission of CAEV through sexual contact (Ali Al Ahmad et al. 2008a, b). The possible carriers of the virus were identified to be the non-spermatic cells in the semen (Peterson et al. 2008). These cells, which include monocytes and macrophages, were tested positive for CAEV proviral DNA (Ali Al Ahmad et al. 2008b).

Lentiviruses such as HIV and FIV were shown to be transferred via semen. This was also reported in the case of BLV, wherein the virus was detected in semen. Therefore, semen or artificial insemination could transfer the virus and that artificial insemination (AI) poses a great risk in the rapid transmission of the virus (Ali Al Ahmad et al. 2012; de Souza et al. 2013). Semen contaminated with viral agents can infect numerous areas in a short period of time. The presence of CAEV in semen and genital tract of naturally infected bucks was reported (Andrioli et al. 1999; Ali Al Ahmad et al. 2008b; Peterson et al. 2008) and, thus, supports the theory that CAEV may be transmitted sexually.

Transmission of CAEV was also investigated from semen samples intended for AI. The result of this study confirmed the use of AI with infected semen or mating with infected bucks may result in the transmission of CAEV to the does' genital tract 7 days after insemination. However, it is still possible to obtain CAEV-free early embryos usable for embryo transfer irrespective of the medical status of the semen or the recipient doe (Ali Al Ahmad et al. 2012).

A study conducted by Fieni et al. (2003) identified CAEV proviral DNA in the genital tissues such as the uterus, oviduct, and ovary and even in the cumulus oophorus cells in the oocytes. In bucks, CAEV proviral DNA was detected in the preputial cells, seminal fluids, and non-spermatic cells that were found in the ejaculate of an experimentally infected buck. The CAEV proviral DNA was also detected in the uterine and uterine flushing does inseminated with infected buck's semen, whereas it was not detected in CAEV-free semen. This result suggests that CAEV can be potentially transmitted to the does via mating or AI (Ali Al Ahmad et al. 2012; de Souza et al. 2013).

In goats, it was also reported that there is higher incidence of infection in does serviced with AI compared to natural mating. This could be due to absence or lack of natural defensive barriers (the cervix and vagina) and higher concentration of infectious articles in the semen during AI (Lamara et al. 2001).

Despite detection of CAEV proviral DNA in the uterine and uterine flush as well as in the semen of goats used in AI, the embryo remained or was tested negative for the virus itself (Ali al Ahmad et al. 2008a). The postulated reason behind was the presence of zona pellucida (ZP), which surrounds the early embryonic cell (Eggert-Krusse et al. 2000). The same was reported by Lamara et al. (2002) and Ali Al Ahmad et al. (2008), which only supports that the ZP is a very efficient protective barrier against CAEV infection.

It should still be considered that semen or AI poses a risk in transmitting CAEV (de Souza et al., 2013). Further studies are still needed to assess whether transmission of CAEV via AI or natural mating results in infection of does or even in the developing embryo (Ali Al Ahmad et al. 2008b).

7.4.2 Host Response to Caprine Arthritis-Encephalitis

Infection induces a strong humoral and cell-mediated immune response, but neither is protective. Ponti et al. (2008) investigated the phenotypic composition of leukocyte in blood and milk during lactation. It was observed that the $CD8^+$ cells were more numerous in CAEV-positive goats compared to negative ones. Dias et al. (2012) reported that the seropositive animals showed morphological changes in red blood cells, with significant decrease in MCV, coupled with a significant increase of MCH and MCHC. The absolute values of monocytes and eosinophil of the seropositive animals were significantly higher than the seronegative ones and above the normal range of goats. Infection with CAEV stimulates a chronic, intense, inflammatory response affecting the brain, lungs, joints, and mammary gland, which is ineffective in eliminating the virus (Narayan and Cork 1985).

The antibodies against CAEV are produced after infection but seroconversion may require few weeks to several months (Elfahal et al. 2013). This explains why animals affected with the virus may or may not show clinical signs of the infection. Viral load also varies from animal to animal, which may affect the replication of the virus. It was also observed that the lesions depend on the viral load on a specific site during infection. Genetics, on the other hand, is also being considered as it can affect the susceptibility and/or resistance of the animal to CAEV infection (Elfahal et al. 2013).

The development and manifestations of clinical signs in CAEV infection is highly variable. Most goats become infected when they are young and develop disease months or years later (CSFPH 2007).

7.4.3 Caprine Arthritis-Encephalitis: Clinical Signs and Lesions

The infection with CAEV produces multi-organ diseases characterized by long incubation period and persistent infection. Most infected goats remain asymptomatic, and only about 20 % of animals with CAE show clinical signs of the disease. The common clinical signs of CAEV infection are polyarthritis and mastitis in adult goats and leukoencephalitis or progressive paresis among kids. Pneumonia can also be observed in infected animals (Clements and Zink 1996; Smith and Sherman 2009).

7.4.3.1 Arthritic Form of Caprine Arthritis-Encephalitis

The arthritic form is the most common manifestation of the infection. This is generally observed among sexually mature goats (> 6 months of age) (Pugh 2002; Smith and Sherman 1994). The arthritis in CAEV-infected animal tends to be chronic and progressive. The carpal, tarsal, stifle, fetlock, atlanto-occipital, and coxofemoral joints are commonly affected with the infection. Signs of arthritis in CAEV-infected animal include stiffness, shifting leg lameness, decreased ambulation, weight loss, reluctance to rise, and abnormal posture. More severe signs of arthritis include acute swelling without pain upon palpation but eventually lead to painful arthritis (Knight and Joniken 1982; Mathews 1999; Pugh 2002; Smith and Sherman 1994).

De Souza et al. (2014) performed a comparative study of articular cartilage of normal goat and that of infected chronically with CAEV. It was observed that surface of the head of the humerus of positive animals had reduced joint space, increased bone density, and signs of degenerative joint diseases (DJD). It was noted that tarsus is the most affected bone, which is characterized by severe DJD, absence of joint space, increased periarticular soft tissue density, edema, and bone sclerosis. Histological analysis of chronic tissue lesions in the CAE-positive animals revealed complete loss of surface zone, the absence of proteoglycan in transition and radial zones, and destruction of the cartilage surface (Fig. 7.2).

7.4.3.2 Encephalitic Form of Caprine Arthritis-Encephalitis

The encephalitic form of CAEV commonly affects kids between 2 and 6 months of age. The kid may show incoordination, gradual paresis, and paralysis, which commonly affect the hind limbs. Usually, the animal remains alert and responsive but then eventually shows signs of neurologic disorder such as depression, nystagmus, abnormal pupillary response, blindness, head tilt, head tremor, dysphagia, and torticollis (Knight and Joniken 1982; Mathews 1999; Pugh 2002; Smith and Sherman 1994). Invasion of the virus and inflammatory lesions of neuroparenchyma causes neurological signs seen in CAEV-infected animal. The encephalitic form of this infection was documented in a Boer breed of goat, which is less than 1 year of age. The CAEV infection was detected through cELISA (Fig. 7.3).

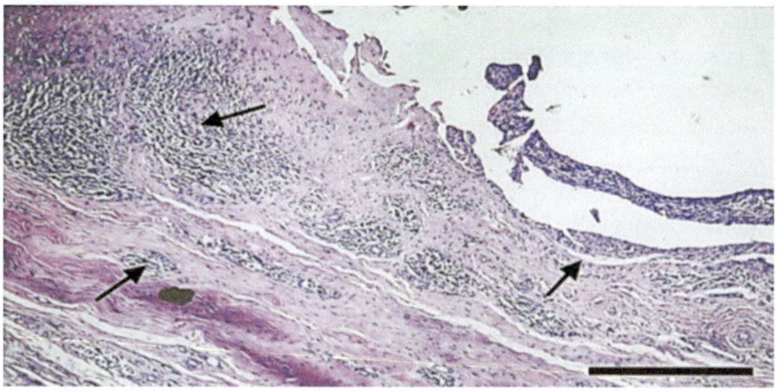

Fig. 7.2 Cross section of carpal joint showing mild to moderate synovial and subsynovial mononuclear cell reaction (*black arrows*) (100 × magnification)

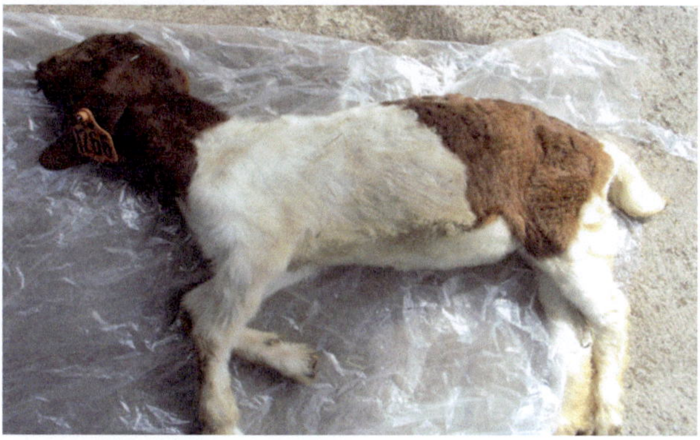

Fig. 7.3 Encephalitic from of CAEV. It is commonly observed among kids. This animal was first observed to have signs of neurological disorder and, eventually, became paralytic. cELISA kit was used to confirm the CAE infection in this animal

7.4.3.3 Mastitis and Respiratory Disease in Caprine Arthritis-Encephalitis

Interstitial mastitis is observed especially in high producing dairy goats. Clinical features of mastitis in CAEV-infected animals are firm and distended udder, swollen mammary gland, and agalactia at the time of parturition. Histopathological changes showed perivascular and gland parenchyma infiltration by mononuclear cells, with lymph proliferation and hyperplasia of lymphoid follicles that would eventually evolve to alveoli calcification and necrosis (Zwahlen et al. 1983; Gonzales et al. 1987; Cheevers and McGuire 1988; Perk 1988; Gregory et al. 2006). In general, the mammary glands of mastitic goats caused by CAEV are characterized by fibrosis of the lactiferous ducts, presence of inflammatory cells, and calcification foci (Gregory et al. 2009) (Fig. 7.4).

In a study conducted by Birgel Jr. et al. (2005), it was noted that healthy goats produce 25–31 % more milk, and lactation was 17 % longer as compared to infected goats. Hardening of the mammary gland was also reported in 19.68 % of the goats infected with CAEV. The substitution of gland parenchyma by connective tissue and fibrosis explains the decrease in milk yield and the quality of milk.

The respiratory syndrome was first believed to only occur in sheep but eventually recorded in goats. The clinical signs syndrome is caused by interstitial pneumonia

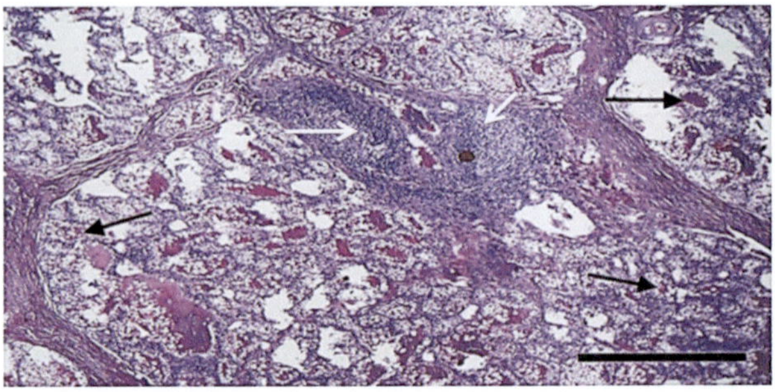

Fig. 7.4 Cross section of mammary gland showing moderate intralobular mononuclear cell reaction (*black arrows*) and focal lymphoid follicle formation (*white arrows*) (40 × magnification; H & E)

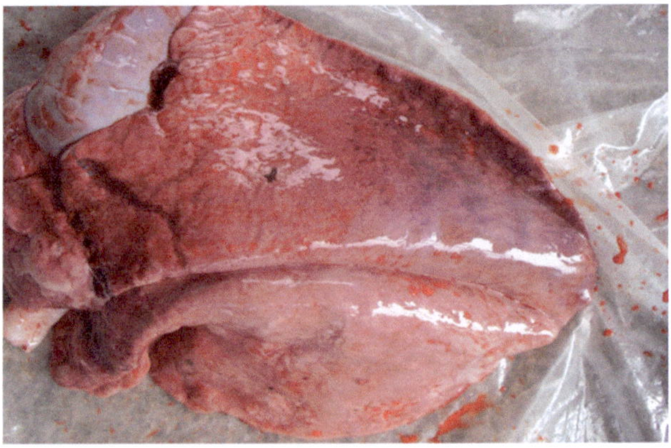

Fig. 7.5 Interstitial pneumonia in goats is one of the signs of CAEV infection. The lung is usually grayish pink in color, heavy, and filled with pleura

that leads to progressive dyspnea. On gross examination, lungs of the CAEV-infected goats are firm and are gray pink in color with multiple foci (Fig. 7.5).

7.5 Management and Control of Caprine Arthritis-Encephalitis

7.5.1 Methods of Detection of Caprine Arthritis-Encephalitis

CAEV screening and detection is done by a wide array of techniques that include virus isolation, serological tests, and molecular methods. Currently, there is no specified "gold standard" in the detection of CAEV, since most of the tests that are being used are efficient and have been valuable in the success of control programs and in reducing the infection prevalence (Peterhans et al. 2004). However, the OIE (2012) has prescribed ELISA and AGID tests for animals meant for international trade. The most practical and reliable approach in the detection of CAEV as indicated by the OIE is the combination of serology and clinical evaluation (OIE 2012).

7.5.1.1 Serological-Based Detection of Caprine Arthritis-Encephalitis

Serological diagnosis of CAEV infection in animals is considered convenient in the detection of the virus (de Andre's et al. 2005). Serological tests that are employed are RIPA, AGID, ELISA, and Western blot (Reddy et al. 1993; Peterhans et al. 2004; OIE 2008; OIE 2012).

AGID Test It is a very valuable serological-based test in the detection of CAEV infection. This test is specific but less sensitive compared to ELISA (Rimstad et al. 1994; Varea et al. 2001). According to Adams and Gorham (1986) and Knowles et al. (1994) as cited by the OIE (2012), the sensitivity of the AGID test is dependent on both viral strain and viral antigen used. The viral surface envelope glycoproteins gp135 and p28 are commonly used in the preparation of anti-CAEV antibody (Cutlip et al. 1977). Adams and Gorham (1986) demonstrated that AGID test utilizing the CAEV gp135 has greater sensitivity compared to p28. The AGID test has a high sensitivity (92 %) and specificity (100 %) when appropriate antigen is used in the test (Herman et al. 2003; OIE 2012). This test is described as simple but the interpretation of results is subjective and requires technical experts (Vitu et al. 1982; Rosati et al. 1994; Starick et al. 1995; Peterhans et al. 2004; OIE 2012).

ELISA Test It is more suitable whenever large numbers of samples are to be tested (Peterhans et al. 2004). This serological test is more sensitive compared to AGID test. Recombinant antigens, whole virus, recombinant proteins, and peptides used in ELISA are usually derived from the gag or env genes. The antigen preparation should contain gp135 and p28 proteins. Samples other than serum can also be subjected to ELISA test. Some studies investigated the agreement of ELISA when serum and milk or colostrums are used in the test (OIE 2012). ELISA is a convenient test compared to other serological tests since quantitative analysis of the raw data can be done using computer-based softwares (Peterhans et al. 2004; OIE 2012). AGID test is the most widely used even if ELISA is more convenient (OIE 2012) due to the reliability of the test results. According to Zanoni et al. (1994) as cited in OIE (2012), there is only one ELISA that has met the standards for serological testing of CAEV. Peterhans et al. (2004) emphasized the selection of appropriate peptides in the development of a reliable ELISA-based test.

Western Blotting and RIPA They are not routine diagnostic methods for CAEV. However sensitivity and specificity of these tests are comparable to ELISA and AGID tests (Peterhans et al. 2004; CFSPH 2007; OIE 2012).

All these tests were proven to be useful in the detection and control of the disease spread. But they have their own limitations and drawbacks (CFSPH 2007). The sensitivity of the serological-based tests is being compromised by the delay or slow seroconversion of infected animals. The seroconversion typically occurs several weeks to months after infection. Screening of CAEV may lead to false negative result, especially in animals with very low antibody titer. Detection of antibody in adult animals means persistent infection (CFSPH 2007). On the contrary, the maternal antibody transfer could lead to false-positive result (CFSPH 2007) especially in animals less than 6 months of age. In the development of serological-based test, it is important to consider the heterogeneity of viral strains. The serological test should be based on the strain present in the population or region (Peterhans et al. 2004).

7.5.1.2 Molecular-Based Methods for Detection of Caprine Arthritis-Encephalitis

The CAEV can be isolated in both live clinical and subclinical cases by obtaining peripheral blood or milk samples (Peterhans et al. 2004; CSFPH 2007; OIE 2012; Minguijon et al. 2015). During necropsy, samples such as the lung, choroid plexus, synovial membrane, and udder can be utilized for virus isolation (OIE 2012). These samples are cocultivated with fibroblasts derived from the synovial membrane cells, which are specific for CAEV (OIE 2012; Minguijon et al. 2015). Microscopic examination of the cytopathic effect, which appears as stellate cells, and presence of syncytia confirm CAEV infection (CFSPH 2007; OIE 2012).

The critical factors affecting virus isolation are source of samples, viral load at the time of obtaining the samples, and the biological and genetic characteristics of the virus (Peterhans et al. 2004).

With the advent of molecular diagnostics, PCR-based assay for the detection of CAEV is becoming more common (Herrman-Hoesing 2010) as rapidity of the assay aids in detection of infected goats even before antibody is detected using ELISA (Brinkof et al. 2007; CFSPH 2007). PCR has been widely used by different research institutions and became a routine diagnostic tool in laboratories. Molecular-based detection protocol aids in rapid detection, quantitation, and identification of the CAEV strain (OIE 2012). The CAEV detection using PCR utilizes DNA isolated from peripheral blood cells.

In a study by Brinkof et al. (2008), it was mentioned that the PCR, though a powerful molecular tool, has its own limitations. The sensitivity of PCR is being affected by viral load and genetic characteristics of the virus (Peterhans et al. 2004; Ali Al Ahmad et al. 2008b; Cruz et al. 2009; Minguijon et al. 2015). The PCR assay requires certain level of virus in the host to be detected using blood sample, while genomic heterogeneity among CAEV could explain why primers based from published articles sometimes do not work when used (Gjerset et al. 2006; Ali Al Ahmad et al. 2008b; Kaba et al. 2009; Gomez-Lucia et al. 2013). The use of other sources of samples such as milk, semen, and synovial fluid may vary the sensitivity of the test (Minguijon et al. 2015). A study by Cruz et al. (2009) reported that the presence of virus was not necessary concordant in semen and peripheral blood samples. It was reported that virus reservoir in semen can be distinct from that in the blood and probably no correlation at all. This contends the diagnostic reliability of PCR when it comes to detection of CAEV in sample collected from blood and semen.

Notomi et al. (2000) developed a molecular technique called loop-mediated isothermal amplification (LAMP). With this technique, amplification and detection of the target gene can be completed in a single step, and DNA can be amplified in 15–60 min with high efficiency and specificity (Notomi et al. 2000; Parida et al. 2006); hence, it is an alternative technique that requires short time and inexpensive machine. Recently, LAMP for the detection of CAEV has been developed targeting p25 gene region of the CAEV provirus (Huang et al. 2012). Balbin et al. (2014) also utilized LAMP for the detection of CAEV circulating in the Philippines. The conserved region in the *gag* gene was used in designing the specific LAMP primers for CAEV (Balbin et al. 2014).

Increasing the sensitivity and specificity of the PCR test involves choosing or designing of primers from conserved region of the virus. Having knowledge on the circulating local strain is also significant while using molecular-based methods (Peterhans et al. 2004; Minguijon et al. 2015; Padiernos et al. 2015).

7.5.2 Treatment and Control of Caprine Arthritis-Encephalitis

There is no treatment specifically indicated to CAEV infection. However, supportive treatment may benefit individual goats. Nonsteroidal anti-inflammatory drugs (NSAIDs) may be used in management of polysynovitis-arthritis. Secondary bacterial infection can be prevented by administration of antibiotics. Other management practices that may improve the condition of CAEV-infected goat include regular foot trimming and provision of additional bedding.

Unlike other virus, infection cannot be prevented using biologicals because of the absence of CAEV vaccine (Reina et al. 2013; Minguijon et al. 2015). There have been efforts to develop a vaccine against CAEV, but researchers were not successful. In one experiment, killed vaccine was used, but developed antibodies did not have protection against arthritis (APHIS 2007). The management and control of CAEV infection relies on different control measures recommended by health authorities.

Management practices greatly affect the prevalence of CAEV in a herd. Control measures have been implemented in many countries since SRLVs, particularly CAEV, were detected in their herd. These methods provided relevant information regarding control and management of CAEV. The following practices are recommended to be implemented in CAEV-infected farms: (1) permanent isolation of kids starting at birth; (2) feeding of heat-treated colostrum (45°C for 60 min); (3) frequent serologic testing of herd (semiannually), with segregation of seronegative and seropositive goats; and (4) culling of seropositive goats. With regard to segregation, it is recommended that the seropositive and seronegative animals must have a minimum distance of 1.8 m between pens during segregation (APHIS 2007; CPSPH 2007).

These control measures were initially implemented among European countries where prevalence was reported to be high during the 1980s. The Netherlands is one of the countries that adapted these protocols and became successful in the eradication of the virus. Switzerland started to eradicate CAEV since 1984, and the seroprevalence was reduced from 60–80 % to 1 %.

In areas where CAEV has not yet been reported, any new goat that has to be added to an uninfected herd must be screened prior to movement. The animal must also be quarantined before addition to the herd (CSFPH 2007). Mixing of sheep and goat in herd must also be avoided since cross-species infection has been reported. Lentiviruses were also found to be susceptible to lipid solvents, periodate, phenolic disinfectants, formaldehyde, and chemicals with low pH (<4.2).

7.5.3 Impact and Progress in the Management of the Disease

7.5.3.1 Economic Impact of the Caprine Arthritis-Encephalitis

In general, significant economic losses were recorded due to infection with CAEV. Peterhans et al. (2004) enumerated some factors that might influence economic losses with SRLV infection: (1) Clinical disease caused by SRLV infection develops slowly. (2) Only about 30 % of infected animals develop clinical disease. (3) Disease signs and economic losses are related to seroprevalence with no signs of evidence in herd with low prevalence of infection. (4) Genetic factor influences the extent of the disease. (5) Certain management can influence the rate of viral transmission. (6) Disease signs may be influenced by concurrent infections, geographical area, and management practices.

It is well established that subclinical mastitis has a big impact on the production performance in terms of the quantity and quality of meat and milk produced (Leitner et al. 2004). The estimated decrease in milk production varies from 70 to 150 kg per animal per year (Smith and Cutlip 1998). In intensive dairy farms, an average of 35 % of animals are replaced yearly, as a result of disease, death, infertility, udder health, and low milk yield (Bergonier et al. 2003; Mahler et al. 2001).

According to Birgel Jr. et al. (2005), CAE virus greatly affects the physical-chemical and characteristics of milk, wherein the chloride content and somatic cell counts are far greater in CAEV-infected goat milk. The protein and fat contents were also noted to be lower compared to goats, which are not affected by CAE virus. In a retrospective study done in Spain, it was found that compared to seronegative animals, seropositive does had significantly shorter lactations. These animals produce less milk with reduced milk fat, lactose, and dry extract and had higher somatic cell counts in the milk. Regression models confirmed the association between CAEV seropositivity and reduced milk production. It was noted that seropositive goats has 10 % lesser milk production compared to seronegative goats. This only proves that infection with CAEV can be a major cause of reduction of milk yield among goat counterparts (Martinez-Navlon et al. 2007). Therefore, management and control strategies should be considered as a part of dairy goat herd health schemes.

Since CAEV causes mastitis in infected animals, infection with this virus is usually associated with reduced lactation length, milk fat, and protein yield as well as incidence of other diseases and reduced birth weight. Martinez-Navlon et al. (2007) found out that the reduction in yield increases with lactation number. With regard to weight gain, a significant loss was indicated in SRLV infection in sheep (Peterson et al. 2004). Mortality in CAEV infection is not directly related to death of animals affected. Decreased average life span of the animal due to infection, culling due to decreased productivity, and increased in mortality and survivability among kids are related to decrease in milk being produced.

7.5.3.2 Progress in the Diagnosis and Management of Caprine Arthritis-Encephalitis

CAEV infection is indeed widespread and has become a major concern among dairy goats. Continuous monitoring on the status of CAEV infection is significant to establish an effective program to control or eradicate this disease.

Most of the present control programs are focused on the elimination of virus in the herd, which were proven to be very effective especially in a herd, which has high CAEV prevalence. Development of more sensitive and specific diagnostic methods, on the other hand, became the focus of most researchers. Though the present diagnostic methods have shown very efficient, most of these techniques require expensive equipment and technical experts. A more convenient, less expensive, rapid yet sensitive and specific test is being developed. The use of nanotechnology, for example, is now being employed in the development of new diagnostic test that can be used on-site. In the development of molecular- and serological-based tests, it is important to consider the local strain of the virus circulating in an area. The knowledge on the virus strain present in the area may increase the sensitivity and specificity of the test being developed.

One area of interest that is being studied is the influence of genetics on the resistance and/or susceptibility of goats to diseases. Certain breed of goats has been reported to be resistant to CAEV infection; therefore scientists are searching for markers that may prove the resistance and/or susceptibility to CAEV. In the future, if proven that resistance to CAEV exists, these markers can be used in selective breeding programs.

With regard to movement of animals from one country to another, disease screening prior to movement should be strictly implemented in countries that import or export animals.

References

Abelson ML, Schoborg RV (2003) Characterization of the caprine arthritis encephalitis virus (CAEV) rev N-terminal elements required for efficient interaction with the RRE. Virus Res 2:23–35

Adams DS, Crawford TB, Klevjer-Anderson P (1980) A pathologentic study of the early connective tissue lesions of caprine arthritis encephalitis. Am Pathol 99:257–278

Adams DS, Klevjer-Anderson P, Carlson JL et al (1983) Transmission and control of caprine arthriris-encephalitis virus. Am J Vet Res 44:1670–1675

Adams DS, Gorgham JR (1986) The gp135 of caprine arthritis encephalitis virus affords greater sensitivity than the p28 in immunodiffusion serology. Res Vet Sci 40:157–160

Andrioli A, Gouveia AMG, Martins AS, Pingheiro RR et al (1999) Detection of caprine lentivirus proviral DNA in semen of naturally infected goats. Rev Bras Rep Anim 23:420–421

Al-Qudah K, Al-Majali AM, Ismail ZB (2006) Epidemiological studies on caprine arthritis-encephalitis virus infection in Jordan. Small Rumin Res 66:181–186

Ali Al Ahmad MZ, Chebloune Y, Bouzar BA, Baril G, Bouvier F, Chatagnon G et al (2008a) Lack of risk of caprine-arthritis encephalitis virus (CAE-V) after an appropriate embryo transfer procedure. Theriogenology 69:408–415

Ali Al Ahmad MZ, Fieni F, Pellerin JL, Guigen F, Chere Y, Chatagnon G (2008b) Detection of viral genome of caprine arthritis encephalitis virus (CAEV) in semen and in genital tract tissues of male goat. Theriogenology 69:473–480

Ali Al Ahmad MZ, Dubreil L, Chatagnon G, Khayli Z, Therst M, Martignat L et al (2012) Goat uterine epithelial cells are susceptible to infection with caprine arthritis encephalitis virus (CAE-V) in vivo. Vet Res 43:1–7

APHIS (2007) Caprine arthritis encephalitis virus. Animal and Plant Health Inspection Service, United States Department of Agriculture, Maryland, USA

Balbin MM, Belotindos LP, Abes NS, Mingala CN (2014) Caprine arthritis encephalitis virus detection in blood by loop-mediated isothermal amplification (LAMP) assay targeting the proviral *gag* region. Diagn Microbiol Infect Dis 79:37–42

Bandeira DA, de Castro RS, Azevedo EO et al (2009) Seroprevalence of caprine arthritis encephalitis virus in goats in the Cariri region, Paraiba state. Brazil Vet J 180:399–401

Bergonier D, De Cremoux R, Rupp R, Lagriffoul G, Berthelot X (2003) Mastitis of dairy small ruminants. Vet Res 34:689–716

Belino ED, Ezeifeka GO (1984) Maedi-Visna antibodies in sheep and goats in Nigeria. Vet Rec 114:570

Bertolitti L, Mazzei M, Puggionni D, Carroza ML et al (2011) Characterization of new small ruminant lentivirus subtype B3 suggests animal trade within Mediterranean Basin. J Gen Virol 92:1923–1929

Birgel Junior EH, Cestari V, Sampaio RM, Birgel DB, et al (2005) Influencia de infeccao pelo virus da artrite encefalite caprina nas caracteristicas fisico-quimicas, celulares e microbiologicas do leite de caprinos. In: CONGRESSO BRASILEIRO DE BUIATRIA, 5. 2005, Buzios-RJ Anais, Editora, Rio de Janeiro. (Resumo)

Blacklaws BA, Berriatua E, Torsteinsdottir S, Watt NJ, de Andre's D, Klein D, Harkiss GD (2004) Transmission of small ruminant lentiviruses. Vet Microbiol 101(3):199–208

Bolea R, Manleon F, Carasso L, Vargas A, de Andre's D et al (2006) Meadi-visna virus infectiob of ovine mammary epithelial cells. Vet Res 37:133–144

Brajon G, Mandas D, Liciardi M, Taccori F, Meloni M, Corrias F et al (2012) Development and field testing of a real-time PCR assay for Caprine Arthritis-Encephalitis-Virus (CAEV). Open Virol J 6:82–90

Brinkof JM, van Maanen C (2007) Evaluation of five enzyme-linked immunosorbent assays and agar gel immunodiffusion test for detection of antibodies to small ruminant lentiviruses. Clin Vaccine Immunol 14:1210–1214

Brinkhof JMA, Van Maanen C, Wigger R, Peterson K, Houwers DJ (2008) Specific detection of small ruminant lentiviral nucleic acid sequences located in the proviral long terminal repeat and leader gag regions using real-time polymerase chain reaction. J Virol Methods 147:338–344

Burgu I, Akca Y, Alkan F, Ozkuln A, Karaglu T, Cabalar M (1994) antibody prevalence of caprine arthritis virus (CAEV) in goats inturkey. Dtsch Tierartl Wochenschr 10:390–391

CFSPH (2007) Caprine Arthritis Encephalitis, Small Ruminant lentivirus Infection. 2007. The Center for Food Security and Public Health. College of Veterinary Medicine, Iowa State University, Ames, Iowa 50011

Cheevers WP, McGuire TC (1988) The lentiviruses:maedi visna, caprine arthritis-encephalitis, and equine infectious anemia. Adv Virus Res 34:189–215

Clements J, Zink C (1996) Molecular biology and pathogenesis of animal lentivirus infections. Clin Microbiol Rev 9:100–117

Concha-Bermejillo C, De La A (2003) Caprine arthritis encephalitis: an update. Sheep Goat Res J 18:68–78

Cork LC, Hadlow WJ, Crawford TB, Gorham JR, Piper RC (1974) Infectious leukoencephalomyelitis of young goats. J Infect Dis 129(2):134–114

Cork LC, Narayan O (1980) The pathogenesis of viral leukoencephalitis-arthritis in goats: I. persistent viral infection with progressive pathologic changes. Lab Invest 42:596–602

Cutlip RC, Jackson TA, Laird OA (1977) Immunodiffusion test for the ovine progressive pneumonia. Am J Vet Res 38:997–999

Cutlip RC, Lehmukuhl HD, Sack JM, Weaver AL (1992) Prevalence of antibody to caprine arthritis encephalitis virus in goats in United States. JAVMA 200:802–805

Cruz JCM, Gouveia AMG, Souza KC, Braz KC (2009) Caprine arthritis-encephalitis virus (CAEV) detection in semen of endangered goat breeds by nested polymerase chain reaction. Small Rumin Res 85:149–152

Crawford TB, Adams DS, Cheevers WP, Cork LC (1980) Chronic arthritis in goats caused by retrovirus. Science 207:997–999

Crawford TB, Adams DS (1981) Caprine arthritis encephalitis: clinical features and presence of antibody in selected goat populations. J Am Vet Med Assoc 178:713–719

Dawson M, Wilesmith JW (1985) Serological survey of lentivirus (maedi-visna/caprine arthritis encephalitis) infection in British goat herds. Vet Rec 117:86–89

de Andre's D, Klein D, Watt NJ, Berriatua E, Torsteinsdottir S, Blacklaws BA, Harkiss GD (2005) Diagnostic tests for small ruminant lentiviruses. Vet Microbiol 107:49–62

De Boer GF, Terpstra C, Houwers DG, Hendricks J (1979) Studies in the epidemiology of maedi-visna in sheep. Res Vet Sci 26:202–208

de Souza KC, Pinheiro RR, Santos DO, de Brito RLL et al (2013) Transmission of the caprine arthritis-encephalitis virus through artificial insemination. Small Rumin Res 109:193–198

de Sousa VR, Sousa FDCA, Filho OFDS, Rici REG et al (2014) Comparative study by computes radiography, histology, and scanning electron microscopy of the articular cartilage of normal goat and in chronic infection with caprine arthritis-encephalitis virus. Microsc Res Tech 77:11–16

Dias RP, de Brito RLL, Rodrigues AS, Alves LAO, Andrioli A, Pinheiro RR, Teixeira MFS (2012) Influence of seropositivity for caprine arthritis encephalitis virus in the hemogram of lactating goats. Arq Inst Biol 79(4):503–509

East NE, Rowe JD, Dahlberg JE, Theilen G, Pederson NC (1993) Modes of transmission of caprine arthritis-encephalitis virus infection. Small Rumin Res 10(3):251–262

Eggert-Krusse W, Botz I, Pohl S, Rohr G, Strowitzki T (2000) Antimicrobial activity of human cervical mucus. Hum Reprod 15:778–784

Elfahal AM, Hussein MO, Enan KA, Taha KM, Salih DA, Halfawi RJ, Mohammed ZA, El-Hussein AM (2013) Investigation of caprine arthritis encephalitis virus in Sudan using competitive enzyme linked immunosorbent assay. Vet World 6(8):558–562. doi:10.5455/vetworld.2013.558-562

Fieni F, Rowe J, Van Hoosear K, Buruoca C, Oppenheim S, Anderson G et al (2003) Presence of caprine arthritis-encephalitis virus (CAEV) proviral DNA in genital tract tissues of super ovulated dairy goat does. Theriogenology 59:1515–1523

Fieni F, Pellerin JL, Roux C, Puolin N et al (2012) Can caprine arthritis encephalitis virus (CAEV) be transmitted by in-vitro fertilization with experimentally infected sperm? Theriogenology 77(3):644–651

Gendelman HE, Narayan O, Molineaux S, Clements JE, Ghotbi Z (1985) Slow, persistent replication of lentiviruses: role of tissue macrophages and macrophage precursors in bone marrow. Proc Natl Acad Sci U S A 82(20):7086–7090

Ghanem YM, El-Khodery SA, Saad AA, Elragaby SA, Abdelkader AH, Heybe A (2009) Prevalence and risk factors of caprine arthritis encephalitis virus infection (CAEV) in North Somalia. Small Rumin Res 85:142–148

Gjerset B, Storset AK, Rimstad E (2006) Genetic diversity of small ruminant lentiviruses: characterization of Norweigian isolates of caprine arthritis encephalitis virus. J Gen Virol 87:573–580

Glaria I, Reina R, Crespo H, de Andre's X, Ramirez H et al (2009) Phylogenetic analysis of SRLV sequences from an arthritic sheep outbreak demonstrates the introduction of CAEV-like viruses among Spanish sheep. Vet Microbiol 138:156–162

Grego E, Bertolotti L, Quasso A, Profiti M, Lacerenza D, Muz D, Rosati S (2007) Genetic characterization of small ruminant lentivirus in Italian mixed flocks: evidence for a novel genotype circulating in a local goat population. J Gen Virol 88:3423–3427

Gomez-Lucia E, Rowe J, Collar C, Murphy B (2013) Diversity of caprine arthritis encephalitis virus promoters isolated from goat milk and passaged in-vitro. Vet J 196:431–438

Gonzales JVC, Domingo CYJ, Abes NS, Gutierrez CA, Villanueva MA, Mingala CN (2013) Concordance of competitive enzyme linked immunosorbent assay and nested-polymerase chain reaction in the detection of caprine arthritis-encephalitis virus. Small Rumin Res 115:134–139

Gonzales L, Gelabert JE, Marlo JC, Saez de Okariz C (1987) Caprine arthritis encephalitis in Basque country. Spain Vet Rec 120:102–109

Greenwood PL, North RN, Kirkland PD (1995) Prevalence, spread and control of caprine arthritis-encephalitis virus in dairy goat herds in New South Wales. Aust Vet J 72(9):341–345

Gregory L, Angelini M, Lara MC, Benesi FJ et al (2006) Clinical evaluation of viral indurative mastitis caused by caprine arthritis-encephalitis. In: XXIV World Buiatrics Congress. Nice, France. Anals of XXIV World Buiatrica Congress

Gregory L, Birgel Junior EH, Lara MC, Angelini M, Araujo WP et al (2009) Clinical features of indurative mastitis caused by caprine arthritis encephalitis virus. Braz J Vet Pathol 2(2):64–68

Grewal AS (1986) Comparison of two gel precipitation tests in the serodiagnosis of caprine arthritis encephalitis virus infection in goats. Aust Vet J 63:341–342

Gufler H (2004) CAEV: clinical and serological findings and the economical losses in a goat herd of 'Passeirer Gebirgsziege. Tierarztl Praxis 32:263–268

Gufler H, Gasteiner J et al (2007) Serology study of small ruminant lentiviirus in goats in Italy. Small Rumin Res 73(1–3):169–173

Hermann LM, Cheevers WP, Mcguire TC et al (2003) A competitive-inhibition enzyme-linked immunosorbent assay (cELISA) for detection of serum antibodies to caprine arthritis encephalitis virus (CAEV): a diagnostic tool for successful eradication. Clin Diagn Lab Immunol 10:267–271

Herrmann-Hoesing LM (2010) Diagnostic assays used to control small ruminant lentiviruses. J Vet Diagn Investig 22:843–855

Huang J, Sun Y, Liu Y, Xiao H, Zhuang S (2012) Development of a loop-mediated isothermal amplification method for rapid detection of caprine arthritis-encephalitis virus proviral DNA. Arch Virol 157:1463–1469

Kaba J, Rola M, Materniak M, Nowicki M (2009) Isolation and characteriztion of caprine arthritis encephalitis virus in goats from Poland. Pol J Vet Sci 12:183–188

Kaba J, Winnicka A, Zaleska M, Nowicki M, Bagnicka E (2011) Influence of chronic caprinne arthriris-encephalitis virus infection on the population of peripheral blood leukocytes. Pol J Vet Sci 14(4):585–590

Knight AP, Joniken MP (1982) Caprine arthritis encephalitis. Compend Cont Educ Pract Vet 4:263–269

Knowles DP, Evermann JF, Schropshire C, Vander Schalie J et al (1994) Evaluation of agar gel immunodiffusion serology using caprine and ovine lentiviral antigens for detection of antibody to caprine arthritis encephalitis virus. J Clin Microbiol 32:243–245

Konishi M, Tsuduku S, Haritani M, Murakami K, Tsuboi T, Kobayashi C, Yoshikawa K, Kimura K, Sentsui H (2004) An epidemic of caprine arthritis encephalitis in Japan: isolation of the virus. J Vet Med Sci 66(8):911–917

Kuzmak J, Rola M et al (2007) Molecular characterization of lentiviruses from goats from Poland based on gag gene sequence analysis. Comp Immunol Microbiol Infect Dis 30(4):211–223

Krieg A, Peterhans E (1990) Caprine arthritis encephalitis in Switzerland: epidemiological and clinical studies. Schweiz Arc Tierheilkd 132(7):345–352

L'Homme Y, Ourdan M, Leuesque V, Bertoni G, Simard C, Pisoni G (2011) Molecular characterisation and phylogenetic analysis of small ruminant lentiviruses isolated from Canadian sheep and goats. J Virol 8:1–7

Lamara A, Fieni F, Mselli-Lakhal L, Tainturier D, Chebloune Y (2001) Efficient replication of caprine arthritis – encephalitis virus in goat granulose cells. Virus Res 87:69–77

Lamara A, Fieni F, Mselli-Lakhal L, Chatagnon G et al (2002) Early embryonic cells from in-vivo produced embryos transmit the caprine arthritisencephalitis virus (CAEV). Theriogenology 58(6):1153–1163

Lara MCCSH, Birgel Junior EH, Birgel EH (2005) Possibility of vertical transmission of caprine arthritis-encephalitis virus in neonate kids. Arq Bras Vet Zootec 57(4):553–555

Leitner G, Merin U, Silanikove N (2004) Changes in milk composition as affected by subclinical mastitis in goats. J Dairy Sci 87:1719–1726

Lilenbaum W, de Souza GN, Ristow P, Moreira MC et al (2007) A serological study on Brucella abortus, caprine arthritis encephalitis virus and Leptospira in dairy goats in Rio de Jeneiro. Braz Vet J 173:408–412

Lluz A, Orais A, Dumilon R, Orais A, Cerbito W (2011) First detection of caprine arthritis encephalitis infection in goats using ELISA test in Eastern Visayas, Philippines. J Vet Anim Sci 37(2):159–166

Mahin L, Chadli M, Houwers DJ (1984) A preliminary report on the occurrence of maedi-visna in sheep in Morocco. Vet Q 6:104

Malher X, Seegers H, Beaudeau F (2001) Culling and mortality in large dairy goat herds managed under intensive conditions in western France. Livestock Prod Sci 71:75–86

Martins G, Penna B et al (2012) Leptospirosis as the most frequent infectious diseases impairing productivity in small ruminat inRio de Jeneiro, Brazil. Trop Anim Health Prod 44(4):773–777

Martinez-Navlon B, Peris C, Gomez EA, Peris B, Roche ML, Caballero C, Goyena E, Berriatua E (2007) Quantitative estimation of the impact of caprine arthritis-encephalitis virus infection on milk production by dairy goats. Vet J 197:311–317

Mathews JG (1999) Diseases of goat, 2nd edn. Blackwell Science, Chemsford, pp 80–87

Minguijon E, Reina R, Perez M, Polledo L, Villoria M, Ramirez H, Leginagoikoa I, Badiola JJ, Garcia-Marin JF, de Andres D, Lujan L, Amorena JRA (2015) Small ruminat lentivirus infections and diseases. Vet Microbiol http://dx.doi.org/10.1016/j.vwtmic.2015.08.007

Murphy FA, Gibbs EPJ, Horzinek MC, Studdert MJ (1999) Veterinary virology, 3rd edn. Academic Press, San Diego

Muz D, Oguzoglu TC, Rosati S, reina R, Bertolotti L, Burgu I (2012) First molecular characterization of isna/maedi viruses from naturally infected goats from Turkey. Arch Virol 158:559–570

Narayan O, Wolinsky JS, Clements JE, Strandberg JD, Griffin DE, Cork LC (1982) Slow virus replication: the role of macrophages in the persistence and expression of visna viruses of sheep and goats. J Gen Virol 59:345–356

Narayan O, Kennedy-Stoskopf S, Sheffer D, Griffin DE, Clements JE (1983) Activation of caprine arthritis-encephalitis virus expression during maturation of monocytes to macrophages. Infect Immun 4:67–73

Narayan O, Cork LC (1985) Lentiviral diseases of sheep and goats: chronic pneumonia leukoencephalomyelitis and arthritis. Rev Infect Dis 7:89–88

Nord K, Rimstad E, Storset AK, LokenT (1998) Prevalence of antibodies against caprine arthritis-encephalitis virus infection in goat herds in Norway. Small Rumin Res 28:115–121

Notomi T, Okayama H, Masubuchi H, Yonekawa T, Watanabe K, Amino N et al (2000) Loop-mediated isothermal amplification of DNA. Nucleic Acids Res 28:e63

Oem J, Chung J, Byun J, Kim H, Kwak D, Jung B (2012) Large-scale survey of caprine arthritis-encephalitis virus (CAEV) in Korean black goats (*Capra hircus aegagrus*). J Vet Med Sci 74(12):1657–1659

Office International des Epizootes (2008) Manual of Diagnostic Tests and Vaccines for Terrestrial Animals. 6th Edn. Paris, France, pp 983–991

Office International des Epizootes (2012) Manual of Diagnostic Tests and Vaccines for Terrestrial Animals (manual, birds and bees). 7th Edn, Vol. 2. Paris, France, pp 978–986

Padiernos RBCP, Balbin MM, Parayao AM, Mingala CN (2015) Molecular characterization of gag gene of caprine arthritis encephalitis virus from goats in the Philippines. Arch Virol 160(4):969–978

Parida MM, Santhosh SR, Dash PK, Tripathi NK, Saxena P, Ambuj S et al (2006) Development and evaluation of reverse transcription loop mediated isothermal amplification for rapid and real-time detection of Japanese encephalitis virus. J Clin Microbiol 44:4172–4178

Pereira CL, Baule C, Costa R, Langa A (1989) Occurrence of caprine erthriris encephalitis in Mozambique. Trop Anim Health Prod 21:237–238
Perk K (1988) Ungulate lentiviruses: pathogenesis and relationship to AIDS. Adv Vet Sci Comp Med 32:97–127
Peterhans E, Greenland T, Badiola J, Harkiss J, Bertoni G, Amorena B, Eliaszewicz M, Juste RA, Krassnig R, Lafont JP, Lenihan P, Petursson G, Pritchard G, Thorley J, Vitu C, Mornex JF, Pepin M (2004) Routes of transmission and consequences of small ruminant lentiviruses (SRLVs) infection and eradication schemes. Vet Res 35(3):257–274
Peterson K, Brinkof J, Houwers D, Colenbrander B, Gadella B (2008) Presence of pro-lentiviral DNA in male sexual organs and ejaculats of small ruminants. Theriogenology 69:433–442
Petursson G, Andresdottir V, Andrresson OS, Georgsson G, Palsson PA, Rafnar B, Torsteinsdottir S (1992) Lentivirus diseases of sheep and goats: maedi visna and caprine arthritis encephalitis. In: Speedy AW (Ed), Progress in sheep and goat research. CAEB International, Oxford, 107–129.
Pisoni G, Bertoni G, Manarolla G, Vogt H-R, Scaccabarozzi L, Locatelli C, Moroni P (2010) Genetic analysis of small ruminant lentiviruses following lactogenic transmission. Virology 407:91–99
Ponti W, Paape M, Bronzo V, Pisoni G, Pollera C, Moroni P (2008) Phenotypic alteration of blood and milk leukocytes in goats naturally infected with caprine arthritis-encephalitis virus (CAEV). Small Rumin Res 78:176–180
Pugh DG (2002) Sheep and goat medicine. W. B. Saunders and Company, Philadelphia, pp 126, 239–240, 296, 388
Qu J, Liu H et al (2005) Current situation of research on caprine arthritis encephalitis. Chin J Prev Med 27(5):431–434
Ravazzolo A, Nenci C, Vogt H, Waldvogel A et al (2006) Viral load, organ distribution, histopathological lesions, and cytokine mRNA expression in goats infected with a molecular clone of the caprine arthritis encephalitis virus. Virology 350:116–127
Reddy PG, Sapp WJ, Heneine W (1993) Detection of caprine arthritis encephalitis virus by polymerase chain reaction. J Clin Microbiol 31:3042–3043
Reina R, Mora I, Glaria I, Garcia C, Solano I, Lujan JJ et al (2006) Molecular characterisation and phylogenetic study of maedi visna and caprine rthritis encephalitis viral sequences in sheep and goats from Spain. Virus Res 121:189–198
Reina R, Grego E, Bertolotti L, de Meneghi D, Rosati S (2009) Genome analysis of small-ruminant lentivirus genotype E: a Caprine Lentivirus with Natural Deletiona of the dUTOase Subunit, vpr-Like Accessory Gene, and 70-Base-Pair Repeat of the U3 Region. J Virol 83(2):1152–1155
Reina R, de Andre's D, Amorena B (2013) Immunization against small ruminant lentivirus. Viruses 5:1948–1963
Rimstad E, East N, Higgins J, Pedersen NC (1994) Detection of antibodies to caprine arthritis encephalitis virus using recombinant gag protein. Arch Virol 134:345–356
Rosati S, Kwang T, Tolari F, Keen JE (1994) A comparison of a whole virus and recombinant transmembrane ELISA and immunodiffusion for detection of ovine lentivirus antibodies in Italian sheep flocks. Vet Res Commun 18:73–80
Rowe JD, East NE (1997) Risk factors for transmission and methods for control of caprine arthritis-encephalitis virus infection. Vet Clin North Am Food Anim Pract 13:33–53
Sanchez A, Conteras A et al (2001) Relationship between infection with caprine arthritis encephalitis virus, intramammary bacterial infection and somatic cell counts in dairy goats. Vet Rec 148:711–714
Santry LA, De Jong J, Gold AC, Walsh SR, Menzies PI, Wootton SK (2013) Genetic characterization of small ruminant lentiviruses circulating in naturally infected sheep and goats in Ontario, Canada. Virus Res 175(1):30–44

Shah C, Huder JB, Boni J, Schonmann M, Muhlherr J, Lutz H, Schupbach J (2004) Direct evidence for natural transmission of small ruminant lentiviruses of subtype A4 from goats to sheep and vice versa. J Virol 78(14):7518–7522

Smith MC, Cutlip R (1998) Effects of infection with caprine arthritis-encephalitis virus on milk production in goats. J Am Vet Med Assoc 193:63–67

Smith MC, Sherman DM (1994) Goat Medicine, 1st Edn., Lea and Febriger, Philadephia. pp. 73–79, 135–138

Smith MC, Sherman DM (2009) Goat medicine, 2nd edn. Wiley and Blackwell, Ames, pp 96–106

Starick E, Enke KH (1995) Comparative studies of maedi-visna diagbosis: immunodiffusion test, enzyme immunoassay, immunoblot. Berl Munch Tierarztl Wochenschr 108:138–142

Synge BA, Ritchie CM (2010) Eliminationn of small ruminant lentivirus infection from sheep flocka and goat herds aided by health svhemes in Great Britain. Vet Rec 167(19):739–743

Thant N, saroch N, Kanisak O, Prachin O, Mongkol T (2011) Seroprevalence and risk factors associated with caprine arthritis encephalitis virus infection in goats in the western part of Thailand. Thai J Vet Med 41:353–360

Torres-Acosta JFJ, Gutierrez-Ruiz EJ, Butler V, Schmidt A, Evans J et al (2003) Serological survey of caprine arthritis-encephalitis virus in 83 goat herds in Yucatan, Mexico. Small Rumin Res 49:207–211

Travassos C, Benoit C, Valas S, da Silva A, Perin G (1999) Caprine erthritis-encephalitis virus in semen of naturally infected bucks. Small Rumin Res 32:101–106

Varea R, Monleon E, Pacheco C, Lujan L et al (2001) Early detection of maedi visna (ovine progressive pneumonia) virus seroconversion in field sheep samples. J Vet Diagn Investig 13:301–307

Villoria M, Leginagoikoa I, Lujan L, Perez M, Salazar E, Berriakia E et al (2013) Detection of small ruminant lentivirus in environmental sample of air and water. Small Rumin Res 110:155–160

Vitu C, Russo P, Filippi P, Vigne R et al (1982) An ELISA test for detection of maedi-visan antibodies: comparative study with gel immunodiffusion and complement fixation test. Comp Immunol Micro Infect Dis 5:469–481

Wander JG (1970) Progresisve interstitial pneumonia in Kenya. Vet Rec 86:434–438

Waseem A, Pawaiya RVS, Singh R, Gupta VK, Rajukumar K, Mir MS, Aamir S (2015) Seroprevalence of caprine arthritis encephalitis virus infection (CAEV) in Indian goats. Indian J Vet Pathol 39(1):15–11

Zanoni RG, Vogt HR, Pohl B, Bottcher J et al (1994) An ELISA based on whole virus for the detection of antibodies to small-ruminant lentiviruses. J Vet Med B 41:662–669

Zanoni RG (1998) Phylogenetic analysis of small ruminant lentiviruses. J Gen Virol 79(Pt 8):1951–1961

Zwahlen M, Aeschbacher M, Balcer TH et al (1983) Lentivirusinfectionen bei Ziegen mit Carpitis und intertieller Mastitis. Schweiz Arch Tierheil 125:281–299

Equine Infectious Anemia

8

Praveen Malik, Harisankar Singha, and Sanjay Sarkar

8.1 Introduction

Equine infectious anemia (EIA), a persistent infectious disease of horses and all other equids, is caused by a macrophage tropic lentivirus, equine infectious anemia virus (EIAV), under the family of *Retroviridae* (Cordes and Issel 1996; Cook et al. 2001, 2013; Issel et al. 1982). The disease is distributed worldwide and also known as "swamp fever" (Cook and Issel 2004). EIAV infection can be represented in three forms, namely, acute, chronic, and clinically inapparent infection (Issel et al. 1982; Craigo and Montelaro 2011). Although the clinical outcome of EIA depends largely on the genetics of host as well as infecting virus strain, the acute form of the disease is characterized by the symptoms of pyrexia, lethargy, anorexia, dependent edema, hemolytic anemia, and death in severe cases (Sellon et al. 1994; Issel et al. 2014). If the animal survives from the acute infection, the disease becomes chronic and subclinical. The chronic form of EIA is characterized by the recurrent episodes of pyrexia, lethargy, anorexia, rapid weight loss, thrombocytopenia, anemia, dependent edema, and occasionally neurologic signs such as ataxia and encephalitis (Sellon et al. 1994; McGuire et al. 1990). The acute form of the disease often goes

P. Malik (✉)
Department of Animal Husbandry, Dairying and Fisheries, Ministry of Agriculture and Farmers Welfare, Government of India, CCS National Institute of Animal Health, Baghpat 250609, Uttar Pradesh, India
e-mail: Praveen.Malik.niah@gov.in; malikphisar@hotmail.com

H. Singha (✉)
ICAR-National Research Centre on Equines, Hisar 125001, Haryana, India
e-mail: hsssankarbty@gmail.com

S. Sarkar (✉)
Maxwell H Gluck Equine Research Centre, University of Kentucky, Lexington, KY 40546-0099, USA
e-mail: sanjaysarkarvet@gmail.com

J. Bayry (ed.), *Emerging and Re-emerging Infectious Diseases of Livestock*, DOI 10.1007/978-3-319-47426-7_8

unnoticed; however, chronic form is the most frequent clinically diagnosed form of the disease (Cook et al. 2013; Sellon et al. 1994). Like other retroviruses, once a horse is infected with EIAV, it remains infected for life. Because of the stringent immunologic control over virus replication, more than 90 % of the EIAV-infected animals progress to inapparent carrier stage from chronic stage despite the tremendous antigenic variation and continuous viral replication making EIAV unique among other lentiviruses (Craigo and Montelaro 2013; Montelaro et al. 1993). Interestingly, EIAV-infected inapparent carriers are also resistant to secondary infection with various other EIAV strains implicating the presence of a strong prophylactic immune control in these carrier animals and also serving as a natural model for the immunologic control of a lentivirus infection and disease (Craigo and Montelaro 2011, 2013; Montelaro et al. 1993).

In nature, the infection is mechanically transmitted from an infected animal to another susceptible animal by the bite of hematophagous insect vectors, but in the field condition, the virus can also be transmitted parenterally with contaminated syringes and blood products (Cook et al. 2013; Issel et al. 1982; Craigo and Montelaro 2013). The clinical signs associated with the disease were first described in 1843 in France (Lignee 1843). Later in 1904, EIA was shown to be caused by a "filterable agent" which could be transmissible from one horse to others through blood (Valle and Carre 1904). Consequently, EIA was credited to be the first animal disease under viral etiology (Craigo and Montelaro 2011). The EIAV was first assigned to be in the family of *Retroviridae*, subfamily of *Lentivirus* in 1970 along with other lentiviruses such as maedi-visna virus (MVV) of sheep and caprine arthritis encephalitis virus (CAEV) of goat (Craigo and Montelaro 2011). Human immunodeficiency virus (HIV) after its discovery in 1983 was classified as lentivirus based on the similarity in morphology with EIAV and serological cross-reactivity of EIAV-infected horse serum (Montagnier et al. 1984). Currently, the disease has been reported all over the world including Europe, Asia, Australia, and the Americas (Pearson 1972).

8.2 Etiology

8.2.1 The Virus

EIAV is a positive-sense, single-stranded ribonucleic acid (RNA) virus, which is classified under the genus *Lentivirus*, subfamily of *Orthoretrovirinae*, and family of *Retroviridae*. EIAV, the "country cousin" of HIV, is very similar in morphology as well as in antigenicity with other lentiviruses such as MVV, CAEV, bovine immunodeficiency virus (BIV), feline immunodeficiency virus (FIV), simian immunodeficiency virus (SIV), and HIV (Cook et al. 2013; Leroux et al. 2004). Electron microscopic studies revealed that, morphologically, the EIAV is oval or circular, is 115 nm in diameter, and contains two copies of single-stranded, positive-sense RNA genome (Fig. 8.1a) (Weiland et al. 1977). The two copies of RNA genome are tightly associated with other viral proteins including viral integrase, reverse

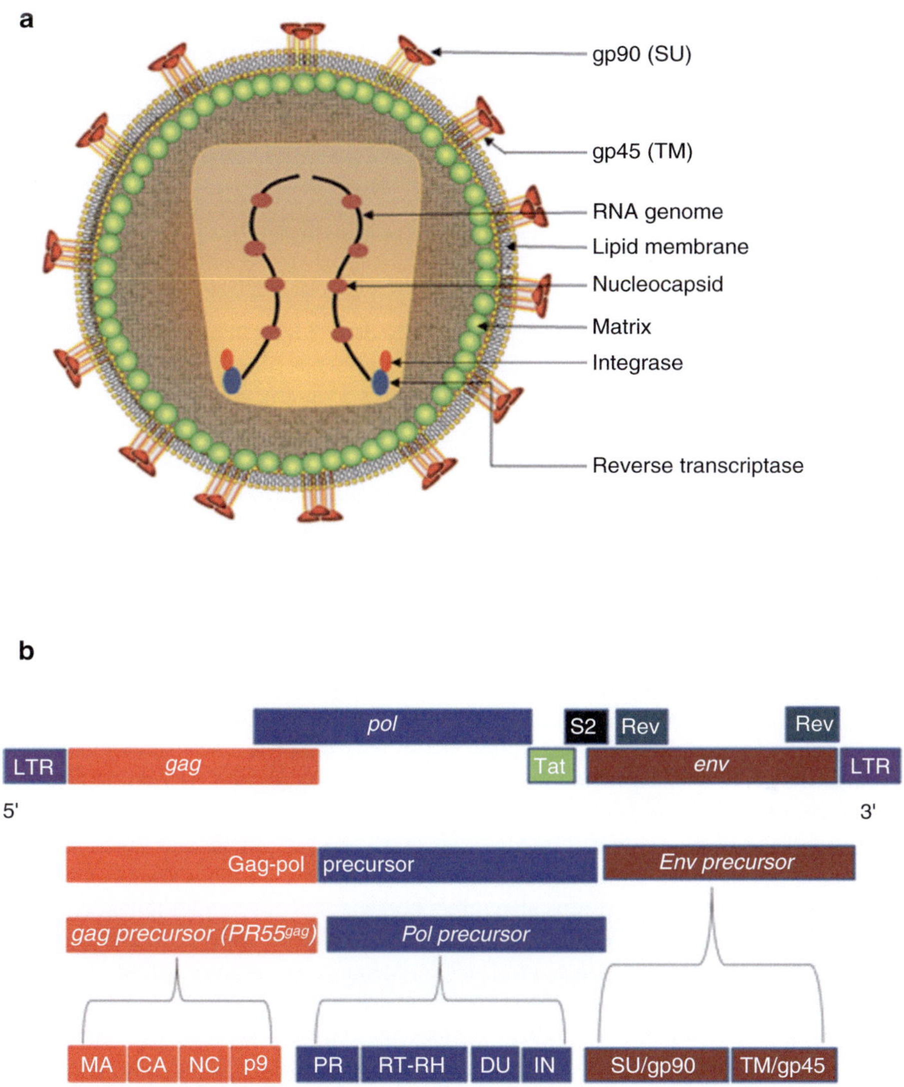

Fig. 8.1 Schematic diagram and genome organization of EIA virus. (**a**) Schematic representation of EIAV: The two copies of RNA genome are associated with reverse transcriptase, integrase, and nucleocapsid. Nucleocapsid is covered by a matrix protein which is again surrounded by a lipid bilayer membrane (envelope). Glycoproteins gp90 and gp45 are embedded in the envelope membrane. (**b**) Genome organization of EIA virus

transcriptase, and nucleocapsid proteins in a virion particle (Weiland et al. 1977). The genome of the EIAV makes a conical-shaped core, which is encased in a proteinaceous matrix that is surrounded by a lipid bilayer membrane derived from the host cells (Weiland et al. 1977). The lipid bilayer membrane has several 6–8 nm projections containing the trimers of envelope proteins consisting of the surface unit and transmembrane domain (Fig. 8.1a) (Weiland et al. 1977; Matheka et al. 1976). The virus is readily inactivated by heating at 60 °C for 60 min, or by formalin and ether treatment but resistant to phenol or trypsin (Pearson 1972; Dreguss and Lombard 1954; Kono et al. 1970).

8.2.2 Genome Organization and Virus Replication

The EIAV has a very simple genome of about 8.2 Kbp length making EIAV to have the smallest genome of all known lentiviruses to date (Liang et al. 2006; Dong et al. 2013). The genome contains three major structural genes (*gag*, *pol*, and *env*), which are flanked by two long terminal repeats (LTR) at 5′ and 3′ regions of the genome (Fig.8.1b). Besides the major structural genes, the EIAV genome also has three other open reading frames (ORFs) (Tat, Rev, and S2) encoding the accessory proteins involved in the regulatory functions, countering of host defenses, and enhancing pathogenicity (Fig.8.1b) (Cook et al. 2013). Although ORFs Tat and Rev are common to all lentiviruses, ORF S2 is unique to EIAV (Dong et al. 2013). Another unique characteristic to EIAV is the absence of Viforthologue, which is present in all other lentiviruses (Issel et al. 2014). *Gag* gene encodes p15 matrix (MA), p26 capsid antigen (CA), p11 nucleocapsid (NC), and p9 "late-domain" protein, and *pol* gene encodes RT, integrase, dUTPase (DU), and other enzymes such as proteinase involved in EIAV replication (Fig. 8.1a and b) (Cook et al. 2013). Protein products of *gag* gene form the core of the EIAV virion where p11 remains associated with viral RNA genome, p26 forms the conical core structure, and p15 surrounds the core forming the matrix (Fig. 8.1a) (Issel et al. 2014).

Though the EIAV primarily infects mature tissue macrophages in vivo for virus replication, peripheral blood monocytes are also infected. In laboratory conditions, cell culture-adapted EIAV strains also infect horse leukocytes, endothelial cells, and fibroblast cells from equines as well as from heterologous species like canine and felines (Oaks et al. 1998; Maury et al. 1998; Brindley and Maury 2008). After the virus is transmitted through contaminated blood, it enters into the host cells such as macrophages by receptor-mediated endocytosis which is low pH dependent (Brindley and Maury 2005, 2008; Jin et al. 2005). The EIAV binds with the EIAV receptor present on the host cells, known as equine lentivirus receptor-1 (ELR1), which is a member of the tumor necrosis factor receptor (TNFR) family of proteins, and this virion binding is mediated by the complex discontinuous sequences present in the C-terminal two-thirds of EIAV envelope glycoprotein gp90 (Zhang et al. 2005; Sun et al. 2008). After the receptor-virus binding, the wild-type EIAV strain is generally internalized by clathrin-mediated and low-pH-dependent endocytosis (Brindley and Maury 2008; Jin et al. 2005). Some superinfecting EIAV strain like $EIAV_{vMA\text{-}1c}$ enters into the target cells pH-independently by direct membrane fusion of viral envelope and host plasma membrane (Brindley and Maury 2008). Once the virus is inside the host cell, it uncoats its genetic material and releases into the cytosol. The replication of the EIAV genome starts with the conversion of RNA genome into a single-stranded complementary deoxyribonucleic acid (cDNA) catalyzed by the virally encoded enzyme, reverse transcriptase (RT) (Rubinek et al. 1994). The lentivirus RT enzyme has three very characteristic intrinsic properties including ribonuclease activity (RNase H), RNA-dependent DNA polymerase activity (RDDP), and DNA-dependent DNA polymerase activity (DDDP) (Rubinek et al. 1994). The RT enzyme of EIAV has an approximate sequence similarity of 42 % and 40.6 % at amino acid level with HIV-1 and HIV-2, respectively, and also very similar in many functional properties including the *Km* values for the DNA polymerase activities, the heat stability of the DNA polymerase functions, and the specific activity of the RNase H function (Rubinek et al. 1994). The RDDP activity of RT copies the plus-stranded RNA genome forming an RNA-DNA hybrid. The RNase H activity degrades the viral RNA specifically in the RNA-DNA hybrid (Rubinek et al. 1994). Synthesis of the second DNA strand (complementary to the first strand) is mediated by the DDDP activity of RT resulting in the production of linear double-stranded viral DNA (dsDNA) (Rubinek et al. 1994). Though EIAV RT is error prone, it catalyzes nucleotide mismatches with higher specificity for A:C. The enzyme also cannot correct the errors for the initial viral replication which occurs in the cytosol where proofreading ability of the cell is absent. Moreover, the enzyme also lacks the 3'–5'exonuclease activity, like other RTs (reviewed in Cook et al. 2013). The error rate of EIAV RT has been estimated to be at least one nucleotide substitution per replication cycle of the virus (Issel et al. 2014). In addition to the absence of proofreading capacity, frequent recombination events occurring between two copies of EIAV genomic RNA during the replication cycle causes an accumulation of mutations in the viral genome resulting in the emergence of many new variants of viral genotypes known as quasispecies (Issel et al. 2014). The dsDNA of EIAV is then translocated to the nucleus where it is integrated into the host cell chromosomal DNA by pre-integration complex (PIC) mediated by virally encoded enzyme integrase and other cellular proteins (Debyser et al. 2015; Hacker et al. 2006). This viral dsDNA integrated inside the host cell genome is now called provirus. The EIAV DNA is often integrated in the AT-rich region of the genome avoiding transcription start site, and integration favors long interspersed elements (LINEs) in contrast to short interspersed elements (SINEs) for HIV-1 (Hacker et al. 2006; Liu et al. 2015). Incorporation of provirus into the host cell chromosome makes the provirus an integral part of the host chromosome, and this aids the provirus in the evasion of cellular immune surveillance resulting in the persistent infection of EIAV in the equine hosts for life (Issel et al. 2014 Liu et al. 2015). Subsequent to integration, the provirus utilizes cellular machineries to replicate its proviral DNA which then undergoes transcription to produce multiple messenger RNAs (mRNAs) from alternate splicing. The EIAV mRNAs are then exported into the cytoplasm where they are translated into different proteins. The viral accessory protein Rev plays an important role in the nuclear export of full-length genomic as well as single spliced viral RNAs and the Tat functions in the elongation of nascent viral RNA transcripts by RNA polymerase II (Carvalho and Derse 1991; Martarano et al. 1994). The EIAV mRNAs are then translated into viral proteins in the cytoplasm utilizing host cell ribosomal machineries. Genomic full-length mRNAs are translated into polyprotein precursor (PR55^{gag}) which gets cleaved by virally encoded proteinase and produces p15 MA, p26 CA, p11 NC, and p9 "late-domain" protein (Cook et al. 2013). Another polyprotein precursor, PR180$^{gag/pol}$, translated from the genomic full-length EIAV mRNA is cleaved by virally encoded proteinase to produce RT, integrase, dUTPase (DU), and other enzymes such as proteinase involved in EIAV replication (Cook et al. 2013). The newly translated viral proteins along with two copies replicated viral RNA genome assembles into a viral particle. The NC protein mediates the packaging of genomic RNA into progeny virions (Cook et al. 2013). The MA protein initiates the EIAV particle assembly, while CA protein controls packaging by polymerizing capsid shell around the NC protein coupled with genomic RNA (Cook et al. 2013). The progeny virions then bud from the host cell membrane retaining a portion of the lipid bilayer membrane as envelope and infect a new host cell.

8.3 Epidemiology

Although EIAV infection is worldwide in distribution and endemic in many countries, the virus is not endemic in Southern Africa (Barnard 1997; DeKock 1923). The disease has been reported in most of the states in the USA but is more prevalent in the Gulf Coast Region with a favorable climate for vector-mediated transmission of EIAV (Mealey 2014). After the incorporation of the EIA, control program involving identification of EIA-positive animals by serological tests followed by their humane killing has reduced the cases from 4 % to 0.003 % from 1972 to 2010, respectively (reviewed in Mealey 2014). In India, the first EIA case was diagnosed in 1987 from a Thoroughbred horse, and after adopting strict EIA control measures, the prevalence of EIA-positive animals has remarkably reduced from 4 % in 1987 to practically zero in 1999 (Malik et al. 2013). In Europe also, the disease is very common and endemic in Italy and Romania. EIA cases have also been reported in Ireland in 2006, and very recently, EIA outbreaks have been observed in many other parts of the Europe including Hungary, France, Greece, Belgium, Germany, and Great Britain (Office International des Épizooties, OIE databases; Cruz et al. 2015; Mooney et al. 2006). The 2006 EIA outbreak in Ireland indicated the involvement of aerosol transmission, which is an unusual method for EIAV transmission besides usual iatrogenic means or by insect vectors (More et al. 2008; Bolfa et al. 2013). Strict control measures were taken by the Government of Ireland for successful eradiation of EIA costing more than €1 million, and no more cases have been reported to date (More et al. 2008; Bolfa et al. 2013).

In some countries, the disease has been reported to reemerge after a long gap. For example, two unrelated cases of EIA have been reported in India, one each in 2010 and 2012, in a recent report on the continuous serological surveillance over a period of 13 years (1999 to 2012) consisting more than 67,000 equine sera (including Thoroughbred and indigenous equines) (Malik et al. 2013). Though no case of EIA has been found in India till date, after these unusual cases (personal communication), the threat of potential reemergence of the EIAV infection in equids is not implausible. Similarly, the EIA cases have been reported to reemerge after a 17-year long gap in Japan (Murakami et al. 2012). In recent past, the disease incidences have also been detected in other Southeast Asian countries such as China, Mongolia, Thailand, Uzbekistan, the Philippines, and Malaysia (Murakami et al. 2012; Pagamjav et al. 2011). In India, the disease is strictly notifiable to the international regulating authority such as OIE like other countries including the USA besides reporting to national regulatory authorities like the Department of Animal Husbandry, Dairying and Fisheries of Ministry of Agriculture and Farmers Welfare, Government of India (Malik et al. 2013).

All the members of the Equidae family are susceptible to the virus, and clinical cases of EIA have been reported in horses, ponies, mules, zebras, and donkeys. Clinical symptoms depend on a number of factors including the viral factors (virus strain, titer of infecting virus in the blood of animal) and host factors (individual host genetics and also host species) (Cook et al. 2001). For example, virulence of the virus strain can be increased with serial in vivo passage of the virus (Kemeny et al. 1971; Rwambo et al. 1990). On the other hand, under identical experimental conditions, donkeys exhibit reduced viremia and mild or no clinical symptoms if infected with horse-adapted EIAV and also mount slower and lower humoral immune responses compared to horses (Cook et al. 2001).

EIAV is the classical example of a blood-borne pathogen, which is usually transmitted from one infected animal to another susceptible animal mechanically by the blood-sucking insect vectors, mainly tabanids (horse flies and deer flies) besides stable flies (*Stomoxys calcitrans*) (Hawkins et al. 1973, 1976; Issel and Foil 1984) (Fig. 8.2). Horse flies are more efficient vectors in transmitting EIAV because of their large mouthparts and carry more blood compared to deer flies or stable flies (Issel and Foil 1984). Under favorable environment, a single horse fly is sufficient to transmit the virus from an acutely infected horse to another susceptible horse (Hawkins et al. 1976). On the other hand, a group of 25 medium-sized horse flies (*Tabanus fuscicostatus*) can transmit the virus from an inapparent carrier horse to a susceptible horse (Issel et al. 1982). In a comparative study, it has been shown experimentally that a single horse fly or at least six deer flies or at least 52 stable flies are required to transmit EIAV from an acutely infected animal to another uninfected healthy animal (Hawkins et al. 1976; Issel and Foil 1984). For the effective transmission, the distance between infected and separating animals is considered to be a critical factor and should be less than 200 yards (~182 meters) (Hawkins et al. 1976). In nature, the painful bite of a tabanid on an infected animal leads to the defensive movement by the animal and this interrupts the blood feeding of the fly. The interrupted feeding further stimulates the fly to complete its meal, and the fly returns to bite again either on the same infected host or a new susceptible host, thus potentially transmitting the virus into the new susceptible host (Hawkins et al. 1973; Issel and Foil 1984; Issel et al. 1988). Besides the close proximity between animals, several other factors such as the number and the feeding habits of insects, the number of times the insect bites to the same and other hosts, the viral load in the initial blood meal, and the amount of infected blood transferred between hosts decide the fate of the transmission (Cordes and Issel 1996). Climate plays an important role in

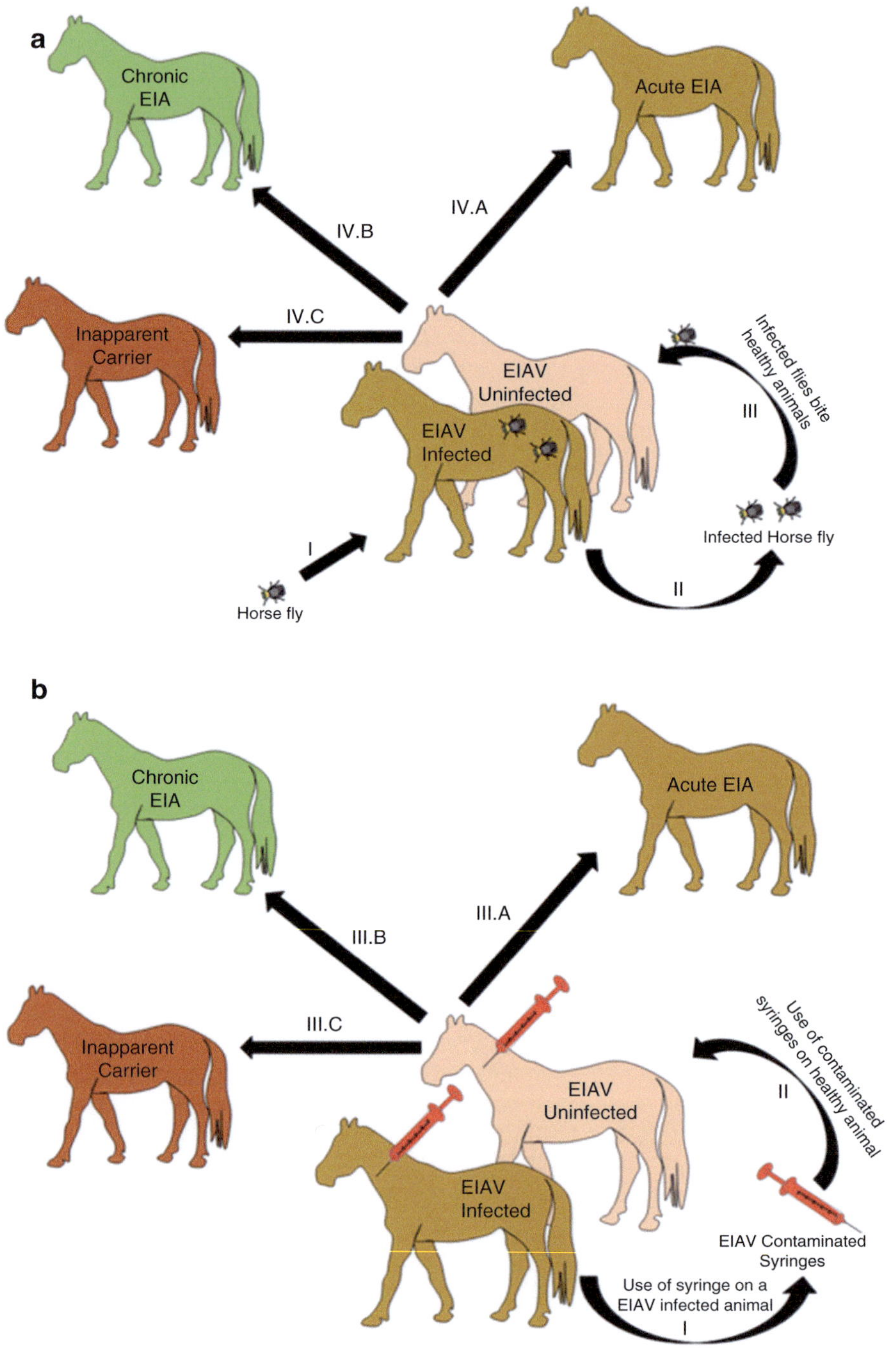
a
Chronic EIA
Acute EIA
IV.A
IV.B
Inapparent Carrier
IV.C
EIAV Uninfected
EIAV Infected
Infected flies bite healthy animals
III
Infected Horse fly
I
II
Horse fly
b
Chronic EIA
Acute EIA
III.A
III.B
Inapparent Carrier
III.C
EIAV Uninfected
EIAV Infected
Use of contaminated syringes on healthy animal
II
EIAV Contaminated Syringes
Use of syringe on a EIAV infected animal
I

the natural transmission since the insect vectors transmitting the EIAV are more prevalent in the warm climates compared to the cold climates, and that is the reason why in countries like Brazil, infection rate in some of the equine herds is nearly 50 % (Mealey 2014). The presence of hot-humid climate and hematophagous vectors remains the most important risk factor that poses a potential threat for transmission of EIAV in India too (Malik et al. 2013). The EIAV is found in the secretions and excretions of the infected animals, and iatrogenic transmission mediated by humans through the use of EIAV-contaminated needles, syringes, veterinary equipment, and plasma transfusions has also been reported in EIA outbreaks (Sellon et al. 1994; Issel et al. 1988; Cheevers and McGuire 1985; Stein et al. 1944). In a free-ranging herd of equids with EIA, stallions have been reported to be at more risk to acquiring EIAV infection because of their fighting behavior among themselves (Cook and Issel 2004). Vertical transmission may also be possible. Pregnant mare with the presence of clinical signs of EIA may transmit the virus to her foal through the perinatal transfer or transplacental route (McConnico et al. 2000). For transplacental transmission, pregnant mares need to suffer from clinical disease; febrile episodes and high-titer viremia are prerequisite factors (Sellon et al. 1994). However, high-tittered EIAV-positive mares with the absence of clinical signs of EIA may deliver uninfected healthy foals (Cook and Issel 2004; Issel et al. 1985). Venereal transmission of EIAV may rarely occur. Experimental subcutaneous inoculation of EIAV-infected semen from a stallion shows the transfer of the virus to an uninfected horse (Stein et al. 1944).

8.4 Pathogenesis

Pro-inflammatory cytokines such as tumor necrosis factor-α (TNF-α), interleukin-1α (IL-1α), interleukin-1β (IL-1β), interleukin-6 (IL-6), interleukin-10 (IL-10), and transforming growth factor-β (TGF-β) play the major role in shaping the clinical signs of EIA (Lim et al. 2005). Being the macrophage-tropic virus, majority of the EIAV strains primarily replicate in the tissue macrophages in vivo. However, some

Fig. 8.2 Transmission of EIA in equids. (**a**) Natural transmission. Horse fly (*Tabanus fuscicostatus*) bites an EIAV-infected animal and carries the virus within the blood meal on its mouth part (*I*). Because of defensive behavior of animal, EIAV infected fly has interrupted feeding (*II*). The fly comes back to complete its meal and bites a new uninfected susceptible animal in close proximity (*III*). The newly infected animal may develop either one of the three conditions: acute (*IV.A*), chronic (*IV.B*), and inapparent (*IV.C*) carrier. (**b**) Iatrogenic transmission. Hypodermic needles, syringes, or veterinary equipment get contaminated with EAIV-infected blood (*I*). Contaminated needles, syringes, or veterinary equipment if used on healthy EIAV-negative animal, the animal gets infected (II) and may become either acute case (*III.A*), chronic (*III.B*), or inapparent (*III.C*) carrier

strains of EIAV also replicate well in the endothelial cells. Viral burden, virus strain, and host species are critical in determining the outcome of the EIAV infection. When the viral load crosses a critical threshold level (EIAV RNA level of 5×10^7 to 1×10^8 copies/ml of plasma in experimental horse challenge model), the pro-inflammatory cytokines are released by the host cells, especially tissue macrophages (Lim et al. 2005; Cook et al. 2003), yet it is not known at this point if the cytokines are secreted by the infected or the bystander cells. Once secreted, pyrogenic cytokines such as TNF-α, IL-1α, IL-1β, and IL-6 trigger the activation of arachidonic acid pathways resulting in the production of prostaglandin E2 (PGE_2) leading to febrile episodes (Lim et al. 2005; Dinarello 1996; Sellon et al. 1999; Costa et al. 1997). In addition to fever, these cytokines may also contribute to EIA symptoms such as anorexia and hypermetabolism leading to wasting disease as has also been reported in HIV AIDS (Chang et al. 1998). It is thought that cytokines such as TNF-α and TGF-β are also the major contributors in the pathogenesis of thrombocytopenia seen in acute clinical cases of EIA by suppressing megakaryocyte function. Although EIAV does not infect megakaryocyte, it has been experimentally observed that EIAV infection causes elevated levels of TNF-α, IFN-α, and TGF-β in the plasma and bone marrow of EIAV-infected SCID foals (Tornquist and Crawford 1997; Crawford et al. 1996). Both TNF-α and TGF-β suppress the megakaryocyte colony proliferation in vitro and are thought to suppress the platelet formation resulting in thrombocytopenia (Tornquist and Crawford 1997). Besides inhibiting the platelet formation, TNF-α has also been shown to cause severe thrombocytopenia by inducing secretion of platelet agonists such as thrombin, plasmin, and serotonin in mice and humans (Michelmann et al. 1997; Tacchini-Cottier et al. 1998). IL-1 and IL-6 also activate platelets, and activated platelets in thrombocytopenic ponies infected with EIAV are removed from circulation quickly causing a reduction in platelets in the circulation (Russell et al. 1998). Furthermore, TNF-α also downregulates erythropoiesis and may be partly responsible for EIAV-associated anemia in equids (Moldawer et al. 1989; Dufour et al. 2003). In addition to the pro-inflammatory cytokines, adaptive immune responses also contribute to the pathogenesis of the EIA, mainly mediated by antibody and complement system. Platelets of EIAV-infected horses undergo immune-mediated destruction (either by complement-mediated lysis or phagocytosis) after binding with antibody (IgG or IgM) leading to thrombocytopenia, splenomegaly, and hepatomegaly (Banks and Henson 1972; Clabough et al. 1991). In chronic cases of EIA, the C3 component of the complement systems is deposited on the erythrocytes and in the glomeruli of the kidneys of horses (Perryman 1971). It has been observed that EIAV can trigger agglutination of erythrocytes in vitro, and therefore, it is thought that virus or virus-containing immune complex on the membrane of erythrocyte induces the activation of complement cascade via the classical pathway (Sentsui and Kono 1987a, b). Once the erythrocytes are coated with C3 component, it is readily destructed by phagocytosis causing anemia (Sentsui and Kono 1987a). This hypothesis is supported by the significant shortened life span (136 days vs 28 days or in some animals 136 days vs 87 days) of the erythrocytes found in the animals with acute EIAV infection compared to uninfected animals (Sentsui and Kono 1987b; McGuire et al.

1969). EIAV-associated destruction of erythrocytes also results in the deposition of hemosiderin granules in the macrophages of the liver, spleen, and lymph nodes (Perryman 1971; Sentsui and Kono 1987a). Deposition of immune complexes and C3 components of the complement on the glomeruli causes thickened glomerular tufts within the kidney of chronic EIAV-infected equids (Henson and McGuire 1971). Moreover, EIAV infection of horses has been shown to induce oxidative stress resulting from the modulation in the glutathione peroxidase and uric acid level (Bolfa et al. 2012). This increased oxidative stress may partly contribute to the pathogenesis of EIA by increasing inflammatory responses, facilitating viral replication while reducing the immune cell proliferation (Cook et al. 2013; Bolfa et al. 2012).

8.5 Clinical Symptoms

The clinical signs of EIA vary greatly and in many cases, the EIAV-seropositive horses do not show any clinical signs. However, depending on the virulence of the infecting strain, viral load in the blood, and genetics of the host, EIAV-infected animals may show one of the three forms of clinical syndromes – acute, chronic, and inapparent carrier (Sellon et al. 1994; McGuire et al. 1990). Although exact distinction among acute, chronic, and inapparent carrier is not always very clear, the acute infection represents the first febrile episode in the infected animals (Sellon et al. 1994). Animals infected with a virulent strain of EIAV may develop acute form of EIA with high viremia within an incubation period of 1–4 weeks (Cordes and Issel 1996; Sellon et al. 1994). The most common clinical signs in the acute infection is pyrexia (>102.5 °F), which can be associated with thrombocytopenia and depression (Cordes and Issel 1996; Sellon et al. 1994; McGuire et al. 1990). The febrile episodes usually last for short period (1–2 days) and can often go unnoticed. Some of the EIAV-infected horses may show a prolonged febrile episode that lasts for 3–6 days or until their death (Cook and Issel 2004; Sellon et al. 1994). Thrombocytopenia is one of the earliest and most consistent clinical signs found in acutely infected horses with EIAV (Sellon et al. 1994). Horses in acute EIA also become lethargic and may develop moderate to severe hemolytic anemia, anorexia, weight loss, and dependent edema. Petechiation and epistaxis may also be seen in severely infected animals (Issel et al. 2014). Death is rare in the first febrile episodes of EIA but may occur in animals infected with high virulent strains of EIAV (Sellon et al. 1994). If the animal recovers from the acute infection, the disease becomes chronic, and the animal may be afebrile for a period of time (days to months) followed by a recurrent febrile episode associated with other clinical signs such as lethargy, anorexia, rapid weight loss, thrombocytopenia, anemia, dependent edema, and occasionally neurologic signs such as ataxia and encephalitis (Sellon et al. 1994; Issel et al. 2014; Issel and Coggins 1979; McClure et al. 1982). If the animals do not die from the recurrent disease episodes, the frequency and intensity of the clinical signs usually reduce within 1–2 years. In the absence of frequent and severe clinical disease, the infected animals become inapparent carriers (Cook and Issel

2004). These clinically inapparent carriers may not show any recognizable signs until the detection of EIAV antibody during routine surveillance (Issel and Adams 1979). However, as with other lentiviruses, EIAV-infected animals remain infected for their lifetime (Sellon et al. 1994). Being the potential reservoir for an outbreak, the carrier animals always pose a threat to the uninfected healthy animals. Recrudescence of the clinical disease in inapparent carriers may occur during stress as has been shown by experimental administration of immune suppressive drugs as corticosteroids (Kono et al. 1976).

8.6 Pathological Lesions

Depending on the clinical signs, the pathological lesions also vary. At necropsy gross pathological changes found in EIAV-infected animals include generalized lymphadenopathy, splenomegaly, hepatomegaly, accentuated hepatic lobular structure, mucosal and visceral hemorrhages, ventral edema, evidence of anemia, icterus, and thrombosis (Sellon et al. 1994; Konno and Yamamoto 1970; Kemen and Coggins 1972). On histopathological examination, lymphocytic proliferation is observed in the spleen and lymph node (Pearson 1972). Other microscopic lesions include infiltration of lymphocytes as well as macrophages in different organs such as the liver (in periportal areas), spleen, bone marrow, adrenal gland, heart, kidney, and meninges (McGuire et al. 1990; Pearson 1972; Henson and McGuire 1971; Ishii and Ishitani 1975). There is hyperplasia of Kupffer cells containing hemosiderin aggregates (Sellon et al. 1994; Henson and McGuire 1971; Ishii and Ishitani 1975). Accumulation of hemosiderin is also found in the infiltrated macrophages in the lymph nodes, spleen, and bone marrow (Sellon et al. 1994). In about 75 % of EIAV-infected animals, there may be macroscopic lesions of glomerulitis with increased cellularity and thickening of glomerular tufts (Sellon et al. 1994; McGuire et al. 1990).

8.7 Diagnosis

As there is no specific pathognomonic clinical sign in EIA (acute or chronic) or no clinical sign at all (inapparent carrier), accurate diagnosis of EIA is not possible only on the basis of clinical history of the animal. However, clinical history of EIA such as fever (acute) or recurrent febrile episodes (chronic) accompanied with thrombocytopenia may sometimes aid in the diagnosis of EIA. But differential diagnosis from other infectious diseases such as equine viral arteritis, *Anaplasma phagocytophilum,* and other causes of edema, fever, anemia, or thrombocytopenia/ ecchymoses should be made (OIE 2013). As per the OIE guidelines, reliable diagnosis of EIA is based on the demonstration of EIAV-specific antibody present in the serum of the suspected animals. Currently, a combination of serological tests is advised for the confirmatory diagnosis of EIA infection. Other diagnostic tests targeting viral antigen such as polymerase chain reaction (PCR) have also been developed.

Virus Isolation Isolation of virus can be done by inoculating the blood of the suspected animal on the leukocyte cultures from the uninfected healthy animals (OIE 2013). However, virus isolation is not recommended for the diagnosis of EIA as the technique lacks the sensitivity, and also preparing the leukocyte culture itself is difficult and requires an expert technician (Issel et al. 2014; OIE 2013).

Serological Tests Because of the persistence of the virus in EIAV-infected animals, the OIE recommends demonstration of EIAV-specific precipitating antibodies by serological tests for the diagnosis of EIA (OIE 2013). The most reliable serological test used to date is the agar gel immunodiffusion (AGID) test or Coggins test, which is the gold standard for the diagnosis of EIA. This test which was developed by Leroy Coggins and coworkers in 1970 is still used routinely in the laboratory for the diagnosis of EIA (Coggins and Norcross 1970). AGID test detects the antibodies against the major core protein of EIAV, p26, where a specific reaction is characterized by the presence of a precipitin line between the EIAV antigen and the suspected test serum. The test is read by comparing with the precipitin reaction between antigen and positive control serum. Previously the test antigen for AGID test was prepared from the spleen of acutely infected horse, but antigens prepared on cultured equine cells or by recombinant DNA technology are now widely accepted (Coggins et al. 1972; Malmquist et al. 1973; Kong et al. 1997). In fact, majority of the commercially available test kits now use recombinant viral protein (p26) as the test antigens because of the low cost for antigen preparation and clearer and sharper precipitation lines seen in the AGID test (Cook et al. 2013). Antibodies detected by the AGID test greatly correlates with the detection of EIAV in the experimentally infected horses (Kemeny et al. 1971).

Although the AGID test is relatively rapid, inexpensive, simple, and highly specific, it also lacks the sensitivity and hence may provide false-negative results. As a result of this, many countries including the USA have developed enzyme-linked immunosorbent assay (ELISA)-based assays for the diagnosis of EIA. The US Department of Agriculture, Animal and Plant Health Inspection Service (USDA-APHIS) has approved the use of multiple ELISAs as equivalent diagnosis method to AGID test (Shen et al. 1984; Shane et al. 1984; Pare and Simard 2004; Singha et al. 2013). Majority of the ELISAs developed to date use EIAV core protein, p26, as coating antigen, but viral protein gp45 has also been used as antigen in the ELISA (Pare and Simard 2004). In countries where ELISAs are not available and AGID is the only approved test, the shortcomings of the AGID test can be managed by testing paired serum samples from acute and convalescent phase serum (Cook and Issel 2004).

In addition to AGID test and ELISA, other serological tests such as Western blot have also been developed in the late 1980s where sucrose gradient-purified EIAV is used as test antigen. The test can detect the antibodies against viral proteins p26, gp90, and gp45 (Cook et al. 2013; Hussain et al. 1987). The Western blot results show that serum from EIAV-infected horses with low antibodies to the core protein, p26 (or nearly negative by AGID test), generally has higher reactivity to envelope proteins, gp45 and gp90. Because of these discrepancies among different serological assays, a three-tiered EIA diagnostic strategy is recommended where serum samples are first screened with ELISAs followed by the confirmatory AGID test. If there is any discrepancy between ELISA and AGID test results, Western blot test is performed. The USA approved the use of Western blot technique as supplemental test provided the test is performed in the specific reference laboratories including National Veterinary Science Laboratories and the University of Kentucky. A survey report has shown that the three-tiered EIA diagnostic strategy dramatically increases the detection of EIA cases by about 17 % when compared to diagnosis by AGID test only (Issel et al. 2013, 2014).

Polymerase Chain Reaction None of the serological assays available at this moment for the diagnosis of EIA can detect the virus immediately or even within few days after the exposure of the animal to the virus because it takes a considerable time before the humoral immune response is initiated. But during the time between the exposure and the presence of antibody in the serum, the recently exposed animals can be a potential risk for transmission of the virus to other healthy animals (Issel et al. 2014). Since the current recommended serodiagnostic strategies cannot detect the recent infection, additional diagnostic procedures are required to be developed to detect either infectious virus or viral RNA or proviral DNA immediately after infection. As discussed earlier that isolation of infectious virus particle is not a practical and viable approach, many PCR-based assays, both nested reverse-transcriptase PCR (RT-nPCR) and quantitative real-time RT-PCR (qRT-PCR), have been developed to detect the viral nucleic acids. Majority of the PCR-based assays have targeted gag gene for amplification (Capomaccio et al. 2012; Cappelli et al. 2011; Cook et al. 2002; Nagarajan and Simard 2001; Dong et al. 2012; Langemeier et al. 1996). A nested PCR assay, the currently OIE-recommended PCR assay, developed by Nagarajan and Simard (2001), has been used in Canada to successfully detect EIA proviral DNA from the peripheral blood of horses. Other PCR assays are also available for detection of EIAV nucleic acids from the plasma of infected equines at the early stage of EIAV infection as well as from the inapparent carrier animals (Cappelli et al. 2011; Dong et al. 2012; Langemeier et al. 1996; Quinlivan et al. 2007; Harrold et al. 2000). In spite of the availability of multiple PCR-based assays, none of them can detect all strains of EIAV currently circulating all over the globe. The major challenge in designing a universal PCR assay is the ability of the virus to mutate resulting in the difficulty to find a conserved region for amplification. EIAV sequences vary greatly between geographical areas. Even in the same geographical area, the sequences can vary significantly within a particular host (Cook et al. 2013). If there is any mutation in the primer binding sequences or in the probe (for TaqMan qRT-PCR assay), then the specificity of the assay will be compromised. Another problem in developing a PCR-based assay is the lack of sensitivity during low viral burden in the blood especially in inapparent carriers (Harrold et al. 2000).

8.8 Treatment, Development of Vaccines, and Prophylaxis Measures

Currently, no specific treatment for EIA is available. American Association of Equine Practitioners (AAEP) along with other regulatory authorities recommends the euthanasia of the EIAV-seropositive animals (Issel et al. 2014; OIE 2013). The palliative therapy through symptomatic and supportive treatment has no effect on controlling the virus itself (Issel et al. 2014). Corticosteroids are contraindicated as it may stimulate the recurrence of the disease symptoms. Regulatory guidelines in several countries prohibit treatment of EIA-infected animals.

Several approaches have been undertaken to develop a successful vaccine that can prevent EIAV infection and clinical disease in equids. So far, none of them including inactivated whole-virus vaccine, modified live virus vaccine, subunit vaccines, vaccines using recombinant viral proteins, or peptide vaccines with T-helper epitopes have been successful in preventing the clinical EIA disease against all strains (homologous and heterologous strains) (Cook et al. 2013; Craigo et al. 2005; Issel et al. 1992; Li et al. 2003). Trials with inactivated virus, subunit, or DNA vaccines have shown a limited protection against the clinical disease. Inactivated whole-virus vaccine has protected equids from EIA disease only from the homologous virus challenge but not from heterologous virus challenge (Issel et al. 1992). The major challenges in the development of a universal EIAV vaccine are the ability of the virus to mutate resulting in the antigenic variation (antigenic drift), emergence of new strain in different geographical locations, resistance to neutralizing antibodies, ability of the virus to be latent, and ability of the virus to integrate its genome into the host genetic material (Issel et al. 2014). A modified live-attenuated virus vaccine developed in the Republic of China developed in the 1970s by serially passing the virulent virus on donkey leukocyte culture was extensively used in China between 1970 and 1990 over 60 million horses (Shen 1983, 1986). Although the Chinese live-attenuated vaccine claimed to protect the vaccinated horses from clinical disease by EIAV infection, detailed reports on the vaccine such as the genetic mechanism of this live attenuation, sequence information, and long-term effect of this vaccination were not widely accessible to the scientific community and were only mentioned in the Chinese literatures (Cook and Issel 2004; Shen 1983; Cohen 2004). In addition to this, the live-attenuated vaccine cannot protect all the vaccinated animals especially from heterologous virus challenge, and the vaccine strains also can persist in the animal (Meng et al. 2011). Besides, using a live-attenuated vaccine with retrovirus like EIAV itself carries other risks including the possibility of the reversion of the attenuated virus to wild type due to the tremendous ability of EIAV to mutate, to recombine, and to integrate into host genome (Issel et al. 2014; OIE 2013). Due to the low prevalence of disease after 1990 and also to avoid the interference of diagnostic testing by vaccine antibodies, the vaccine was discontinued in China and the EIA control strategy now changed from vaccination to quarantine of EIAV-seropositive animals (OIE 2013). Hence, development of safe and effective universal vaccines that can provide protection of EIAV-infected animals from clinical disease from a broad spectrum of EIAV strains is yet to be achieved. Use of vaccines against EIA is not recommended in many countries where official control program is in place, to elude interference in diagnostic testing due to vaccination.

Because there are no treatments or prophylactic measures for EIA, effective control of the disease depends entirely on the successful breaking of the transmission cycle of the virus by interfering with the EIAV transmission by mechanical vectors (tabanids) or blood contamination by human intervention. To break the transmission cycle, proper identification followed by quarantine or euthanasia of the EIAV-positive animals from the susceptible population must be done (Issel et al. 2014). Although EIA was first described many years ago, screening of equids for detection of EIA cases was not possible until the discovery of AGID or Coggins test in 1972. International regulatory bodies including OIE also recommend the sero-surveillance of susceptible equine populations for the international trade and imposition of restrictions on the trade or movement of the serologically positive equids (Cook and Issel 2004; Issel et al. 2014; OIE 2013). In many countries, the only control measure is the identification of EIAV-seropositive equids and their euthanasia. In other countries where quarantine of the seropositive equids is an option, the animals must be segregated from the seronegative horses by the distance of more than 180 meters (~200 meters) to break the transmission of the virus by tabanids. During the interrupted feeding, the hematophagous fly will return to the same host and will not infect the second host to complete its meal if the distance between two animals is kept about 200 meters (Hawkins et al. 1976). Often following this restriction is impractical for many animal owners, and they decide for human killing of the infected animals (Issel et al. 2014).

As discussed earlier that in addition to the vector-mediated transmission, contaminated blood by human interventions also plays a major role in the transmission of EIA. In fact, during feeding on host, horse and deer flies carry very little amount of contaminated blood in their mouthparts usually less than one hundred thousandth of a milliliter which is much smaller compared to the residual volume of blood in a hypodermic needle (0.05–0.1 ml) used during the collection of blood by venipuncture (Cook et al. 2013; Foil et al. 1987). Also the viremia in an inapparent carrier horse is generally less than 50 % horse-infective dose per milliliter (HID_{50}/ml), and flies will carry very little amount of this infected blood which may not be sufficient to induce clinical disease (Issel et al. 1982). Hence, human intervention in the transmission of EIAV from inapparent carrier animal to an uninfected animal may not be ruled out. Indeed, in controlled experimental condition, the role of iatrogenic transmission of EIAV has been demonstrated (Cook et al. 2013). However, iatrogenic transmission of EIAV can be controlled by following proper sanitary measures designed to prevent transmission of blood-borne pathogens (Issel et al. 1982; Montelaro et al. 1993). Constant monitoring and surveillance associated with quarantine and elimination of seropositive animals under the appropriate legal provisions followed by regulation of movement of horses coupled with other sanitary and phytosanitary measures under the national policy on control of EIA resulted in successfully controlling EIA in India. Veterinarians also need to report properly to their respective regulatory authorities if they diagnose any new case of EIA. In the EIA endemic countries where horses are used as vital agricultural animals, provision of proper compensation before adopting the identification and destruction policy may aid in controlling EIA (Issel et al. 2014). Otherwise, this may be impractical to remove EIA-infected animals and may interrupt the goal of EIA control program. In summary, horses should be tested annually for EIAV, and if found positive, proper control measures should be taken to stop further transmission of the virus. New horses in herd should always be tested for EIA before inclusion in the herd to minimize the accidental outbreak of EIA in a herd.

References

Banks KL, Henson JB (1972) Immunologically mediated glomerulitis of horses. II. Antiglomerular basement membrane antibody and other mechanisms in spontaneous disease. Lab Invest 26:708–715

Barnard BJ (1997) Antibodies against some viruses of domestic animals in southern African wild animals. Onderstepoort J Vet Res 64:95–110

Bolfa P, Nolf M, Cadore JL et al (2013) Interstitial lung disease associated with equine infectious anemia virus infection in horses. Vet Res 44:113

Bolfa PF, Leroux C, Pintea A et al (2012) Oxidant-antioxidant imbalance in horses infected with equine infectious anaemia virus. Vet J 192:449–454

Brindley MA, Maury W (2005) Endocytosis and a low-pH step are required for productive entry of equine infectious anemia virus. J Virol 79:14482–14488

Brindley MA, Maury W (2008) Equine infectious anemia virus entry occurs through clathrin-mediated endocytosis. J Virol 82:1628–1637

Capomaccio S, Willand ZA, Cook SJ et al (2012) Detection, molecular characterization and phylogenetic analysis of full-length equine infectious anemia (EIAV) gag genes isolated from Shackleford Banks wild horses. Vet Microbiol 157:320–332

Cappelli K, Capomaccio S, Cook FR et al (2011) Molecular detection, epidemiology, and genetic characterization of novel European field isolates of equine infectious anemia virus. J Clin Microbiol 49:27–33

Carvalho M, Derse D (1991) Mutational analysis of the equine infectious anemia virus Tat-responsive element. J Virol 65:3468–3474

Chang HR, Dulloo AG, Bistrian BR (1998) Role of cytokines in AIDS wasting. Nutrition 14:853–863

Cheevers WP, McGuire TC (1985) Equine infectious anemia virus: immunopathogenesis and persistence. Rev Infect Dis 7:83–88

Clabough DL, Gebhard D, Flaherty MT et al (1991) Immune-mediated thrombocytopenia in horses infected with equine infectious anemia virus. J Virol 65:6242–6251

Coggins L, Norcross NL (1970) Immunodiffusion reaction in equine infectious anemia. Cornell Vet 60:330–335

Coggins L, Norcross NL, Nusbaum SR (1972) Diagnosis of equine infectious anemia by immunodiffusion test. Am J Vet Res 33:11–18

Cohen J (2004) HIV/AIDS in China. Vaccine development with a distinctly Chinese flavor. Science 304:1437

Cook FR, Issel, CJ (2004) Equine infectious anemia. In: Coetzer JAW, Tustin RC (ed), Infectious diseases of livestock, vol 2. Oxford University Press South Africa, Cape Town, pp 747–752

Cook RF, Cook SJ, Berger SL et al (2003) Enhancement of equine infectious anemia virus virulence by identification and removal of suboptimal nucleotides. Virology 313:588–603

Cook RF, Cook SJ, Li FL et al (2002) Development of a multiplex real-time reverse transcriptase-polymerase chain reaction for equine infectious anemia virus (EIAV). J Virol Methods 105:171–179

Cook RF, Leroux C, Issel CJ (2013) Equine infectious anemia and equine infectious anemia virus in 2013: a review. Vet Microbiol 167:181–204

Cook SJ, Cook RF, Montelaro RC et al (2001) Differential responses of *Equus caballus* and *Equus asinus* to infection with two pathogenic strains of equine infectious anemia virus. Vet Microbiol 79:93–109

Cordes T, Issel, CJ (1996) Equine infectious anemia: a status report on its control. USDA-APHIS

Costa LR, Santos IK, Issel CJ et al (1997) Tumor necrosis factor-alpha production and disease severity after immunization with enriched major core protein (p26) and/or infection with equine infectious anemia virus. Vet Immunol Immunopathol 57:33–47

Craigo JK, Li F, Steckbeck JD et al (2005) Discerning an effective balance between equine infectious anemia virus attenuation and vaccine efficacy. J Virol 79:2666–2677

Craigo JK, Montelaro RC (2011) Equine infectious anemia virus infection and immunity: lessons for AIDS vaccine development. Futur Virol 6:139–142

Craigo JK, Montelaro RC (2013) Lessons in AIDS vaccine development learned from studies of equine infectious, anemia virus infection and immunity. Viruses 5:2963–2976

Crawford TB, Wardrop KJ, Tornquist SJ et al (1996) A primary production deficit in the thrombocytopenia of equine infectious anemia. J Virol 70:7842–7850

Cruz F, Fores P, Ireland J et al (2015) Freedom from equine infectious anaemia virus infection in Spanish Purebred horses. Vet Rec Open 2:e000074

De Kock GVDW (1923) A contribution to the study of the virus, hematology, and pathology of equines under South African conditions. 9th and 10th reports of the Director of Veterinary Education and Research, pp 253–313

Debyser Z, Christ F, De Rijck J et al (2015) Host factors for retroviral integration site selection. Trends Biochem Sci 40:108–116

Dinarello CA (1996) Biologic basis for interleukin-1 in disease. Blood 87:2095–2147

Dong JB, Zhu W, Cook FR et al (2012) Development of a nested PCR assay to detect equine infectious anemia proviral DNA from peripheral blood of naturally infected horses. Arch Virol 157:2105–2111

Dong JB, Zhu W, Cook FR et al (2013) Identification of a novel equine infectious anemia virus field strain isolated from feral horses in southern Japan. J Gen Virol 94:360–365

Dreguss MN, Lombard LS (1954) Experimental studies in equine infectious anemia. University of Pennsylvania Press, Philadelphia

Dufour C, Corcione A, Svahn J et al (2003) TNF-alpha and IFN-gamma are overexpressed in the bone marrow of Fanconi anemia patients and TNF-alpha suppresses erythropoiesis in vitro. Blood 102:2053–2059

Foil LD, Adams WV, McManus JM et al (1987) Bloodmeal residues on mouthparts of *Tabanus fuscicostatus* (Diptera: Tabanidae) and the potential for mechanical transmission of pathogens. J Med Entomol 24:613–616

Hacker CV, Vink CA, Wardell TW et al (2006) The integration profile of EIAV-based vectors. Mol Ther 14:536–545

Harrold SM, Cook SJ, Cook RF et al (2000) Tissue sites of persistent infection and active replication of equine infectious anemia virus during acute disease and asymptomatic infection in experimentally infected equids. J Virol 74:3112–3121

Hawkins JA, Adams WV Jr, Wilson BH et al (1976) Transmission of equine infectious anemia virus by *Tabanus fuscicostatus*. J Am Vet Med Assoc 168:63–64

Hawkins JA, Adams WV, Cook L et al (1973) Role of horse fly (*Tabanus fuscicostatus* Hine) and stable fly (*Stomoxys calcitrans* L.) in transmission of equine infectious anemia to ponies in Louisiana. Am J Vet Res 34:1583–1586

Henson JB, McGuire TC (1971) Immunopathology of equine infectious anemia. Am J Clin Pathol 56:306–313

Hussain KA, Issel CJ, Schnorr KL et al (1987) Antigenic analysis of equine infectious anemia virus (EIAV) variants by using monoclonal antibodies: epitopes of glycoprotein gp90 of EIAV stimulate neutralizing antibodies. J Virol 61:2956–2961

Ishii S, Ishitani R (1975) Equine infectious anemia. Adv Vet Sci Comp Med 19:195–222
Issel CJ, Adams WV Jr (1979) Serologic survey for equine infectious anemia virus in Louisiana. J Am Vet Med Assoc 174:286–288
Issel CJ, Adams WV Jr, Foil LD (1985) Prospective study of progeny of inapparent equine carriers of equine infectious anemia virus. Am J Vet Res 46:1114–1116
Issel CJ, Adams WV Jr, Meek L et al (1982) Transmission of equine infectious anemia virus from horses without clinical signs of disease. J Am Vet Med Assoc 180:272–275
Issel CJ, Coggins L (1979) Equine infectious anemia: current knowledge. J Am Vet Med Assoc 174:727–733
Issel CJ, Cook RF, Mealey RH et al (2014) Equine infectious anemia in 2014: live with it or eradicate it? Vet Clin North Am Equine Pract 30:561–577
Issel CJ, Foil LD (1984) Studies on equine infectious anemia virus transmission by insects. J Am Vet Med Assoc 184:293–297
Issel CJ, Horohov DW, Lea DF et al (1992) Efficacy of inactivated whole-virus and subunit vaccines in preventing infection and disease caused by equine infectious anemia virus. J Virol 66:3398–3408
Issel CJ, Rushlow K, Foil LD et al (1988) A perspective on equine infectious anemia with an emphasis on vector transmission and genetic analysis. Vet Microbiol 17:251–286
Issel CJ, Scicluna MT, Cook SJ et al (2013) Challenges and proposed solutions for more accurate serological diagnosis of equine infectious anaemia. Vet Rec 172:210
Jin S, Zhang B, Weisz OA et al (2005) Receptor-mediated entry by equine infectious anemia virus utilizes a pH-dependent endocytic pathway. J Virol 79:14489–14497
Kemen MJ Jr, Coggins L (1972) Equine infectious anemia: transmission from infected mares to foals. J Am Vet Med Assoc 161:496–499
Kemeny LJ, Mott LO, Pearson JE (1971) Titration of equine infectious anemia virus. Effect of dosage on incubation time and clinical signs. Cornell Vet 61:687–695
Kong XG, Pang H, Sugiura T et al (1997) Application of equine infectious anemia virus core proteins produced in a baculovirus expression system to serological diagnosis. Microbiol Immunol 41:975–980
Konno S, Yamamoto H (1970) Pathology of equine infectious anemia. Proposed classification of pathological types of disease. Cornell Vet 60:393–449
Kono Y, Hirasawa K, Fukunaga Y et al (1976) Recrudescence of equine infectious anemia by treatment with immunosuppressive drugs. Natl Inst Anim Health Q (Tokyo) 16:8–15
Kono Y, Yoshino T, Fukanaga Y (1970) Growth characteristics of equine infectious anemia virus in horse leukocyte cultures. Arch Gesamte Virusforsch 30:252–256
Langemeier JL, Cook SJ, Cook RF et al (1996) Detection of equine infectious anemia viral RNA in plasma samples from recently infected and long-term inapparent carrier animals by PCR. J Clin Microbiol 34:1481–1487
Leroux C, Cadore JL, Montelaro RC (2004) Equine Infectious Anemia Virus (EIAV): what has HIV's country cousin got to tell us? Vet Res 35:485–512
Li F, Craigo JK, Howe L et al (2003) A live attenuated equine infectious anemia virus proviral vaccine with a modified S2 gene provides protection from detectable infection by intravenous virulent virus challenge of experimentally inoculated horses. J Virol 77:7244–7253
Liang H, He X, Shen RX et al (2006) Combined amino acid mutations occurring in the envelope closely correlate with pathogenicity of EIAV. Arch Virol 151:1387–1403
Lignee M (1843) Memoire et observations sur ure maladie de sarg, connue sous le nom d'anhemie hydrohemie. Cachexie acqueuse du cheval. Recueil de Medecine Veterinaire 20:30
Lim WS, Payne SL, Edwards JF et al (2005) Differential effects of virulent and avirulent equine infectious anemia virus on macrophage cytokine expression. Virology 332:295–306
Liu Q, Wang XF, Ma J et al (2015) Characterization of equine infectious anemia virus integration in the horse genome. Viruses 7:3241–3260
Malik P, Singha H, Goyal SK et al (2013) Sero-surveillance of equine infectious anemia virus in equines in India during more than a decade (1999-2012). Indian J Virol 24:386–390

Malmquist WA, Barnett D, Becvar CS (1973) Production of equine infectious anemia antigen in a persistently infected cell line. Arch Gesamte Virusforsch 42:361–370
Martarano L, Stephens R, Rice N et al (1994) Equine infectious anemia virus trans-regulatory protein Rev controls viral mRNA stability, accumulation, and alternative splicing. J Virol 68:3102–3111
Matheka HD, Coggins L, Shively JN et al (1976) Purification and characterization of equine infectious anemia virus. Arch Virol 51:107–114
Maury W, Oaks JL, Bradley S (1998) Equine endothelial cells support productive infection of equine infectious anemia virus. J Virol 72:9291–9297
McClure JJ, Lindsay WA, Taylor W et al (1982) Ataxia in four horses with equine infectious anemia. J Am Vet Med Assoc 180:279–283
McConnico RS, Issel CJ, Cook SJ et al (2000) Predictive methods to define infection with equine infectious anemia virus in foals out of reactor mares. J Equine Vet 20:387–392
McGuire TC, Henson JB, Quist SE (1969) Viral-induced hemolysis in equine infectious anemia. Am J Vet Res 30:2091–2097
McGuire TC, O'Rourke KI, Perryman LE (1990) Immunopathogenesis of equine infectious anemia lentivirus disease. Dev Biol Stand 72:31–37
Mealey RH (2014) Equine infectious anemia. In: Sellon DC, Long MT (ed) Equine infectious diseases, 2nd ed. Saunders, St. Louis, pp 232–238
Meng Q, Lin Y, Ma J et al (2011) A pilot study on an attenuated Chinese EIAV vaccine inducing broadly neutralizing antibodies. Arch Virol 156:1455–1462
Michelmann I, Bockmann D, Nurnberger W et al (1997) Thrombocytopenia and complement activation under recombinant TNF alpha/IFN gamma therapy in man. Ann Hematol 74:179–184
Moldawer LL, Marano MA, Wei H et al (1989) Cachectin/tumor necrosis factor-alpha alters red blood cell kinetics and induces anemia in vivo. FASEB J 3:1637–1643
Montagnier L, Dauguet C, Axler C et al (1984) A new type of retrovirus isolated from patients presenting with lymphadenopathy and acquired immune deficiency syndrome: structural and antigenic relatedness with equine infectious anaemia virus. Annales de l'Institut Pasteur/Virologie 135:119–134
Montelaro RC, Ball JM, Rushlow KE (1993) Equine retroviruses. In: Levy JA (ed) The retroviridae, vol 2. Plenum Press, New York, pp 257–360
Mooney J, Flynn O, Sammin D (2006) Equine infectious anaemia in Ireland: characterisation of the virus. Vet Rec 159:570
More SJ, Aznar I, Bailey DC et al (2008) An outbreak of equine infectious anaemia in Ireland during 2006: investigation methodology, initial source of infection, diagnosis and clinical presentation, modes of transmission and spread in the Meath cluster. Equine Vet J 40:706–708
Murakami K, Konishi M, Kameyama K et al (2012) Detection of equine infectious anaemia virus in native Japanese ponies. Vet Rec 171:72
Nagarajan MM, Simard C (2001) Detection of horses infected naturally with equine infectious anemia virus by nested polymerase chain reaction. J Virol Methods 94:97–109
Oaks JL, McGuire TC, Ulibarri C et al (1998) Equine infectious anemia virus is found in tissue macrophages during subclinical infection. J Virol 72:7263–7269
OIE (2013) Equine infectious anaemia, chapter 2.5.6. OIE Terrestrial Manual 2013. pp 1–6
Pagamjav O, Kobayashi K, Murakami H et al (2011) Serological survey of equine viral diseases in Mongolia. Microbiol Immunol 55:289–292
Pare J, Simard C (2004) Comparison of commercial enzyme-linked immunosorbent assays and agar gel immunodiffusion tests for the serodiagnosis of equine infectious anemia. Can J Vet Res 68:254–258
Pearson JE (1972) Equine Infectious Anemia Diagnosis and Prevention. Iowa State University Veterinarian 34(2), Article 4
Perryman LE, McGuire TC, Banks KL et al (1971) Decreased C3 levels in a chronic virus infection: equine infectious anemia. J Immunol 106:1074–1078

Quinlivan M, Cook RF, Cullinane A (2007) Real-time quantitative RT-PCR and PCR assays for a novel European field isolate of equine infectious anaemia virus based on sequence determination of the gag gene. Vet Rec 160:611–618

Rubinek T, Loya S, Shaharabany M et al (1994) The catalytic properties of the reverse transcriptase of the lentivirus equine infectious anemia virus. Eur J Biochem 219:977–983

Russell KE, Walker KM, Miller RT et al (1998) Hyperglobulinemia and lymphocyte subset changes in naturally infected, inapparent carriers of equine infectious anemia virus. Am J Vet Res 59:1009–1015

Rwambo PM, Issel CJ, Hussain KA et al (1990) In vitro isolation of a neutralization escape mutant of equine infectious anemia virus (EIAV). Arch Virol 111:275–280

Sellon DC, Fuller FJ, McGuire TC (1994) The immunopathogenesis of equine infectious anemia virus. Virus Res 32:111–138

Sellon DC, Russell KE, Monroe VL et al (1999) Increased interleukin-6 activity in the serum of ponies acutely infected with equine infectious anaemia virus. Res Vet Sci 66:77–80

Sentsui H, Kono Y (1987a) Complement-mediated hemolysis of horse erythrocytes treated with equine infectious anemia virus. Arch Virol 95:53–66

Sentsui H, Kono Y (1987b) Phagocytosis of horse erythrocytes treated with equine infectious anemia virus by cultivated horse leukocytes. Arch Virol 95:67–77

Shane BS, Issel CJ, Montelaro RC (1984) Enzyme-linked immunosorbent assay for detection of equine infectious anemia virus p26 antigen and antibody. J Clin Microbiol 19:351–355

Shen DT (1986) Equine infectious anemia vaccine. In: Salzman LA (ed) Animal models of retrovirus infection and their relationship to AIDS. Academic Press, Orlando, pp 387–392

Shen DT, Gorham JR, McGuire TC (1984) Enzyme-linked immunosorbent assay for detection of equine infectious anemia antibody to purified P26 viral protein. Am J Vet Res 45:1542–1543

Shen R (1983) Development and use of an equine infectious anemia donkey leucocyte attenuated vaccine. In: Proceedings of the international symposium on immunity to equine infectious anemia. Harbin Veterinary Research Institute, Chinese Academy of Agricultural Sciences, Beijing, pp 34–53

Singha H, Goyal SK, Malik P et al (2013) Development, evaluation, and laboratory validation of immunoassays for the diagnosis of equine infectious anemia (EIA) using recombinant protein produced from a synthetic p26 gene of EIA virus. Indian J Virol 24:349–356

Stein CD, Osteen OS, Mott LO (1944) Experimental transmission of equine infectious anemia by contact and body secretions and excretions. Vet Med 39:46–52

Sun C, Zhang B, Jin J et al (2008) Binding of equine infectious anemia virus to the equine lentivirus receptor-1 is mediated by complex discontinuous sequences in the viral envelope gp90 protein. J Gen Virol 89:2011–2019

Tacchini-Cottier F, Vesin C, Redard M et al (1998) Role of TNFR1 and TNFR2 in TNF-induced platelet consumption in mice. J Immunol 160:6182–6186

Tornquist SJ, Crawford TB (1997) Suppression of megakaryocyte colony growth by plasma from foals infected with equine infectious anemia virus. Blood 90:2357–2363

Valle H, Carre H (1904) Sur la natur infectieuse de l'anenie du cheval. C R Acad Sci 139:331–333

Weiland F, Matheka HD, Coggins L et al (1977) Electron microscopic studies on equine infectious anemia virus (EIAV). Brief report. Arch Virol 55:335–340

Zhang B, Jin S, Jin J et al (2005) A tumor necrosis factor receptor family protein serves as a cellular receptor for the macrophage-tropic equine lentivirus. Proc Natl Acad Sci U S A 102:9918–9923

Duck Tembusu Virus Infection

9

Lijiao Zhang and Jingliang Su

9.1 Introduction

Tembusu virus (TMUV) was first detected in mosquitoes in Malaysia during surveillance studies of arthropod-borne pathogenic diseases in 1955 and was subsequently isolated in Thailand (cited by Mackenzie and Williams 2009). In 2000, a TMUV-related virus, designated Sitiawan virus, was isolated from diseased 4-week-old broiler chicken flocks in Perak State, Malaysia, that had clinical signs of encephalitis, growth retardation, and increased blood glucose levels (Kono et al. 2000). The disease was reproduced experimentally with isolates from chicks, providing direct evidence of its association with animal infections. The most serious outbreak of TMUV infection in domestic flocks occurred in April 2010 in Southeast China, where a highly infectious disease in laying ducks, characterized by reduced feed uptake and severe egg drop, was observed and spread rapidly to most farms in the duck-producing regions of the country by the end of the year (Cao et al. 2011; Su et al. 2011; Yan et al. 2011a, b). This epidemic was responsible for significant economic losses arising from the consequent reduced egg production in laying flocks, especially breeder flocks. During this period, infections were also frequently detected in young ducks and geese with neurological symptoms, and the etiological agent was identified as a Tembusu-related virus or duck Tembusu virus (Huang et al. 2013; Li et al. 2012a; Liu et al. 2012a; Yun et al. 2012a, b). An outbreak of infection with this virus in young duck flocks has recently been reported in Malaysia (Homonnay et al. 2014). These data indicate that Tembusu virus infection poses a threat to the poultry industry in Southeast Asia, where more than 80 % of the world's waterfowl are produced per annum.

L. Zhang • J. Su (✉)
Key Laboratory of Animal Epidemiology and Zoonosis, Ministry of Agriculture,
College of Veterinary Medicine, China Agricultural University, Beijing 100193, China
e-mail: suzhang@cau.edu.cn

J. Bayry (ed.), *Emerging and Re-emerging Infectious Diseases of Livestock*,
DOI 10.1007/978-3-319-47426-7_9

9.2 Etiology

Duck Tembusu virus (DTMUV) is a member of the Ntaya antigenic group in the genus *Flavivirus*, family *Flaviviridae*. It has also been referred to as Baiyangdian virus (Su et al. 2011) and duck egg drop syndrome virus when it was isolated from diseased ducks.

DTMUV has a nucleocapsid enclosed in a lipid bilayer envelope, and its infectivity is destroyed by treatment with lipid solvents (ether or chloroform) (Su et al. 2011). An electron-microscopic examination of in vitro-infected cells revealed viral particles with a diameter of 40–50 nm localized within cytoplasmic vesicles (Fig. 9.1). Its genome consists of a positive-sense, single-stranded RNA of 10,990 nucleotides (nt), with an open reading frame (ORF) of 10,278 nt, flanked by 5′- and 3′-noncoding regions of 94 and 618 nt, respectively (Liu et al. 2012b). A phylogenetic analysis showed that all DTMUV isolates are highly homologous, with strong similarity to the chicken-origin Sitiawan virus and the Tembusu prototype strain MM1775, but are less closely related to the Bagaza virus (BAGV) (Fig. 9.2) (Yu et al. 2013). BAGV was recently shown to be the same viral species as Israel turkey meningoencephalitis virus in this antigenic group (Fernández-Pinero et al. 2014).

9.3 Laboratory Host Systems

DTMUV can be propagated in embryonated duck, chicken, and goose eggs after allantoic sac inoculation (Su et al. 2011; Yan et al. 2011b). Viral infection may result in embryo death 3–5 days after injection, and lesions can be seen in the

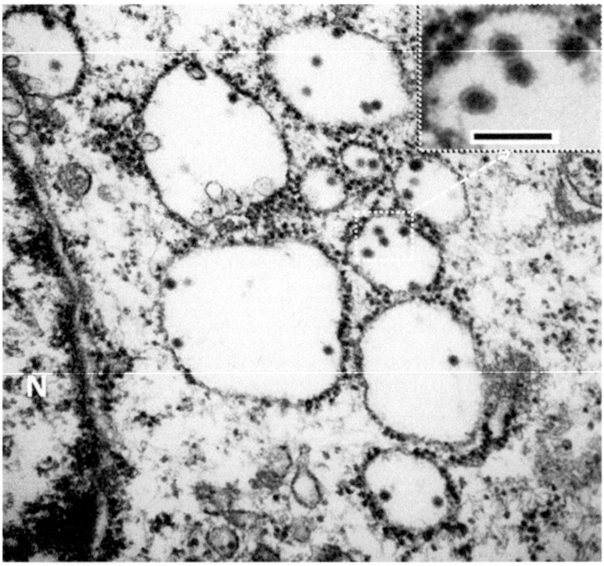

Fig. 9.1 Electron micrograph of a BHK-21 cell infected for 28 h with duck *Tembusu* virus. Viral particles are apparent in the cytoplasmic vesicles. *N* nucleus (*black bar* = 100 nm)

dead embryos, including body hemorrhage and focal necrosis in the liver. The virus can also replicate in a variety of primary cells and vertebrate and mosquito cell lines, including duck embryo fibroblasts (DEF), a chicken fibroblast cell line (DF-1), a baby hamster kidney fibroblast cell line (BHK-21), and an African green monkey kidney epithelial cell line (Vero). Cytopathic effects are readily produced in both primary and continuous cell cultures, and the infectious viral titer can be increased by serial passage. The *Aedes albopictus* cell line (C6/36) is susceptible to DTMUV and is frequently used to isolate the virus from mosquito samples (Pandey et al. 1999). However, some DTMUV isolates may not be well adapted to Vero cells, with significant reductions in their titers after 3–5 passages. Ducks are the animals used most frequently in laboratory studies to assess the pathogenicity of DTMUV and to study its pathogenesis. One-day-old chicks have been shown to develop the usual disease course, with mortality, after experimental infection (O'Guinn et al. 2013), and viremia is detected in 3–6-day-old chicks between 1 and 4 days after infection (O'Guinn et al. 2013). Three to six-week-old BALB/c mice are susceptible to intracerebral infection with duck TMUV (Li et al. 2013). Affected mice show significant weight loss and neurological signs; the mortality rate decreases with increasing age (Li et al. 2013). The intracerebral inoculation of suckling mice with duck TMUV causes typical encephalitis and approximately 100 % mortality (Li et al. 2013). The sucrose/acetone-extracted antigen can be prepared from infected mouse brains and used for the hemagglutination inhibition test, as has been described for *Japanese encephalitis virus* B (http://www.oie.int/fileadmin/Home/eng/Health_standards/tahm/2.01.07_JEV.pdf).

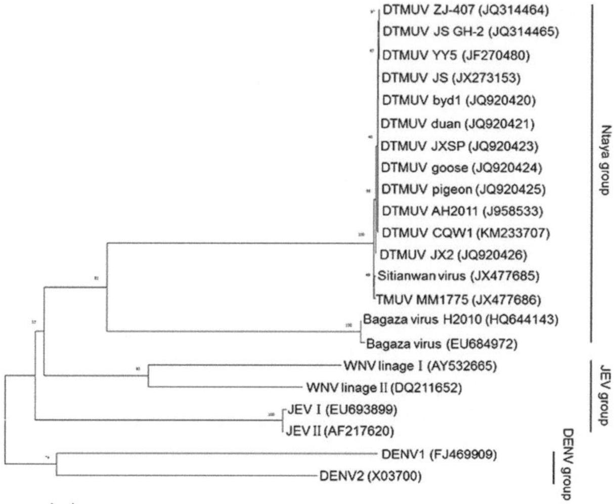

Fig. 9.2 Phylogenetic analysis of representative flaviviruses based on the ORF region using the neighbor-joining method with 1000 bootstrap replications. GenBank accession numbers are shown after each isolate name

9.4 Pathobiology and Epizootiology

9.4.1 Incidence and Distribution

The recognized disease caused by TMUV was first detected in Malaysian chicken flocks in 2000 (Kono et al. 2000), and severe outbreaks in farm-raised ducks and geese have been identified in China since 2010 (Cao et al. 2011; Huang et al. 2013; Li et al. 2012a; Liu et al. 2012a; Su et al. 2011; Yun et al. 2012a, b). Recently, TMUV infections in broiler ducks have been reported in some flocks in Perak, Malaysia (Homonnay et al. 2014). In Thailand, TMUVs have been detected in the sera of sentinel ducks that display no sign of illness (O'Guinn et al. 2013).

9.4.2 Natural and Experimental Hosts

TMUV is commonly detected in mosquito samples (Grubaugh et al. 2013; O'Guinn et al. 2013; Petz et al. 2014; Sudeep 2014) and has been isolated from *Culex tritaeniorhynchus*, *Cx. vishnui*, *Cx. gelidus*, *Cx. pseudovishnui*, *Cx. sitiens*, *Aedes linneatopennis*, and *Anopheles philippinensis* in Southeast Asia. Viral isolation and serological evidence suggest that wild and domestic birds are the principal vertebrate hosts of TMUV (Wolfe et al. 2001). Domestic ducks and geese of all ages are highly susceptible to infection under both field and laboratory conditions. Although field cases of TMUV infection in chickens have been reported (Kono et al. 2000; Chen et al. 2014b), outbreaks of TMUV infection are most commonly recorded in duck and goose flocks (Chen et al. 2013; Huang et al. 2013; Wan et al. 2012; Yun et al. 2012b; Zeng et al. 2015; Zhu et al. 2012; 2015;). The virus has also been isolated from house sparrows (*Passer domesticus*) and domestic pigeon (*Columba livia domestica*) collected near duck farms during the period of a disease outbreak (Liu et al. 2012b; Tang et al. 2013), but the pathogenicity of TMUV in these birds requires further investigation. Birds that show no significant clinical signs after infection may act as the maintenance and amplifying hosts in the transmission cycle of TMUV.

9.4.3 Transmission and Vectors

Because most flaviviruses can be transmitted through the bites of arthropod vectors, mosquitos may play an important role in the transmission of TMUV (Vaidya et al. 2012). TMUV transmission among chickens by mosquito bites has been confirmed under laboratory conditions (O'Guinn et al. 2013). Chicks survived and developed viremia after experimental infection at 3 or 5 days, which was high enough to infect mosquitos and transmit the virus to susceptible chickens. However, the disease occurred in the late winter in Northern China, when the activity of insect vectors was minimal, and the dramatic drop in egg production during the infection of laying duck flocks was greater than expected if arthropod-borne transmission was the principal route of infection. During disease outbreaks in domestic poultry flocks, the virus appears to be transmitted primarily by direct contact between infected and susceptible birds or by indirect contact through virus-contaminated fomites and aerosols. DTMUV is readily detectable in the respiratory tracts and feces of infected ducks (Yan et al. 2011a), and experiments have shown that the virus can be transmitted efficiently among ducks by direct contact and through aerosols (Li et al. 2015).

Vertical transmission seems to be another route of infection. Duck TMUV invades and replicates in the reproductive systems of infected adult ducks. It has been frequently detected in samples of the ovary follicular envelope during disease outbreaks (Cao et al. 2011; Huang et al. 2013; Su et al. 2011; Wu et al. 2014). A pathogenic TMUV strain isolated from a duck embryo provided direct proof of vertical transmission (Zhang et al. 2015).

More than 80 % of the world's domestic waterfowl are produced in China and Southeast Asia (FAO 2013), and most duck and goose flocks are raised in semi-open or open systems in these regions. Birds raised outdoors or that have access to the outdoors are exposed to wild birds and blood-feeding arthropods. Therefore, it is highly likely that TMUV is introduced into these commercial flocks by one or several routes. Once the infection is established, the virus can spread rapidly to different flocks and farms in other geographic areas when duck eggs are transported and hatched ducklings are distributed. Contaminated transportation vehicles are an important agent of viral dispersal.

9.5 Clinical Signs and Pathology

The incubation period ranges from 3 to 5 days in experimentally infected domestic ducks. The greatest incidence of TMUV infection is in laying duck flocks in the field. Reduced feed uptake is usually first observed in affected flocks, followed by a marked decline in egg production, varying from 30 % to more than 90 % within 3–5 days. Ducks with neurological signs, including uncoordinated gait, reluctance to move, and wing and leg paralysis, may be observed later, and most of these birds will die because they are unable to reach food or water (Fig. 9.3a). The mortality rate ranges from 5 to 15 %, depending on the coinfection rate with other pathogens and any supportive treatment provided by the owner. Feed uptake starts to recover in an affected flock when the clinical signs disappear, and egg production resumes 3–4 weeks after the outbreak. However, both the fertility and hatchability rates of eggs from breeding stock are lower than those from normal flocks.

Various degrees of neurological involvement are prominent in young TMUV-affected ducks and geese. These birds first display depression and inappetence, and 40–50 % reductions in feed intake have been reported (Homonnay et al. 2014). As the infection progresses, signs of incoordination become pronounced (Fig. 9.3b). Torticollis and opisthotonus may be seen in some birds when disturbed, and birds may fall while attempting to stand. Most ducks that have difficulty walking die from being trampled or from starvation or are culled by the

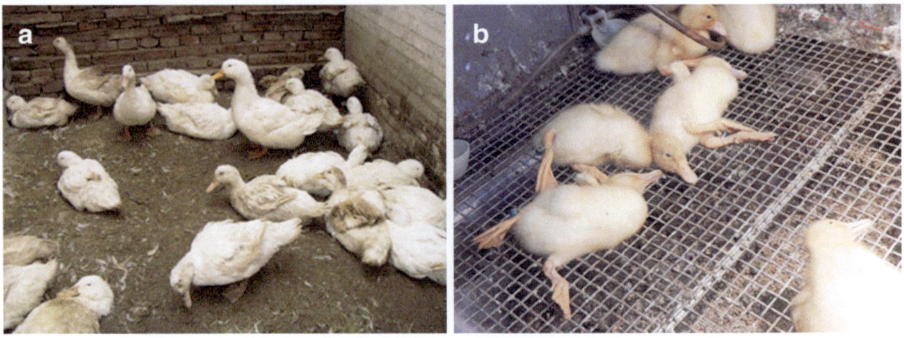

Fig. 9.3 Signs in Tembusu,virus-infected ducks. (**a**) Laying ducks appear depressed, with torticollis and paralysis. (**b**) Ducklings are reluctant to move or unable to stand 3–4 days after experimental infection

owner. Losses of up to 20 % of birds are common in some young broiler duck flocks concurrently infected with *Riemerella anatipestifer* or *Escherichia coli*. Geese infected with TMUV present with clinical symptoms similar to those of young ducks (Huang et al. 2013).

As with the clinical signs, the organs that are affected and the lesions that form depend on many factors. A replication kinetics study showed that TMUV replicated in most internal organs of the adult duck, including the brain, after subcutaneous infection (Wu et al. 2014). The viral load decreased significantly on day 7 and no infectious virus was detected 18 days after infection (Wu et al. 2014). In laying ducks, hemorrhage of the ovary follicles is consistent with the lesions in female ducks euthanized in the early stage of a disease outbreak (Fig. 9.4), and follicle atrophy and degeneration are evident as the disease progresses. Egg yolk peritonitis is occasionally observed. Some ducks may have enlarged spleens (Fig. 9.4). Ovarian hemorrhage and follicle atresia and rupture are the most obvious and consistent microscopic changes in affected female ducks (Fig. 9.5). Reduced sperm production, spermatocyte swelling, and vacuolar degeneration occur in the testes of infected male ducks, with focal lymphocytic infiltration in the later stages (Wu et al. 2014). Nonpurulent encephalitis can be found in the brains of some ducks, characterized by multifocal gliosis, and perivascular lymphocyte infiltration under the cranial arachnoid can also be seen (Fig. 9.5). Lesions of other internal organs have been described in several reports, but have not been consistently reproduced.

As in the adult duck, TMUV invades and replicates in most internal duckling organs after subcutaneous infection, with higher viral loads in the spleen, kidney, lung, and cloacal bursa. Viremia peaks on day 2 after infection and lasts until day 6. However, no consistent gross lesions are found on the postmortem examination of young infected ducks, except the enlargement of the spleen, which becomes round and dark. Swelling of the cloacal bursa is seen in some affected ducklings. The typical microscopic changes associated with nonpurulent encephalitis can be seen in most diseased ducklings.

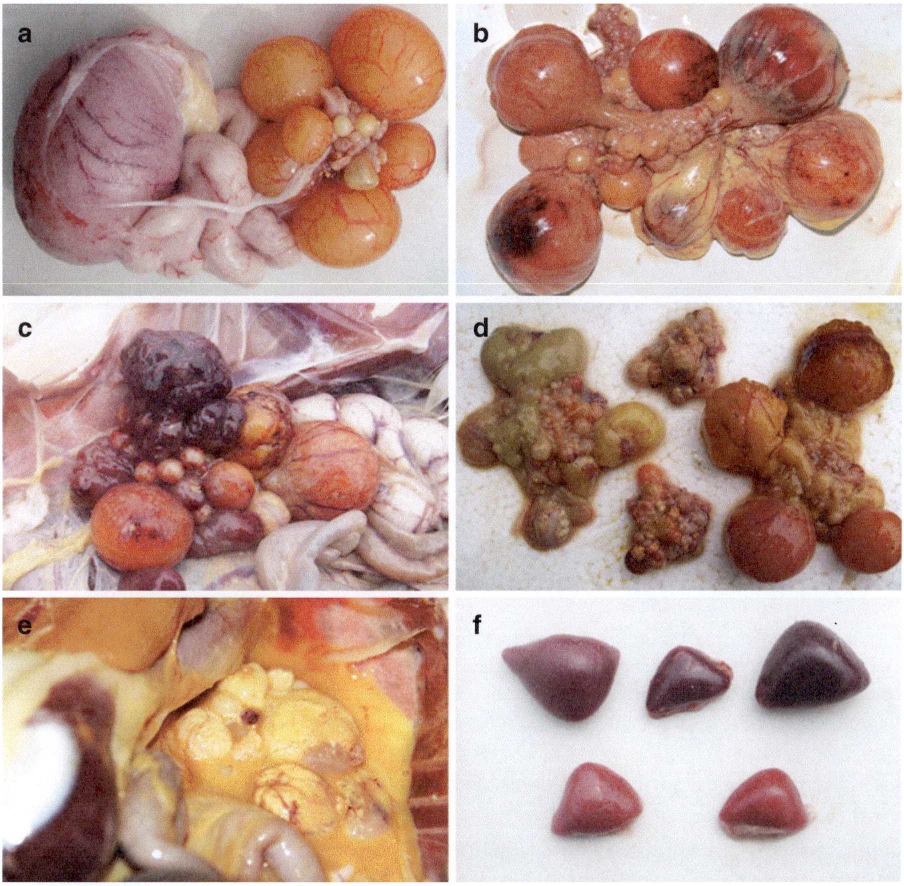

Fig. 9.4 Gross lesions in laying duck infected with Tembusu virus. (**a**) Ovary of uninfected duck, with an egg inside the oviduct. (**b–d**) Hemorrhage and degeneration of the follicles and ovaries of an infected duck. (**e**) Fluid yolk material in the abdominal cavity of an infected laying duck. (**f**) Enlarged spleens of experimentally infected ducks (*upper panel*); spleens in the lower panel were collected from uninfected control ducks

9.6 Diagnosis

The presumptive diagnosis of duck TMUV infection is based on the clinical manifestations in laying ducks, including reduced feed intake and a significant decline in egg production, with hemorrhage of the ovary follicles on postmortem examination. The demonstration of TMUV in clinical material is required to confirm the cause. A laboratory diagnosis can be made with viral isolation, the detection of TMUV-specific nucleic acids with reverse transcription–polymerase chain reaction (RT–PCR), or the demonstration of viral antigens with immunohistochemistry.

The virus can be isolated by inoculating 9–11-day-old embryonated duck eggs via the allantoic yolk sac route or a monolayer of primary DEF. The brain, spleen,

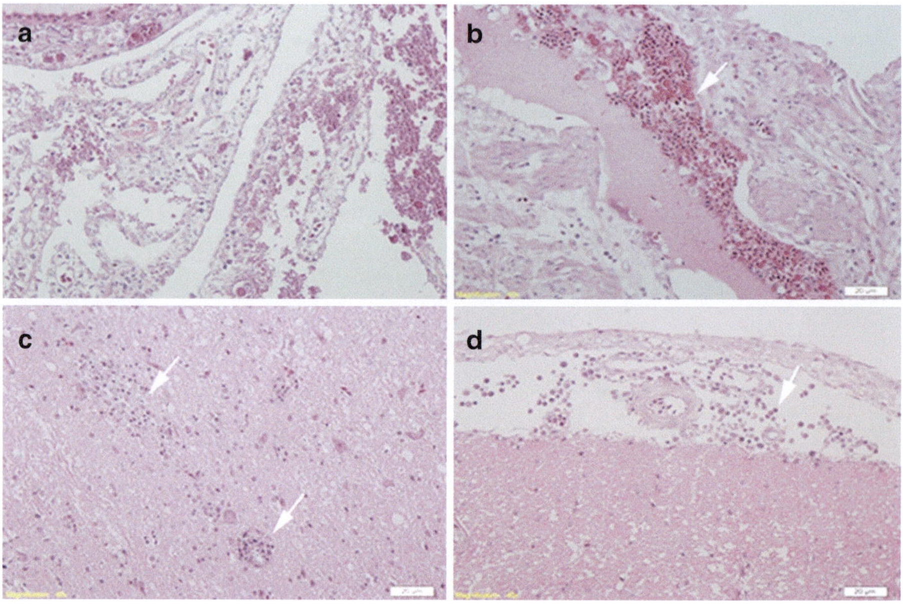

Fig. 9.5 Microscopic lesions of laying duck infected with Tembusu virus. (**a, b**) Hematoxylin–eosin (H&E)-stained ovary section showing hemorrhage and follicle rupture with red-stained bodies. (**c, d**) H&E-stained brain section showing focal gliosis (*arrow*) and lymphocyte and mononuclear cell infiltration under the cranial arachnoid

and ovary tissues from birds suspected of infection are the preferred materials for viral isolation (Su et al. 2011). The tissues are homogenized in phosphate-buffered saline (pH 7.4), clarified by centrifugation, and filtered through a 0.2-μm filter. One or more passages in embryonated duck eggs or DEF may be required, although most duck TMUV isolates can cause embryo body hemorrhage, liver necrotic foci, or embryo death 3–5 days after the first injection. A readily recognized cytopathic effect appears in DEF 48–72 h after infection. The inoculation of suckling mice can also be used to isolate the virus. The mice show neurological signs 5–6 days after intracerebral infection and die.

An RT–PCR assay targeting the conserved region of the TMUV E and NS5 genes has been used to detect the virus in infected tissues (Petz et al. 2014). Real-time RT–PCR methods have also been established for the rapid and quantitative detection of TMUV in field and laboratory samples and are more sensitive than conventional RT–PCR (Jiang et al. 2012; Liu et al. 2013; Yan et al. 2012; Yun et al. 2012a). A reverse-transcription loop-mediated isothermal amplification (RT–LAMP) assay has also been developed to detect the viral nucleic acids (Tang et al. 2012; Wang et al. 2011).

Serological tests are useful for determining the prevalence of TMUV infection in duck flocks and for evaluating the antibody response to vaccination.

Antibody-based methods include viral neutralization assays and enzyme-linked immunosorbent assays (ELISAs). Diagnosis requires a significant increase in the antibody titers in paired sera collected during the acute and convalescent phases. Although an antigen epitope common to members of the genus *Flavivirus* has been detected in TMUV (Liu et al. 2012b), the plaque reduction neutralization test is highly specific and sensitive for TMUV when BHK-21 cell cultures are used. Several ELISAs for antibody detection have been reported (Fu et al. 2015; Li et al. 2012b; Yin et al. 2013). A blocking ELISA based on a duck TMUV-specific monoclonal antibody has proved useful for antibody detection (Li et al. 2012b), and the results correlate positively with those of the serum neutralizing antibody test. A viral E protein-based ELISA and a solid-phase competition ELISA have also been described as sensitive and rapid tools for the serological detection of DTMUV (Fu et al. 2015; Yin et al. 2013). Recently, an antigen-capture ELISA and a double-antibody-sandwich ELISA have been developed to detect viral antigen in clinical samples, based on monoclonal antibody specifically directed against the DTMUV E protein (Bai et al. 2015; Chen et al. 2014a).

9.7 Differential Diagnosis

Nervous symptoms in young ducks and geese may be caused by a variety of bacteria, including *Riemerella anatipestifer*, *Escherichia coli*, *Salmonella* spp., and *Streptococcus* spp., and a diagnosis can be made by direct bacterial isolation from fresh tissues, especially the brain.

Highly pathogenic avian influenza viruses, including the H5N1, H5N2, and H5N8 subtype viruses, can reduce feed uptake and cause significant egg drop in laying duck flocks with only a slight increase in the mortality rate (Swayne et al. 2013). The situation differs in chicken flocks, in which the mortality rates are very high and can reach 50–100 % (Swayne et al. 2013). However, domestic ducks< 5 weeks old and geese of all ages are highly sensitive to highly pathogenic avian influenza virus infections in the field (Pantin-Jackwood et al. 2007; Perkins and Swayne 2002). The affected birds display depression, anorexia, and neurological disorders, including recumbency to one wing, tremors of the head, torticollis, opisthotonus, and an inability to stand (Capua and Mutinelli 2001). The mortality rate can exceed 30 %. Devastating infections in unvaccinated goose flocks are not uncommon. Severe egg drop caused by *Avian metapneumovirus* infection has recently been reported in Muscovy duck flocks in China (Sun et al. 2014b), and the clinical manifestations of the disease are similar to those of TMUV infection.

The *West Nile virus* is considered endemic to many countries in Africa, Asia, Europe, and North America (Chancey et al. 2015). Although the infection of domestic geese and ducks by this virus has not been reported in China, nervous symptoms have been described in young geese and ducks (Shirafuji et al. 2009). Laboratory studies should be conducted to clarify the differential diagnosis of DTMUV infection.

9.8 Prevention and Control

Because most ducks and geese are raised in semi-intensive accommodation or in the open in China, establishing biosecurity measures to control the disease is difficult. Flock vaccination appears to be the best option for the active prevention of the disease. The successful use of *Flavivirus* vaccines, such as those against Japanese encephalitis virus B and *West Nile virus*, in domestic animals provides an excellent precedent for the development of a DTMUV vaccine (Tomohiro et al. 2014). A live attenuated candidate vaccine designed against DTMUV, FX2010-180P, was successfully developed by serial passage of the virus in chicken embryo fibroblasts (Li et al. 2014), and a virulent strain of DTMUV, designated Du/CH/LSD/110128, was attenuated after 90 passages in embryonated chicken eggs (Sun et al. 2014a). Both strains are nonpathogenic in ducks but retain the immunogenicity of the parental strain and provide effective protection to challenged ducks under laboratory conditions (Li et al. 2014; Sun et al. 2014a). In addition to these conventional measures, recombinant vaccines using an attenuated duck enteritis virus as the vector to express the DTMUV E or prM protein have been developed (Zou et al. 2014). In laboratory experiments, immunized ducks produced high levels of neutralizing antibodies and were protected against viral challenge.

Public Health Although antibodies directed against TMUV have been detected in humans, no disease caused by TMUV has yet been reported in humans or mammals (Wolfe et al. 2001). Because most flaviviruses are arthropod-borne and can infect humans, including the yellow fever virus, Japanese encephalitis virus, *Tick-borne meningoencephalitis virus*, and *Dengue virus*, the potential threat posed by DTMUV to public health must be considered.

References

Bai X, Shao Z, Zhang Q et al (2015) Characterization of monoclonal antibodies against duck Tembusu virus E protein: an antigen-capture ELISA for the detection of Tembusu virus infection. Arch Virol 160(3):757–764. doi:10.1007/s00705-014–2312–z

Cao Z, Zhang C, Liu Y et al (2011) Tembusu virus in ducks, china. Emerg Infect Dis 17(10):1873–1875. doi:10.3201/eid1710.101890

Capua I, Mutinelli F (2001) Mortality in Muscovy ducks (Cairinamoschata) and domestic geese (Anseranser var. domestica) associated with natural infection with a highly pathogenic avian influenzavirusofH7N1subtype.AvianPathol30(2):179–183.doi:10.1080/03079450120044597

Chancey C, Grinev A, Volkova E et al (2015) The global ecology and epidemiology of West Nile virus. Biomed Res Int 2015:376230. doi:10.1155/2015/376230

Chen H, Liu X, Tang Y et al (2013) Complete genome sequences of two waterfowl-origin tembusu virus strains isolated in shandong province, china. Genome Announc 1(6) pii: e00789–e00713. doi:10.1128/genomeA.00789–13

Chen H, Ou Q, Tang Y et al (2014a) Development and evaluation of a DAS-ELISA for rapid detection of Tembusu virus using monoclonal antibodies against the envelope protein. PLoS One 9(5):e96366. doi:10.1371/journal.pone.0096366

Chen S, Wang S, Li Z et al (2014b) Isolation and characterization of a Chinese strain of Tembusu virus from Hy-Line Brown layers with acute egg-drop syndrome in Fujian, China. Arch Virol 159(5):1099–1107. doi:10.1007/s00705-013-1931–0

Fernández-Pinero J, Davidson I, Elizalde M et al (2014) Bagaza virus and Israel turkey meningoencephalomyelitis virus are a single virus species. J Gen Virol 95(Pt 4):883–7. doi:10.1099/vir.0.061465–0

Food and Agriculture Organization of the United Nations (2013) The FAO register. http://faostat3.fao.org/browse/Q/QA/E

Fu Y, Ji Y, Liu B et al (2015) Development of a solid-phase competition ELISA to detect antibodies against newly emerged Duck Tembusu virus. J Virol Methods 224:73–76. doi:10.1016/j.jviromet.2015.08.017

Grubaugh ND, McMenamy SS, Turell MJ et al (2013) Multi-gene detection and identification of mosquito-borne RNA viruses using an oligonucleotide microarray. PLoS Negl Trop Dis 7(8):e2349. doi:10.1371/journal.pntd.0002349 eCollection 2013

Homonnay ZG, Kovács EW, Bányai K et al (2014) Tembusu-like flavivirus (Perak virus) as the cause of neurological disease outbreaks in young Pekin ducks. Avian Pathol 43(6):552–560. doi:10.1080/03079457.2014.973832

Huang X, Han K, Zhao D et al (2013) Identification and molecular characterization of a novel flavivirus isolated from geese in China. Res Vet Sci 94(3):774–780

Jiang T, Liu J, Deng YQ et al (2012) Development of RT-LAMP and real-time RT-PCR assays for the rapid detection of the new duck Tembusu–like BYD virus. Arch Virol 157(12):2273–2280. doi:10.1007/s00705-012–1431–7

Kono Y, Tsukamoto K, Hamid M et al (2000) Encephalitis and retarded growth of chicks caused by sitiawan virus, a new isolate belonging to the genus flavivirus. Am J Trop Med Hyg 63(1, 2):94–101

Li L, An H, Sun M et al (2012a) Identification and genomic analysis of two duck-origin Tembusu virus strains in southern China. Virus Genes 45(1):105–112. doi:10.1007/s11262-012-0753-6

Li X, Li G, Teng Q et al (2012b) Development of a blocking ELISA for detection of serum neutralizing antibodies against newly emerged duck Tembusu virus. PLoS One 7(12):e53026. doi:10.1371/journal.pone.0053026

Li S, Li X, Zhang L et al (2013) Duck Tembusu virus exhibits neurovirulence in BALB/c mice. Virol J 10:260. doi:10.1186/1743–422X-10-260

Li G, Gao X, Xiao Y et al (2014) Development of a live attenuated vaccine candidate against duck Tembusu viral disease. Virology 450-451:233–242. doi:10.1016/j.virol.2013.12.028

Li X, Shi Y, Liu Q et al (2015) Airborne Transmission of a Novel Tembusu Virus in Ducks. J Clin Microbiol 53(8):2734–2736. doi:10.1128/JCM.00770–15

Liu M, Chen S, Chen Y et al (2012a) Adapted Tembusu-like virus in chickens and geese in China. J Clin Microbiol 50(8):2807–2809. doi:10.1128/JCM.00655–12

Liu P, Lu H, Li S et al (2012b) Genomic and antigenic characterization of the newly emerging Chinese duck egg-drop syndrome flavivirus: genomic comparison with Tembusu and Sitiawan viruses. J Gen Virol 93(Pt 10):2158–2170. doi:10.1099/vir.0.043554–0

Liu Z, Fu Y, Ji Y et al (2013) Development and validation of one-step SYBR green real-time RT-PCR for the rapid detection of newly emerged duck Tembusu virus. Avian Dis 57(3):595–601

Mackenzie JS, Williams DT (2009) The zoonotic flaviviruses of southern, south-eastern and eastern Asia, and Australasia: the potential for emergent viruses. Zoonoses Public Health 56:338–356

O'Guinn ML, Turell MJ, Kengluecha A et al (2013) Field detection of Tembusu virus in western Thailand by rt–PCR and vector competence determination of select culex mosquitoes for transmission of the virus. Am J Trop Med Hyg 89(5):1023–1028. doi:10.4269/ajtmh.13–0160

Pandey BD, Karabatsos N, Cropp B et al (1999) Identification of a flavivirus isolated from mosquitos in Chiang Mai Thailand. Southeast Asian J Trop Med Public Health 30(1):161–165

Pantin-Jackwood MJ, Suarez DL, Spackman E et al (2007) Age at infection affects the pathogenicity of Asian highly pathogenic avian influenza H5N1 viruses in ducks. Virus Res 130(1–2):151–161

Perkins LE, Swayne DE (2002) Pathogenicity of a Hong Kong-origin H5N1 highly pathogenic avian influenza virus for emus, geese, ducks, and pigeons. Avian Dis 46(1):53–63

Petz LN, Turell MJ, Padilla S et al (2014) Development of conventional and real-time reverse transcription polymerase chain reaction assays to detect Tembusu virus in Culextarsalis mosquitoes. Am J Trop Med Hyg 91(4):666–671. doi:10.4269/ajtmh.13–0218
Shirafuji H, Kanehira K, Kubo M et al (2009) Experimental West Nile virus infection in aigamo ducks, a cross between wild ducks (Anasplatyrhynchos) and domestic ducks (Anasplatyrhynchos var. domesticus). Avian Dis 53:239–244
Su J, Li S, Hu X et al (2011) Duck egg-drop syndrome caused by BYD virus, a new Tembusu-related flavivirus. PLoS One 6(3):e18106. doi:10.1371/journal.pone.0018106
Sudeep AB (2014) Culexgelidus: an emerging mosquito vector with potential to transmit multiple virus infections. J Vector Borne Dis 51(4):251–258
Sun L, Li Y, Zhang Y et al (2014a) Adaptation and attenuation of duck Tembusu virus strain Du/CH/LSD/110128 following serial passage in chicken embryos. Clin Vaccine Immuno 21(8): 1046–53. doi:10.1128/CVI.00154–14
Sun S, Chen F, Cao S et al (2014a) Isolation and characterization of a subtype C avian metapneumovirus circulating in Muscovy ducks in China. Vet Res 45:74. doi:10.1186/s13567–014-0074–y
Swayne DE, Suarez DL, Sims LD (2013) Influenza. In Swayne DE (ed) Disease of poultry, 13th edn. Wiley, New Jersey, pp 181–218
Tang Y, Diao Y, Yu C et al (2012) Rapid detection of Tembusu virus by reverse-transcription, loop-mediated isothermal amplification (RT-LAMP). Transbound Emerg Dis 59(3):208–213. doi:10.1111/j.1865–1682.2011.01257
Tang Y, Diao Y, Yu C et al (2013) Characterization of a Tembusu virus isolated from naturally infected house sparrows (*Passer domesticus*) in Northern China. Transbound Emerg Dis 60(2):152–158. doi:10.1111/j.1865–1682.2012.01328
Tomohiro I, Atsushi Y, Eiji K (2014) A review of successful flavivirus vaccines and the problems with those flaviviruses for which vaccines are not yet available. Vaccine 32:1326–1337
Vaidya NK, Wang FB, Zou X et al (2012) Transmission dynamics of the recently-identified BYD virus causing duck egg-drop syndrome. PLoS One 7(4):e35161. doi:10.1371/journal.pone.0035161
Wan C, Huang Y, Fu G et al (2012) Complete genome sequence of avian tembusu-related virus strain WR isolated from White Kaiya ducks in Fujian, China. J Virol 86(19):10912. doi:10.1128/JVI.01582–12
Wang Y, Yuan X, Li Y et al (2011) Rapid detection of newly isolated Tembusu-related Flavivirus by reverse-transcription loop-mediated isothermal amplification assay. Virol J 8:553. doi:10.1186/1743–422X–8–553
Wolfe ND, Kilbourn AM, Karesh WB et al (2001) Sylvatic transmission of arboviruses among Bornean orangutans. Am J Trop Med Hyg 64(5–6):310–316
Wu L, Liu J, Chen P et al (2014) The sequential tissue distribution of duck Tembusu virus in adult ducks. Biomed Res Int 2014:703930. doi:10.1155/2014/703930
Yan L, Yan P, Zhou J et al (2011a) Establishing a TaqMan-based real-time PCR assay for the rapid detection and quantification of the newly emerged duck Tembusu virus. Virol J 8:464. doi:10.1186/1743–422X–8–464
Yan P, Zhao Y, Zhang X et al (2011b) An infectious disease of ducks caused by a newly emerged Tembusu virus strain in mainland China. Virology 417(1):1–8. doi:10.1016/j.virol.2011.06.003
Yan L, Peng S, Yan P et al (2012) Comparison of real-time reverse transcription loop-mediated isothermal amplification and real-time reverse transcription polymerase chain reaction for duck Tembusu virus. J Virol Methods 182(1–2):50–55
Yin X, Lv R, Chen X et al (2013) Detection of specific antibodies against tembusu virus in ducks by use of an E protein-based enzyme-linked immunosorbent assay. J Clin Microbiol 51(7):2400–2402. doi:10.1128/JCM.00361–13
Yu K, Sheng ZZ, Huang B et al (2013) Structural, antigenic, and evolutionary characterizations of the envelope protein of newly emerging Duck Tembusu Virus. PLoS One 8(8):e71319. doi:10.1371/journal.pone.0071319

Yun T, Ni Z, Hua J et al (2012a) Development of a one-step real-time RT-PCR assay using a minor-groove-binding probe for the detection of duck Tembusu virus. J Virol Methods 181(2):148–154. doi:10.1016/j.jviromet.2012.01.019

Yun T, Ye W, Ni Z et al (2012b) Identification and molecular characterization of a novel flavivirus isolated from Pekin ducklings in China. Vet Microbiol 157(3–4):311–319. doi:10.1016/j.vetmic.2012.01.013

Zeng T, Xie Z, Xie L et al (2015) Identification and whole-genome sequence analysis of Tembusu Virus GX2013G, isolated from a Cherry Valley Duckling in Southern China. Genome Announc 3(1) pii: e00007–e00015. doi:10.1128/genomeA.00007-15

Zhang Y, Li X, Chen H et al (2015) Evidence of possible vertical transmission of Tembusu virus in ducks. Vet Microbiol 179(3–4):149–154. doi:10.1016/j.vetmic.2015.06.004

Zhu W, Chen J, Wei C et al (2012) Complete genome sequence of duck Tembusu virus, isolated from Muscovy ducks in southern China. J Virol 86(23):13119. doi:10.1128/JVI.02361–12

Zhu K, Huang J, Jia R et al (2015) Identification and molecular characterization of a novel duck Tembusu virus isolate from Southwest China. Arch Virol 160(11):2781–2790. doi:10.1007/s00705–015–2513–0

Zou Z, Liu Z, Jin M (2014) Efficient strategy to generate a vectored duck enteritis virus delivering envelope of duck Tembusu virus. Viruses 6(6):2428–43. doi:10.3390/v6062428

Aujeszky's Disease

10

Ewelina Czyżewska Dors and Małgorzata Pomorska Mól

10.1 History

The history of PRV goes back to the nineteenth century, but at that time, the disease was not linked with swine. In 1813, Dr. Hildreth, a physician from Marietta, Ohio, reported the first case of what later known as Aujeszky's disease (AD). He described a case of "mad itch" in one of his client's cows. The clinical symptoms of the disease included rubbing of the head, twitching neck muscles, scratching and mutilating itself. The cow died in agony 12–14 h from the onset of clinical signs. Due to similarities of the clinical pattern to rabies, the name "pseudorabies" was given to the disease observed in cattle.

In the first decades of the twentieth century, AD was mainly observed in cattle and dogs in Hungary, Romania, France, Russia, Brazil and the USA (Wittmann and Rziha 1989). In pigs, a first description of AD was made in 1902 by the Hungarian veterinary surgeon Aladár Aujeszky. He also revealed that the disease has a viral background. The first enzootic occurrence was reported in 1931 in the Netherlands (Wittmann and Rziha 1989). In general, at that time AD was observed rarely in pigs and was associated with central nervous system disorders. The strong intensification of pig production occurred during 1960–1970 probably contributed to an increase in case numbers and change in disease patterns (respiratory disorder, fever, anorexia, high mortality in piglets, abortion in sows) (Glorieux 2009). Similar changes of disease patterns have been observed in the USA, Europe and in Northern Ireland (Pensaert et al. 1987; Pol 1990; Wittmann and Rziha 1989; Nauwynck et al. 2007).

The rapid spread of infection and huge losses in swine sector forced the introduction of eradication strategy focused on culling of infected herds and vaccination programme. Thereafter, to distinguish infected herds vs. vaccinated herds, marker

E. Czyżewska Dors • M. Pomorska Mól (✉)
National Veterinary Research Institute, Department of Swine Diseases,
57 Patryzantów Str, 24-100 Pulawy, Poland
e-mail: malgorzata.pomorska@piwet.pulawy.pl

J. Bayry (ed.), *Emerging and Re-emerging Infectious Diseases of Livestock*,
DOI 10.1007/978-3-319-47426-7_10

vaccine (deletion of *gE* gene) and appropriate ELISA test were developed. Introduction of this marker vaccine contributed to the eradication of AD in the USA and significant part of Europe (Mettenleiter 2000). AD is a notifiable disease by the World Organisation for Animal Health (OIE) (list B).

10.2 Aetiology of AD

Aujeszky's disease virus (ADV), the causative agent of AD, is a DNA virus of 150–180 nm in diameter belonging to the subfamily *Alphaherpesvirinae* within the family *Herpesviridae*, genus *Varicellovirus*.

The virion of ADV consists of four morphologically distinguishable structures: the core containing the linear double-stranded DNA of approximately 142 kilo base pairs (kbp), capsid, tegument and envelope. The capsid consists of 162 capsomers and 150 hexons that surround and protect the genome. Together, genome and capsid form nucleocapsid. Between the icosahedral capsid and envelope, a proteinaceous layer is situated – called as a tegument. The tegument protein plays an important role in viral life cycle (Luxton et al. 2006; Guo et al. 2010). The outer layer of ADV is a bilayered phospholipid membrane containing 11 glycoproteins (gB, gC, gD, gE, gG, gH, gI, gK, gL, gM and gN) and three nonglycosylated proteins (UL20, UL43, US9). The envelope proteins play important roles in viral entry, egress, cell-to-cell spread, induction of protective immunity and immune evasion (Mettenleiter 2000; Pomeranz et al. 2005; Nauwynck et al. 2007).

Herpesvirus genomes are split into six classes, A to F, and the ADV genome belongs to D class (Roizman and Pellet 2001). The genome of ADV contains 72 open reading frames (ORFs), which encode 70 different proteins. The characteristic feature of D class is that the genome is composed of two unique regions, a long (unique long (UL)) and a short one (unique short (US)). These two regions are bracketed by inverted repeat sequences, an internal repeat sequence (IRS) and a terminal repeat sequence (TRS) (Nauwynck et al. 2007). The whole genome of ADV has been sequenced and published by Klupp et al. (2004).

10.3 Distributions of AD

Due to trade restrictions imposed by AD-free countries, national AD eradication programmes have been conducted in many countries. Thereafter, ADV has been eliminated from domestic pig population in Denmark, Finland, France, Germany, Hungary, Slovakia, Luxembourg, the Netherlands, Sweden, Switzerland, Great Britain (England, Scotland and Wales), Canada, the USA and New Zealand. Till date, ADV has never been detected in Australia. In ADV-free countries, vaccination is forbidden. Nowadays, AD is endemic in commercial pigs in Eastern and Southeastern Europe, Latin America, Africa and Asia. Despite AD has been eradicated from many countries, the ADV continues to circulate among feral swine in the USA and wild boars in many European countries (Müller et al. 2011; Pedersen et al.

2013). In the USA, 18 % of feral swines were seropositive in 25 of 35 states where samples were collected (Pedersen et al. 2013). In Europe, serosurveys revealed the occurrence of ADV infections in wild boar in several European countries, including Belgium, Croatia, the Czech Republic, the former Yugoslavia, France, Germany, Italy, Poland, Romania, Russia, Slovenia, Spain and Switzerland. The average seroprevalence in these countries for the investigated periods was ranging from 0.57 to more than 60 % at national or regional level (Szweda et al. 1998; Zupancic et al. 2002; Lari et al. 2006; Vengust et al. 2006; Köppel et al. 2007; Sedlak et al. 2008; Kukushkin et al. 2009; Montagnaro et al. 2010; Boadella et al. 2012; Pannwitz et al. 2012; Meier et al. 2015). Wild swine are known reservoirs of ADV and may serve as a potential source of virus reintroduction to domestic swine and the other susceptible mammals. Thus, due to the risk of spill over infections to domestic pigs, attention should be paid to control AD in populations of wild swine.

10.4 Replication Cycle

Replication cycles are almost identical among alphaherpesviruses. First step of ADV infection is attachment to the host cell membrane (Fig. 10.1). This process consists of labile binding of the viral envelope glycoprotein gC (and less gB) to cellular heparan sulphate proteoglycans found at the cell surface (Mettenleiter et al. 1990). Afterwards, glycoprotein gD interacts with cellular receptors, herpesvirus entry mediator B (HveB), nectin-2, poliovirus receptor-related protein 2 (PRR2), HveC (nectin-1, PRR1) and HveD (poliovirus receptor (PVR), CD155), which leads to stable binding of virus to the host cell (Nixdorf et al. 1999; Spear 2000; Mettenleiter 2002). Attachment is followed by the viral envelope fusion with host plasma membrane. The glycoproteins gB, gD and gH/gL are involved in this process (Rauh and Mettenleiter 1991; Mettenleiter 1994; Klupp et al. 1997). After penetration to the host cell, capsid is transported to the nucleus via interaction with microtubules (Guo et al. 2010). Upon released into the nucleus, the ADV genome circularises and is transcribed in a tightly regulated cascade-like manner in the order immediate-early (IE), early (E) and late (L) gene expression. ADV encodes only one IE protein termed as IE180. The ADV IE180 gene codes immediate-early transcriptional activator of viral genes required for DNA replication and RNA transcription (Pomeranz et al. 2005). Subsequently, E genes are activated, which are mainly involved in replication of the viral DNA (Huang and Wu 2004). Finally, the L genes are transcribed, encoding mainly structural proteins (capsid, tegument and envelope (glyco)proteins) involved in virion assembly. Capsid assembly occurs inside the nucleus. Nucelocapsid is formed by pulling monomeric units of DNA through a cylindric entry pore to new capsid. Nucleocapsid leaving to the cytosol consists of two steps: primary envelopment (budding of capsids at the inner leaflet of the nuclear membrane) and de-envelopment (primary envelope fusion with the outer leaflet of the nuclear membrane). The primary envelopment is mediated by two viral proteins encoded by UL34 and UL31 (Mettenleiter et al. 2009). Thus, US3 protein is involved in de-envelopment

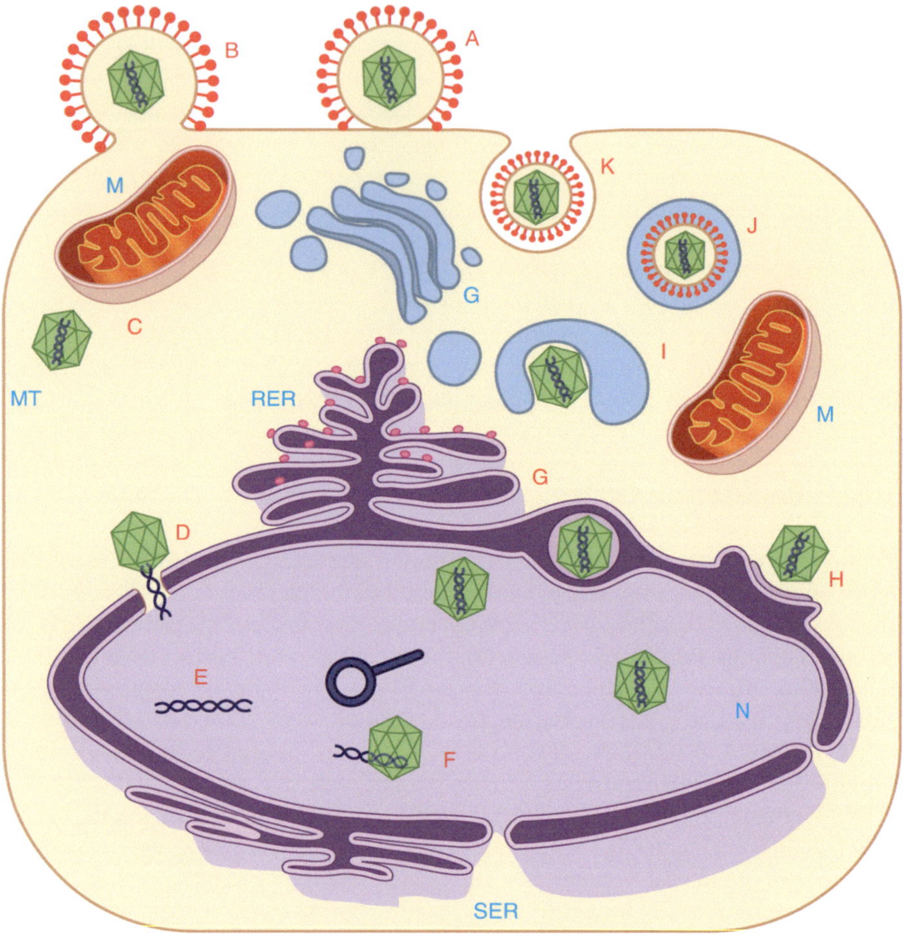

Fig. 10.1 Schematic replication cycle of the ADV in the host cell. ADV attachment (*A*) and penetration (*B*) to the host cell, capsid transport (*C*) along microtubules (*MT*) to the nucleus (*N*), release of the viral genome into the nucleus (*D*), DNA transcription and replication (*E*), encapsidation of viral genome (*F*), primary envelopment (*G*), de-envelopment (*H*), secondary envelopment (*J*) and egress of mature and enveloped virion (*K*). *G* Golgi apparatus, *M* mitochondrion, *MT* microtubules, *N* nucleus, *RER* rough endoplasmic reticulum, *SER* smooth endoplasmic reticulum (Adapted from Mettenleiter 2000 with own modification)

at the outer leaflet of the nuclear membrane (Granzow et al. 2004). Thereafter, the nucleocapsids are transported to the trans-Golgi area in which different viral glycoproteins are expressed. The acquisition of final tegument and envelope is done by budding of nucleocapsids into trans-Golgi-derived vesicles. During this process, different tegument and envelope proteins are assembled into the mature virion. The ADV proteins such as UL11, glycoprotein E (gE) and gM are involved in secondary envelopment of tegumented nucleocapsids in the cytoplasm (Kopp et al. 2004). Release of newly formed virion into the extracellular space occurs by fusion of the vesicle with the plasma membrane. Spread of ADV infection may occur via free virions and by cell-associated way of spreading. Several different mechanisms of the cell-associated spread have been described (for details, see Nauwynck et al. 2007).

10.5 Pathogenesis

The nasal mucosa acts as a main entry gate of ADV. After intranasal/peroral exposure, ADV primary replicates in the respiratory tract including the nasal cavity, tonsils, pharynx and lungs (Sabo et al. 1969; Baskerville 1973; Miry and Pensaert 1989). Replication in lower parts of the respiratory tract is restricted, except when the virus is directly administrated in these locations (intratracheal or aerosol inoculation) (Baskerville 1973; Miry and Pensaert 1989). Within 24 h, ADV crosses the basement membrane barrier in the nasal respiratory epithelium and infects all sorts of cells including fibrocytes, endothelial cells and mononuclear cells in the lamina propria (Kritas et al. 1994; Glorieux 2009).

In ADV invasion through the basement towards the lamina propria, trypsin-like serine protease activity is involved (Glorieux et al. 2011). After reaching blood vessels, nerves and lymphatics localised in lamina propria, ADV spreads to internal organs wherein the secondary replication takes place. Viraemia can be observed at 24 h after challenge and for up to 9 days, whereby monocytes are the most susceptible mononuclear cells to ADV infection (Nauwynck and Pensart 1995). Replication in secondary site can be observed starting from 48 h postinoculation (Sabo et al. 1969; Wittmann et al. 1980).

One of the important secondary sites of ADV replication is the central nervous system (CNS). ADV travels from the nasal mucosa to the CNS via some of the cranial nerves such as the olfactory and trigeminal nerves (McFerran and Dow 1965; Sabo et al. 1969; Wittmann et al. 1980). Like the other alphaherpesviruses, ADV is capable of moving efficiently within axons to distant cell body (retrograde) and away from cell bodies (anterograde) (Smith et al. 2001). After axonal entry, virus particle moves along microtubules, which are oriented with the plus end at the cell periphery or axon terminal and the minus end towards the cell body. The kinesin family of motor proteins generally mediates plus-end-directed anterograde transport, and dynein motor complexes (recruited by virus U36 protein) generally mediate minus-end-directed retrograde transport (Kapitein and Hoogenraad 2011; Zaichick et al. 2013). A recent study indicated that for efficient ADV retrograde transport and infection of cell bodies, local translation of host proteins including proteins involved in cytoskeletal remodelling, intracellular trafficking, signalling and metabolism is required (Koyuncu et al. 2013b). Another important site of secondary replication is the genital tract. It is suggested that ADV reaches genital organs via blood or nerves (McFerran and Dow 1965) and via ADV-contaminated semen (Bolin et al. 1985). Replication of ADV has been observed in the ovaries and the uterus of sows (Hsu et al. 1980) and in the mesorchium of boars (Miry and Pensaert 1989).

10.6 Virulence Factor

Individual ADV strains show various biological features concerning their virulence as expressed by the different severities of disease in pigs, the quantity and duration of virus shedding. For example, one of the recent studies comparing the differences in pathogenicity between ADV JS-2012 strain (emerging ADV strain isolated in Bartha-K61-vaccinated piglets in 2012, China) and classical virulent ADV SC strain indicated higher pathogenicity of JS-2012 (the earlier onset of clinical signs, higher mortality and more severe pathological lesions) compared with SC strain (Tong et al. 2015).

ADV virulence is determined by viral membrane glycoproteins, virus-encoded enzymes and non-essential capsid-associated proteins (Mettenleiter 2000). Among ADV membrane glycoproteins, five (gB, gD, gH, gK, gL) are essential for viral replication in cell culture. Thus, gC, gE, gG, gI, gM and gN are non-essential but have been involved in important viral functions such as target cell recognition and virulence (Mettenleiter 1996; Pomeranz et al. 2005). From the non-essential glycoprotein, gE is important for the expression of ADV virulence. There is evidence that deletion of gE can induce a considerable impairment in the neuroinvasiveness of ADV, either directly or through its association with gI (Babic et al. 1996). Another study indicated that deletion of two amino acids (valine (125) and cysteine (126)) in gI strongly decreases virulence and neurotropism of ADV for pigs (Jacobs et al. 1993). Besides glycoproteins, thymidine kinase (TK) enzyme (non-essential for virus replication in cell culture) is also an important determinant of ADV virulence. Removal of the TK function leads to attenuation of ADV (McGregor et al. 1985; Prieto et al. 1991; Pan et al. 2001). Studies on viral components affecting the ADV virulence have important significance for the development of efficient engineered vaccines.

10.7 Latency

The hallmark of the herpesviruses infection is its ability to establish a latent state within specific tissues and its characteristic feature for each virus: *Alphaherpesvirinae* (PNS neurons), *Betaherpesvirinae* (monocytes and macrophage precursors) and *Gammaherpesvirinae* (B and T cells)) (Koyuncu et al. 2013a). The viral latency is the ability of the virus to remain dormant (latent) inside the neurons, which allow evading the host's immune system. In the domestic swine, ADV mainly hides in the trigeminal ganglia and other sites, such as the olfactory bulb, medulla oblongata and less frequently the brainstem and spinal cord (Rziha et al. 1986). While in feral swine, ADV typically shows latent infection of the sacral ganglia (Romero et al. 2003). During latency, viral DNA persists, but infectious virus is not produced. Under this state, expression of ADV gene is limited to a small region of the viral genome termed latency-associated transcript (LAT) and to several microRNAs (miRNAs) (Cheung 1989; Farrell et al. 1991; Umbach et al. 2008). There is evidence that miRNAs are involved in blocking apoptosis of infected neurons and in the control of reactivation from latency (Koyuncu et al. 2013a). Moreover, it is speculated that IFN alpha (IFN-α) represents a key immune component involved in the efficient establishment of *Alphaherpesvirus* latency in sensory neurons (De Regge et al. 2010). Reactivation from latency takes place after stress stimuli (pregnancy, parturition, fever, extreme temperatures, transportation) and is associated with virus production, excretion and spread in the environment (Davies and Beran 1980; Tanaka and Mannen 2003, 2008).

10.8 Co-infections with Other Pathogens

ADV is one of the primary viral aetiologic agents of porcine respiratory disease complex (PRDC). Co-infection of ADV with other swine pathogens exacerbates clinical signs and pathological lesions in the respiratory tract. For example, co-infections of specific pathogen-free (SPF) pigs with porcine reproductive and respiratory syndrome virus (PRRSV) and ADV increase and prolong the clinical signs and pathologic lesions in the lungs (Shibata et al. 2003). Other studies also indicated that pigs dually infected with ADV and *Mycoplasma hyopneumoniae* or ADV and *Actinobacillus pleuropneumoniae* show very severe clinical symptoms and pneumonia and lung lesions as compared to pigs infected only with one agent (Sakano et al. 1993; Shibata et al. 1998).

10.9 Other Animals

Most mammals and birds including dogs, cats, cattle, sheep, goats, foxes, hedgehogs, polecats, opossums, jackals, some nonhuman primates (rhesus macaques, marmosets), chickens, pigeons, geese, ducks, buzzards, sparrow, hawks, rabbits, guinea pigs, rats and mice are susceptible for ADV infection. However, only members of the family *Suidae* (true pigs) are known to survive an acute infection and in addition have subclinical and latent infections. Therefore, swine represents natural reservoir of ADV and the source of virus transmission (Mettenleiter 1996). In non-porcine species, ADV typically causes fatal neurologic disease (respiratory distress, salivation, muscular stiffness), often associated with localised pruritus resulting from self-trauma (Glass et al. 1994; Zanin et al. 1997; Cramer et al. 2011; Verpoest et al. 2014). There are two possible explanations with regard to the fatal infection in nonnatural host: (a) the immune system fails to control the infection and viral replication destroys cells and tissues, and (b) the immune system responds too strongly and a systemic inflammatory response overwhelms the host (Brittle et al. 2004).

Among non-porcine species, deviation from hallmark clinical patterns is presented in minks. In this species, pruritus is usually not observed, and the inflammatory reaction in the mink nervous system is minimal or even absent, and the most prevalent lesions are haemorrhages and ischaemia associated with a systemic vasculopathy, which may be the consequence of an increased endotheliotropism of ADV in this species (Marccacini et al. 2008).

10.10 Viral Transmission

ADV is transmitted most efficiently by direct contact between pigs. The virus is not very contagious. The high quantities of virus are usually necessary to infect animals ($>1 \times 10^{4\text{-}5}$ $TCID_{50}$), except piglets (1×10^2 $TCID_{50}$) (Wittmann 1991). In vaccinated pigs, a dose needed to infect pigs has been reported to be 10–1000 times higher (Visser 1997; Wiśniewski and Siemionek 1988).

Virus can also spread by oral route through contact with contaminated materials (including water, bedding, meat products, other contaminated feedstuffs and contaminated carcasses of rats and raccoons) (Donaldson et al. 1983). Transmission of virus by the aerosol route over short and long distances may occur (Allepuz et al. 2009; Kluge et al. 1999). Also, venereal transmission by contaminated vaginal mucosa or semen is possible. Virus can be transmitted vertically (transplacental), and foetuses may be infected in utero in the last third of gestation. ADV can also spread via colostrum and milk from an infected sow to suckling piglets (Beran 1991). Other animals such as dogs, cats, raccoons, skunks and rats are considered potential carriers within an endemic area (Kirkpatrick et al. 1980). Common sources of viral transmission into herds are animals with long-term or recrudescent shedding.

Typically, ADV enters into a susceptible swine via the nasal mucosa (when they inhale the virus) or via the tonsils or oral/digestive tract mucosa (when they ingest the virus) (Donaldson et al. 1983).

10.11 Duration and Routes of Shedding

Infected swine can shed ADV in high concentrations from almost all body secretions, excretions and aerosol. At the height of the disease, pigs can excrete $10^{5.8}$–$10^{8.3}$ $TCID_{50}\ ml^{-1}$ nasally (Pensaert and Kluge 1989). Virus shedding starts 1–2 days after infection, with nasal and oral excretion, and in adult swine vaginal, preputial and/or milk secretion prior to the onset of viraemia and clinical signs. Peak of shedding occurs at 2–5 days of infection and it will last up to 17 days (Müller et al. 2001; Wittmann 1991). Persistence of ADV in infected swine has been observed from 6 weeks to 13 months (Beran et al. 1980). Transplacental transmission leads to considerable virus shedding during abortion and birth (Beran 1991; Blaha 1989).

Lifelong latent infection commonly follows clinical recovery or inapparent infection, with virus remaining in trigeminal ganglia and tonsils (White et al. 1994; Balasch et al. 1998). Stress such as subsequent disease, farrowing, crowding, mingling with unfamiliar animals or transport can lead to recrudescence of shedding for 3–4 days (Schoenbaum et al. 1990a).

10.12 Persistence of Virus in the Environment

ADV is one of the most environmentally persistent animal viruses. However, virus survival depends on many environmental factors, such as temperature, pH and humidity.

Virus inactivation rates increase with increasing temperature (Turner et al., 2000). The virus is inactivated at 5 and 20 °C in 15 weeks and 2 weeks, respectively. At 35 °C (mesophilic conditions), the virus is inactivated in 5 h and at 55 °C (thermophilic conditions); no virus could be detected after 10 min (Bøtner 1991). In another study, persistence of ADV was determined in liquid cattle manure stored at temperatures of 4 and 20 °C for up to 26 weeks (Biermann et al. 1990). Infectivity of ADV in an aerated slurry at pH 9.6 and 44 °C declines in 1–3 weeks (Wittman 1991). Temperatures during composting process effectively and quickly kill ADV (Morrow et al. 1995; Garcia-Siera et al. 2001; Paluszak et al. 2012). In food waste, the virus is inactivated within 24 h at 20 and 30 °C but remains infectious at 10 °C for 48 h and at 5 °C for 96 h (Donaldson et al. 1983). Maturation of pig meat at 4 °C does not inactivate the virus. ADV in pig muscle and bone marrow is killed by freezing (−18 °C) for 35 days (Durham et al. 1980) and after heat treatment of meat at 80 °C (Donaldson et al. 1983).

ADV is stable at pH 4–12 and even at extreme pH values of 2.0 and 13.5; complete inactivation can take 2–4 h (Benndorf and Hantschel 1963).

The best survival of ADV was shown to occur at intermediate (50 %) relative humidity (half-life of 36.1 min) (Schoenbaum et al. 1990b). Survival times of this virus were lower and similar at high and low humidity, with half-lives of 17.4 and 18.8 min, respectively.

Davies and Beran (1981) have shown that at 37 °C and pH 6–8, ADV outside the living host is inactivated at a rate of 0.6 $\log_{10}$ per day. ADV becomes inactive at pH of 4.3 and temperature of 39 °C between 1 and 7 days (Kluge et al. 1999).

ADV has been shown to be persistent in well water for up to 7 days and in chlorinated water for less than 1 day. Studies examining the survival of ADV in buffered solutions found that the virus was stable for at least 10 days in PBS and saline solution. On polypropylene, vinyl, loam soil and meat and bone meal, virus inactivation rate was more rapid with 7 $\log_{10}$ reduction within 7 days. For straw, concrete and pelleted feed, virus dropped below detectable levels within 4 days. For wood, faeces, green grass, alfalfa and denim fabric, virus dropped below detectable levels in 1–2 days (Schoenbaum et al. 1990b).

Susceptibility to disinfectant ADV like other enveloped viruses can be inactivated by many disinfectants, including orthophenylphenate compounds, formalin, quaternary ammonium compound, iodine compound, sodium hydroxide, hypochlorides and chlorine solutions (Beran 1991; Dvorakova et al. 2008).

10.13 Clinical Signs

Clinical picture of AD depends on the virulence of the ADV isolate, the dose and route of infection and the age of the affected swine. Infection can cause neonatal mortality due to neurological disease, respiratory disease in growing and adult pigs and reproductive disorders in breeding animals. The most sensitive age groups to virus infection are newborn pigs. In piglets less than 3 weeks old, sudden death can occur with few, if any, clinical signs. More often, fever, lethargy, loss of appetite, weakness, lack of coordination and convulsions precede death.

Vomiting and diarrhoea can be present. Pigs less than 2 weeks old usually die. Suckling piglets can be infected before birth in the uterus. After birth, they die within 2 days, occasionally having manifested violent shaking and shivering (shaker pig syndrome). Piglets infected immediately after birth show clinical signs within the first 2 days and usually die before they are 5 days old. In older pigs, the symptoms start with fever followed by loss of appetite, listlessness, agony, somnolence, laboured breathing, vomiting, rambling and, in some animals, lack of coordination and weakness occurring in the hindquarters. Death is usually preceded by convulsions. Involvement of the respiratory tract becomes apparent with sneezing, coughing and nasal discharge. Recovered pigs have significant loss of weight. The intensity of the clinical signs impairs with rising age. Hence, in adult pigs, the disease is usually not severe. Fever is always present, and nasal discharge, coughing, agony and somnolence frequently occur, whereas typical nervous symptoms can be observed only occasionally. Usually, no marked pruritus develops in pigs of any age (Pensaert et al. 1982; Wittmann 1986; Kluge et al. 1999; Mettenleiter et al. 2012).

In sows infected in the first trimester of pregnancy, resorption of foetuses can occur. ADV infection in the second or third trimester results in abortion or stillborn pigs (Kluge et al. 1999). In boars, virus replication occurs in the mesorchium, which leads to exudative periorchitis and increased scrotal fluid, resulting in a severely swollen scrotal region. Moreover, morphologic alteration and lowered sperm cell concentration can be observed, leading to infertility (Miry et al. 1987; Miry and Pensaert 1989; Kluge et al. 1999; Mettenleiter et al. 2012).

10.14 Pathological Examination

Gross pathological changes in necropsied pigs are often subtle, absent or difficult to find. Many pigs have serous or fibrinonecrotic rhinitis, but this may be visible only if the head is split and the nasal cavity opened. Pulmonary oedema, congestion or consolidation is sometimes found, and secondary bacterial pneumonia can result in more severe gross lesions. In the lymph nodes, small haemorrhages may be observed. Affected pigs may also have necrotic tonsillitis or pharyngitis, congested meninges or necrotic placentitis. Necrotic foci can occur in the liver; this is particularly common in very young piglets.

Microscopic examination of the white and grey matter typically reveals non-suppurative meningoencephalitis with relatively mild myelitis. Mononuclear perivascular cuffing and neuronal necrosis may be seen, and the meninges are usually thickened from mononuclear cell infiltration. Additional microscopic findings may include necrotic tonsillitis, bronchitis, bronchiolitis and alveolitis. Very often, in the liver, spleen, adrenal glands and lymph nodes of affected foetuses, focal necrosis is present (Narita et al. 1985; Mettenleiter et al. 2012; Wang et al. 2015).

10.15 Immunity

The viral envelope glycoproteins (gB, gC, gE) are reported as major immunogenic components involved in the induction of protective immunity against ADV. Among the mentioned glycoproteins, gB is important for the induction of cell-mediated immunity (CMI), including cytotoxic T-lymphocyte (CTL) responses, but not for inducing virus-neutralising antibody. The gD is important for inducing the virus-neutralising antibody, but not for inducing CMI. Thus, gC elicits only a weak CMI response and virus-neutralising antibody (van Rooij et al. 2000). In contrast, other reports have shown that gC offers complete protection against lethal ADV challenge (Gerdts et al. 1997). Another finding indicated that gB produces the strongest ADV-specific IgG response in the sera. gB and gD induce contrast pattern of immunity with a bias to Th1 and Th2 types, respectively. Moreover, it has been shown that gC is a potent inducer of $CD8^+$ T-cell-mediated CTL activity against ADV (Yoon et al. 2006). In-depth knowledge on the immunological parameters relevant for protection is a pivotal step for improving the effectiveness of vaccines against ADV.

Infection of pigs with ADV elicits specific antibody and T-cell response against ADV. CMI may be detected from day 4 after infection. In immune pigs, secondary lymphoproliferation responses were detectable from day 2 post-challenge, and both $CD4^+$ and $CD8^+$ cells contribute to lymphoproliferation responses. Humoral immunity appears few days later than CMI. Neutralising antibodies may be detected at day 7, reaching optimal titres at day 14. Interestingly, challenge of previously infected pigs does not result in a secondary B-lymphocyte response. In contrast, the proliferative T-lymphocyte response is enhanced after challenge, which suggests that humoral immune response may be of minor importance in the control of ADV infection (Wittman et al. 1976; Kimman et al. 1995; De Bruin et al. 1998).

10.16 Colostral Immunity

Passively acquired maternally derived antibodies (MDA) prevent newborn pigs against transmission and invasion of ADV in the nervous system and decrease clinical manifestation of the disease (McFerran et al. 1979; Wittmann and Jakubik 1979; Bouma et al. 1997). However, MDA did not provide protection against establishment of latent infection in piglets after exposure to low doses of virulent ADV (McCaw et al. 1997). The amount of passive immunity depends on antibody titre in sow, colostrum intake by the newborn piglets and the number of suckling piglets (Andries et al. 1978). MDA can be detected in ELISA as long as 27 weeks postpartum, while in the serum neutralisation test, up to 15 weeks postpartum. The half-life of maternal neutralising antibodies was determined to be 11.3 and 21 days in domestic swine and wild boars, respectively (Tenhagen et al. 1995; Müller et al. 2005).

The disadvantage of MDA is that they may reduce or abolish antibody development after vaccination (Andries et al. 1978; Bouma et al. 1997; Pomorska-Mol et al. 2010b).

Although early vaccination (at 7 days) against AD in the presence of high levels of MDA did not impair T-cell responses, the intensity and duration of recall proliferative response were dependent on the time of booster vaccination (Bouma et al. 1998; Pomorska-Mól and Markowska-Daniel 2010a).

10.17 Evasion Strategy

ADV has developed several strategies to counter and evade host's antiviral responses. One of the strategies is evasion of antibody-dependent cell lysis. It has been reported that gE-gI expressed on the viral envelope acts as receptor for Fc, and by binding to the constant portion of IgGs (Fc) at the surface of virus-infected cells, they inhibit downstream immune functions of IgG (Walle 2003). Moreover, gE-gI can participate in creating antibody bipolar bridging (ABB) by simultaneous binding of the Fab sides of an antibody to viral proteins in the viral envelope or on the surface of infected cells and of the Fc side of the same antibody to viral Fc receptors (Ndjamen et al. 2014). Another ability of the ADV is evasion of major histocompatibility complex (MHC) class I-dependent cell lysis. A downregulation of the MHC I antigen presentation pathway allows ADV to avoid the CTL response. The downregulation process is associated with the ability of ADV to interfere with peptide transport activity (transporter associated with antigen presentation), due to interactions of viral gN protein encoded by the UL49.5 gene (Ambagala et al. 2000; Flori et al. 2008). Furthermore, it has been shown that the US3 of ADV is required but not sufficient for MHC I downregulation process. However, the mechanisms of ADV-mediated MHC I cell surface downregulation are highly cell type dependent, with variable roles for US3, UL49.5 and, additionally, unidentified early viral proteins (Deruelle et al. 2009). Immune circumvent is also associated with anti-apoptotic activity. Apoptosis, known as programmed cell death, is a fundamental cellular process required for embryonic development, organogenesis and the elimination of damaged or aged cells during maintenance of cellular homoeostasis (Fuentes-González et al. 2013). Many cells undergo apoptosis in response to viral infection, which results in reduction in the release of progeny virus (Thomson 2001). It is known that ADV possesses the ability to block apoptosis of infected cells, which allow to complete the replication cycle in host cell (Aleman et al. 2001). In relation to ADV, there is evidence that virus US3 protein kinase possesses anti-apoptotic activity that protects cells from undergoing apoptosis during infection (Geenen et al. 2005). The next immune evasion mechanism is suppression of expression of the most IFN-β-stimulated genes that have potent antiviral effects (Brukman and Enquist 2006). Also, gG inhibits chemokine responses that mediate the migration of immune cells to sites of injury and infection (Viejo-Borbolla et al. 2010). Moreover, it has been reported that after infection of ADV, the interleukin-18 (IL-18) protein expression in the tracheobronchial lymph node is impaired. Importantly, IL-18 is involved in the induction of strong Th1-type immune response (IL-2 and IFN-γ) (Miller et al. 2010). Furthermore, one of the latest studies pointed out the role of ADV in impairing natural killer (NK) cell functions. It was reported that expression of gD protein leads to degradation of CD112/nectin-2, a ligand for the NK-activating receptor DNAX accessory molecule 1 (DNAM-1). This impairs binding of DNAM-1 to the cell surface, thereby suppressing NK-mediated killing of virus-infected (or gD-transfected) cells (Grauwet et al. 2014).

Identification of immune evasion mechanism helps in the design and/or improved herpesvirus vaccines and herpesvirus-based therapeutic vectors.

10.18 Diagnosis

The clinical manifestation of ADV can resemble other infectious and noninfectious diseases including rabies, polioencephalomyelitis (teschovirus), Nipah virus infection, classical and African swine fever, Japanese encephalitis, haemagglutinating encephalomyelitis, *Streptococcus suis* infection, salt poisoning and hypoglycaemia, which should be excluded before the start of ADV laboratory confirmatory test. In sow, diseases which are able to cause reproductive disorders such as porcine parvovirus (PPV), porcine reproductive and respiratory syndrome virus (PRRSV), porcine circovirus type 2 (PCV2), swine influenza virus (SIV) and leptospirosis should be taken into account for the differential diagnosis.

In non-porcine species, other diseases causing neurological symptoms (rabies, scrapie (sheep), bovine spongiform encephalopathy (BSE), hepatic encephalopathy (dogs, cats)) and itching need to be considered.

Various methods are used for the laboratory diagnosis of ADV infections. AD can be confirmed by virus isolation, DNA detection by polymerase chain reaction (PCR) and indirectly by serological methods. The oropharyngeal fluid, nasal fluid (swabs) or tonsil swabs are appropriate specimens for the virus isolation from living pigs. From dead pigs, the preferred specimens are brain, tonsil and lung samples. In latently infected pigs, the trigeminal ganglia is the most consistent site for virus isolation, although latent virus is usually non-infective unless reactivated, making it difficult to recover in culture. In non-porcine species, the section of the spinal cord that innervates the pruritic area should be collected. The pruritic area of the skin, together with subcutaneous tissues, should also be submitted (The Center for Food Security and Public Health 2006; OIE Terrestial Manual 2012).

The collected samples are homogenised in normal saline or cell culture medium with antibiotics. The resulting suspension is next clarified by low-speed centrifugation. The virus particle-containing supernatant is used to inoculate sensitive cell culture system for virus passage and cytopathic effect (CPE) examination. The typical CPE is usually observed 24–72 h after inoculation and is characterised as the infected cells detached from each other, became round and form an approximate circular or oval plaque. In addition, large syncytia can also be formed (Ren et al. 2012). However, virus identity should be confirmed by immunofluorescence, immunoperoxidase or neutralisation using a specific antiserum (OIE Terrestial Manual 2012).

Serological tests such as VN, a variety of ELISA tests, latex agglutination test (LAT) and agar gel immunodiffusion test (AGID) can be used to diagnose AD (Corn et al. 2004; Yong et al. 2005; OIE Terrestial Manual 2012; Serena et al. 2015). It can be carried out only in pigs (feral swine, wild boars), as other susceptible animals (herbivores and carnivores) die before mounting the humoral response.

Among serology methods, VN has been recognised as the reference method and is considered as a gold standard test for ADV infection. In general, VN has been widely replaced by ELISA assay because of its suitability for large-scale testing. Moreover, VN is very expensive and time-consuming and both require expensive equipment and trained technicians. Due to these conditions, VN is not used in routine laboratory diagnostics. Moreover, some ELISA tests (i.e. gE differential ELISA test) are able to distinguish naturally infected and the glycoprotein E (gE)-deleted modified live virus vaccine-immunised pigs. This step is a pivotal part of successful eradication of AD in domestic swine.

Any serological technique used in AD diagnosis should be sensitive enough to give a positive result with the OIE international standard reference serum. This serum can be obtained from the OIE reference laboratory for AD in France. For international trade purposes, the test should be sensitive enough to detect the standard serum diluted one half. To authorise pig movement from an area where deleted gE vaccines are used to a free area, serological assays should be able to detect gE of the OIE reference standard serum at one eighth dilution as prescribed by the European Commission (2008) (OIE Terrestial Manual 2012).

At present, real-time PCR, a high-throughput test system is widely used in AD diagnosis in domestic and feral swine, as well as non-porcine species. The advantages of real-time PCR are the rapid and specific ADV detection (acute phase of infection – before seroconversion) and differentiation of gE-deleted vaccine viruses from field ADV (Ma et al. 2008; Zanella et al. 2012).

10.19 Vaccination

In 1977, Norden Laboratories (Lincoln, Nebraska) licensed the first conventional, commercial modified live vaccine (MLV) and inactivated ADV vaccine. The first gene-deleted PRV vaccine (gG deletion) with a companion differential serological diagnostic test was licensed in 1988. In 1990, a new vaccine based on a virus with a gE glycoprotein deletion and a companion differential ELISA diagnostic serology test has been introduced to the market (USDA 2008).

In several studies, it has been documented that both live and inactivated vaccines induce protective immunity after wild-type ADV challenge (Alva-Valdes 1981; Cook et al. 1990; Vannier and Cariolet 1991; Vilnis et al. 1998; Mikulska-Skupień et al. 2005). Inactivated vaccines are capable of mounting equal or higher titre of VN antibodies such as MLV, but it is less effective in inducing virus-specific IFN-γ-producing cells (Zuckermann et al. 1998; Zuckermann 2000). It has been reported that CMI response is a major component of early protection against ADV infection and that ADV-induced IFN-γ responses may serve as a suitable indicator for assessing the immune status of vaccinated pigs (Hoegen et al. 2004; van Rooij et al. 2004). An additional important issue associated with MLV ADV vaccine is the ability of vaccine virus to prevent or reduce the establishment of latent infection by the wild-type virus. Moreover, it has been reported that efficacy of MLV depends on the MLV strain and route of administration (Vilnis et al. 1998).

The important findings reported recently indicated that simultaneous administration of therapeutic doses of antibiotics (doxycycline, enrofloxacin, ceftiofur) and vaccination (MLV) against ADV might affect the development of post-vaccinal humoral and cellular immune responses (Pomorska-Mól et al. 2014, 2015a, 2015b). Whereas, development of humoral and CMI response after vaccination against ADV in pigs with subclinical infection, e.g. PCV2, was altered (Díaz et al. 2012).

10.20 Prevention and Control

Severe economic consequences in pig husbandry are related to AD, mainly due to international trade restrictions, as well as the costs of vaccination and/or slaughtering of infected pigs. For example, an AD epidemic in West Germany in 1980–1982 caused indirect losses of 61 million Deutsche Marks in compensation to farmers for slaughtered animals (Vicente-Rubiano et al. 2014). In response to huge economic losses, most countries have launched strict AD control and eradication programmes mainly based on vaccination and elimination of infected herds.

Nowadays, due to rigorous AD control and eradication programmes, several countries reach the AD-free status. However, ADV still circulates globally in non-domestic swine population and might reintroduce the disease to commercial swine as well as other domestic animals and wildlife. To prevent the reintroduction, proper surveillance and strict biosecurity measures are compulsory especially in areas where wild swine population may have a direct contact with commercial swine (free-roaming pig husbandry). In Spain where open-air pig production coexists with a large wild boar population (ADV seropositive), a new approach for rapidly assessing the risk of AD reintroduction into a disease-free Spanish territory by analysing the movement of live pigs and potential contacts with wild boar has been proposed. This new approach may help in designing risk-based surveillance and control strategies to rapidly eradicate the disease from endemic areas and to prevent further reintroduction of the virus into free territories (Vicente-Rubiano et al. 2014).

References

Aleman N, Quiroga MI, Lopez-Pena M et al (2001) Induction and inhibition of apoptosis by pseudorabies virus in the trigeminal ganglion during acute infection of swine. J Virol 75:469–479

Allepuz A, Seaz M, Solymosi N et al (2009) The role of spatial factors in the success of an Aujeszky's disease eradication programme in a high pig density area (Northeast Spain, 2003-2007). Prev Vet Med 91:153–160

Alva-Valdes R (1981) Pseudorabies (Aujeszky's disease): effect of vaccination on lesion development and immune response in challenged swine. Dissertation, Iowa State University

Ambagala AP, Hinkley S, Srikumaran S (2000) An early pseudorabies virus protein down-regulates porcine MHC class I expression by inhibition of transporter associated with antigen processing (TAP). J Immunol 164:93–99

Andries K, Pensaert MB, Pandeputte J (1978) Effects of experimental infection with pseudorabies (Aujeszky's disease) on pigs with maternal immunity from vaccinated sows. Am J Vet Res 39:1282–1285

Babic N, Kluup B, Brack A et al (1996) Deletion of glycoprotein gE reduces the propagation of pseudorabies virus in the nervous system of mice after intranasal inoculation. Virology 219:279–284

Balasch M, Pujols J, Segalés J (1998) Study of the persistence of Aujeszky's disease (pseudorabies) virus in peripheral blood mononuclear cells and tissues of experimentally infected pigs. Vet Microbiol 62:171–183

Baskerville A (1973) The histopathology of experimental pneumonia in pigs produced by Aujeszky's disease virus. Res Vet Sci 14:223–228

Benndorf E, Hantschel H (1963) Zum Verhalten des Aujeszkyvirus bei veschiedenen Wasserstoffkonzentrationen. Arch Exp Vet Med 17:1357–1362

Beran GW (1991) Transmission of Aujeszky's disease virus. In: Proceedings of the 1st International Symposium on the Eradication of Pseudorabies (Aujeszky's) Virus. St. Paul, Minnesota, pp 93–111

Beran GW, Davies EB, Arambulo PV et al (1980) Persistence of pseudorabies virus in infected swine. J Am Vet Med Assoc 176:988–1000

Biermann U, Herbst W, Schliesser T (1990) The persistence of bovine enterovirus and pseudorabies virus in liquid cattle manure at different storage temperatures. Berl Munch Tierarztl Wochenschr 103:88–90

Blaha T (1989) Aujeszky's disease (pseudorabies). In: Blaha T (ed) Applied veterinary epidemiology. Elsevier, Amsterdam, pp 83–87

Boadella M, Gortázar C, Vicente J et al (2012) Wild boar: an increasing concern for Aujeszky's disease control in pigs? BMC Vet Res 8:7

Bolin CA, Bolin SR, Kluge JP et al (1985) Pathologic effects of intrauterine deposition of pseudorabies virus on the reproductive tract of swine in early pregnancy. Am J Vet Res 46:1039–1042

Bøtner A (1991) Survival of Aujeszky's disease virus in slurry at various temperatures. Vet Microbiol 29:225–235

Bouma A, De Jong MC, Kimman TG (1997) The influence of maternal immunity on the transmission of pseudorabies virus and on the effectiveness of vaccination. Vaccine 15:287–294

Bouma A, De Jong MDM, Kimman TG (1998) The influence of maternal immunity on the development of the in vitro lymphocyte proliferation response against pseudorabies virus in pigs. Res Vet Sci 64:167–171

Brittle EE, Reynolds AE, Enquist LW (2004) Two modes of pseudorabies virus neuroinvasion and lethality in mice. J Virol 78:12951–12963

Brukman A, Enquist LW (2006) Suppression of the interferon-mediated innate immune response by pseudorabies virus. J Virol 80:6345–6356

Cheung AK (1989) Detection of pseudorabies virus transprits in trigeninal ganglia of latently infected swine. J Virol 63:2908–2913

Cook D, Hill HT, Kinker DR (1990) Efficacy of a killed gpX deleted pseudorabies virus vaccine. Can J Vet Res 54:438–445

Corn JL, Stallknecht DE, Mechlin NM et al (2004) Persistence of pseudorabies virus in feral swine populations. J Wildl Dis 40:307–310

Cramer SD, Campbell GA, Njaa BL et al (2011) Pseudorabies virus infection in Oklahoma hunting dogs. J Vet Diagn Invest 23:915–923

Davies EB, Beran G (1980) Spontaneous shedding of pseudorabies virus from a clinically recovered postparturient sow. J Am Vet Med Assoc 176:1345–1347

Davies EB, Beran GW (1981) Influence of environmental factors upon the survival of Aujeszky's disease virus. Res Vet Sci 31:32–36

De Bruin MG, De Visser YE, Kimman TG et al (1998) Time course of the porcine cellular and humoral immune response in vivo against pseudorabies virus after inoculation and challenge: significante of in vitro antigenic restimulation. Vet Immunol Immunopathol 65:75–87

De Regge N, Van Opdenbosch N, Nauwynck HJ et al (2010) Interferon alpha induces establishment of alphaherpesvirus latency in sensory neurons in vitro. PLoS One 5(9):e13076. doi:10.1371/journal.pone.0013076
Deruelle MJ, Van den Broeke C, Nauwynck HJ et al (2009) Pseudorabies virus US3- and UL49.5-dependent and -independent downregulation of MHC I cell surface expression in different cell types. Virology 395:172–181. doi:10.1016/j.virol.2009.09.019
Díaz I, Cortey M, Darwich L et al (2012) Subclinical porcine circovirus type 2 infection does not modulate the immune response to an Aujeszky's disease virus vaccine. Vet J 194:84–88. doi:10.1016/j.tvjl.2012.02.014
Donaldson AI, Wardley RC, Martin S et al (1983) Experimental Aujeszky's disease in pigs: excretion, survival and transmission of the virus. Vet Rec 113:490–494
Durham PJK, Gow A, Poole WSH (1980) Survival of Aujeszky's disease virus in frozen pig meat. Res Vet Sci 28:256–258
Dvorakova H, Prodelalova J, Reichelova M (2008) Comparative inactivation of Aujeszky's disease virus, porcine teschovirus and vesicular stomatitis virus by chemical disinfectants. Vet Med (Praha) 53:236
European Commission (2008). Commission Decision of 21 February 2008 on additional guarantees in intra-Community trade of pigs relating to Aujeszky's disease and criteria to provide information on this disease 2008/185/EC: Official Journal of the European Communities L 316, 5–35
Farrell MJ, Dobson AT, Feldman LT (1991) Herpes simplex virus latency associated transcript is a stable intron. Proc Natl Acad Sci U S A 88:790–794
Flori L, Rogel-Gaillard C, Cochet M et al (2008) Transcriptomic analysis of the dialogue between Pseudorabies virus and porcine epithelial cells during infection. BMC Genomics 9:123. doi:10.1186/1471-2164-9-123
Fuentes-González AM, Contreras-Paredes A, Manzo-Merino J et al (2013) The modulation of apoptosis by oncogenic viruses. Virol J 10:182. doi:10.1186/1743-422X-10-182
Garcia-Siera J, Rozeboom DW, Straw BE et al (2001) Studies on survival of pseudorabies virus, Actinobacillus pleuropneumoniae, and Salmonella serovar Choleraesuis in composted swine carcasses. J Swine Health Prod 9:225–231
Geenen KH, Favoreel W, Nauwynck HJ (2005) Higher resistance of porcine trigeminal ganglion neurons towards pseudorabies virus-induced cell death compared with other porcine cell types in vitro. J Gen Virol 86:1251–1260
Gerdts V, Jons A, Makoschey B et al (1997) Protection of pigs against Aujeszky's disease by DNA vaccination. J Gen Virol 78:2139–2146
Glass C, McLean R, Katz J et al (1994) Isolation of pseudorabies (Aujeszky's disease) virus from a Florida panther. J Wildl Dis 30:180–184
Glorieux S (2009) Invasion of pseudorabies virus in porcine nasal respiratory mucosa explants. Dissertation, Ghent University
Glorieux S, Favoreel HW, Steukers L et al (2011) A trypsin-like serine protease is involved in pseudorabies virus invasion through the basement membrane barrier of porcine nasal respiratory mucosa. Vet Res 42:58. doi:10.1186/1297-9716-42-58
Granzow H, Klupp BG, Mettenleiter TC (2004) The pseudorabies virus US3 protein is a component of primary and of mature virions. J Virol 78:1314–1323
Grauwet K, Cantoni C, Parodi M et al (2014) Modulation of CD112 by the alphaherpesvirus gD protein suppresses DNAM-1-dependent NK cell-mediated lysis of infected cells. Proc Natl Acad Sci U S A 111:16118–16123. doi:10.1073/pnas.1409485111
Guo H, Shen S, Wang L et al (2010) Role of tegument proteins in herpesvirus assembly and egress. Protein Cell 1:987–998
Hoegen B, Saalmuller A, Rottgen M et al (2004) Interferon-gamma response of PBMC indicates productive pseudorabies virus (PRV) infection in swine. Vet Immunol Immunopathol 102:389–397
Hsu FS, Chu RM, Lee RC et al (1980) Placental lesions caused by pseudorabies virus in pregnant sows. J Am Vet Med Assoc 177:636–641
Huang C, Wu CY (2004) Characterization and expression of the pseudorabies virus early gene UL54. J Virol Methods 119:129–136

Jacobs L, Mulder WA, Van Oirschot JT et al (1993) Deleting two amino acids in glycoprotein gI of pseudorabies virus decreases virulence and neurotropism for pigs, but does not affect immunogenicity. J Gen Virol 74:2201–2206

Kapitein LC, Hoogenraad CC (2011) Which way to go? Cytoskeletal organization and polarized transport in neurons. Mol Cell Neurosci 46:9–20

Kimman TMG, De Bruin JJM, Voermans BPH et al (1995) Development and antigen specificity of the lymphoproliferation response of pigs to pseudorabies virus: dichotomy between secondary B- and T-cell response. Immunology 86:372–378

Kirkpatrick CM, Kanitz CL, McCrocklin SM (1980) Possible role of wild mammals in transmission of pseudorabies to swine. J Wildl Dis 16:601–614

Kluge JP, Beran GW, Hill HT et al (1999) Pseudorabies (Aujeszky's disease). In: Straw BE, D'allaire S, Mengeling WL, Taylor DJ (eds) Disease of Swine, 8th edn. Blackwell Science, Ames, pp 233–246

Klupp BG, Fuchs W, Weiland E et al (1997) Pseudorabies virus glycoprotein L is necessary for virus infectivity but dispensable for virion localization of glycoprotein H. J Virol 71:7687–7695

Klupp BG, Hengartner CJ, Mettenleiter T (2004) Complete, annotated sequence of the pseudorabies virus genome. J Virol 78:424–440

Kopp M, Granzow H, Fuchs W et al (2004) Simultaneous deletion of Pseudorabies virus tegument protein UL11 and glycoprotein M severely impairs secondary envelopment. J Virol 78:3024–3034.

Köppel C, Knopf L, Ryser MP et al (2007) Serosurveillance for selected infectious disease agents in wild boars (*Sus scrofa*) and outdoor pigs in Switzerland. Eur J Wildl Res 53:212–220

Koyuncu OO, Hogue IB, Enquist LW (2013a) Virus infections in the nervous system. Cell Host Microbe 13:379–393

Koyuncu OO, Perlman DH, Enquist LW (2013b) Efficient retrograde transport of pseudorabies virus within neurons requires local protein synthesis in axons. Cell Host Microbe 13:54–66

Kritas SK, Pensaert MB, Mettenleiter TC (1994) Invasion and spread of single glycoprotein deleted mutants of Aujeszky's disease virus (ADV) in the trigeminal nervous pathway of pigs after intranasal inoculation. Vet Microbiol 40:323–334

Kukushkin S, Baborenko E, Baybikov T et al (2009) Seroprevalence of antibodies to main porcine infectious pathogens in wild boar in some regions of Russia. Acta Silvatica et Lignaria Hungarica 5:147–152

Lari A, Lorenzi D, Nigrelli D et al (2006) Pseudorabies virus in European wild boar from Central Italy. J Wildl Dis 42:319–324

Luxton GW, Lee JI, Haverlock-Moyns S et al (2006) The pseudorabies virus VP1/2 tegument protein is required for intracellular capsid transport. J Virol 80:201–209

Ma W, Lager KM, Richt JA et al (2008) Development of real-time polymerase chain reaction assays for rapid detection and differentiation of wild-type pseudorabies and gene-deleted vaccine viruses. J Vet Diagn Invest 20:440–447

Marcaccini A, López Peña M, Quiroga MI et al (2008) Pseudorabies virus infection in mink: a host-specific pathogenesis. Vet Immunol Immunopathol 124:264–273. doi:10.1016/j.vetimm.2008.03.013

McCaw MB, Osorio FA, Wheeler J et al (1997) Effect of maternally acquired Aujeszky's disease (pseudorabies) virus-specific antibody in pigs on establishment of latency and seroconversion to differential glycoproteins after low dose challenge. Vet Microbiol 55:91–98

McFerran JB, Dow C (1965) The distribution of the virus of Aujeszky's disease (pseudorabies virus) in experimentally infected swine. Am J Vet Res 26:631–635

McFerran JB, Dow C, McCracken RM (1979) Experimental studies in weaned pigs with three vaccines against Aujeszky's disease. Comp Immunol Microbiol Infect Dis 2:327–334

McGregor S, Easterday BC, Kaplan AS et al (1985) Vaccination of swine with thymidine kinase-deficient mutants of pseudorabies virus. Am J Vet Res 46:1494–1497

Meier RK, Ruiz-Fons F, Ryser-Degiorgis M-P (2015) A picture of trends in Aujeszky's disease virus exposure in wild boar in the Swiss and European contexts. BMC Vet Res 11:277. doi:10.1186/s12917-015-0592-5

Mettenleiter TC (1994) Initiation and spread of α-herpesvirus infections. Trends Microbiol 2:2–4
Mettenleiter TC (1996) Immunobiology of pseudorabies (Aujeszky's disease). Vet Immunol Immunopathol 54:221–229
Mettenleiter TC (2000) Aujeszky's disease (pseudorabies) virus: the virus and molecular pathogenesis – State of the art. Vet Res 31:99–115
Mettenleiter TC (2002) Brief overview on cellular virus receptors. Virus Res 82:3–8
Mettenleiter TC, Ehlers B, Muller T et al (2012) In: Zimmerman JJ, Karriker LA, Ramírez A, Schwartz KJ, Stevenson GW (eds) Diseases of swine, 10th edn. Wiley-Blackwell Publishing, Ames, pp 421–455
Mettenleiter TC, Klupp BG, Granzow H (2009) Herpesvirus assembly: an update. Virus Res 143:222–234
Mettenleiter TC, Zsak L, Zuckermann F et al (1990) Interaction of glycoprotein gIII with a cellular heparinlike substance mediates adsorption of pseudorabies virus. J Virol 64:278–286
Mikulska-Skupień E, Szweda W, Procajło Z (2005) Evaluation of specific humoral immune response in pigs vaccinated intradermally with deleted Aujeszky's disease vaccine and challenged with virulent strain of Herpesvirus suis type 1. Pol J Vet Sci 8:11–16
Miller LC, Zanella EL, Waters WR et al (2010) c. Clin Vaccine Immunol 17:728–734
Miry C, Pensaert MB (1989) Sites of virus replication in the genital organs of boars inoculated in the cavum vaginale with pseudorabies virus. Am J Vet Res 50:345–348
Miry C, Pensaert MB, Bonte P et al (1987) Effect of intratesticular inoculation with Aujeszky's disease virus on genital organs of boars. Vet Microbiol 14:355–363
Montagnaro S, Sasso S, De Martino L et al (2010) Prevalence of antibodies to selected viral and bacterial pathogens in wild boar (*Sus scrofa*) in Campania Region, Italy. J Wildl Dis 46:316–319
Morrow WEM, O'Quinn P, Barker J et al (1995) Composting as a suitable technique for managing swine mortalities. Swine Health Prod 3:236–243
Müller T, Hahn EC, Tottewitz F et al (2011) Pseudorabies virus in wild swine: a global perspective. Arch Virol 156:1691–1705. doi:10.1007/s00705-011-1080-2
Müller T, Teuffert J, Staubach C et al (2005) Long-term studies on maternal immunity for Aujeszky's disease and classical swine fever in wild boar piglets. J Vet Med B Infect Dis Vet Public Health 52:432–436
Müller TF, Teuffert J, Zellmer R et al (2001) Experimental infection of European wild boars and domestic pigs with pseudorabies viruses with differing virulence. Am J Vet Res 62:252–258
Narita M, Haritani M, Moriwaki M et al (1985) Pseudorabies virus in dexamethasone-treated pigs. Vet Pathol 22:417–419
Nauwynck H, Glorieux S, Favoreel H et al (2007) Cell biological and molecular characteristics of pseudorabies virus infections in cell cultures and in pigs with emphasis on the respiratory tract. Vet Res 38:229–241
Nauwynck HJ, Pensaert MB (1995) Cell-free and cell-associated viremia in pigs after oronasal infection with Aujeszky's disease virus. Vet Microbiol 43:307–314
Ndjamen B, Farley AH, Lee T et al (2014) The herpes virus Fc receptor gE-gI mediates antibody bipolar bridging to clear viral antigens from the cell surface. PLoS Pathog 10:e1003961. doi:10.1371/journal.ppat.100396
Nixdorf R, Schmidt J, Karger A et al (1999) Infection of Chinese hamster ovary cells by pseudorabies virus. J Virol 73:8019–8026
OIE Terrestrial Manual (2012) Aujeszky's disease, Chapter 2.1.2. http://www.oie.int/fileadmin/Home/fr/Health_standards/tahm/2.01.02_AUJESZKYS_DIS.pdf
Paluszak Z, Lipowski A, Ligocka A (2012) Survival rate of Suid herpesvirus (SuHV-1, Aujeszky's disease virus, ADV) in composted sewage sludge. Pol J Vet Sci 15:51–54
Pan Z, Zhang C, Ding J et al (2001) Thymidine kinase gene mutation leads to reduced virulence of pseudorabies virus. Chin Sci Bull 46:1972–1975
Pannwitz G, Freuling C, Denzin N et al (2012) A long-term serological survey on Aujeszky's disease virus infections in wild boar in East Germany. Epidemiol Infect 140:348–358
Pedersen K, Bevins SN, Baroch JA et al (2013) Pseudorabies in feral swine in the United States, 2009–2012. J Wildl Dis 49:709–713

Pensaert MB, Kluge JB (1989) Pseudorabies virus (Aujeszky's disease). In: Pensaert MB (ed) Virus infection of porcine. Elsevier, Amsterdam, pp 39–64
Pensaert MB, Miry C, Biront P et al (1987) Evolution of Aujeszky's disease in Belgium (1974-1987). Tijdschr Diergeneeskd 56:425–432
Pensaert MB, Vandeputte J, Andries K (1982) Oronasal challenge of fattening pigs after vaccination with an inactivated Aujeszky's disease vaccine. Res Vet Sci 32:12–16
Pol JMA (1990) Interferons affect the morphogenesis and virulence of pseudorabies virus. Dissertation, Rijksuniversiteit
Pomeranz LE, Reynolds AE, Hengartner CJ (2005) Molecular biology of pseudorabies virus: impact on neurovirology and veterinary medicine. Microbiol Mol Biol Rev 69:462–500. doi:10.1128/MMBR.69.3.462-500.2005
Pomorska-Mól M, Czyżewska-Dors E, Kwit K et al (2015a) Immune response in pigs treated with therapeutic doses of enrofloxacin at the time of vaccination against Aujeszky's disease. Res Vet Sci 100:68–74. doi:10.1016/j.rvsc.2015.04.003
Pomorska-Mól M, Czyżewska-Dors E, Kwit K et al (2015b) Ceftiofur hydrochloride affects the humoral and cellular immune response in pigs after vaccination against swine influenza and pseudorabies. BMC Vet Res 11:268. doi:10.1186/s12917-015-0586-3
Pomorska-Mól M, Kwit K, Markowska-Daniel I et al (2014) The effect of doxycycline treatment on the postvaccinal immune response in pigs. Toxicol Appl Pharmacol 278:31–38. doi:10.1016/j.taap.2014.04.006
Pomorska-Mól M, Markowska-Daniel I (2010a) Interferon-γ secretion and proliferative responses of peripheral blood mononuclear cells after vaccination of pigs against Aujeszky's disease in the presence of maternal immunity. FEMS Immunol Med Microbiol 58:405–411. doi:10.1111/j.1574-695X.2010.00651.x
Pomorska-Mól M, Markowska-Daniel I, Pejsak Z (2010b) Evaluation of humoral and antigen-specific T-cell responses after vaccination of pigs against pseudorabies in the presence of maternal antibodies. Vet Microbiol 144:45–454. doi:10.1016/j.vetmic.2010.01.015
Prieto J, Hernandez AMM, Tabares E (1991) Loss of pseudorabies virus thymidine kinase activity due to a single base mutation and amino acid substitution. J Gen Virol 72:1435–1439
Rauh I, Mettenleiter TC (1991) Pseudorabies virus glycoproteins gII and gp50 are essential for virus penetration. J Virol 65:5348–5356
Ren Y, Li G, Su D et al (2012) Characterization of pseudorabies viruses produced in mammalian cells by viral genome transfection. Afr J Biotechnol 11:4098–4102
Roizman B, Pellet PE (2001) The family herpesviridae: a brief introduction. In: Knipe DM, Howley PM (eds) Fields virology, vol 2, 4th edn. Lippincott William&Wilkins, Philadelphia, pp 2381–2397
Romero CH, Meade PN, Homer BL (2003) Potential sites of virus latency associated with indigenous pseudorabies viruses in feral swine. J Wildl Dis 39:567–575
Rziha HJ, Mettenleiter TC, Ohlinger V et al (1986) Herpesvirus (pseudorabies virus) latency in swine: occurrence and physical state of viral DNA in neural tissues. Virology 155:600–613
Sabo A, Rajcani J, Blaskovic D (1969) Studies on the pathogenesis of Aujeszky's disease virus. III. The distribution of virulent virus in piglets after intranasal infection. Acta Virol 13:407–714
Sakano T, Shibata I, Samegai Y et al (1993) Experimental pneumonia of pigs infected with Aujeszky's disease virus and Actinobacillus pleuropneumoniae. J Vet Med Sci 55:575–579
Schoenbaum MA, Beran GW, Murphy DP (1990a) Pseudorabies virus latency and reactivation in vaccinated swine. Am J Vet Res 51:334–338
Schoenbaum MA, Zimmerman JJ, Beran GW et al (1990b) Survival of pseudorabies virus in aerosol. Am J Vet Res 51:331–333
Sedlak K, Bartova E, Machova J (2008) Antibodies to selected viral disease agents in wild boars from the Czech Republic. J Wildl Dis 44:777–780
Serena MS, Metz GE, Panei CJ et al (2015) Development of an AGID based on baculovirus expressed Pseudorabies virus glycoprotein B. In: Front Immunol Conference Abstract: ImmunoColombia 2015 - 11th Congress of the Latin American Association of Immunology-10o. Congreso de la Asociación Colombiana de Alergia, Asma e Inmunología. doi:10.3389/conf.fimmu.2015.05.00236

Shibata I, Okada M, Urono K et al (1998) Experimental dual infection of cesarean-derived, colostrum-deprived pigs with Mycoplasma hyopneumoniae and pseudorabies virus. J Vet Med Sci 60:295–300

Shibata I, Yazawa S, Ono M et al (2003) Experimental dual infection of specific pathogen-free pigs with porcine reproductive and respiratory syndrome virus and pseudorabies virus. J Vet Med B Infect Dis Vet Public Health 50:14–19

Smith GA, Gross SP, Enquist LW (2001) Herpesviruses use bidirectional fast-axonal transport to spread in sensory neurons. Proc Natl Acad Sci U S A 98:3466–3470

Spear PG, Eisenberg RJ, Cohen GH (2000) Three classes of cell surface receptors for alphaherpesvirus entry. Virology 275:1–8.

Szweda W, Lipowski A, Ciecierski H et al (1998) European wild boar (*Sus scrofa* L.) as a reservoir of Herpesvirus suis 1. Med Weter 54:541–544

Tanaka S, Mannen K (2003) Effect of mild stress in mice latently infected pseudorabies virus. Exp Anim 52:383–386

Tanaka S, Mannen K (2008) Pregnancy and parturition of mice latently infected with Pseudorabies virus. Exp Anim 57:149–152

Tenhagen BA, Bollwahn W, Seidler MJ (1995) Vaccination trial against Aujeszky's disease: development of antibodies in sow serum, colostrum and the serum of suckling piglets and the influence of maternal antibodies on the serologic vaccination reaction of weaned piglets. Dtsch Tierärztl Wschr 102:86–90

The Center for Food Security and Public Health (2006) Iowa State University. http://www.cfsph.iastate.edu/Factsheets/pdfs/aujeszkys_disease.pdf

Thomson BJ (2001) Viruses and apoptosis. Int J Exp Pathol 82:65–76

Tong W, Liu F, Zheng H et al (2015) Emergence of a pseudorabies virus variant with increased virulence to piglets. Vet Microbiol 181(3-4):236–240. doi:10.1016/j.vetmic.2015.09.021

Turner C, Williams SM, Cumby TR (2000) The inactivation of foot and mouth disease, Aujeszky's disease and classical swine fever viruses in pig slurry. J Appl Microbiol 89:760–767

Umbach JL, Kramer MF, Jurak I et al (2008) MicroRNAs expressed by herpes simplex virus 1 during latent infection regulate viral mRNAs. Nature 454:780–783

USDA (2008) Pseudorabies (Aujeszky's disease) and its eradication. Animal and plant health inspection service. Technical Bulletin No. 1923. https://www.aasv.org/documents/pseudorabiesreport.pdf

Van De Walle G (2003) Pseudorabies virus-specific antibodies mask infected monocytes from immune recognition, induce a quiescent infection, and allow virus transmission to endothelial cells. Dissertation, Gent University

Van Rooij EM, de Bruin MG, de Visser YE et al (2004) Vaccine-induced T cell-mediated immunity plays a critical role in early protection against pseudorabies virus (suid herpesvirus type 1) infection in pigs. Vet Immunol Immunopathol 99:113–125

Van Rooij EM, Haagmans BL, Glansbeek HL et al (2000) A DNA vaccine coding for glycoprotein B of pseudorabies virus induces cell-mediated immunity in pigs and reduces virus excretion early after infection. Vet Immunol Immunopathol 74:121–136

Vannier P, Cariolet R (1991) Vaccination of pigs against Aujeszky's disease by the intradermal route using live attenuated and inactivated virus vaccines. Vet Microbiol 26:11–23

Vengust G, Valencak Z, Bidovec A (2006) A serological survey of selected pathogens in wild boar in Slovenia. J Vet Med Ser B Infect Dis Vet Public Health 53:24–27

Verpoest S, Cay AB, Bertrand O et al (2014) Isolation and characterization of pseudorabies virus from a wolf (*Canis lupus*) from Belgium. Eur J Wildl Res 60:149–153

Vicente-Rubiano M, Martınez-Lopez B, Sanchez-Vizcaıno F et al (2014) A new approach for rapidly assessing the risk of Aujeszky's Disease reintroduction into a disease-free Spanish territory by analysing the movement of live pigs and potential contacts with wild boar. Transbound Emerg Dis 61:350–361. doi:10.1111/tbed.12041

Viejo-Borbolla A, Munoz A, Tabares E et al (2010) Glycoprotein G from pseudorabies virus binds to chemokines with high affinity and inhibits their function. J Gen Virol 91:23–31

Vilnis A, Sussman MD, Thacker BJ et al (1998) Vaccine genotype and route of administration affect pseudorabies field virus latency load after challenge. Vet Microbiol 62:81–96

Visser N (1997) Vaccination strategies for improving the efficacy of programs to eradicate Aujeszky's disease virus. Vet Microbiol 55:61–74
Wang T, Xiao Y, Yang Q et al (2015) Construction of a gE-deteted psudorabies virus and its efficacy to the new-emerging variant PRV challenge in the form of killed vaccine. Biomed Res Int 2015:684945. doi:10.1155/2015/684945
White AK, Ciacci-Zanella J, Galeota-Wheeler J et al (1994) Detection of latent pseudorabies virus infections. In: 37th Annual Meeting of the American Association of Veterinary Laboratory Diagnosticians. Grand Rapids
Wiśniewski J, Siemionek J (1988) Chosen problems of immunoprophylaxis of Aujeszky's disease in swine. Med Weter 44:81–84
Wittmann G (1986) Aujeszky's disease. Rev Sci Tech Off Int Epiz 5:959–977
Wittmann G (1991) Spread and control of Aujeszky's disease (AD). Comp Immunol Microbial Infect Dis 14:165–173
Wittmann G, Bartenbach G, Jakubik J (1976) Cell-mediated immunity in Aujeszky's virus infected pigs. Arch Virol 50:215–222
Wittmann G, Jakubik J (1979) Colostral immunity in piglets from sows vaccinated with inactivated Aujeszky disease virus vaccine. Arch Virol 60:33–42
Wittmann G, Jakubik J, Ahl R (1980) Multiplication and distribution of Aujeszky's disease (pseudorabies) virus in vaccinated and non-vaccinated pigs after intranasal infection. Arch Virol 66:227–240
Wittmann G, Rziha HJ (1989) Aujeszky's disease (pseudorabies) in pigs. In: Wittmann G (ed) Herpesvirus diseases of cattle, horses, and pigs. Kluwer Academic Publ, Boston/Dordrecht/London, pp 230–325
Yong T, Huanchun C, Shao-Bo X et al (2005) Development of a latex agglutination test using the major epitope domain of glycoprotein E of pseudorabies virus expressed in *E. coli* to differentiate between immune responses in pigs naturally infected or vaccinated with pseudorabies virus. Vet Res Commun 29:487–497
Yoon HA, Aleyas AG, George JA et al (2006) Differential segregation of protective immunity by encoded antigen in DNA vaccine against pseudorabies virus. Immunol Cell chain reaction assay for Pseudorabies virus surveillance purposes. J Vet Diagn Invest 24:739–745
Zaichick SV, Bohannon KP, Hughes A et al (2013) The herpesvirus VP1/2 protein is an effector of dynein-mediated capsid transport and neuroinvasion. Cell Host Microbe 13:193–203
Zanella EL, Miller LC, Lager KM et al (2012) Evaluation of a real-time polymerase chain reaction assay for Pseudorabies virus surveillance purposes. J Vet Diagn Invest 24:739–745. doi:10.1177/1040638712447279
Zanin E, Capua I, Casaccia C et al (1997) Isolation and characterization of Aujeszky's disease virus in captive brown bears from Italy. J Wildl Dis 33:632–634
Zuckermann FA (2000) Aujeszky's disease virus: opportunities and challenges. Vet Res 31:121–131
Zuckermann FA, Husmann RJ, Schwartz R (1998) Interleukin-12 enhances the virus-specific interferon gamma response of pigs to inactivated pseudorabies virus vaccine. Vet Immunol Immunopathol 63:57–67
Zupancic Z, Jukic B, Lojkic M et al (2002) Prevalence of antibodies to classical swine fever, Aujeszky's disease, porcine reproductive and respiratory syndrome, and bovine viral diarrhoea viruses in wild boars in Croatia. J Vet Med B Infect Dis Vet Public Health 49:253–256

Porcine Epidemic Diarrhea

11

Ju-Yi Peng, Cai-Zhen Jian, Chia-Yu Chang, and Hui-Wen Chang

11.1 Porcine Epidemic Diarrhea Virus Structure and Genome

Porcine epidemic diarrhea (PED) first appeared in England and Belgium in the 1970s. The etiological agent of the disease is porcine epidemic diarrhea virus (PEDV), which belongs to the order *Nidovirales*, genus *Alphacoronavirus*, and family *Coronaviridae* (EN 1977; Bridgen et al. 1993; Duarte et al. 1993). It consists of a single-stranded positive-sense RNA genome of approximately 28 kb in size with gene order 5′-replicase (1a/1b)-S-ORF3-E-M-N-3′. The viral genome encodes three nonstructural protein, Pol 1a/1b and ORF3, and four major structural proteins, the spike (S) glycoprotein, nucleocapsid (N) protein, membrane (M) glycoprotein, and envelope (E) protein. The S protein of PEDV is a type 1 transmembrane envelope protein and consists of the S1 and S2 domains. The S protein is responsible for the viral entry via specific binding of the S1 domain with the cellular receptor, fusion and interaction of the S2 domain with host cellular membrane, and for induction of neutralizing antibodies in the host (Bosch et al. 2003). The M and E proteins are associated with virus assembly via interacting with S and N proteins (Klumperman et al. 1994). The primary role of N protein is to pack the viral genomic RNA into viral particles (Spaan et al. 1983).

J.-Y. Peng • C.-Z. Jian • C.-Y. Chang • H.-W. Chang (✉)
Graduate Institute of Molecular and Comparative Pathobiology,
School of Veterinary Medicine, National Taiwan University, Taipei 10617, Taiwan
e-mail: huiwenchang@ntu.edu.tw

J. Bayry (ed.), *Emerging and Re-emerging Infectious Diseases of Livestock*,
DOI 10.1007/978-3-319-47426-7_11

11.2 PEDV Transmission

PEDV is mainly transmitted via oral-fecal route, though aerosolized PEDV is also infectious (Alonso et al. 2014). The major transmission source of PEDV may be from the feces or vomitus. Other possible carriers for PEDV may be asymptomatic pigs or persons that carry contaminated fomites from farm to farm (Lowe et al. 2014). Besides horizontal transmission, potential route for vertical transmission of PEDV via sow milk is also suggested (Li et al. 2012; Sun et al. 2012).

11.3 PEDV Pathogenesis

PEDV establishes its infection majorly in porcine villous enterocytes, which express the cellular receptor, porcine aminopeptidase N (pAPN; CD13) (Li et al. 2007). PEDV replicates in the cytoplasm of villous epithelial cells in the small intestine and sometimes in the colon resulting in severe villous atrophy and leading to malabsorptive diarrhea (Straw et al. 2006).

In the past, the prevalence of PED was low. It only caused endemic infection with very few mini-outbreaks. Suckling piglets are protected from the disease via maternal antibodies and immunity (Bandrick et al. 2014). The disease majorly appeared in postweaning pigs as the maternal antibody titer drops. Possibly due to the fast turnover time (2–3 days) of enterocytes from crypt stem cells in postweaning pigs as well as the low virulence of traditional PEDV strains, the affected piglets usually show transient diarrhea with low or without mortalities.

Since late 2010, however, new PEDV variants with evidence of increased virulence have been isolated in several countries. The novel PEDV variants attack neonatal piglets regardless of their vaccination status or maternal immunity, which derived from CV777-based vaccination or preexisting historic PEDV infection (Sun et al. 2012; Chiou 2015; Stevenson et al. 2013). Factors underlying the potential pathogenesis of the PEDV outbreaks and the high mortality in piglets include the mutation of the virus (Chiou et al. 2015; Pasick et al. 2014), the lacking of maternal antibodies for protection of the piglets, and the slower turnover rate of enterocytes (5–7 days) of the neonatal piglets as compared to postweaning pigs (2–3 days) (Jung and Saif 2015a; Straw et al. 2006).

11.4 Clinical Signs and Lesions in PEDV Infection

The major clinical signs of PED are watery diarrhea and/or vomiting. Piglets might die from dehydration and electrolyte imbalance due to severe diarrhea and vomiting. At necropsy, gross lesions majorly include distension of small intestine with

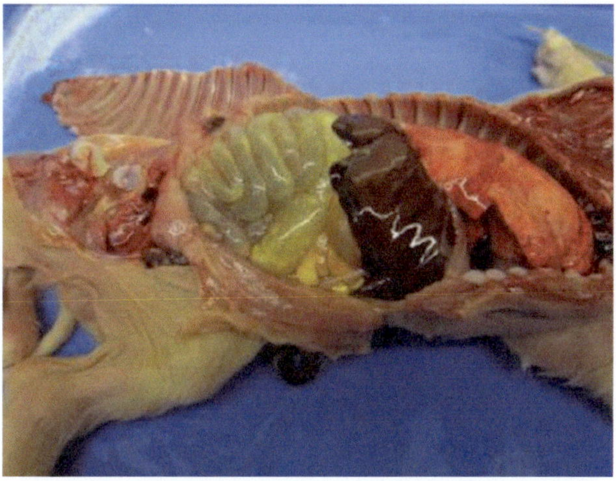

Fig. 11.1 Gross lesions in a PED-affected 1-day-old piglet. Distension of small intestine with yellowish fluid content and thin and transparent intestinal walls were noted

yellowish fluid, thin and transparent intestinal walls (Fig. 11.1), and the stomach filled with curdle milk. Congestion of mesenteric vessels and edema of mesenteric lymph nodes are often seen. Under microscopic examination, an acute, diffuse, severe atrophic enteritis characterized by reduction in the villous height and crypt depth ratio, villous blunting and fusion, and cell exfoliation on the tips of villous enterocytes are often seen (Straw et al. 2006; Jung and Saif 2015a).

11.5 Differential Diagnosis Between Diarrheal Pathogens

Several viruses can cause diarrhea in pigs with similar clinical signs and pathologic features to PED. These viruses include porcine deltacoronavirus (PDCoV), transmissible gastroenteritis virus (TGEV), and porcine rotavirus. As listed in Table 11.1, these viral infections exhibit similar clinical signs, age tropisms, replication sites, gross lesions, and microscopic lesions. Therefore, a definitive diagnosis of PED majorly depends on molecular methods.

11.6 Diagnostic Methods of PEDV

Several laboratory diagnostic methods are available for the definitive diagnosis of PEDV infection, such as immunofluorescence assay (IFA) and immunohistochemistry (IHC) by using specific antibodies for the detection of PEDV antigen in paraffin-embedded tissues (Madson et al. 2014) (Fig. 11.2.), enzyme-linked immunosorbent assay (ELISA) for detecting virus or serum antibodies (Okda

Table. 11.1 Comparison of clinical and pathologic features among PEDV, PDCoV, PEDV, and TGEV infections

	PEDV	PDCoV	TGEV	Rotavirus
Etiology	Coronavirus	Coronavirus	Coronavirus	Rotavirus
Ages	All ages (Traditional PED: postweaning pigs; novel PED: neonatal piglets)	Young nursing piglets	All ages	Neonates
Replication site	Villous epithelium cells			
Clinical signs	Vomiting and profuse watery, yellowish diarrhea			
Gross lesions	1. The small intestine is distended with yellow, foamy fluid, and the wall is thin and transparent 2. The stomach is distended with curdled milk			
Microscopic lesions	Reduction in villous height to crypt depth ratio; cell exfoliation; vacuolation of superficial epithelial cells	Acute diffuse severe atrophic enteritis and vacuolation of superficial epithelial cells in cecum and colon	Reduction in villous height to crypt depth ratio; cytoplasmic vacuoles in villous enterocytes and Peyer's patches	Degeneration of villous epithelial cells

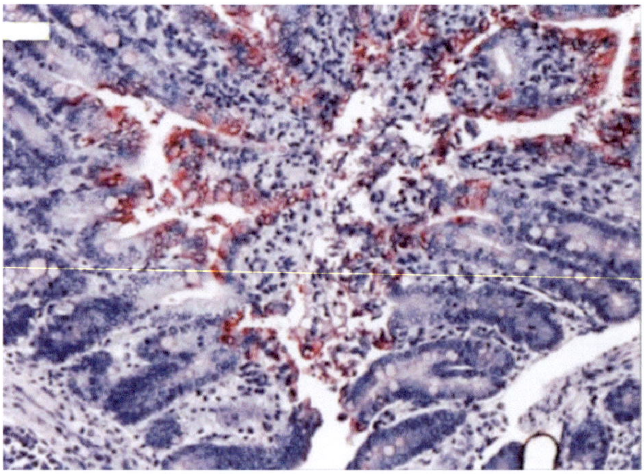

Fig. 11.2 Detection of PEDV antigen in formalin-embedded tissues by the immunohistochemistry staining

et al. 2015; Gerber et al. 2014; Ren et al. 2011; Carvajal et al. 1995; Knuchel et al. 1992; van Nieuwstadt and Zetstra 1991), reverse transcription polymerase chain reaction (RT-PCR) for detecting the viral genome (Ben Salem et al. 2010; Ishikawa et al. 1997; Kweon et al. 1997), electron microscopy (EM) for demonstration of PEDV particles (Fig.11.3) in the diarrheal feces (Straw et al. 2006; Jung and Saif 2015a), and viral isolation together with immunocytochemistry (ICC) performed in Vero cells for propagation and in vitro characterization of the virus (Fig.11.4).

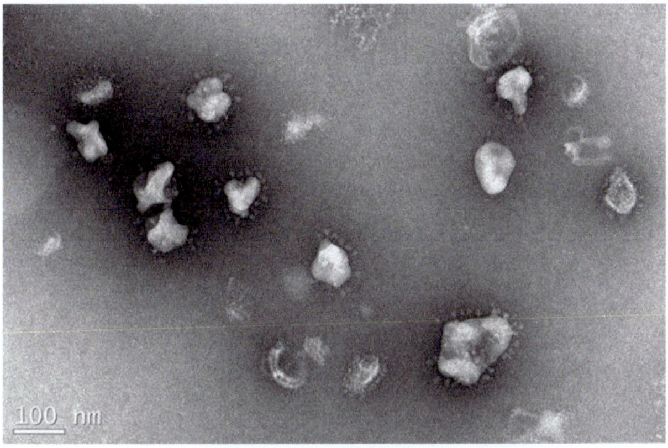

Fig. 11.3 Transmission electron micrograph (TEM) of PEDV particles (*virions*, *round*) in a tissue culture sample

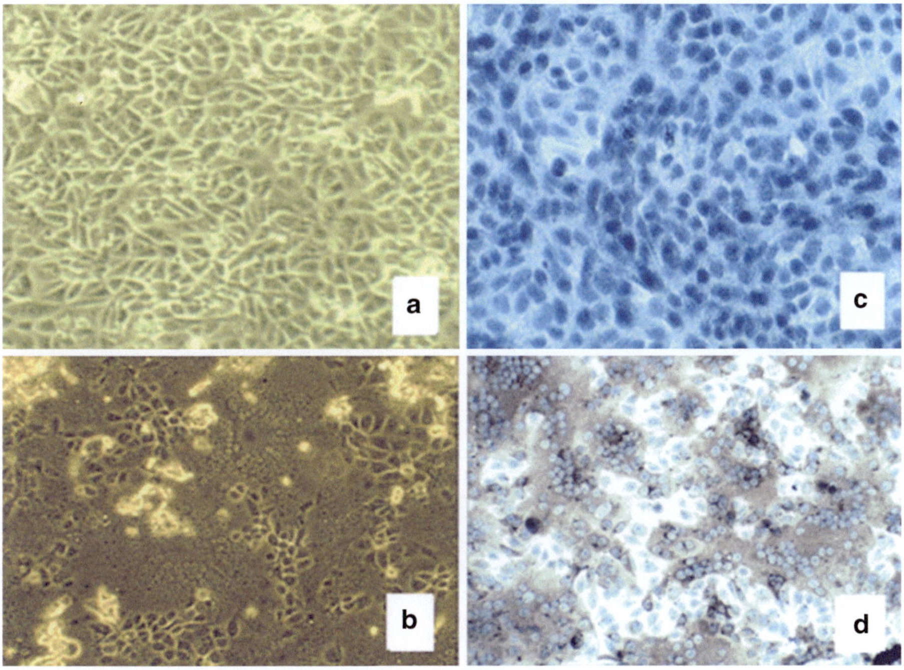

Fig. 11.4 Isolation and detection of PEDV in Vero cells. As compared with mock-infected Vero cells (**a, c**), typical CPEs in PEDV-infected Vero cells are characterized by cell fusion and syncytial formation under the light microscopy (**b**) and distinct cytoplasmic signals under the immunocytochemistry staining (**d**)

11.7 Epidemiology of PEDV Worldwide

Molecular characteristics of the S gene of PEDV strains have been often investigated to better understand the genetic diversity of PEDVs. After the new PEDV outbreaks, the global PEDVs have been divided into two major groups according to the phylogenetic analysis of S gene of PEDV. Group 1 (G1) comprises global new variants of PEDVs and US S non-INDEL strains in North America; group 2 (G2) is composed of CV777 and DR13 vaccine strains, historical PEDVs isolated in Asia and Europe before the recent outbreaks, and the emerging US S INDEL PEDVs. The G1 strains isolated from the USA, China, and Asia PED-affected piglets show high sequence identities to each other (Chiou et al. 2015; Chen et al. 2014; Pasick et al. 2014; Temeeyasen et al. 2014; Vui et al. 2014). These novel G1 PEDV strains are suggested to share a common ancestor. The G2 PEDV strains are suggested being less virulent and are reported circulating in Europe, the USA, and Asia (Vlasova et al. 2014; Dennis Hanke et al. 2015).

11.7.1 Epidemiology of PEDV in Asia

In Asia, historic PEDs are not uncommon in the field. Most of PEDV infections caused transient diarrhea in postweaning piglets and rarely any disease in suckling piglets. The new variants of PEDV, however, cause severe diseases and serious economic loss of the neonatal piglets in many Asian countries, including China (Tian et al. 2013; Li et al. 2012), Taiwan (Chiou et al. 2015; Lin et al. 2014), South Korea (Cho et al. 2014; Choi et al. 2014), Thailand (Temeeyasen et al. 2014), and Japan, raising the importance of PED in Asia. It has been demonstrated that new PEDV strains that emerged in Asia are distinct from previous historic PEDV strains. They cause higher morbidity and mortality in neonatal piglets than previous strains (Jung and Saif 2015b).

11.7.1.1 China

In the past, the infection of PEDV was endemic, and the prevalence of PED was low with few mini-outbreaks reported in China. The CV777-based vaccine is frequently used in the field. Since late 2010, PEDV-infected piglets, regardless of vaccination or immune status, showed large-scale outbreaks of PED. These variants caused severe watery diarrhea in pigs of all ages and high death rate in neonatal and suckling piglets. Based on the phylogenetic analysis of full-length sequences of the S gene of PEDVs isolated in China, both G1 and G2 strains are concurrently isolated. While China G2 PEDV strains share 96–99 % homologies to the CV777-based vaccine strain, the China G1 PEDV strains only share 93–95 % similarity to the CV777-based vaccine strain as well as China historic PEDV strains. These China G1 PEDV strains present several nucleotides insertions and deletions in the S gene (Li et al. 2012). Comparing the G1 PEDV strains with the CV777-based vaccine strain, there are two to three amino acid mutations in the neutralizing epitope SS6 and one to eight amino acid differences in the CO-26 K equivalent epitope (COE) domain (499-638 aa) in the S protein. These mutations have been speculated to represent the viral evolution through escaping from the antibodies derived from the CV777-based vaccine or the historic PEDV strains in several studies (Chen et al. 2013; Hao et al. 2014).

11.7.1.2 Taiwan

Before outbreaks of new variants of PED, historically PED was endemic in Taiwan. PEDV infection only caused mild transient diarrhea in postweaning piglets. PEDV vaccines, such as the CV777-based or DR13-based vaccines, are not routinely used in Taiwan. Nearly at the same time as in North America, new variants of PEDV infection leading to dramatic outbreaks and losses of suckling piglets reemerged in late 2013 in Taiwan. Phylogenetic analysis of the S gene showed that new variants of Taiwan PEDV strains belong to G1 and are closely related to other G1 strains from the USA, Canada, and China. These new Taiwan PEDV strains share 99.2 to 99.3 % nucleotide sequence identity to China CH/ZMDZY/11 and 94.1 to 94.2 % homologies to historical Taiwan PEDV strains. Similar to other new global variants of PEDV G1 strains, three serine amino acid substitutions (A522S, A554S, and G599S) in the COE and two serine amino acid substitutions (L769S and D771S) in SS6 are also observed in the reemerging Taiwan PEDVs as compared with the CV777-based vaccine strain (Chiou et al. 2015).

11.7.1.3 Korea

In Korea, the endemic PED has been in the field for decades (Yeo et al. 2003; Kim and Chae 2000; Kubota et al. 1999), and live PEDV vaccine strains are also available (Song et al. 2007). In addition to the reemerging of new variants of G1 PEDV strains resulting in severe PED in suckling piglets (Lee and Lee 2014; Park et al. 2013, 2014; Cho et al. 2014), a novel G1 PEDV variant, MF3809, is identified containing numerous sequence variations in the S protein, including a large (204-aa) deletion at positions 713–916 and a 2-aa (D/NI) deletion at positions 163–164. The antigenicity/immunogenicity alteration of MF3809 is still unknown (Park et al. 2014).

11.7.2 Epidemiology of PEDV in North America

The first PED case in the USA was reported in April–May, 2013. There was no previous description of the disease in the country before that time. No PEDV vaccines were available in the USA as well. After the first outbreak, PEDV quickly spread throughout the USA, then to nearby countries, including Canada (Song et al. 2015; Pasick et al. 2014). The US PEDV strains distributed both in G1 and G2 groups. The strains causing severe outbreaks exhibit high similarity to G1 PEDVs and are designed as S non-INDEL strains. The S non-INDEL PEDV strains were most closely related to China PEDV strains of CH/ZMZDY/11 and AH2012 in G1 group. Later after the outbreak, variants of US PEDV strains are isolated and are clustered in G2 in the phylogenetic analysis. As compared with the original US G1 PEDV strains, the US G2 PEDV strains contain insertions and deletions in S protein and are designed as S INDEL strains. Based on the full-length S gene analysis of PEDVs, the US S non-INDEL strains in G1 shared 99.8–100 % homology to each other and shared 96.6–97.1 % homologies to other PEDV G2 strains from North America. The S INDEL and original S non-INDEL strains are suggestive co-circulating and could have been introduced simultaneously in North America (Anastasia et al. 2014).

11.7.3 Epidemiology of PEDV in Europe

The first PEDV was reported in Belgium in 1978. The virus then spread over Europe until the end of the 1990s. However, there were no well-documented reports for PEDV in Europe during the past decades; most of the countries have not implemented active monitoring for this particular disease (23). Recently, several European PEDVs variants, such as PEDV/GER/L00719/214 and PEDV/GER/L00721/2014 from Germany, FR/001/2014 from France, BEL/15 V010/2015 from Belgium, and some isolates from Portugal and Italy, have been isolated and studied. They shared high identity (99 %) with the USA/OH851 strain (S INDEL; G2) and the Chinese AH2012 strain, but less similar (97.1 %) to the historic isolate, CV777, which isolated from Europe (Grasland et al. 2015; Theuns et al. 2015; Dennis Hanke et al. 2015).

Conclusion

The recent global outbreaks of PED in pigs of all ages with the high mortality in neonatal piglets may be contributed by several factors, including the mutation of the virus, the lacking of maternal antibodies, and the slower turnover of enterocytes (5–7 days) of neonatal piglets (Jung and Saif 2015a; Straw et al. 2006). It had been demonstrated that current available vaccines, CV777 and the attenuated PEDV DR13 vaccines, might not be able to fully protect piglets against the infection or control disease progression due to the high genetic diversity of new variants of global PEDVs. A new generation of PEDV vaccine is therefore urgently in need.

References

Alonso C, Goede D, Morrison R, Davies P, Rovira A, Marthaler D, Torremorell M (2014) Evidence of infectivity of airborne porcine epidemic diarrhea virus and detection of airborne viral RNA at long distance form infected herd. Vet Res 45:73

Anastasia N, Vlasova DM, Wang Q, Culhane MR, Rossow KD, Rovira A, Collins J, Saif LJ (2014) Distinct characteristics and complex evolution of PEDV strains, North America, May 2013–February 2014. Emerg Infect Dis 20(10):1620–1628

Bandrick M, Theis K, Molitor T (2014) Maternal immunity enhances Mycoplasma hyopneumoniae vaccination induced cell-mediated immune responses in piglets. BMC Vet Res 10:124

Ben Salem AN, Chupin Sergei A, Bjadovskaya Olga P, Andreeva Olga G, Mahjoub A, Prokhvatilova Larissa B (2010) Multiplex nested RT-PCR for the detection of porcine enteric viruses. J Virol Methods 165(2):283–293. doi:10.1016/j.jviromet.2010.02.010

Bosch B, van der Zee R, de Haan C, Rottier P (2003) The coronavirus spike protein is a class I virus fusion protein: Structural and functional characterization of the fusion core complex. J Virol 77:8801–8811

Bridgen A, Duarte M, Tobler K, Laude H, Ackermann M (1993) Sequence determination of the nucleocapsid protein gene of the porcine epidemic diarrhoea virus confirms that this virus is a coronavirus related to human coronavirus 229E and porcine transmissible gastroenteritis virus. J Gen Virol 74(Pt 9):1795–1804

Carvajal A, Lanza I, Diego R, Rubio P, Carmenes P (1995) Evaluation of a blocking ELISA using monoclonal antibodies for the detection of porcine epidemic diarrhea virus and its antibodies. J Vet Diagn Invest Off Publ Am Assoc Vet Lab Diagn Inc 7(1):60–64

Chen J, Liu X, Shi D, Shi H, Zhang X, Li C, Chi Y, Feng L (2013) Detection and molecular diversity of spike gene of porcine epidemic diarrhea virus in China. Viruses 5(10):2601–2613. doi:10.3390/v5102601

Chen Q, Li G, Stasko J, Thomas JT, Stensland WR, Pillatzki AE, Gauger PC, Schwartz KJ, Madson D, Yoon KJ, Stevenson GW, Burrough ER, Harmon KM, Main RG, Zhang J (2014) Isolation and characterization of porcine epidemic diarrhea viruses associated with the 2013 disease outbreak among swine in the United States. J Clin Microbiol 52(1):234–243. doi:10.1128/JCM.02820-13

Chiou HY, Huang YL, Deng MC, Chang CY, Jeng CR, Tsai PS, Yang C, Pang VF, Chang HW (2015) Phylogenetic analysis of the spike (S) gene of the new variants of porcine epidemic diarrhoea virus in Taiwan. Transbound Emerg Dis. doi:10.1111/tbed.12357

Cho YY, Lim SI, Kim YK, Song JY, Lee JB, An DJ (2014) Complete Genome Sequence of K14JB01, a novel variant strain of porcine epidemic diarrhea virus in South Korea. Genome Announc 2(3):e00505–e00514. doi:10.1128/genomeA.00505-14

Choi JC, Lee KK, Pi JH, Park SY, Song CS, Choi IS, Lee JB, Lee DH, Lee SW (2014) Comparative genome analysis and molecular epidemiology of the reemerging porcine epidemic diarrhea virus strains isolated in Korea. Infect Genet EvolJ Mol Epidemiol Evol Genet Infect Dis 26:348–351. doi:10.1016/j.meegid.2014.06.005

Dennis Hanke MJ, Petrov A, Ritzmann M, Stadler J, Valerij Akimkin SB, Pohlmann A, Schirrmeier H, Beer M, Höper D (2015) Comparison of porcine epidemic diarrhea viruses from Germany and the United States, 2014. Emerg Infect Dis 21(3):493–496

Duarte M, Gelfi J, Lambert P, Rasschaert D, Laude H (1993) Genome organization of porcine epidemic diarrhoea virus. Adv Exp Med Biol 342:55–60

EN W (1977) An apparently new syndrome of porcine epidemic diarrhea. Vet Rec 100:243–244

Gerber PF, Gong Q, Huang YW, Wang C, Holtkamp D, Opriessnig T (2014) Detection of antibodies against porcine epidemic diarrhea virus in serum and colostrum by indirect ELISA. Vet J 202(1):33–36. doi:10.1016/j.tvjl.2014.07.018

Grasland B, Bigault L, Bernard C, Quenault H, Toulouse O, Fablet C, Rose N, Touzain F, Blanchard Y (2015) Complete genome sequence of a porcine epidemic diarrhea s gene indel strain isolated in france in december 2014. Genome Announc 3(3):e00535–e00515. doi:10.1128/genomeA.00535-15

Hao J, Xue C, He L, Wang Y, Cao Y (2014) Bioinformatics insight into the spike glycoprotein gene of field porcine epidemic diarrhea strains during 2011-2013 in Guangdong, China. Virus Genes 49(1):58–67. doi:10.1007/s11262-014-1055-y

Ishikawa K, Sekiguchi H, Ogino T, Suzuki S (1997) Direct and rapid detection of porcine epidemic diarrhea virus by RT-PCR. J Virol Methods 69(1-2):191–195

Jung K, Saif L (2015a) Porcine epidemic diarrhea virus infection: etiology, epidemiology, pathogenesis and immunoprophylaxis. Vet J 204:134–143

Kim O, Chae C (2000) In situ hybridization for the detection and localization of porcine epidemic diarrhea virus in the intestinal tissues from naturally infected piglets. Vet Pathol 37(1):62–67

Klumperman J, Locker J, Meijer A, Horzinek M, HJ G, Rottier P (1994) Coronavirus M proteins accumulate in the Golgi complex beyond the site of virion budding. J Virol 68:6523–6534

Knuchel M, Ackermann M, Muller HK, Kihm U (1992) An ELISA for detection of antibodies against porcine epidemic diarrhoea virus (PEDV) based on the specific solubility of the viral surface glycoprotein. Vet Microbiol 32(2):117–134

Kubota S, Sasaki O, Amimoto K, Okada N, Kitazima T, Yasuhara H (1999) Detection of porcine epidemic diarrhea virus using polymerase chain reaction and comparison of the nucleocapsid protein genes among strains of the virus. J Vet Med Sci/Japanese Soc Vet Sci 61(7):827–830

Kweon CH, Lee JG, Han MG, Kang YB (1997) Rapid diagnosis of porcine epidemic diarrhea virus infection by polymerase chain reaction. J Vet Med Sci/Japanese Soc Vet Sci 59(3):231–232

Lee S, Lee C (2014) Outbreak-related porcine epidemic diarrhea virus strains similar to US strains, South Korea, 2013. Emerg Infect Dis 20(7):1223–1226. doi:10.3201/eid2007.140294

Li B, Ge J, Li Y (2007) Porcine aminopeepdidase N is a functional receptor for the PEDV coronavirus. Virology 365:166–172

Li W, Li H, Liu Y, Pan Y, Deng F, Song Y, Tang X, He Q (2012) New variants of porcine epidemic diarrhea virus, China, 2011. Emerg Infect Dis 18:1350–1353

Lin CN, Chung WB, Chang SW, Wen CC, Liu H, Chien CH, Chiou MT (2014) US-like strain of porcine epidemic diarrhea virus outbreaks in Taiwan, 2013-2014. J Vet Med Science/Japanese Soc Vet Sci 76(9):1297–1299

Lowe J, Gauger P, Harmon K, Zhang J, Connor J, Yeske P, Louia T, Levis L, Dufresne L, Main R (2014) Role of transpotation in spread of porcine epidemic diarrhea virus infection, United States. Emerg Infect Dis 20:872–874

Madson DM, Magstadt DR, Arruda PH, Hoang H, Sun D, Bower LP, Bhandari M, Burrough ER, Gauger PC, Pillatzki AE, Stevenson GW, Wilberts BL, Brodie J, Harmon KM, Wang C, Main RG, Zhang J, Yoon KJ (2014) Pathogenesis of porcine epidemic diarrhea virus isolate (US/Iowa/18984/2013) in 3-week-old weaned pigs. Vet Microbiol 74(1-2):60–68. doi:10.1016/j.vetmic.2014.09.002

Okda F, Liu X, Singrey A, Clement T, Nelson J, Christopher-Hennings J, Nelson EA, Lawson S (2015) Development of an indirect ELISA, blocking ELISA, fluorescent microsphere immunoassay and fluorescent focus neutralization assay for serologic evaluation of exposure to North American strains of Porcine Epidemic Diarrhea Virus. BMC Vet Res 11:180. doi:10.1186/s12917-015-0500-z

Park S, Kim S, Song D, Park B (2014) Novel porcine epidemic diarrhea virus variant with large genomic deletion, South Korea. Emerg Infect Dis 20(12):2089–2092. doi:10.3201/eid2012.131642

Park SJ, Song DS, Park BK (2013) Molecular epidemiology and phylogenetic analysis of porcine epidemic diarrhea virus (PEDV) field isolates in Korea. Arch Virol 158(7):1533–1541. doi:10.1007/s00705-013-1651-5

Pasick J, Berhane Y, Ojkic D, Maxie G, Embury-Hyatt C, Swekla K, Handel K, Fairles J, Alexandersen S (2014) Investigation into the role of potentially contaminated feed as a source of the first-detected outbreaks of porcine epidemic diarrhea in Canada. Transbound Emerg Dis 61(5):397–410. doi:10.1111/tbed.12269

Ren X, Suo S, Jang YS (2011) Development of a porcine epidemic diarrhea virus M protein-based ELISA for virus detection. Biotechnol Lett 33(2):215–220. doi:10.1007/s10529-010-0420-8

Song D, Moon H, Kang B (2015) Porcine epidemic diarrhea: a review of current epidemiology and available vaccines. Clin Exp vaccine Res 4(2):166–176. doi:10.7774/cevr.2015.4.2.166

Song DS, Oh JS, Kang BK, Yang JS, Moon HJ, Yoo HS, Jang YS, Park BK (2007) Oral efficacy of Vero cell attenuated porcine epidemic diarrhea virus DR13 strain. Res Vet Sci 82(1):134–140. doi:10.1016/j.rvsc.2006.03.007

Spaan WDH, Skinner M, Armstrong J, Rottier P, Smeekens S, van der Zeijst BA, Siddell SG (1983) Coronavirus mRNA synthesis involves fusion of non-contiguous sequences. EMBO J 2(10):1839–1844

Stevenson GW, Hoang H, Schwartz KJ, Burrough ER, Sun D, Madson D, Cooper VL, Pillatzki A, Gauger P, Schmitt BJ, Koster LG, Killian ML, Yoon KJ (2013) Emergence of Porcine epidemic diarrhea virus in the United States: clinical signs, lesions, and viral genomic sequences. J Vet Diagn Invest Off Publ Am Assoc Vet Lab Diagn Inc 25(5):649–654. doi:10.1177/1040638713501675

Straw B, Zimmerman J, D'Allaire S, Taylor D (2006) Diseases of swine. 9th edn. Blackwell Publishing, Ames, Iowa.

Sun R, Cai R, Chen Y, Liang P, Chen D, Song C (2012) Outbreak of porcine epidemic diarrhea in suckling piglets, China. Emerg Infect Dis 18:161–163

Temeeyasen G, Srijangwad A, Tripipat T, Tipsombatboon P, Piriyapongsa J, Phoolcharoen W, Chuanasa T, Tantituvanont A, Nilubol D (2014) Genetic diversity of ORF3 and spike genes of porcine epidemic diarrhea virus in Thailand. Infect, Genet Evol J Mol Epidemiol Evol Genet Infect Dis 21:205–213. doi:10.1016/j.meegid.2013.11.001

Theuns S, Conceicao-Neto N, Christiaens I, Zeller M, Desmarets LM, Roukaerts ID, Acar DD, Heylen E, Matthijnssens J, Nauwynck HJ (2015) Complete genome sequence of a porcine epidemic diarrhea virus from a novel outbreak in Belgium, January 2015. Genome Announc 3(3):e00506–e00515. doi:10.1128/genomeA.00506-15

Tian Y, Su D, Zhang H, Chen RA, He D (2013) Complete genome sequence of a very virulent porcine epidemic diarrhea virus strain, CH/GDGZ/2012, isolated in Southern China. Genome Announc 1(4):00645–00613. doi:10.1128/genomeA.00645-13

van Nieuwstadt AP, Zetstra T (1991) Use of two enzyme-linked immunosorbent assays to monitor antibody responses in swine with experimentally induced infection with porcine epidemic diarrhea virus. Am J Vet Res 52(7):1044–1050

Vlasova AN, Marthaler D, Wang Q, Culhane MR, Rossow KD, Rovira A, Collins J, Saif LJ (2014) Distinct characteristics and complex evolution of PEDV strains, North America, May 2013-February 2014. Emerg Infect Dis 20(10):1620–1628. doi:10.3201/eid2010.140491

Vui DT, Tung N, Inui K, Slater S, Nilubol D (2014) Complete genome sequence of porcine epidemic diarrhea virus in Vietnam. Genome Announc 2(4):e00753–e00714. doi:10.1128/genomeA.00753-14

Yeo SG, Hernandez M, Krell PJ, Nagy EE (2003) Cloning and sequence analysis of the spike gene of porcine epidemic diarrhea virus Chinju99. Virus Genes 26(3):239–246

Nipah Virus Infection

12

Diwakar D. Kulkarni, Chakradhar Tosh, Sandeep Bhatia, and Ashwin A. Raut

12.1 Introduction

Among all the emerging viral infections recorded since last 25 years, majority is of zoonotic importance. Many of them are transmitted by bats. One of those infections is Nipah virus (NiV) infection and is caused by a virus under newly classified genus of *paramyxovirus*. Nipah virus causes highly fatal encephalitis in humans. In Malaysia, during the last trimester of 1998, an outbreak of acute febrile encephalitis with high mortality in the pig-rearing farmers was reported in Kinta district of Perak followed by more outbreaks in three districts of Negeri Sembilan province including Sungai Nipah that continued till February 1999. The later outbreaks were due to the movement of infected pigs showing respiratory illness and migration of persons from Kinta district pig farms to newer areas. This infection, unknown till then, caused a relatively mild disease in pigs, but in humans, it caused 300 cases and over 100 deaths. To halt the outbreak, more than a million pigs were slaughtered (Chua 2003; 2010). In March 1999, a similar illness was reported among 11 abattoir workers in Singapore, with one death. This incidence was also traced back to Malaysia, importation of infected pigs (Paton et al. 1999). Investigation into this sequence of events led to the discovery of Nipah virus in March 1999, a highly virulent virus of the *Paramyxoviridae* family, as the etiological agent of theses fatal outbreaks. It was named as Nipah virus due to its first isolation reporting from a human patient in Sungai Nipah District (Wong et al. 2002). Although reported from only four countries on the globe, viz., Malaysia, Singapore, India, and Bangladesh, it has a vast potential for spread due to involvement of fruit bats as a carrier host.

D.D. Kulkarni (✉) • C. Tosh • S. Bhatia • A.A. Raut
ICAR-National Institute of High Security Animal Diseases,
OIE Reference Laboratory for Avian Influenza, Bhopal, India
e-mail: ddkulkar@gmail.com

J. Bayry (ed.), *Emerging and Re-emerging Infectious Diseases of Livestock*,
DOI 10.1007/978-3-319-47426-7_12

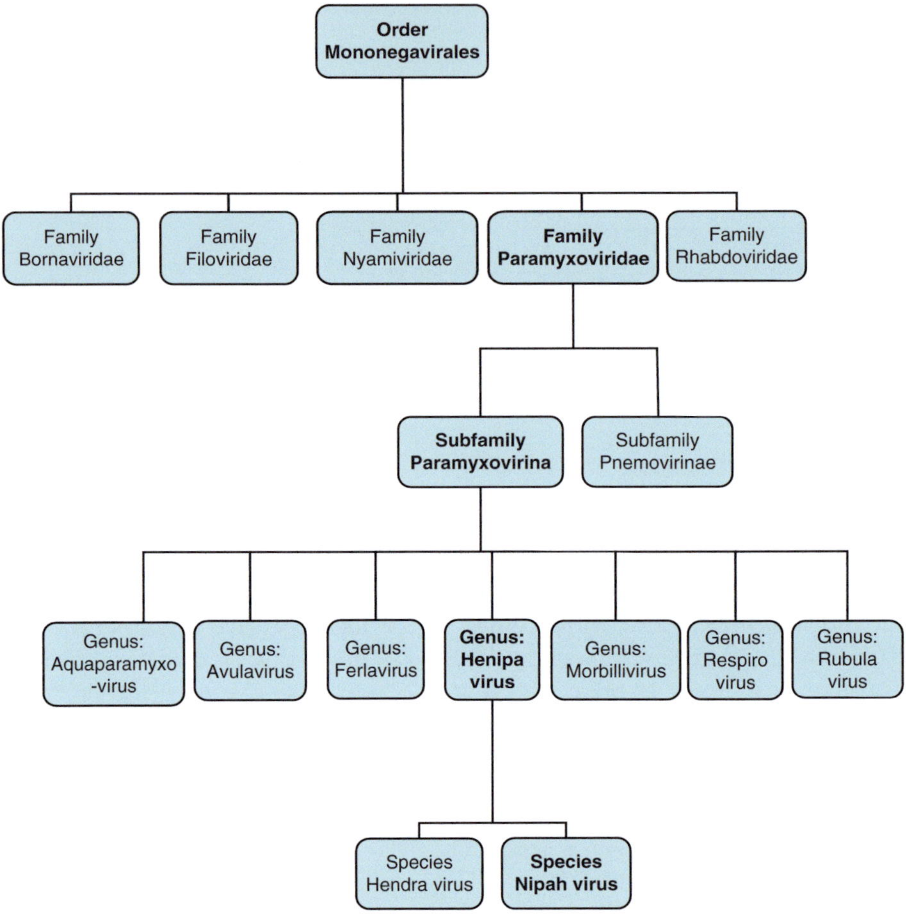

Fig. 12.1 Virus classification of paramyxoviruses in relation to Nipah virus

12.2 Etiology

NiV is an enveloped virus that has non-segmented single-stranded negative-sense RNA genome. The virus is classified as a member of genus *Henipavirus* in the subfamily *Paramyxovirinae* of the family *Paramyxoviridae.* The detailed classification is given in Fig. 12.1. Nucleocapsid of NiV is 18 nm in diameter, 1 μm in length and has a pitch of 5.5 nm. The genome is comprised of six genes, viz., N, P, M, F, G, and L yielding the proteins – nucleoprotein (np), phosphoprotein (php), matrix protein (mp), fusion (fp), glycoprotein (gp), and large RNA polymerase (rnp), respectively (Chan et al. 2001). The virion nucleocapsid core consists of np, php, and large rnp helping the initiation of the virus replication in the host cell cytoplasm. The mp is present between the envelope and the nucleocapsid that is critical to the virion structure and is released when virus enters host cells. The gp and fp of NiV are supposed to mediate the viral entry into the cell and also play a role in inducing neutralizing antibodies. The infection of the host cell requires coordination of these two glycoproteins. Upon attachment to cell surface receptors, gp triggers the fp to execute membrane fusion. In addition to the php, the P gene encodes at least three nonstructural proteins (C, V, and W) (Harcourt et al. 2000, 2001). Nonstructural C protein is supposed to regulate viral RNA synthesis and expression of pro-inflammatory cytokines, thereby coordinate the chemokine-induced immune response and may play a role as a virulence factor by controlling the lethal outcome of the infection. The V protein of NiV plays an important role in downregulating IFN signalling and also known to form high molecular weight complexes in cytoplasm and thus inhibiting host cell signalling functions. The W protein of NiV also has been reported to have immune suppression function. Php is the only essential gene product for genome replication; the additional gene products are not required for in vitro virus replication but may play a role as virulence factors in vivo (Lamb et al. 2005; Lamb and Parks 2007).

There is a slight difference between genomes of the Malaysian isolates and the Bangladesh isolates of NiV; the former has 18,246 nucleotides, the latter with 18,252 nucleotides. The sequences obtained from all the human isolates in Malaysia which were nearly identical, whereas the Bangladesh strains are found to be more divergent as compared to Malaysian strains as observed by the differences in np nucleotide homology and multiple introductions of the virus into human populations even within a single year (Harcourt et al. 2005).

NiV is so closely related to Hendra virus (HeV) that NiV can even be neutralized with anti-HeV antibodies. NiV shares 70–78 % nucleotide homology with HeV in three major genes – N, P, and M (Daniels et al. 2001).

NiV is highly pathogenic to developing chicken embryos (DCE) and that the chicken embryo may represent a useful model for studying the vascular and neuronal tropisms of NiV (Tanimura et al. 2006). However, from biosafety point of view, owing to its high pathogenicity, biosafety level 4 containment laboratory is essential to work with live NiV (Lo and Rota 2008). Therefore, handling of virus in DCE is more risky and to be carried out with appropriate personal protection and due caution. The virus readily grows in many mammalian cells with variable results. Mostly the virus is grown in Vero cells, where initially cytopathic effect (CPE) is recorded between 3 and 6 days postinoculation but adapted isolates grow within 24–48 h and produce high viral titer at full CPE. The electron microscopy during the viral growth shows nucleocapsids inclusions in the host cell cytoplasm, budding of the nucleocapsid at the plasma membrane and pleomorphic extracellular virus particles that are enveloped. In electron microscopic studies, negative stain preparations reveal nucleocapsids with the typical "herringbone" appearance (Chow et al. 2000; Goldsmith et al. 2000; Daniels et al. 2001; Hyatt et al. 2001).

The survivability of the virus in the environment is not clearly understood; however, unpublished experiments suggest that it can survive for days in fruit juice or fruit bat urine. NiV is easily inactivated by soaps, detergents, and disinfectants. Routine cleaning and disinfection with any good quality commercial disinfectants is effective (Lam and Chua 2002). Sodium hypochlorite was successfully used in pig farms in Malaysia (Nor et al. 2000).

12.3 Epidemiology

12.3.1 Host Range

The virus seems to have the ability to infect many mammalian species. Apart from fruit bat as a natural host (Yob et al. 2001), NiV host range thus includes swine, humans, cattle, goat, cat, and dog, as well as horses as indicated by serological studies (Tamin et al. 2009). Only pigs were primarily infected in initial NiV outbreaks in Malaysia before transmitting the infection to humans (Chua et al. 1999). It was later revealed that other animals such as dogs, cats, and horses could also be infected when they come in contact with infected pigs. Experimental infection is shown to be produced in cat (Middleton et al. 2002), golden hamsters (Wong et al. 2003), guinea pig (Torres-Velez et al. 2008), African green monkey *(Chlorocebus aethiops)* (Geisbert et al. 2010), and squirrel monkey (*Saimiri sciureus*) (Marianneau et al. 2010).

After the initial Nipah virus infection reported from Malaysia during 1998–1999 (CDC 1999; Chua et al. 1999) and one incidence in Singapore in 1999 (Paton et al. 1999), several outbreaks of NiV have been reported in human beings in Bangladesh from 2001 to 2015 (Kulkarni et al. 2013; Anonym 2015) and two outbreaks in India (Chhadha et al. 2006; Arankalle et al. 2011)

12.3.2 Geographical Distribution

The disease in typical clinical form has been reported in humans from Malaysia (1998–1999), Singapore (1999), India (2001 and 2007), and Bangladesh (2001–2015). In Bangladesh, several outbreaks have been reported every year since 2001 with significant human morbidity and nearly 70 % case fatality. No country other than the abovementioned list has reported the disease. Except for mortality in pigs in Malaysia and one incidence of transfer of few infected pigs to Singapore slaughterhouse from Malaysia, all other incidences were reported only in human beings. However, the bats carrying NiV are present in several countries right from Western part of Africa to South Asia, Southeast and East Asian countries (Kulkarni et al. 2013). Although the closely related Hendra virus-carrying bats are present in Queensland area, NiV has not been found in Australia (Breed et al. 2013; Anonym 2016).

12.3.3 Role of Bats in Spread of the Disease

Fruit bats are considered as agricultural pests by orchard owners and are hunted for sport, for food, or for medicinal purpose. Bats are the only mammals having tremendous capacity to fly hundreds of kilometers and may travel over 2000 km in a year. This may have significant implications for disease spread since they are the carriers of many dangerous pathogens including NiV (Breed et al. 2006). With variations,

Pteropus bats tested from Southeast Asian or South Asian countries do have antibodies to *Henipavirus*. East African bats have also shown significant proportion of seropositivity to NiV. Studies indicated that 23 or more species of bats from 10 genera carry NiV. These bats are widely spread among many countries including Bangladesh, China, India, Cambodia, Thailand, Indonesia, Papua New Guinea, Madagascar, Gulf of Guinea, Cameroon, Nigeria, and some West African countries. NiV was isolated from urine specimens of *P. vampyrus* and *P. hypomelanus* fruit bats as well as from partially eaten fruits in Malaysia (Chua et al. 2002); Lyle's flying fox (*P. lylei*) in Cambodia (Reynes et al. 2005) and urine of *P. hypomelanus* and *P. lylei* as well as from saliva of *P. lylei* in Thailand (Wacharapluesadee et al. 2005). Of the several species of *Pteropus* bats living in South and Southeast Asia, *P. giganteus* is widely distributed across India and Bangladesh (Bates and Harrison 1997) and is supposed to be mainly responsible for spread of all the NiV infection in India-Bangladesh belt. NiV RNA was detected in a liver homogenate of *P. giganteus* from West Bengal in India (Yadav et al. 2012). Similarly, the presence of NiV antibodies has been detected in bats in India (Epstein et al. 2008), Indonesia (Sendow et al. 2013), Madagascar (Lehlé et al. 2007), China (Li et al. 2008), and Vietnam (Hasebe et al. 2012), and NiV RNA has been reported in *Eidolon helvum* in Ghana (Drexler et al. 2009).

Bats neither naturally suffer from the NiV infection nor become ill with experimental infection by the virus indicating possible coevolution of NiV with its *Pteropus* hosts over a period of centuries. They are the symptomless carriers but have potential to excrete the virus through their excretions and secretions intermittently. It has been postulated that the long and short distance travelling of these bats is not only responsible for disease spread but also results in intercolony or intra-/interspecies transmission of the pathogens (Middleton et al. 2007; Chong et al. 2009; FAO 2011).

12.3.4 Transmission

NiV appears to be transmitted from bats to other hosts in a specific pattern. The virus shedding from bats through saliva and urine contaminate food and water sources and consumption of such NiV-contaminated material results in setting up of infection. From several observations in Malaysian outbreak in 1998, it was concluded that fruit bats eat fruits on the plants and partially eaten saliva-laden fruits falling down from these orchids are consumed by pigs leading to infection of pigs. The disease spreads rapidly among pigs through direct contact with excretions and secretions such as urine, saliva, and pharyngeal and respiratory secretions of infected pigs. The clinical signs and lesions in pigs during Malaysian episode revealed involvement of respiratory system indicating aerosol spread of NiV from pig to human through respiratory route (CDC 1999; Chua et al. 2000). Although the dogs can be infected, the serologic status in the dogs indicated that in the absence of infected pigs, dogs were not a secondary reservoir for NiV and could not be considered a threat for disease transmission. (Mills et al. 2009).

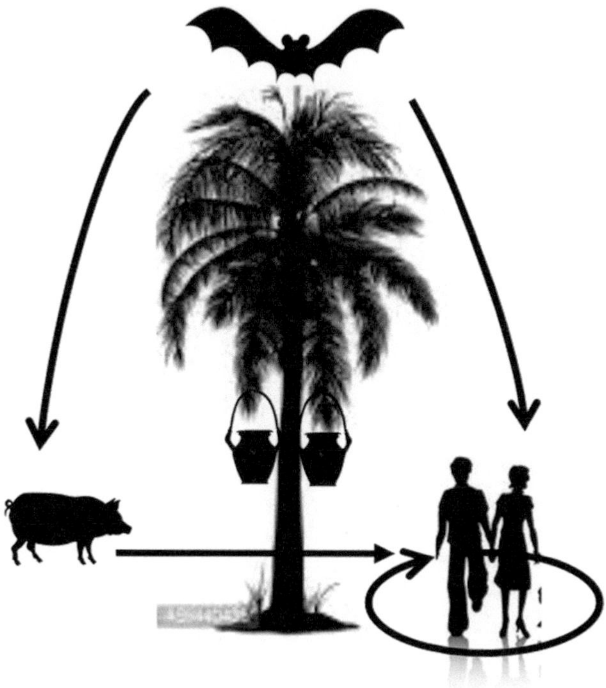

Fig. 12.2 Transmission pathways for Nipah virus

In comparison with the Malaysian NiV outbreak, the disease transmission in Bangladesh seems to be different and without involvement of pigs. Commonly observed NiV transmission routes were contact with NiV-infected persons, climbing trees or contact with sick animals (Hsu et al. 2004; Montgomery et al. 2008). In most part of Bangladesh and in neighboring West Bengal state of India, fresh, uncooked date palm sap is consumed by human being as a popular drink. About 1–2 l of palm sap slowly oozes from scratched palm trunk in one night that is collected in earthen pots tied to the tree below the drain. Since these earthen pots containing sap are mostly open and are easily accessible to fruit bats, the sap gets contaminated with the virus-laden saliva and other excretions of bat and the NiV is transmitted to the person drinking this contaminated sap (Fig. 12.2) (Luby et al. 2006; Yadav et al. 2012).

It is not clearly understood about how wide is the circulation of NiV in bat populations; however, detection of NiV, or its RNA or antibodies to NiV is recorded from many countries where no clinical cases in humans or animals are reported. Not all seroreactor bats in a colony shed the virus continuously. It is postulated that bats excrete the virus during breeding season. However, there is uncertainty over the seasonal excretion of the virus by bats (Anonym 2016; Wacharapluesadee et al. 2010). The domesticated animals can be infected by contact with infected pigs. Experimental infections have also been established in cats by intranasal and oral inoculation. Horizontal transmission has not been demonstrated between cats, but it is theoretically possible since NiV has been found in feline respiratory secretions, urine, the placenta, and embryonic fluids of the infected cat. Transmission in cats via semen is also possible. Experimentally, the high virus titers in the placenta and uterine fluid in a NiV-infected pregnant cat with virus shedding has been demonstrated indicating vertical transmission. In utero transmission has been demonstrated in this species (Mungall et al. 2007; Anonym 2016). Although experimental studies have not been published in dogs, serological surveys in Malaysia suggest that the virus did not spread horizontally in dogs during this outbreak (Middleton et al. 2002). Humans can be infected by direct contact with infected swine, probably through the mucous membranes but possibly also through skin abrasions. Humans could theoretically be infected by contact with domesticated animals other than pigs, but this has not been demonstrated. Direct or indirect bat-to-human transmission was apparently responsible for several outbreaks. Ingestion of virus in contaminated, unpasteurized date palm juice seems to be the source of the initiation factor in majority of the outbreaks in Bangladesh. However, once it infects a human, then it can be transmitted to another human with direct or indirect contact (Gurley et al. 2007). In humans, NiV can be shed in saliva, upper respiratory secretions, and urine (Chua et al. 2000; Harcourt et al. 2005). Nosocomial transmission has been demonstrated, the virus getting transmitted through fomites and handling inanimate objects infected with virus. Handlings of respiratory secretions, nursing of the human patients and relatives of the patients in the same family have all been shown as potential transmission mechanism of NiV (Tan and Tan 2001; Hsu et al. 2004). Person-to-person transmission has also been reported after close direct contact with either NiV patients or even with cadavers of the patients died of Nipah (Sazzad et al. 2013).

12.4 Pathology and Pathogenesis

There is lack of clear understanding over NiV pathogenesis, since Malaysia-Singapore and Bangladesh-India episodes pose different clinical pictures. For the convenience of understanding, the nervous and respiratory routes may be explained separately. The outbreaks in Malaysia and Singapore have primarily been associated with the development of severe febrile encephalitis; however, it is not very clear how the virus spreads to the central nervous system. Ephrin B2 is host cell receptor for G glycoprotein of NiV. The widespread presence of Ephrin B2 in vertebrates, particularly in arterial endothelial cells and in neurons, provides an explanation for the wide host range for Henipaviruses and their systemic infection. Cellular tropism of NiV correlates well with the expression pattern of Ephrin B2 as it acts as the entry receptor for NiV on endothelial cells, neurons and smooth muscle cells surrounding small arteries and arterioles. After primary replication of the virus in these cells, viremia occurs, which results in systemic spread of the virus leading to thrombosis, vascular occlusion, ischemia, affecting the central nervous system most severely. The pro-inflammatory cytokines like TNF-α and IL-1β are responsible to increase the permeability of the blood-brain barrier and the induction of neuronal injury, thereby disrupting the blood-brain barrier, after which the neurological signs are apparent. As seen from the lesions of vasculitis, the blood vessels are the early targets of infection, with the central nervous system being the most severely affected, although the lung, kidney, and other organs are also infected (Wong et al. 2002; Negrete et al. 2006; Pernet and Lee 2012). Another characteristic of the NiV infection are the endothelial syncytia. The F and G glycoproteins play an important role in formation of these multinucleated giant-endothelial cells. Experimental inoculation of NiV into golden hamster brain shows the presence of both viral antigen and RNA in the many eosinophilic inclusions in the cytoplasm of the neurons around the site of vasculitis resulting in severe vascular and parenchymal lesions explaining encephalitis observed in humans. However, unlike humans, involvement of bronchial epithelium is not seen in golden hamster, although NiV-positive multinucleated giant cells and inflammatory cells are observed occasionally along with parenchymal inflammation and fibrinoid necrosis (Wong et al. 2003). Acute encephalitis with vasculitis-induced thrombosis in the brain is common, but atypical pneumonia and respiratory involvement is also noticed in other cases. In a very small proportion of human cases, there could be recurrence of encephalitis after several months or even years (Luby et al. 2009).

For respiratory system involvement, experimental animal infections indicate efficient initiation of infection after inhalation of virus, targeting endothelial cells as it is an important first line of defense. In human NiV infection, the virus is demonstrable in bronchiolar epithelium and is shed through nasopharyngeal and tracheal secretions in the early phase of the illness hence such cases are more likely to transmit NiV. Histopathologic changes noticed are hemorrhagic or necrotizing alveolitis, pulmonary edema, and aspiration pneumonia. Intra-alveolar inflammatory cells and occasional multinucleated giant cells are also seen in affected alveolar tissue (Lo et al. 2010).

In a recent finding, it has been shown that human lymphocytes and monocytes are not permissive for NiV but can bind NiV and transfer infection to endothelial and Vero cells without themselves being productively infected. It means that the virus uses the leukocytes as vehicle for its own spread. However, a low level of virus replication occurs in dendritic cells (Mathieu et al. 2011).

12.5 Symptoms and Lesions

12.5.1 Human Beings

Incubation period of NiV infection varies between 4 and 30 days but may be as long as 2 months. NiV causes severe, rapidly progressive encephalitis that carried a high mortality rate. The case fatality rates vary depending upon the quickness of diagnosis and early palliative treatment. During Malaysian outbreaks, the fatality was around 40 % whereas in India-Bangladesh epidemics, it is always higher around 70 %. A small proportion of survivors may have objective neurological dysfunction from a few months to a few years after infection. Symptoms of NiV infection in humans are variable, like influenza. Symptoms include fever and muscular pain or inflammation of the brain leading to disorientation or coma. Nausea and vomiting may occur. Sometimes, respiratory signs with acute respiratory distress may also be seen. Acute encephalitis follows febrile reaction along with drowsiness, confusion leading to coma. Encephalitis may occur at later stage and even takes months after earlier symptoms. In such late onsets, it becomes very difficult to correlate the etiology of encephalitis with NiV infection since exposure may have taken place several months earlier. Further, those who recover once may again suffer from encephalitis. Most of the lesions are noticed in the brain. Viral inclusions are seen in neurons. Vasculitis leading to thrombosis and occluding vascular path with necrotic areas and ischemia are the only prominent lesions. However, vasculitis may also be seen in the lung and kidneys. The other organs may not show significant changes (Goh et al. 2000; Chua et al. 2001; Wong et al. 2002; Chadha et al. 2006).

12.5.2 Pigs

On the basis of clinical signs during Malaysian outbreaks, the disease in pigs was called as porcine respiratory and neurologic syndrome, porcine respiratory and encephalitic syndrome, barking pig syndrome, and one mile cough. The disease was generally associated with high (~80 %) morbidity and low (< 5 %) mortality but relatively higher mortality in piglets. The incubation period runs around 1–2 weeks. Clinical manifestations in pigs depend upon involvement of central nervous system or respiratory system and the age of the pigs. Pigs develop febrile respiratory disease with a loud cough. Mostly adult animals show neurological signs while in piglets and young ones respiratory involvement is more common. Respiratory form shows acute febrile reaction that includes fever, distressed respiration, dry cough, and hemoptysis in severe cases. Neurological form initiates with fever, followed by tremors, muscular twitching, muscular spasms, incoordination of movements and hind leg weakness in the initial stages followed by spastic or flaccid paresis. In some cases, nervous and respiratory symptoms are seen together. In adults, febrile illness with respiratory symptoms is observed with increased salivation, lachrymation, and nasal discharge. Nervous signs include nystagmus, head knocking, clamping of mouth, tetanus-like spasm, and seizures. Pregnant sow may abort. (Nor et al. 2000; Chua 2003; Wong and Ong 2011).

In pigs, lung consolidation with hemorrhages ranging from petechiae to ecchymosis is noticed. The bronchi and trachea may contain frothy or blood-stained fluid. The cut surface showed exudate of varying consistency in the bronchi. The brain and kidneys show generalized congestion and edema. Histologically, the lungs show interstitial pneumonia with hemorrhages and syncytia formations in the endothelial cells of blood vessels. The kidney and brain may also show generalized vasculitis with fibrinoid necrosis, hemorrhages, infiltration of mononuclear cells, and thrombosis. In some cases, meningeal inflammatory infiltrates may also be noted in meninges. Demonstration of high concentration of NiV antigens in the endothelium of the blood vessels in the lung may be taken as a proof for capacity of such infected pigs to excrete a large amount of the virus through respiratory route (Chua et al. 2000; Wong et al. 2003).

12.5.3 Dogs, Cats, and Experimental Animals

The observations on clinical signs in dogs are not recorded in detail as these animals during pig outbreaks in Malaysia could be suspected to have suffered from NiV infection only after their death. Necropsy of dogs died with illness revealed presence of exudates in the trachea and bronchi, vasculitis in lungs, and glomerular and tubular necrosis with syncytia formation with severe hemorrhages in kidneys (Nor et al. 2000). Cats are also susceptible to infection and show lesions of vascular changes and endothelial syncytia in vital organs. The lungs show inflammation with involvement of bronchial epithelium (Middleton et al. 2002). Experimental animals could be infected are guinea pig, hamster, ferret, and nonhuman primates (squirrel monkey and African green monkey). Most of these animals show extensive vascular and parenchymal lesions in central nervous system and other organs like the liver, lungs, kidneys, muscles, lymphoid tissue, etc. Mouse and rat do not respond to experimental infection (Wong et al. 2003; Torres-Velez et al. 2008; Geisbert et al. 2010; Marianneau et al. 2010).

12.6 Diagnosis

The disease is not reported in pigs after 1999; however, the presence of virus in bat population requires that the veterinarians keep them abreast with clinical diagnosis of NiV and its differentiation with other pig diseases, viz., swine fever, porcine reproductive and respiratory syndrome (PRRS), swine enzootic pneumonia, and pseudorabies. The disease in humans is generally diagnosed on the basis of history and signs of an acute fever, altered mental status, cough, and respiratory distress as initiating clinical signs; however, there are several acute encephalitis cases confusing NiV diagnosis that needs to be differentiated. Of particular concern are Japanese encephalitis virus, other arboviruses, and bacterial meningitis.

Clinical specimens like cerebrospinal fluid, throat swabs, and blood and urine samples from such suspected patients are subjected to nucleic acid amplification tests (RT-PCR, real-time RT-PCR, and duplex nested RT-PCR) and also for virus isolation. For NiV isolation and further virus neutralization test, BSL-4 laboratory is essential (Anonym 2016). However, suspected sample processing before virus isolation can be conducted in BSL-3 conditions, and if found positive, all the samples and culture fluids are immediately transferred to BSL-4 laboratory by following transport protocol appropriate for dangerous pathogens. Under such circumstances, adequate biosafety precautions for laboratory workers, sterilization of the laboratory, and equipment used need to be taken. Where such BSL-3 labs also are not available, immunofluorescent test over acetone-fixed infected cells or viral antigen-capture ELISA is carried out (OIE 2010). Under serological tests, at an early stage, positive IgM ELISA and at later stage, IgG ELISA would also be indicative of NiV infection.

12.7 Treatment, Prevention, and Control

As of now, there is no approved or licensed therapeutics or effective treatment for NiV infection (Broder 2010; Border et al. 2012) and vaccine for prevention of disease in human or livestock populations does not exist. The patients are admitted in the hospital only to manage fever, respiratory distress, neurological symptoms, and general supportive care. An early medical attention would increase the chances of survival. The antiviral drug ribavirin is shown to be effective against the virus in vitro, and although in many patients, it has not given assured recovery, it has been helpful for reduction in hospitalization period, dependence on ventilator and alleviating the symptoms of nausea, vomiting, and convulsions (Aljofan et al. 2009; Chong et al. 2001). However, there is a need to have authenticated data with more controlled studies on antiviral trials. It has also been shown that passive immunotherapy with polyclonal or monoclonal antibody specific for the NiV envelope glycoproteins has proved successful in hamsters (Guillaume et al. 2004, 2006). Recombinant canarypox-based NiV vaccine candidates are being tried for pig immunization (Weingartl et al. 2006). A subunit consisting of a recombinant soluble and oligomeric form of the G glycoprotein of Hendra virus (HeV-sG) affords protection against HeV and NiV challenge. A human monoclonal antibody known as m102.4 isolated from a recombinant human phage-Fab library has been shown as potential postexposure therapy against NiV infection (Zhu et al. 2008).

NiV infection can be prevented by avoiding exposure to bats and sick pigs in endemic areas and by avoiding drinking of raw date palm sap. It is transmissible from one human to another through contact or excretions and secretions and hence is an occupational hazard. During medical examination and care, standard infection control practices and proper barrier nursing techniques should be strictly followed for preventing nosocomial transmission. Similarly, in disease prone area, there is a need for adequate public education on the disease risk and biosafety precautions for animal handlers, butchers, and other professional people. In an area where there is possibility of disease, medical and veterinary practitioners, medical attendants, nursing staff, laboratory, and paramedical staff should be trained for handling clinical cases, sampling, and laboratory operations with adequate biosafety procedures. A second area for targeted intervention is reducing the exposure of caretakers of the patients to the saliva/urine or other excretions, cloths, utensils, and other belongings of sick persons or the cadavers of the persons suspected to have died with NiV. When a NiV outbreak is recognized, it is appropriate to implement standard precautions both at field and hospital level to prevent its spread.

It is now known that fruit bat populations that carry the virus exist in several countries right from Western part of Africa to South Asia, Southeast and East Asian countries, although their present spillover events are limited to only three countries. The ever-mixing bat populations carrying NiV has potential for spread in more countries. Climatic changes like global warming, floods, draughts, and others may compel bats to have an impact on the virus circulation and spillover risk in different manner, viz., change in geographical distribution as a direct effect, decreased local food resources or extreme weather conditions putting them under physiological stress, leading to immunosuppression or prolonged virus shedding due to stress, or sharp rise in viral incidence within populations. In some cases, less or nonavailability of natural foraging may divert bats to preferential feeding on human cultivated crops that may increase the risk of viral spillover surrounding human or animal populations (Daszak et al. 2013).

In many such instances, as an immediate reaction to disease occurrence, local people and at times even the administration may favor the general sentiments to eliminate of fruit bats. However, it should be remembered that many plants depend partly or wholly on bats to pollinate their flowers or spread their seeds, while other bats also help in biological control of pest by eating insects. Human being should take care to protect food and water sources getting contaminated from bat excretions and secretions. The date palm sap consumption practice needs a big deterrent. It includes diverting this product in molasses production industry thus making the sap safer after heating. The bat access can also be controlled during the sap collection from date palm trees. Public awareness through mass education would help people understand how palm sap drinking is harmful not only for their personal health but also for the country and region.

References

Aljofan M, Saubern S, Meyer AG, Marsh G, Meers J, Mungall BA (2009) Characteristics of Nipah virus and Hendra virus replication in different cell lines and their suitability for antiviral screening. Virus Res 142(1–2):92–99

Anonym (2015) Nipah disease outbreak – 2015 IEDCR website, http://www.iedcr.org/index.php?option=com_content&view=article&id=27&Itemid=75

Anonym (2016) Queensland Govt Report. https://www.business.qld.gov.au/industry/agriculture/species/diseases-disorders/animals/nipah-virus

Arankalle VA, Bandyopadhyay BT, Ramdasi AY, Jadi R, Patil DR, Rahman M, Majumdar M, Banerjee PS, Hati AK, Goswami RP, Neogi DK, Mishra AC (2011) Genomic characterization of Nipah virus, West Bengal, India. Emerg Infect Dis 17:907–909

Bates PJJ, Harrison DL (1997) Bats of the Indian subcontinent. Harrison Zoological Museum, Kent, pp. 3–18

Breed AC, Field HE, Epstein JH, Daszak P (2006) Emerging henipaviruses and flying foxes – conservation and management perspectives. Biol Conserv 131(2):211–220. doi:10.1016/j.biocon.2006.04.007

Breed AC, Meers J, Sendow I, Bossart KN, Barr JA et al (2013) The distribution of Henipaviruses in Southeast Asia and Australasia: is Wallace's Line a barrier to Nipah virus? PLoS One 8(4):e61316. doi:10.1371/journal.pone.0061316

Broder CC. (2010) Therapeutics and vaccines against Hendra and Nipah viruses. In: Levine MM, Dougan G, Good MF, Liu MA, Nabel GJ, Nataro JP, Rappuoli R, editors. New generation vaccines. 4thed. Informa Healthcare; USA, New York. pp 885–894

Border CC, Geisbert TW, Xu K, Nikolov DB, Wang LF, Middleton D, Pallister J, Bossart KN (2012) Immunization strategies against Henipaviruses. Curr Top Microbiol Immunol 359:197–223

CDC (1999) Outbreak of Hendra-like virus-Malaysia & Singapore 1998–1999. Morb Mortal Wkly Rep 48:265–269

Chadha MS, Comer JA, Lowe L, Rota PA, Rollin PE, Bellini WJ, Ksiazek TG, Mishra A (2006) Nipah virus-associated encephalitis outbreak, Siliguri, India. Emerg Infect Dis 12(2):235–240

Chan YP, Chua KB, Koh CL, Lim ME, Lam SK (2001) Complete nucleotide sequences of Nipah virus isolates from Malaysia. J Gen Virol 82:2151–2155

Chong HT, Abdullah S, Tan CT (2009) Nipah virus and bats. Neurol Asia 14:73–76

Chong HT, Kamarulzaman A, Tan CT, Goh KJ, Thayaparan T, Kunjapan SR, Chew NK, Chua KB, Lam SK (2001) Treatment of acute Nipah encephalitis with ribavirin. Ann Neurol 49(6):810–813

Chow VT, Tambyan PA, Yeo WM, Phoon MC, Howe J (2000) Diagnosis of Nipah virus encephalitis by electron microscopy of cerebrospinal fluid. J Clin Virol 19:143–147
Chua KB (2003) Nipah virus outbreak in Malaysia. J Clin Virol 26(3):265–275
Chua KB (2010) Epidemiology, surveillance and control of Nipah virus infections in Malaysia. Malays J Pathol 32(2):69–73
Chua KB, Bellini WJ, Rota PA, Harcourt BH, Lam SK, Ksiazek TG, Rollin PE, Zaki SR, Shieh WJ, Goldsmith CS, Gubler DJ, Roehrig JT, Eaton BT, Gould A, Olson J, Field H, Daniels P, Ling AE, Peters CJ, Anderson LJ, Mahy WJ (2000) Nipah virus: a recently emergent deadly paramyxovirus. Science 288:1432–1435
Chua KB, Chua BH, Wang CW (2002) Anthropogenic deforestation, El Nin˜o and the emergence of Nipah virus in Malaysia. Malays J Pathol 24:15–21
Chua KB, Goh KJ, Wong KT, Kamarulzaman A, Tan PSK, Ksiazek TG, Zaki SR, Paul G, Lam SK, Tan CT (1999) Fatal encephalitis due to Nipah virus among pig farmers in Malaysia. Lancet 354:1257–1259
Chua KB, Lam SK, Goh KJ, Hooi PS, Ksiazek TG, Kamarulzaman A, Olson J, Tan CT (2001) The presence of Nipah virus in respiratory secretions and urine of patients during an outbreak of Nipah virus encephalitis in Malaysia. J Infect 42(1):40–43.
Daniels P, Ksiazek K, Eaton B T (2001) Laboratory diagnosis of Nipah and Hendra virus infections. Microbes Infect 3:289–295
Daszak P, Torrelio CZ, Bogich TL, Fernandez M, Epstein JH, Murray KA, Hamiltone H (2013) Interdisciplinary approaches to understanding disease emergence: the past, present, and future drivers of Nipah virus emergence. Proc Natl Acad Sci U S A 110(Suppl 1):3681–3688
Drexler JF, Corman VM, Gloza-Rausch F, Seebens A, Annan A, Ipsen A, Kruppa T, Mu¨ller MA, Kalko EK, Adu-Sarkodie Y, Oppong S, Drosten C (2009) Henipavirus RNA in African bats. PLoS One 4(7):e6367. doi:10.1371/journal.pone.0006367
Epstein JH, Prakash V, Smith CS, Daszak P, McLaughlin AB, Meehan A (2008) Henipavirus infection in fruit bats (*Pteropus giganteus*giganteus), India. Emerg Infect Dis 14(8):1309–1311
FAO (2011) Investigating the role of bats in emerging zoonoses: Balancing ecology, conservation and public health interests. In: Newman SH, Field HE, de Jong CE, Epstein JH (eds). FAO Animal Production and Health Manual No. 12. Rome.
Geisbert TW, Daddario-DiCaprio KM, Hickey AC, Smith MA, Chan Y-P et al (2010) Development of an acute and highly pathogenic nonhuman primate model of Nipah virus infection. PLoS One 5(5):e10690. doi:10.1371/journal.pone.0010690
Goh KJ, Tan CT, Chew NK, Tan PSK, Kamarulzaman A, Sarji SA, Wong KT, Abdullah BJ, Chua KB, Lam SK (2000) Clinical features of Nipah virus encephalitis among pig farmers in Malaysia. New Engl J Med 342:1229–1235
Goldsmith CS, Whistler T, Rollin PE, Chua KB, Bellini W, Rota P, Wong KT, Daszak P, Ksiazek TG, Zaki SR (2000) Ultrastructural studies of Nipah virus, a newly emergent paramyxovirus, using thin section, negative stain, immunogold, and in situ hybridization electron microscopy. Microsc Microanal 6:644–645
Guillaume V, Contamin H, Loth P, Georges-Courbot MC, Lefeuvre A, Marianneau P, Chua KB, Lam SK, Buckland R, Deubel V, Wild TF (2004) Nipah virus: vaccination and passive protection studies in a hamster model. J Virol 78(2):834–840
Guillaume V, Contamin H, Loth P, Grosjean I, Courbot MC, Deubel V, Buckland R, Wild TF (2006) Antibody prophylaxis and therapy against Nipah virus infection in hamsters. J Virol 80(4):1972–1978
Gurley ES, Montgomery JM, Hossain MJ, Bell M, Azad AK, Islam MR, Molla MA, Carroll DS, Ksiazek TG, Rota PA, Lowe L, Comer JA, Rollin P, Czub M, Grolla A, Feldmann H, Luby SP, Woodward JL, Breiman RF (2007) Person-to-person transmission of Nipah virus in a Bangladeshi community. Emerg Infect Dis 13(7):1031–1037
Harcourt BH, Lowe L, Tamin A, Liu X, Bankamp B, Bowden N, Rollin PE, Comer JA, Ksiazek TG, Hossain MJ, Gurley ES, Breiman RS, Bellini WJ, Rota PA (2005) Genetic characterization of Nipah viruses isolated during two outbreaks in Bangladesh in 2004. Emerg Infect Dis 11:1594–1597
Harcourt BH, Tamin A, Ksiazek TG, Rollin PE, Anderson LJ, Bellini WJ, Rota PA (2000) Molecular characterization of Nipah virus, a newly emergent paramyxovirus. Virology 271:334–349

Harcourt BH, Tamin A, Halpin K, Ksiazek TG, Rollin PE, Bellini WJ, Rota PA (2001) Molecular characterization of the polymerase gene and genomic termini of Nipah virus. Virology 287:192–201

Hasebe F, Thuy NT, Inoue S, Yu F, Kaku Y, Watanabe S, Akashi H, Dat DT, Mai le TQ, Morita K (2012) Serologic evidence of nipah virus infection in bats, Vietnam. Emerg Infect Dis 18(3):536–537

Hsu VP, Hossain MH, Parashar UD, Ali MM, Ksiazek TG, Kuzmin I, Niezgoda M, Rupprecht C, Bresee J, Breiman RF (2004) Nipah virus encephalitis reemergence, Bangladesh. Emerg Infect Dis 10:2082–2087

Hyatt AD, Zaki SR, Goldsmith CS, Wise TG, Hengstberger SG (2001) Ultrastructure of Hendra virus and Nipah virus within cultured cells and host animals. Microbes Infect 3:297–306

Kulkarni DD, Tosh C, Venkatesh G, Senthil Kumar D (2013) Nipah virus infection: current scenario. Indian J Virol 24(3):398–408

Lam SK, Chua KB (2002) Nipah virus encephalitis outbreak in Malaysia. Clin Infect Dis 34(Suppl 2):S48–S51

Lamb RA, Parks GD (2007) Paramyxoviridae: the viruses and their replication. In: Knipe DM, Griffin DE, Lamb RA, Straus SE, Howley PM (eds) Fields virology. Lippincott, Philadelphia, pp. 1449–1496

Lamb RA, Collins PL, Kolakofsky D, Melero JA, Nagai Y, Oldstone MBA, Pringle CR, Rima BK (2005) Family paramyxoviridae. In: Fauquet CM (ed) Virus taxonomy: the classification and nomenclature of viruses. The eighth report of the International Committee in Taxonomy of Viruses. CPI Group (UK) Ltd. Croydon CR0 4YY pp. 655–668

Lehlé C, Razafitrimo G, Razainirina J, Goodman SM, Faure C, Courbot MCG, Rousset D, Reynes JM, Andriaholinirina N (2007) Henipavirus and Tioman virus antibodies in pteropodid bats, Madagascar. Emerg Infect Dis 13(1):159–161

Li Y, Wang J, Hickey AC, Zhang Y, Li Y, Wu Y, et al. Antibodies to Nipah or Nipah-like viruses in bats, China [letter]. Emerg Infect Dis. 2008; 14(12): 1974–76

Lo MK, Miller D, Aljofan M, Mungall BA, Rollin PE, Bellini WJ, Rota PA (2010) Characterization of the antiviral and inflammatory responses against Nipah virus in endothelial cells and neurons. Virology 404(1):78–88.

Lo MK, Rota PA (2008) Emergence of Nipah virus, a highly pathogenic Paramyxovirus. J Clin Virol 43:396–400

Luby SP, Rahman M, Hossain MJ, Blum LS, Husain MM, Gurley E et al (2006) Foodborne transmission of Nipah virus, Bangladesh. Emerg Infect Dis 12(12):1888–1894

Luby SP, Hossain MJ, Gurley ES, Ahmed B-N, Banu S, Khan SU, Homaira N, Rota PA, Rollin PE, Comer JA, Kenah E, Ksiazek TG, Rahman M (2009) Recurrent zoonotic transmission of Nipah virus into humans Bangladesh 2001–2007. Emerg Infect Dis 15:1229–1235

Marianneau P, Guillaume V, Wong KT, Badmanathan M, Looi RY, Murri S, Loth P, Tordo N, Wild TF, Horvat B, Contamin H (2010) Experimental infection of squirrel monkeys with Nipah virus. Emerg Infect Dis 16(3):507–510

Mathieu C, Pohl C, Szecsi J, Bodennec ST, Devergnas S, Raoul H, Cosset HL, Gerlier D, Wild TF, Horvat B (2011) Nipah virus uses leukocytes for efficient dissemination within a host. J Virol 85(15):7863–7871

Middleton DJ, Westbury HA, Morrissy CJ, van der Heide BM, Russell GM, Braun MA, Hyatt AD (2002) Experimental Nipah virus infection in pigs and cats. J Comp Pathol 126:124–136

Middleton DJ, Morrissy CS, van der Heide BM, Russell GM, Braun MA, Westbury HA, Halpin K, Daniels PW (2007) Experimental Nipah virus infection in Pteropid Bats (*Pteropus poliocephalus*). J Comp Pathol 136:266–272

Mills JN, Alim ANM, Bunning ML, Lee OB, Wagoner KD, Amman BR, Stockton PC, Ksiazek TG (2009) Nipah Virus infection in dogs, Malaysia, 1999. Emerg Infect Dis 15(6):950–952

Montgomery J, Hossain MJ, Gurley E, Carroll DS, Croisier A, Bertherat E, Asgari N, Formenty P, Keeler N, Comer J, Bell MR, Akram K, Molla AR, Zaman K, Islam M, Wagoner RK, Mills JN, Rollin PE, Ksiazek TG, Breiman RF (2008) Risk factors for Nipah virus infection in Bangladesh. Emerg Infect Dis 14(10):1526–1532

Mungall BA, Middleton D, Crameri G, Halpin K, Bingham J, Eaton BT, Broder CC (2007) Vertical transmission and fetal replication of Nipah virus in an experimentally infected cat. J Infect Dis 196(6):812–816

Negrete OA, Wolf MC, Aguilar HC, Enterlein S, Wang W, Mühlberger E, Su SV, Ciarlet AB, Flick R, Lee B (2006) Two key residues in EphrinB3 are critical for its use as an alternative receptor for Nipah virus. PLoS Pathog 2(2):78–86. doi:10.1371/journal.ppat.0020007

Nor MMN, Gan CH, Ong BL (2000) Nipah virus infection of pigs in peninsular Malaysia. Rev Scintech Off Int Epiz 19(1):160–165

OIE (2010) Nipah and Hendra virus diseases. Manual of diagnostic tests and vaccines for terrestrial animals. Chapter 2.9.6. OIE, Paris

Paton NI, Leo YS, Zaki SR, Auchus AP, Lee KE, Ling AE, Chew SK, Ang BSP, Rollin PE, Umapathi T, Sng I, Lee CC, Lim E, Ksiazek TG (1999) Outbreak of Nipah virus infection among abattoir workers in Singapore. Lancet 354:1253–1256

Pernet O, Lee B (2012) Henipavirus receptor usage and tropism. Curr Top Microbiol Immunol 359:59–78. doi:10.1007/82_2012_222

Reynes JM, Counor D, Ong S, Faure C, Seng V, Molia S, Walston J, Georges-Courbot MC, Deubel V, Sarthou JL (2005) Nipah virus in Lyle's flying foxes, Cambodia. Emerg Infect Dis 11:1042–1047

Sazzad HMS, Hossain MJ, Gurley ES, Ameen KMH, Parveen S, Islam MS et al (2013) Nipah virus infection outbreak with nosocomial and corpse-to-human transmission, Bangladesh. Emerg Infect Dis 19(2):210–217. doi:10.3201/eid1902.120971

Sendow I, Ratnawati A, Taylor T, Adjid RMA, Saepulloh M et al (2013) Nipah virus in the fruit bat *Pteropus vampyrus* in Sumatera, Indonesia. PLoS One 8(7):e69544. doi:10.1371/journal.pone.0069544

Tamin A, Harcourt BH, Lo MK, Roth JA, Wolf MC, Lee B, Weingartl H, Audonnet JC, Bellini WJ, Rota PA (2009) Development of a neutralization assay for Nipah virus using pseudotype particles. J Virol Methods 160(1–2):1–6

Tan CT, Tan KS (2001) Nosocomial transmissibility of Nipah virus-correspondence. J Infect Dis 184:1367

Tanimura N, Imada T, Kashiwazaki Y, Sharifah SH (2006) Distribution of viral antigens and development of lesions in chicken embryos inoculated with Nipah virus. J Comp Pathol 135:74–82

Torres-Velez FJ, Shieh WJ, Rollin PE et al (2008) Histopathologic and immunohistochemical characterization of Nipah virus infection in the guinea pig. Vet Pathol 45:576–585

Wacharapluesadee S, Boongird K, Wanghongsa S, Ratanasetyuth N, Supavonwong P, Saengsen D, Gongal GN, Hemachudha T (2010) A longitudinal study of the prevalence of Nipah virus in *Pteropus lylei* bats in thailand: evidence for seasonal preference in disease transmission. Vector Borne Zoonotic Dis 10(2):183–190

Wacharapluesadee S, Lumlertdacha B, Boongird K, Wanghongsa S, Chanhome L, Rollin P, Stockton P, Rupprecht CE, Ksiazek TG, Hemachudha T (2005) Bat Nipah virus, Thailand. Emerg Infect Dis 11:1949–1951

Weingartl HM, Berhane Y, Caswell JL, Loosmore S, Audonnet JC, Roth JA, Czub M (2006) Recombinant nipah virus vaccines protect pigs against challenge. J Virol 80:7929–7938

Wong KT, Ong KC (2011) Pathology of acute Henipavirus infection in humans and animals. Pathol Res Int 2011:ID567248. doi:10.4061/2011/567248

Wong KT, Shieh WJ, Kumar S, Norain K, Abdullah W, Guarner J, Goldsmith CS, Chua KB, Lam SK, Tan CT, Goh KJ, Chong HT, Jusoh R, Rollin PE, Ksiazek TG, Zaki SR (2002) Nipah virus infection: an emerging paramyxoviral zoonosis. Am J Pathol 161:215–228

Wong KT, Grosjean I, Brisson C, Blanquier B, Fevre-Montagne M, Bernard A, Loth P, Georges-Courbot MC, Chevallier M, Akaoka H, Marianneau P, Lam SK, Wild TF, Deubel V (2003) A golden hamster model for human acute Nipah virus infection. Am J Pathol 163(5):2127–2137

Yadav PD, Raut CG, Shete AM, Mishra AC, Towner JS, Nichol ST, Mourya DT (2012) Detection of Nipah virus RNA in fruit bat (*Pteropus giganteus*) from India. Am J Trop Med Hyg 87(3):576–578

Yob JM, Field H, Rashdi AM, Morrissy C, van der Heide B, Rota P, Adzhar A, White J, Daniels P, Jamaluddin A, Ksiazek T (2001) Nipah virus infection in bats (order Chiroptera) in Peninsular Malaysia. Emerg Infect Dis 7:439–441

Zhu Z, Bossart KN, Bishop KA, Crameri G, Dimitrov AS, McEachern JA, Feng Y, Middleton D, Wang LF, Broder CC, Dimitrov DS (2008) Exceptionally potent cross-reactive neutralization of Nipah and Hendra viruses by a human monoclonal antibody. J Infect Dis 197(6):846–853

Bovine Immunodeficiency Virus

13

Sandeep Bhatia and Richa Sood

13.1 Introduction

Bovine immunodeficiency virus (BIV) is a member of the *Lentivirus* genus of the subfamily *Orthoretrovirinae* under *Retroviridae* family (ICTV 2011). Persistent infection of cattle with BIV is reported worldwide. However, the virus has never been found primary cause of any serious illness in cattle unlike human immunodeficiency virus (HIV) in human beings. The virus was first isolated in 1969 in Louisiana, USA, from a Holstein cow (R-29) with clinical signs of mild persistent lymphocytosis, generalized hyperplasia of lymph nodes, central nervous system lesions, weakness, and emaciation (Van der Maaten et al. 1972). Two isolations (named FL491 and FL112) of BIV were made in Florida in 1993 (Suarez et al. 1993). Though, immunodeficiency was not characteristic feature of BIV, it was mainly named so due to its genetic and morphologic similarity with HIV and simian immunodeficiency viruses (SIV). Initially, BIV was designated as "bovine visna-like virus" since it was structurally similar to maedi-visna virus. Its biology went unstudied for a decade and a half until HIV was discovered in 1983 (Barré-Sinoussi et al. 1983). The other two retroviruses discovered were bovine syncytial virus (BSV), a foamy virus or spumavirus; and bovine leukemia virus (BLV), an oncovirus (Malmquist et al. 1969; Miller et al. 1969).

13.2 Molecular Biology

Molecular biology of BIV was elucidated by characterization of original BIV R-29 isolate and infectious cDNA clones BIV106 and BIV127 derived from R-29 isolate (Garvey et al. 1990; Braun et al. 1988). A complete genetic map of BIV was

S. Bhatia (✉) • R. Sood
National Institute of High Security Animal Diseases (NIHSAD), Bhopal, India
e-mail: sandeep.bhatia@hsadl.in

J. Bayry (ed.), *Emerging and Re-emerging Infectious Diseases of Livestock*,
DOI 10.1007/978-3-319-47426-7_13

developed by sequencing of the proviruses from BIV-infected cells (Garvey et al. 1990; Braun et al. 1988) and further genetic characterization of BIV clones (Oberste et al. 1991, 1993; Liu et al. 1992; Pallansch et al. 1992). Based on these experiments and the genetic map of BIV, Gonda et al. (1994) described the genomic organization of BIV. The linear genome of BIV contains 8960 base pairs in the form of proviral DNA, which comprises obligatory retrovirus structural genes *gag*, *pol,* and *env*, flanked on the 5′ and 3′ ends by a complete copy of a LTR. The *gag* (for group antigen-associated gene) gene encodes the Gag precursor protein (Gag Pr53) involved in virus particle assembly. Gag Pr53 is cleaved by the viral protease leading to the production of matrix (MA) p16, capsid (CA) p26, nucleocapsid (NC) p13, and three small proteins p21, p3 and p2 the role of which in the replication cycle of BIV is not clearly determined. The LTRs contain the promoters, enhancers, and terminators of transcription. A complex lentivirus "central region" between and overlapping the *pol* and *env* reading frames contains coding exons of several putative nonstructural accessory genes including *vif* (viral infectivity factor), *tat* (transactivator of transcription), *rev* (regulator of virus expression), *vpw, vpy, and tmx.* The products of the accessory genes "vif" and "Tat" as well as those of the structural genes *gag*, *pol*, and *env* of BIV have some sequence similarity to their counterparts in HIV-1. However, the genomes of BIV and HIV-1 show overall divergence, with the *gag* and *pol* ORFs having the greatest sequence similarity.

13.3 Virus Morphology and Structure

The morphology of BIV is similar to HIV with comparable size (120–130 nm) of virions. BIV particles consist of two positive-sense single-stranded viral RNAs and have a structural capsid which envelops proteins. The bilayer viral envelope consists of the viral surface (SU) gp100 and transmembrane (TM) gp45 proteins and encompasses conical-shaped capsid (CA) and nucleocapsid (NC) structures protecting the BIV genome. The genome is composed of a capped and polyadenylated diploid RNA of 8482 nucleotides (nt) in length closely associated to viral proteins p7 and p14 (St-Louis et al. 2005; Gonda et al. 1994). The BIV genome, like other lentiviruses, contains the structural genes *gag, pol, env,* and several accessory genes, including *tat, rev, vif, vpw, vpy, and tmx* (Avidan et al. 2006; Braun et al. 1988).

13.4 BIV Infection Cycle

BIV has a replication cycle similar to that of other lentiviruses. Free virus particles attach to specific cell receptors and penetrate into the cell. The viral RNA is then released and a complimentary DNA copy is transcribed from the viral RNA by reverse transcriptase enzyme and is incorporated into the host's cell nucleus as a "provirus" by viral integrase. The BIV provirus may remain latent for many years until it gets reactivated into infectious RNA virus in the presence of predisposing factors such as concurrent infection, stress, or age. Viral particle assembly and RNA incorporation occur near the plasma membrane. Cellular machinery and viral proteins contribute to each step of this process.

The expression of a unique enzyme reverse transcriptase (RT) is a characteristic feature of the members of *Retroviridae* family. The reactivation of the virus may initiate pathogenesis inside host leading to various lesions and symptoms. BIV infection cycle has been described by Gonda and Oberste (1992). The BIV genome consists of two positive-sense, single-stranded, protein-encapsidated RNA genomes. During the infection cycle, free BIV particles attach to specific cell surface receptors via the envelop glycoprotein of the virus. Subsequently, the viral envelop fuses with the plasma membrane releasing genomic RNA and mature *pol* gene products from the core of the virus into the cytoplasm. The viral RT transcribes the viral RNA into double-stranded DNA, which is then transported to the nucleus where it is incorporated into the host genome with the help of integrase (IN) enzyme. The integrated provirus remains transcriptionally silent until appropriate cellular signals activate gene expression from the viral long terminal repeat (LTR). Cell-mediated expression from the viral LTR is significantly enhanced by the action of the virally coded Tat protein. Splicing of the primary genome-length viral mRNA into sub-genomic messages and transport to the cytoplasm is carried out by the cellular splicing machinery and another virally encoded protein, "Rev" (regulator of virus expression). Sub-genomic mRNAs are translated on the ribosomes in the cytoplasm of the infected cell. Viral precursors for *gag* (group-specific antigen) and *gag-pol* assemble beneath the plasma membrane and incorporate viral genomic RNA during the process of budding. The viral envelop is studded with surface (SU) and transmembrane (TM) glycoprotein. Following release, *gag*-related precursors in the immature particle are cleaved into their functional subunits by the viral protease (PR) as the virus undergoes morphogenesis into a mature infectious particle. The mature particle can begin the infection cycle again by binding to a naïve cell expressing the appropriate receptor for BIV.

13.5 Pathogenesis of BIV in Cattle

BIV infect cells of the immune system, primarily monocytes /macrophages and lymphocytes in vivo (Gonda et al. 1987). Experimental infection of calves with BIV causes a transient lymphocytosis and lymphadenopathy without any overt clinical signs (Carpenter et al. 1992; Suarez et al. 1993; Onuma et al. 1992; Van der Maaten et al. 1972). Unlike other lentiviruses that cause chronic inflammatory diseases (CAEV and EIAV), natural infection of BIV does not cause any significant effect on health of cattle even though experimental evidences suggest that BIV can cause immune dysfunction and can predispose animals to secondary infections. It is not clear whether BIV induces a specific syndrome or whether it renders animals more susceptible to other infectious agents. A number of reports from the USA have described a clinical syndrome with a variety of signs like mastitis, secondary infections, foot problems, dullness, and stupor. A strong association of these signs with coinfection of BIV with bovine leukosis virus (BLV) has been found. Hematological changes, lymphadenopathy with follicular hyperplasia, skin lesions unresponsive to treatment, and meningoencephalitis have been found in naturally and experimentally infected cattle (Braun et al. 1988; Martin et al. 1991; Carpenter et al. 1992; Onuma et al. 1992; Flamming et al. 1993; Gonda et al. 1994; Rovid et al. 1996). The virus is transcriptionally active in animals showing symptoms and can be isolated for many years after infection (Brownlie et al. 1994; Baron et al. 1995). A number of studies have demonstrated the role of BIV in immune dysfunction including a long-term study (more than 7 years) at Louisiana State University dairy herd, which had high seroprevalence of BIV and had high incidence of common diseases that reduced the economic viability of the dairy (Snider et al. 1997). BIV infection of cattle reduces the responsiveness of various important monocyte functions without a change in CD4/CD8 ratios (Onuma et al. 1992; Zhang et al. 1997a). Other studies have demonstrated either mild or no immunosuppression on the basis of lymphocyte blastogenesis tests, neutrophil function tests, mononuclear subset analysis, and histopathological changes (Martin et al. 1991; Carpenter et al. 1992; Flamming et al. 1993).

There is no evidence that can confirm the mode of transmission of BIV naturally, though experimental transmission through the intravenous inoculation of infected material (blood, cell-free or cell-associated virus) is possible. The virus has been observed in milk (Nash et al. 1995a) and bull semen (Nash et al. 1995b). At present, nothing is known about in utero transmission or the influence of age on an animal's susceptibility to infection.

Long-term studies are necessary to observe BIV pathogenicity since current opinions are based on a limited number of short-term experimental studies. Isolation of new BIV variants shall help understanding its pathogenesis since lentiviruses exhibit variable virulence according to isolates. Though BIV is closely related with pathogenic lentiviruses like JDV and HIV, its nonpathogenic nature is an interesting feature which renders it a good model for lentiviral research especially for understanding the pathogenesis (Gardner and Luciw 1989; Gonda et al. 1990).

13.6 Epidemiology

BIV infection in cattle has been reported from several countries, viz., Southwest USA (Black 1990), Canada (McNab et al. 1994), Germany (Muluneh 1994), Japan (Hirari et al. 1996), Italy (Cavirani et al. 1998), Australia (Burkala et al. 1999), Korea (Cho et al. 1999), Pakistan (Meas et al. 2000), India (Patil et al. 2003, Bhatia et al. 2006, 2008a, b, 2010), Brazil (Meas et al. 2002), and Zambia (Meas et al. 2004). Serologic screening of randomly selected cattle sera has shown a nonuniform distribution in the USA in studies using R-29 as a source of antigen (Ambroski et al. 1989; Black 1990; Whetstone et al. 1990; Cockerell et al. 1992). In some regions of the USA, BIV infection was relatively higher than others. For example, in Louisiana, on an average, 40 % of beef and 60 % of dairy herds were positive (Gonda et al. 1994). Seroprevalence of BIV among cattle herds in Mississippi was reported to be greater than 50 % (St Cyr Coats 1995). On the other hand, approximately 4 % cattle from southern and southwestern part of the USA were positive, while cattle sera from eastern or northeastern part of the USA were rarely positive (Gonda 1992; Gonda et al. 1994). In the central regions of the USA, a 21 % seroprevalence of BIV was reported in a Colorado dairy herd based on an ELISA test (Cockerell et al. 1992).

A chemiluminescence Western blot assay detected anti-BIV antibodies in sera of 5.5 % of 928 adult cows from Ontario (McNab et al. 1994). In Germany, serum samples from 6.6 % of 380 cattle were positive for the presence of anti-BIV antibodies by cell ELISA and immunofluorescence assay (Muluneh 1994). In France, a recombinant 53 kDa BIV R-29 antigen was used, which gave weaker reaction with French sera than with positives from Louisiana indicating the occurrence of distinct French and Louisiana BIV variant (Polack et al. 1996). In Hokkaido, Japan, the seroprevalence of BIV up to 7.5 % in 120 cattle with relatively higher prevalence of BLV (bovine leukemia virus) has been reported (Hirari et al. 1996).

13.7 Diagnostic Methods for BIV Infection

Isolation of BIV is difficult from naturally infected animals and is not a method of choice for diagnosis. Till date only four successful isolations of BIV have been reported. The first reported isolation of BIV was from a cow with persistent lymphocytosis and was named as R-29 (Van der Maaten et al. 1972). The other three were BIVCR1 from Costa Rica (Hidalgo et al. 1995) and FL491 and FL112 from Florida, USA (Suarez et al. 1993). All the four isolates used cocultivation techniques using peripheral blood mononuclear cells (PBMC) from infected animal with either fetal bovine spleen cells, fetal bovine lung cells, or embryonic rabbit embryonic cells (Suarez et al. 1993). Molecular and serological methods have been adopted routinely for diagnosis of BIV due to the difficulties in isolation.

Though detection of BIV in infected cattle can be reliably done by PCR using specific primers, there is no gold standard for BIV diagnosis (Suarez and Whetstone 1998). Sensitive PCR diagnostic methods have been developed for the detection of proviral BIV DNA in mononuclear cells (Miller et al. 1969; Suarez et al. 1995; Zhang et al. 1997b). A nested PCR targeting two separate regions of *pol* and *env* was developed and was found to have a greater sensitivity than serology and virus isolation (Suarez et al. 1995).

For serological studies, the R-29 isolate of BIV was used as a source of antigen initially. Later, the use of recombinant viral proteins became popular for serological studies and facilitated seroepidemiology of BIV infection worldwide (Betemps et al. 1999; Zheng et al. 2000; Abed et al. 1999). Indirect ELISA based on recombinant capsid protein (Zheng et al. 2000) or baculovirus-expressed transmembrane protein (Abed and Archambault 2000) have been in use for serodiagnosis. The recombinant capsid protein has been used extensively by many to perform Western blot and indirect ELISA for detection of serum antibodies against BIV (Bhatia et al. 2006, 2008a, b, 2010).

References

Abed Y, Archambault D (2000) A viral transmembrane recombinant protein-based enzyme-linked immunosorbent assay for the detection of bovine immunodeficiency virus infection. J Virol Methods 85:109–116

Abed Y, St-Laurent G, Zhang H, Jacobs RM, Archambault D (1999) Development of a Western blot assay for detection of bovine immunodeficiency-like virus using capsid and transmembrane envelope proteins expressed from recombinant baculovirus. Clin Diagn Lab Immunol 6:168–172

Ambroski GF, Lo JL, Seger CL (1989) Serological detection of multiple retroviral infections in cattle bovine leukemia virus bovine syncytial virus and bovine visna virus. Vet Microbiol 20:247–253

Avidan O, Bochner R, Hizi A (2006) The catalytic properties of the recombinant reverse transcriptase of bovine immunodeficiency virus. Virology 351(1):42–57

Baron T, Mallet F, Polack B, Betemps D, Belli P (1995) The bovine immunodeficiency-like virus (BIV) is transcriptionally active in experimentally infected calves. Arch Virol 140(8):1461–1467

Barré-Sinoussi F, Chermann JC, Rey F, Nugeyre MT, Chamaret S, Gruest J, Dauguet C, Axler-Blin C, Vézinet-Brun F, Rouzioux C, Rozenbaum W, Montagnier L (1983) Isolation of a T-lymphotropic retrovirus from a patient at risk for acquired immune deficiency syndrome (AIDS). Science 220(4599):868–871

Betemps D, Mallet F, Cheynet V, Baron T (1999) Over expression and purification of an immunologically reactive His BIV capsid fusion protein. Protein Expr Purif 15(3):258–264

Bhatia S, Bhatia AK, Sood R, Pattnaik B, Pradhan HK (2006) Serological evidence of bovine immunodeficiency virus infection in cattle and buffalo through use of recombinant capsid (P26) protein based immunoassay. J Immunol Immunopathol 8(2):128–129

Bhatia S, Patil SS, Sood R, Dubey R, Bhatia AK, Pattnaik B, Pradhan HK (2008a) Prokaryotic expression of a 750-bp capsid region of Bovine Immunodeficiency Virus *gag* gene and development of a recombinant capsid (p26) protein based immunoassay for seroprevalence studies. Indian J Biotechnol 7(1):50–55

Bhatia S, Sood R, Bhatia AK, Pattnaik B, Pradhan HK (2008b) Development of a capsid based competitive inhibition enzyme linked immunosorbent assay for detection of bovine immunodeficiency virus antibodies in cattle and buffalo serum. J Virol Methods 148:218–225

Bhatia S, Gangil R, Gupta DS, Sood R, Pradhan HK, Dubey SC (2010) Single chain fragment variable antibody against the capsid protein of bovine immunodeficiency virus and its use in ELISA. J Virol Methods 167:68–73

Black JW (1990) Bluetongue and bovine retrovirus committee report. In: Proceedings of the 93rd annual meeting of the U.S. Animal Health Association. Carter Printing Co., Richmond, pp 150–152

Braun MJ, Lahn S, Boyd AL, Kost TA, Nagashima K, Gonda MA (1988) Molecular cloning of biologically active proviruses of bovine immunodeficiency-like virus. Virology 167:515–523

Brownlie J, Collins ME, Heaton P (1994) Bovine immunodeficiency-like virus – a potential cause of disease in cattle? Vet Rec 134(12):289–291 Review

Burkala EJ, Ellis TM, Voigt V, Wilcox GE (1999) Serological evidence of an Australian bovine lentivirus. Vet Microbiol 68(1–2):171–177

Carpenter S, Miller LD, Alexandersen S, Whetstone CA, Van Der Maaten MJ, Viuff B, Wannemuehler Y, Miller JM, Roth JA (1992) Characterization of early pathogenic effects after experimental infection of calves with bovine immunodeficiency-like virus. J Virol 66:1074–1083

Cavirani S, Donofrio G, Chiocco D, Foni E, Martelli P, Allegri G, Cabassi CS, De Iaco B, Flamming CF (1998) Seroprevalence to bovine immunodeficiency virus and lack of association with leukocyte counts in Italian dairy cattle. Prev Vet Med 37:147–157

Cho KO, Meas S, Park NY, Kim YH, Lim YK, Endoh D, Lee SI, Ohashi K, Sugimoto C, Onuma M (1999) Seroprevalence of bovine immunodeficiency virus in dairy and beef cattle in Korea. J Vet Med Sci 61(5):549–551

Cockerell GL, Jensen WA, Rovank J, Ennis WH, Gonda MA (1992) Seroprevalence of bovine immunodeficiency-like virus and bovine leukemia virus in a dairy cattle herd. Vet Microbiol 31:109–116

Flamming K, Van Der Maaten M, Whetstone C, Carpenter S, Frank D, Roth J (1993) Effect of bovine immunodeficiency-like virus infection on immune function in experimentally infected cattle. Vet Immunol Immunopathol 36:91–105

Gardner MB, Luciw PA (1989) Animal models of AIDS. FASEB J 3:2593–2606

Garvey KJ, Obsrste MS, Esler JE, Braun MJ, Gonda MA (1990) Nucleotide sequence and genome organization of biologically active proviruses of the bovine immunodeficiency-like virus. Virology 175:391–409

Gonda MA, Oberste MS (1992) AIDS- the human immunodeficiency virus: molecular and structural aspects of its biology. In: Kurstak E (ed) Control of virus diseases. Marcel Dekker, New York, pp 3–31

Gonda MA (1992) Bovine immunodeficiency virus. AIDS. 6(8):759–776

Gonda MA, Braun MJ, Carter SG, Kost TA, Bess JW Jr, Arthur LO, Van Der Maaten MJ (1987) Characterization and molecular cloning of a bovine lentivirus related to human immunodeficiency virus. Nature 330:388–391

Gonda MA, Oberste MS, Garvey KJ, Pallansch LA, Battles JK, Pifat DY, Bess IW Jr, Nagashima K (1990) Development of the bovine immunodeficiency like virus as a model of lentivirus disease. Dev Biol Stand 72:97–110

Gonda MA, Luther DG, Fong SE, Tobin GJ (1994) Bovine immunodeficiency virus molecular biology and virus-host interactions. Virus Res 32:155–181

Hidalgo G, Flores M, Bonilla JA (1995) Detection and isolation of bovine immunodeficiency-like virus (BIV) in dairy herds of Costa Rica. Zentralbl Veterinarmed B 42(3):155–161

Hirari N, Kabeya Ohashi K, Sugimoto C, Onuma M (1996) Detection of antibodies against bovine immunodeficiency like virus in dairy cattle in Hokkaido. J Vet Med Sci 58(5):455–457

King AMQ, Adams MJ, Carstens EB, Lefkowitz EJ, editors. (2011) Virus Taxonomy - Ninth Report of the International Committee on Taxonomy of Viruses. Elsevier/Academic Press: London, United Kingdom

Liu ZQ, Sheridan D, Wood C (1992) Identification and characterization of bovine immunodeficiency like virus tat gene. J Virol 66(8):5137–5140

Malmquist WA, Van der Maaten MJ, Boothe AD (1969) Isolation, immunodiffusion and immunofluorescence and electron microscopy of a lymphosarcomatous and apparently normal cattle. Cancer Res 29:188–200

Martin SJ, Neill PO, Billello JA, Eiseman JL (1991) Lymphocyte transformation abnormalities in bovine immunodeficiency-like virus infected calves. Immunol Lett 27:81–84

McNab WB, Jacobs RM, Smith HE (1994) A serological survey for bovine immunodeficiency like virus in Ontario dairy cattle and association between test results production records and management practices. Can J Vet Res 58:36–41

Meas S, Seto J, Sugimoto C, Bahksh M, Riaz M, Sato T, Naeem K, Ohashi K, Onuma M (2000) Infection of bovine immunodeficiency virus and bovine leukemia virus in water buffalo and cattle populations in Pakistan. J Vet Med Sci 62(3):329–331

Meas S, Ryas J, Faria NA, Usui T, Teraoka Y, Mulenga A, Chang KS, Masuda A, Madryga CR, Ohash K, Omma M, Ruas Faias J (2002) Seroprevalence and molecular evidence for the presence of bovine immunodeficiency virus in Brazilian cattle. Jpn J Vet Res 50(2–3):145

Meas S, Nakayama M, Usui T, Nakazato Y, Yasuda J, Ohashi K, Onuma M (2004) Evidence for bovine immunodeficiency virus infection in cattle in Zambia. Jpn J Vet Res 52(1):3–8

Miller JM, Miller LD, Olson C, Gillette KG (1969) Virus like particles in phytohaemagglutinin stimulated lymphocyte culture with reference to bovine persistence lymphosarcoma. J Natl Cancer Inst 43:1297–1305

Muluneh A (1994) Seroprevalence of bovine immunodeficiency virus (BIV) antibodies ion cattle population in Germany. Zentralbl Veterinarmed B 41(10):679–684

Nash JW, Hanson LA, St Cyr Coats K (1995a) Detection of bovine immunodeficiency virus in blood and milk-derived leukocytes by use of polymerase chain reaction. Am J Vet Res 56(4):445–449

Nash JW, Hanson LA, St Cyr Coats K (1995b) Bovine immunodeficiency virus in stud bull semen. Am J Vet Res 56(6):760–763

Oberste MS, Greenwood JD, Gonda MA (1991) Analysis of the transcription pattern and mapping of putative rev and *env* splice junctions bovine immunodeficiency like virus. J Virol 65:3932–3937

Oberste MS, Williamson JC, Greenwood JD, Nagashima K, Copeland TD, Gonda MA (1993) Characterization of bovine immunodeficiency virus rev cDNAs and identification and subcellular localization of the Rev protein. J Virol 67(11):6395–6405

Onuma M, Koomto E, Furuyama H, Yasutomi Y, Taniyama H, Iwai H, Kawakami Y (1992) Infection and dysfunction of monocytes induced by experimental inoculation of calves with bovine immunodeficiency like virus. J Acquir Immune Defic Syndr 5:1009–1015

Pallansch LA, Lackman Smith CS, Gonda MA (1992) Bovine immunodeficiency-line virus encodes factors which trans activate the long terminal repeat. J Virol 66:2647–2652

Patil SS, Pattanaik B, Mishra N, Banumathi N, Dubey R, Pradhan HK (2003) Detection of proviral genomic sequence of bovine immunodeficiency virus in Indian cattle. Curr Sci 84:563–566

Polack B, Schwartz I, Berthelemy M, Belloc C, Manet G, Wuillaume A, Baron T, Gonda MA, Levy D (1996) Serologic evidence for bovine immunodeficiency virus infection in France. Vet Microbiol 48(1–2):165–173

Rovid AH, Carpenter S, Miller LD, Flaming KP, Long MJ, Van der Maaten MJ, Frank DE, Roth JA (1996 Jul) An atypical T-cell lymphosarcoma in a calf with bovine immunodeficiency-like virus infection. Vet Pathol 33(4):457–459

Snider TG, Hoyt PG, Jenny BF, Coats KS, Luther DG, Storts RW, Battles JK, Gonda MA (1997) Natural and experimental bovine immunodeficiency virus infection in cattle. Vet Clin North Am Food Anim Pract 13:151–176

St Cyr Coats K (1995 Mar 18) Dual infection with bovine immunodeficiency virus and bovine leukaemia virus in Mississippi dairy cattle. Vet Rec 136(11):269–270

St-Louis MC, Abed Y, Archambault D (2005) The bovine immunodeficiency virus: cloning of a tat/rev cDNA encoding a novel Tat protein with enhanced transactivation activity. Arch Virol 150:1529–1547

Suarez DL, Whetstone CA (1998) PCR diagnosis of the bovine immunodeficiency-like virus. Methods Mol Biol 92:67–79

Suarez DL, Van Der Maaten MJ, Wood C (1993) Isolation and characterization of new wild type isolates of bovine lentivirus. J Virol 67:5051–5055

Suarez DL, Van der Maaten MJ, Whetstone CA (1995) Improved early and long-term detection of bovine lentivirus by a nested polymerase chain reaction test in experimentally infected calves. Am J Vet Res 56(5):579–586

Van der Maaten MJ, Boothe AD, Seger CL (1972) Isolation of a virus from cattle with persistent lymphocytosis. J Natl Cancer Inst 49:1649–1657

Whetstone CA, Van Der Maaten MJ, Black JW (1990) Humoral immune response to the bovine immunodeficiency-like virus in experimentally and naturally infected cattle. J Virol 64:3557–3561

Zhang S, Wood C, Xue W, Krunkenberg SM, Minocha HC (1997a) Immune suppression in calves with bovine immunodeficiency virus. Clin Diagn Lab Immunol 4:232–235

Zhang S, Troyer DL, Kapil S, Zheng L, Kennedy G, Weiss M, Xue W, Wood C, Minocha HC (1997b) Detection of proviral DNA of bovine immunodeficiency virus in bovine tissues by polymerase chain reaction (PCR) and PCR *in situ* hybridization. Virology 236:249–257

Zheng L, Swanson M, Liao J, Wood C, Kapil S, Snider R, Louahin TA, Minocha HC (2000) Cloning of the bovine immunodeficiency virus *gag* gene and development at a recombinant protein based Enzyme-linked immunosorbent assay. Clin Diagn Lab Immunol 7:557–562

14 Lumpy Skin Disease (Knopvelsiekte, Pseudo-Urticaria, Neethling Virus Disease, Exanthema Nodularis Bovis)

Sameeh M. Abutarbush

14.1 Introduction

Lumpy skin disease (LSD) is an acute viral disease of cattle that is caused by lumpy skin disease virus, of the genus *Capripoxvirus*, within the subfamily *Chordopoxvirinae* of the family *Poxviridae* (Buller et al. 2005). The LSDV prototype strain is the Neethling virus and it has only one serotype (Alexander et al. 1957). The virus has close antigenic relationship to sheep pox and goat pox viruses, which are also in the same genus, and the three viruses cannot be distinguished by routine virus neutralization or other serological tests (Radostits et al. 2007; Burdin 1959). The clinical syndrome of LSD was first described in Zambia (formerly Northern Rhodesia) in 1929 and was initially thought to be the result of poisoning or insect bite hypersensitivity (Mac-Donald 1931; Davies 1991). The disease is endemic in most African countries and some of the Middle East countries, with a recent incursion into other territories that were not known to have the disease before. Epizootics interspersed with periods of sporadic occurrence. LSD currently poses a serious threat to Europe and the rest of the world (Davies and Atema 1981; Davies 1991; Tuppurainen and Oura 2012).

LSD is associated with high morbidity and low mortality in cattle. It causes serious economic losses due to severe reduction in milk production, feed intake, and weight conversion. Furthermore, it causes abortion, infertility, and damage to cattle hide. LSD is considered a notifiable disease, and in affected countries, it results in serious restrictions to international trade (Davies 1991; Tuppurainen and Oura 2012). The incubation period is reported to be between 1 and 4 weeks (Haig 1957;

S.M. Abutarbush
Department of Veterinary Clinical Sciences, Faculty of Veterinary Medicine, Jordan University of Science and Technology, Irbid 22110, Jordan
e-mail: sameeh75@hotmail.com

J. Bayry (ed.), *Emerging and Re-emerging Infectious Diseases of Livestock*,
DOI 10.1007/978-3-319-47426-7_14

Coetzer 2004). Not all affected animals show clinical sings. High fever, lachrymation, and enlarged palpable lymph nodes are the first clinical signs reported. Soon after the fever, few to several variable-sized cutaneous nodules appear on different regions of the body. The entire body of the animal can be covered with nodules, and lesions may be seen in the mouth and nose as well as the mucous membranes of the eye in affected animals (Haig 1957; Coetzer 2004; Babiuk et al. 2008).

In postmortem, LSD lesions may be seen throughout the gastrointestinal and respiratory tracts. Pox lesions may be widespread and seen in each organ. In some cases limb edema can be seen in one or more limbs (Babiuk et al. 2008).

Diagnosis of LSD is based on characteristic signs, histopathology, and virus isolation as well as polymerase chain reaction (PCR) (Davies 1991; Tuppurainen and Oura 2012).

In endemic regions, vaccination is the only effective method of control (Hunter and Wallace 2001). There are commercial vaccines that are currently used to control the disease. Broad-spectrum antibiotics are used to control secondary bacterial infections (Salib and Osman 2011).

14.2 Epidemiology

14.2.1 Occurrence

The disease used to be confined to sub-Saharan Africa (Radostits et al. 2007). Over the years, LSDV has spread out to most of the African countries. The disease was reported in Egypt (Ali et al. 1990) and Israel (Yeruham et al. 1994). Outbreaks have been reported in the Middle East since 1990 (Kuwait in 1991, Lebanon in 1993, Yemen in 1995, United Arab Emirates in 2000, Bahrain in 2003, Israel in 2006–2007, and Oman in 2010) (Tuppurainen and Oura 2012). In 2006, the disease was reported again in Egypt and Israel (ProMed report no. 20060421.1167 and no. 20060627.1786, 2006). In 2012, the disease reoccurred again in Israel (ProMed report no. 20120729.1219223, 2012). In 2013, the disease spread in the region and was reported in Lebanon (ProMed report no. 20130118.1505118, 2013), Palestinian authority (ProMed report no. 20130311.1581763, 2013), Jordan (Abutarbush et al. 2013), Turkey (ProMed report no.20130918.1951299, 2013; Wainwrigh et al. 2013), Iraq (ProMed report no.20131128.2079065, 2013; Al-Salihi and Hassan 2015), and Egypt (ProMed report no.20131213.2111947,2013). In 2014, the disease was reported in Iran (ProMed report no.20140623.2561202, 2014), Azerbaijan (ProMed report no.20140719.2621294, 2014), and Cyprus (ProMed report no.20141209.3022793, 2014). In 2015, the disease was reported in Kuwait (ProMed report no. 20150105.3072591, 2015), Saudi Arabia (ProMed report no.20150430.3333997, 2015), Greece (ProMed report no. 20150821.3594203, 2015), and Russia (ProMed report no. 20150904.3622855, 2015). The disease continues to spread into new territories, and current methods of control and prevention do not appear to be efficient enough to stop it.

14.2.2 Morbidity, Mortality, and Case Fatality Rates

Generalized severe infections and a high mortality are seen in some field outbreaks. Other outbreaks are associated with few obviously affected animals and no deaths. In general, outbreaks are more severe with the initial introduction of the infection to a region and then abate, probably associated with the development of widespread immunity (Radostits et al. 2007). Morbidity, mortality, and case fatality rates are influenced by many factors including the immune status of the affected cattle and the abundance of the vectors (Thomas and Mare 1945; Tuppurainen and Oura 2012).

Morbidity rates of 1–2 % may reach 80–90 % in different situations (Davies 1991). Morbidity rates reach 80 % during epizootics, but are close to 20 % in enzootic areas. Morbidity rates of 31 % and 10 % were recorded in some outbreaks in Egypt and in 1994 in Israel, respectively. Much lower morbidity rate and milder disease were seen in Kenya (Radostits et al. 2007). Mortality rates of 10–40 % and even higher may be reported on occasion, but the much lower range of 1–5 % is more usual (Davies 1991). In natural outbreaks, the morbidity and mortality rates were reported to range from 3 to 85 % and from 1 to 3 %, respectively (Thomas and Mare 1945; Coetzer 2004). In recent outbreaks, the overall morbidity and mortality rates in Jordan were 26 % and 1.9 %, respectively (Abutarbush et al. 2013). Case fatality rate average is 2 %, but varies with the outbreak. The case fatality rate in the Jordanian outbreak was 7.5 %. In disease outbreak in Egypt, the case fatality rate was reported to be 1.8 % (Salib and Osman 2011). In Israel in 1994, no direct mortality from the disease was reported.

The disease has been reappearing in South Africa in the past decade probably due to higher rainfall and a decrease in the use of vaccination due to previous low incidence in the area (Hunter and Wallace 2001).

14.2.3 Origin of Infection and Transmission

Cattle can be infected by drinking water, but ingestion and direct contact transmission are not common routes, even though the virus is present in nasal and lacrimal secretions, semen, and milk of infected animals. Experimental seminal transmission of LSDV in cattle has been reported recently (Annandale et al. 2014). Although not commonly seen, the disease can be transmitted by direct contact (cutaneous lesions, saliva, respiratory secretions, milk, and semen) and using of contaminated needles (Davies 1991; Hunter and Wallace 2001). Most cases are believed to result from transmission by an arthropod vector. Historically, LSD virus has been isolated from *Stomoxys calcitrans* and *Musca confiscata* and transmitted experimentally using *S. calcitrans*, but other vectors are also suspected including *Biomyia*, *Culicoides*, *Glossina*, and *Musca* spp. However, in a recent study, despite the detection of virus in mosquitoes (*Anopheles stephensi*, *Culex quinquefasciatus*), the stable fly, and a biting midge (*Culicoides nebeculosis*) after they had fed on cattle with lumpy skin disease, the infection did not transmit to susceptible cattle when these arthropods were allowed to refeed on them (Chihota et al. 2003).

Although the specific vector of the disease was not confirmed yet, there is a strong evidence supporting LSD to be mechanically transmitted by arthropods. Lumpy skin disease virus was detected in saliva samples of both *Amblyomma hebraeum* and *Rhipicephalus appendiculatus* ticks (Lubinga et al. 2013). Previous studies showed successful transmission by *Aedes mosquitoes* (Chihota et al. 2001) and by African brown ear ticks (*Rhipicephalus appendiculatus Neum.*) (Tuppurainen et al. 2012). Another study showed evidence of transstadial and transovarial transmission of LSDV by *Rhipicephalus (Boophilus) decoloratus* ticks and mechanical or intrastadial transmission by *Rhipicephalus appendiculatus* and *Amblyomma hebraeum* ticks (Tuppurainen et al. 2011). Further evidence of mechanical/intrastadial and transstadial transmission of LSDV by *Amblyomma hebraeum* implicating *A. hebraeum* as a possible maintenance host in the epidemiology of the disease was reported (Lubinga et al. 2015).

Some more evidence is derived from a study which found that direct contact between infected and susceptible cattle in an insect-free environment does not lead to infection (Carn and Kitching 1995). Moreover, a recent study used a mathematical model to analyze the development of an LSD outbreak that took place at a dairy farm in Israel was reported. The model described the different forms of contact between the cattle groups and concluded that the virus spread could be attributed mostly to indirect transmission via insect vectors (Magori-Cohen et al. 2012).

14.2.4 Risk Factors

14.2.4.1 Animal Risk Factors

All ages and types of cattle are susceptible to the causative virus, except animals recently recovered from an attack, in which case there is a solid immunity lasting for about 3 months. In outbreaks, very young calves, lactating and malnourished cattle develop more severe clinical disease (Hunter and Wallace 2001). Introduction of a new animal to the herd, herd size, and utilization of communal grazing and watering points were found to be risk factors for contracting the disease in a recent study (Hailu et al. 2014).

Bos taurus cattle, imported from Europe, are far more susceptible than the indigenous *Bos indicus* cattle (Davies 1991). British breeds, particularly *Channel Island* breeds, are much more susceptible than zebu types, both in numbers affected and the severity of the disease. In an Ethiopian study (Gari et al. 2011), the annual cumulative incidence of LSD infection in Holstein Friesian/crossbred (33.93 %) was significantly higher than that of local zebu cattle (13.41 %). Also, the annual mortality was also significantly higher in Holstein Friesian/crossbred (7.43 %) than in local zebu cattle (1.25 %) (Gari et al. 2011).

Wildlife species are not affected in natural outbreaks, although there is a concern that they might be reservoir hosts. Typical skin lesions, without systemic disease, have been produced experimentally with Neethling virus in sheep, goat, giraffe, impala, and Grant's gazelle, but wildebeest were resistant. Serological evidence of naturally acquired infection has been observed only in African buffalo (*Syncerus caffer*) (Davies 1982). Natural occurrence of LSD has also been reported in water buffalo (*Bubalis*).

14.2.4.2 Environmental Risk Factors

Outbreaks tend to follow waterways, and extensive epizootics are associated with high rainfall and concomitant high levels of insect activity with a peak of disease in the late summer and early autumn (Hunter and Wallace 2001).

14.2.4.3 Pathogen Risk Factors

Capripox viruses are generally resistant to drying and survive freezing and thawing. Resistance to heat is variable but most are inactivated at temperatures above 60 °C. LSDV has shown to be very resistant to physical and chemical agents. In necrotic skin, the virus persists for at least 33 days. At ambient temperature, the virus remains viable in lesions in air-dried hides for at least 18 days (Weiss 1968a, b). From intact skin nodules kept at −80 °C, the virus was recovered for 10 years (Weiss 1968a, b). It is stable at pH range between 6.6 and 8.6 and sensitive to chloroform and ether, readily inactivated by dodecyl-sulfate (Weiss 1968a, b; Plowright and Ferris 1959).

14.2.4.4 Experimental Production

Experimental transmission can be accomplished using ground-up nodular tissue and blood or tissue culture virus. Disease is produced following intranasal, intradermal, or intravenous challenge (Prozesky and Barnard 1982). While LSD is characterized by generalized nodular skin lesions, less than 50 % of natural or experimental infections develop generalized skin nodules, and the length of the viremic period does not correlate with the severity of the clinical disease (Radostits et al. 2007). A recent study has shown experimental seminal transmission of LSDV in cattle (Annandale et al. 2014).

14.2.4.5 Economic Importance

Economic loss associated with LSD outbreaks is high. Although the disease is not associated with high mortalities, it results in great economic losses due to decreased feed intake, milk production, weight conversion, abortion and infertility, and damaged cattle hides. Secondary mastitis predisposed by the development of lesions on the teats may occur. In addition, this disease is an important notifiable disease and affects international trade (Abutarbush et al. 2013). Lumpy skin disease was proven to be one of the high-risk diseases for spread out of Africa to the outside world and is a potential agent of agroterrorism.

In an outbreak in Jordan, the cost of supportive antibiotic treatment ranged from 0 to 84.3 (mean = 27.9) British Pound/animal (Abutarbush et al. 2013). In another study, the total cost of treatment and losses per animal (including the total value of the animal in case of death) in the holding, as estimated by the owner, was 27–2210 £ (mean = 486) in non-vaccinated cattle and 0–2210 £ (mean = 78) in vaccinated cattle (Abutarbush 2014).

The financial cost of clinical LSD based on questionnaire survey distributed to livestock farmers, in Oromia region of Ethiopia, was studied (Gari et al. 2011). The annual financial cost included the average production losses, due to morbidity and mortality arising from milk loss, beef loss, traction power loss, and treatment and vaccination costs at the herd level. The financial cost in infected herds was estimated to be USD 6.43 (5.12–8) per head for local zebu and USD 58 (42–73) per head for Holstein Friesian/crossbred cattle (Gari et al. 2011).

14.3 Pathogenesis

Viremia accompanied by a febrile reaction and localization in the skin occurs with development of inflammatory nodules in generalized disease (Radostits et al. 2007). Local lesions can develop at the site of challenge, without viremia and generalization of the infection, in the experimental disease after intradermal inoculation, (Prozesky and Barnard 1982).

14.4 Clinical Findings

LSD is an acute infectious disease of cattle of all ages. There have been five instances of clinical cases of LSD in *Bubalus bubalis*, the Asian water buffalo (Ali et al. 1990). No other domestic ruminant species becomes infected naturally during field outbreaks. The incubation period is reported to be between 1 and 4 weeks (Haig 1957; Coetzer 2004). An incubation period of 2–4 weeks is common in field outbreaks and 7–14 days following experimental challenge (Carn and Kitching 1995). In severe cases there is an initial rise of temperature (40–41.5 °C), which lasts for over a week, sometimes accompanied by depression, lacrimation, nasal discharge, salivation, and a reluctance to move. Shortly afterward, the characteristic skin lumps develop. Multiple nodules appear suddenly about a week later, the first ones usually appearing in the perineum. They are intradermal, round, circumscribed, and firm. In most cases, they are confined to the skin area, varying from 5 to 50 mm in diameter, and are flattened, and the hair on them stands on end. They vary in number from a few to hundreds (Fig. 14.1). They may cover the whole body or be restricted to the head, neck, perineum, udder, genitalia, or limbs (Figs. 14.2 and 14.3). The distribution of the lesion, in a decreasing order, is seen on the body, neck, perineum, udder, head, vulva, and mouth (Abutarbush et al. 2013). They are firm and slightly raised above the surrounding normal skin from which they are often separated by a narrow ring of hemorrhage (Fig. 14.4). The lesions are of full skin thickness involving the epidermis, dermis, and adjacent subcutis. Nodules may disappear, but they may persist as hard lumps or become moist, necrotic, and slough (Radostits et al. 2007). Lesions may harden and remain in situ. They may also slough away to leave a hole of full skin thickness and characteristic lesion of "inverted conical zone" of necrosis, also known as "sitfast" (Abutarbush et al. 2013) (Fig. 14.5).

The regional superficial lymph nodes are enlarged and edematous. Nasal and oropharyngeal secretions are seen and may be associated with the development of lesions on the muzzle and in the mouth (Davies 1991) (Fig. 14.6). Lesions may be

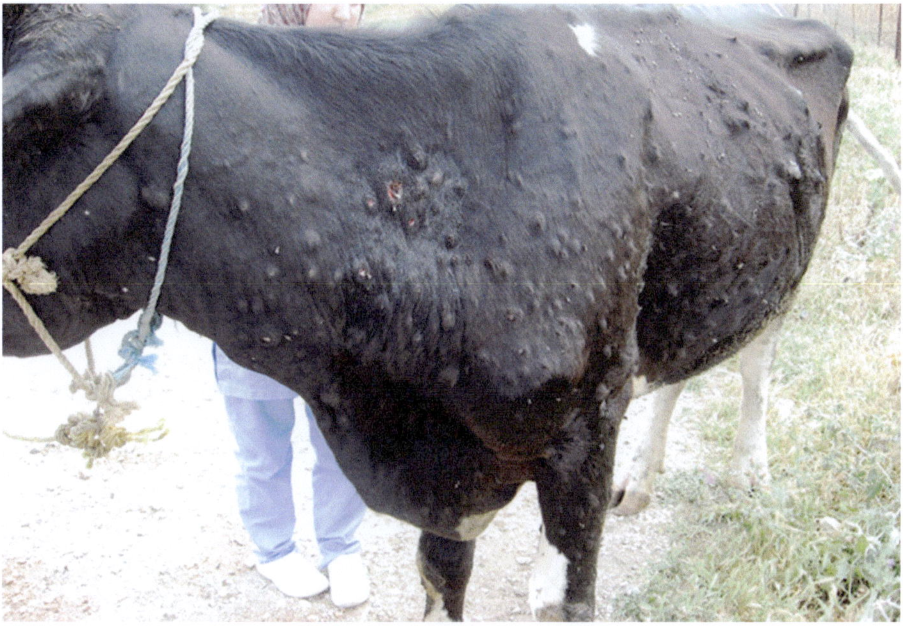

Fig. 14.1 Characteristic skin lumps in cattle infected with LSDV

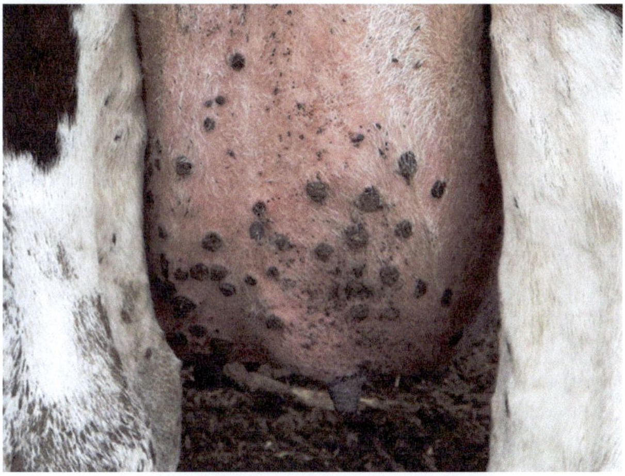

Fig. 14.2 Skin nodules in the udder region of cattle infected with LSDV

found anywhere in the oropharynx, in the upper respiratory tract, throughout the alimentary tract, in the subcutis, in the muscle fascia, and in the muscle itself. The lesions on the mucous epithelium are round (usually 4–40 mm in diameter) and have a ring shape and separated from the normal tissue (Davies 1991). Necrosis follows quickly and the ulcers become infected. Other manifestations that may be observed in severe cases include respiratory obstruction and snoring (Radostits et al. 2007). Conjunctivitis and keratitis may be seen.

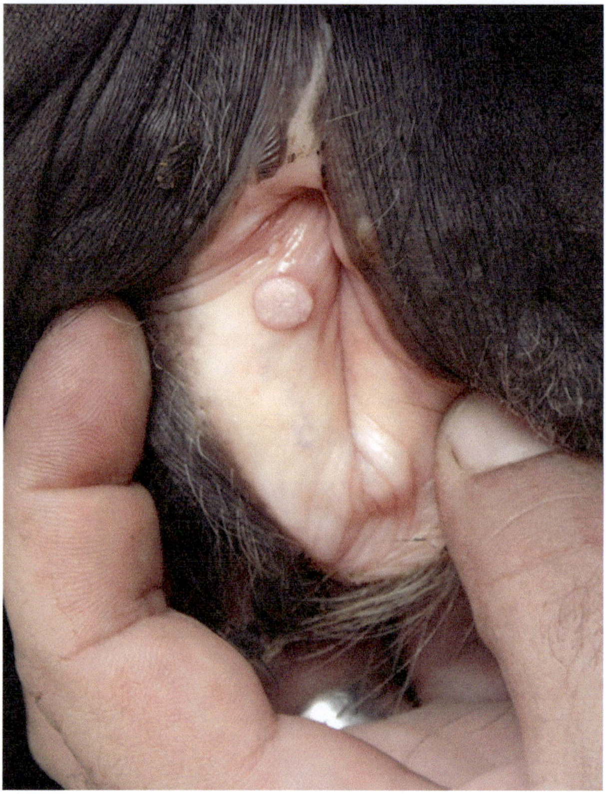

Fig. 14.3 Skin nodules in the genitalia of cattle infected with LSDV

Mouth lesions may interfere with feeding and dehydration and starvation may be seen in some cases. Milk production is reduced and may cease, and udder and teat lesions may result in serious infections with the sloughing of necrotic tissue and mastitis (Davies 1991; Abutarbush et al. 2013). Edematous and inflammatory swellings of the brisket and of one or more limbs may be seen and can severely restrict movement (Fig. 14.7). The sheath of bulls can be similarly affected and may interfere with their ability to serve for many weeks. In addition, estrus may be suppressed during the periods of severe debility (Davies 1991). Secondary bacterial infections and myiasis can develop at the lesion site (Abutarbush et al. 2013) (Fig. 14.8). Pneumonia is a common sequel in animals with lesions in the mouth and respiratory tract. Nodular lesions may have quite extensive surrounding areas of interstitial pneumonia in the lung, and inhalation pneumonia frequently occurs (Radostits et al. 2007; Davies 1991). Pregnant cows may abort, and calves have been born with extensive skin lesions, presumably acquired by intrauterine infection (Radostits et al. 2007; Davies 1991).

Emaciation can be seen and affected cattle may require humane euthanasia. Debility status remains for 1–3 months and sometimes for up to 6 months (Davies 1991). A convalescence of 4–12 weeks is usual (Radostits et al. 2007).

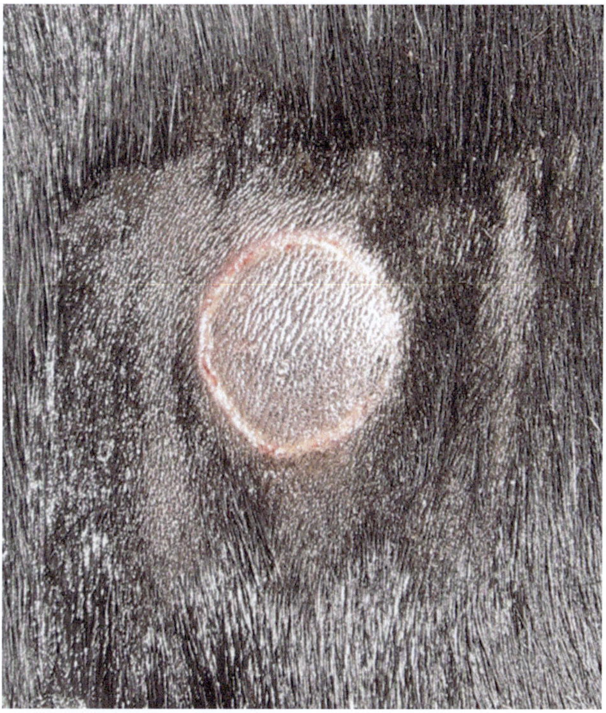

Fig. 14.4 A narrow ring of hemorrhage that separates normal skin from LSD lesion

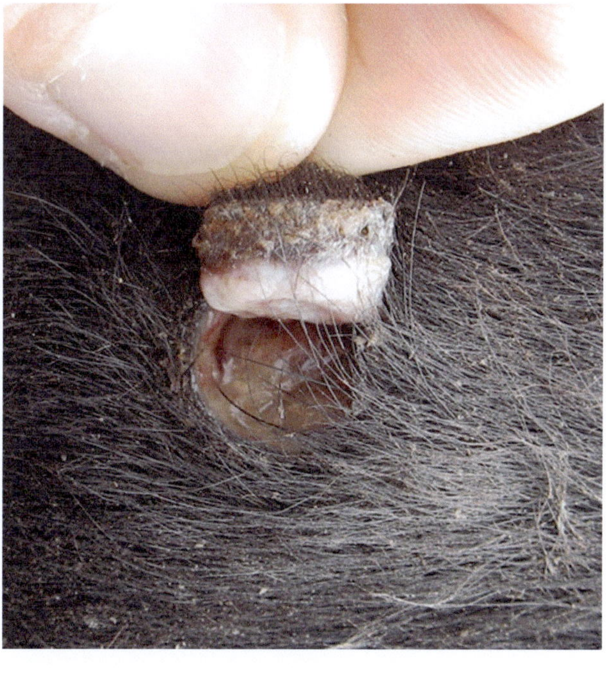

Fig. 14.5 Lesion of "inverted conical zone" of necrosis ("sitfast") on the LSDV-infected cattle skin

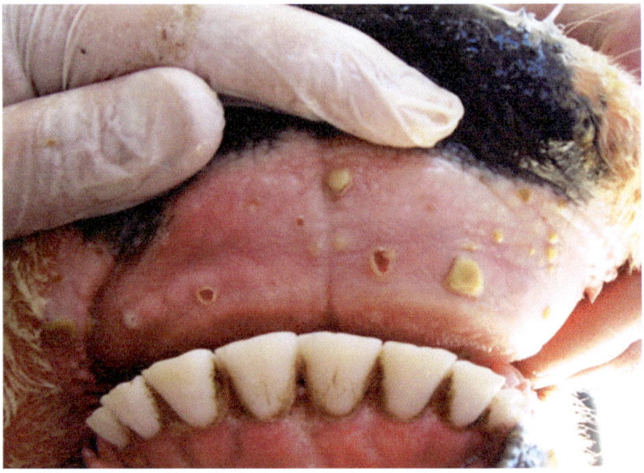

Fig. 14.6 Lesions on the muzzle and mouth of LSDV-infected cattle

Fig. 14.7 Edematous and inflammatory swellings of the brisket and of limbs of LSDV-infected cattle

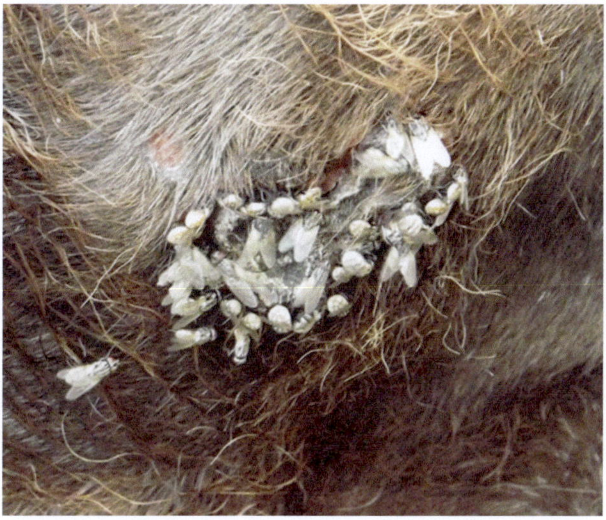

Fig. 14.8 Myiasis of skin lesion in LSDV-infected cattle

14.5 Clinical Pathology

14.5.1 Hematology and Serum Biochemical Analysis

Hematological and serum biochemical findings associated with natural clinical infection of LSD in cattle were recently studied and described (Abutarbush 2015). LSD was reported to be associated with inflammatory leukogram, anemia, thrombocytopenia, hyperfibrinogenemia, hyperproteinemia, decreased creatinine concentration, hyperchloremia, and hyperkalemia. These abnormalities were attributed to the associated severe inflammatory process and disease complications such as anorexia and reduced muscle mass (Abutarbush 2015).

14.5.2 Antigen Detection and Serology

The gold standard methods for the detection of capripox viral antigen and antibody are electron microscopy examination and serum/virus neutralization tests, respectively (Tuppurainen et al. 2011). Typical capripox virions in full thickness skin biopsies or scabs can be seen by electron microscopic examination (Radostits et al. 2007). The clinical diagnosis of LSD can be confirmed using conventional or real-time PCR methods (Tuppurainen et al. 2005; Orlova et al. 2006; Zheng et al. 2007; Tuppurainen et al. 2011). When compared to real-time PCR, gel-based PCR is more time and labor consuming. However, it is a cheap, reliable method and useful in countries with limited resources (Tuppurainen et al. 2011). PCR and immunohistochemistry can be used for the diagnosis of LSD in formalin-fixed paraffin-embedded tissue samples from skin nodules and lymph nodes of affected cattle (Awadin et al. 2006).

In a study to compare the different diagnostic tests in experimentally infected cattle (Tuppurainen et al. 2005), the incubation period in infected animals varied from 4 to 5 days. The PCR was a fast and sensitive method in demonstrating viral DNA in blood and skin samples. Although sensitive and reliable, virus isolation from blood and skin samples was too time consuming to use. However, it is required if infectivity of the LSD virus is to be determined. Virus isolation was successful in detecting viremia from 1 to 12 days, while PCR was successful from 4 to 11 days. The latter could demonstrate viral DNA until day 92 postinfection. The virus was isolated from skin biopsies until 39 days postinfection (Tuppurainen et al. 2005).

Virus can be cultivated from lesions. Antigen can also be detected by antigen detection ELISA with samples taken early in the course of the disease before the development of neutralizing antibodies and by fluorescent antibody tests and PCR. The AGID test can be used but the antigen will also react with parapox virus (Radostits et al. 2007; Tuppurainen et al. 2005).

The host immunity against LSDV is mainly cell mediated, and, therefore, serological testing may not be sensitive enough to detect mild and long-standing infections or antibodies in vaccinated animals (Kitching et al. 1987). Antibody ELISAs have been developed with limited success (Tuppurainen et al. 2011). Indirect fluorescent antibody test can also be used (Gari et al. 2008). AGID tests may be associated with false-positive reactions due to cross-reaction with other viruses such as bovine papular stomatitis and pseudocowpox viruses (Radostits et al. 2007).

14.6 Necropsy Findings

Typical LSD lesions, described previously in the clinical signs section, are seen on the skin, mouth, pharynx, trachea, skeletal muscle, bronchi and stomachs, and there may be accompanying pneumonia (Radostits et al. 2007). Superficial lymph nodes are usually enlarged. Respiratory obstruction by the necrotic ulcers and surrounding inflammation in the upper respiratory tract and/or concurrent aspiration pneumonia may be seen. Widespread vasculitis reflects the viral tropism for endothelial cells is seen histologically. Microscopic examination of the lesion of affected cattle with LSD reveals a granulomatous reaction in the dermis and hypodermis. Intracellular, eosinophilic inclusion bodies can be seen in the earlier acute stages of the disease. Intracytoplasmic viral inclusion bodies may be seen in a variety of cell types (Prozesky and Barnard 1982; Radostits et al. 2007).

14.7 Differential Diagnosis

The clinical signs, rapid spread of the disease, and the sudden appearance of lumps on the skin after an initial fever are characteristics for LSD infection (Radostits et al. 2007). However, pseudo-lumpy skin disease, also known as Allerton virus infection and general infection of cattle with bovine herpesvirus-2, the agent of bovine mammillitis, can be confused with LSD. Pseudo-lumpy skin disease causes circular, up to 2 cm in diameter, multifocal cutaneous lesions involving the superficial layer only and are distributed over the body. These lesions are associated with loss of hair, an intact central area, and raised edges. Sometime, the lesions show a circular ring of necrosis around a central scab, which falls off leaving discrete hairless lesion, which may be depigmented. The duration of the disease is approximately 2 weeks with usually no mortality (Radostits et al. 2007).

Streptotrichosis (*Dermatophilus congolensis* infection), ringworm, *Hypoderma bovis* infection, photosensitization, insect bites, urticaria, bovine papular stomatitis, foot and mouth disease, bovine viral diarrhea, and malignant catarrhal fever are all considered as differential diagnosis of LSD.

14.8 Treatment

Treatment is mainly symptomatic and supportive and no specific treatment is available. The use of antibiotics to prevent secondary bacterial infection is highly recommended (Abutarbush et al. 2013).

14.9 Control

Lumpy skin disease is usually introduced into new territory mainly by the movement of infected cattle or possibly by wind-borne vectors (Yeruham et al. 1995). Further spread is suspected to be via an insect vector. Cattle movement control from uninfected to infected territories is a critical control measure. Beyond that, vaccination is the main control method (Radostits et al. 2007).

Members of the capripox virus family are known to provide cross protection. Live attenuated sheep pox, goat pox, and LSDV vaccines can all be used to protect cattle against LSD infection (Kitching 1983). Available commercial vaccines are live attenuated. Live attenuated capripox virus (CaPV) vaccine strains that are used for cattle to control LSD include lumpy skin disease virus (LSDV) Neethling strain, Kenyan sheep and goat pox virus (KSGPV) O-240 and O-180 strains, Yugoslavian RM65 sheep pox(SPP) strain, Romanian SPP, and Gorgan goat pox (GTP) strains (Kitching et al. 1987; Kitching 2003). In countries previously free of LSD and use sheep pox vaccine to protect sheep against sheep pox, it is recommended to use the same vaccine in LSD outbreaks, because of potential safety issues associated with the live attenuated LSDV vaccine use (Tuppurainen and Oura 2012).

Field trials of the vaccines used in LSD prevention are few in the literature, and their efficacy has been questioned (Brenner et al. 2009; Ayelet et al. 2013). However, new reliable studies have been published recently. In a recent study to assess the value and efficacy of vaccination against a natural outbreak of LSD, epidemiological data were collected from 101 vaccinated and unvaccinated farms in Jordan (Abutarbush 2014). In the unvaccinated holdings, the overall morbidity rate was 42.6 %, mortality rate 10.2 %, and case fatality rate 23.9 %, compared to the overall morbidity rate of 4.7 %, mortality rate of 1 %, and case fatality rate of

22.9 % in vaccinated holdings. Decreased feed intake, decreased milk production, and fever were seen in 100 %, 76.9 %, and 92.3 % of the cattle farms in unvaccinated holdings, respectively, compared to decreased feed intake of 23.8 %, decreased milk production of 21.4 %, and fever of 23.8 % seen in vaccinated holdings. The percentage reduction in milk production in unvaccinated holdings ranged from 0 to 100 % (mean = 38.5 %, SE ± 9.6 %), and the total loss/animal in the farm ranged from £27 to £2210 (mean = 486, SE ± 162), compared to 0 to 100 % (mean = 6 %, SE ± 1.8 %) range of decrease in milk production and total loss/animal that ranged from 0 to 2210 £ (mean = 78, SE ± 29) in the vaccinated holdings (Abutarbush 2014).

In one study, 11.1 % of the cattle vaccinated with RM65 vaccine developed skin lesions after natural exposure to the disease. However, the number of affected cattle with clinical disease was five times higher compared with the unvaccinated cattle (Brenner et al. 2009). In another study, the authors concluded that Kenyan sheep pox vaccine strain used to protect cattle from LSDV infection did not confer good protection against LSDV infection. In this study 22.9 % of animals were clinically affected, while 2.31 % died of the disease (Ayelet et al. 2013). In a recent study three vaccines were evaluated (lumpy skin disease virus (LSDV) Neethling, Kenyan sheep and goat pox (KSGP) O-180 strain vaccines, and a Gorgan goat pox (GTP) vaccine). The Neethling and KSGPO-180 vaccines failed to protect vaccinated cattle against LSDV, whereas the Gorgan GTP vaccinated calves did not show clinical signs of LSD. In addition, incomplete protection of cattle vaccinated by live attenuated KSGP O-240 strain was reported during the 2006 outbreak in Egypt (Marshall 2006). In a randomized controlled field study to compare the efficacy of Neethling lumpy skin disease virus and x10 RM65 sheep pox live attenuated vaccines ($10^{3.5}$TCID50/dose) for the prevention of lumpy skin disease (Ben-Gera et al. 2015). However, enrolled cattle in both groups were vaccinated 2–5 months prior to study onset with a single dose of $10^{2.5}$TCID50 of RM65 attenuated sheep pox vaccine. LSD morbidity rate was significantly lower in the Neethling-vaccinated cattle when compared to that of the x10RM65 vaccinated cattle. In the same study, an incidence of 0.38 % of Neethling-associated disease was reported among Neethling-vaccinated cows, while no such disease occurred in x10RM65 vaccinated cows (Ben-Gera et al. 2015).

Cattle vaccinated with a recombinant capripox-rinderpest vaccine are immune to experimental challenge with both viruses but for a different length of time with each agent (Ngichabe et al. 2002).

Incomplete protection and adverse reactions have been associated with the use of capripox virus vaccines against lumpy skin disease virus (LSDV). The South African Onderstepoort Neethling vaccine strain administration causes a local reaction at the injection site (Weiss 1968a, b). However, no large-scale systemic or generalized reactions have been reported after the use of sufficiently attenuated LSDV vaccines. The Kenyan sheep and goat pox (KSGP) vaccine [also known as Kenya sheep-1 (KS-1)] was derived from the O-240 sheep isolate and is used against LSD. This vaccine has been associated with adverse clinical reactions in certain cattle species (Yeruham et al. 1994; Kitching 2003). The resulting clinical signs in dairy cattle were reported to be similar to those seen with natural LSD infection and included fever, skin lesions, and decreased milk production (Yeruham et al. 1994; Ayelet et al. 2013). However, this strain has been used effectively in sheep and goats in Kenya without severe or generalized reactions being observed (Kitching et al. 1987).

Currently available vaccines in the market against LSD include LSD vaccine by Onderstepoort Biological Products, South Africa (LSDV Neethling strain); Lumpyvax – Merck, Intervet, SA (attenuated LSDV field strain); Herbivac LS – Deltamune, South Africa (LSDV Neethling strain); Yugoslavian SPPV RM-65 (Jovac/Jordan and Abic/Israel) (10 × sheep dose); Bakirköy SPPV strain (Turkey, 3 to 4× sheep dose); and Romanian SPPV strain. KSGP O-240 and O-180 strains have been characterized as LSDV (Tulman et al. 2002; Le Goff et al. 2009; Davies 1976, Davies and Otema 1978), and vaccines using these strains are not recommended for cattle against LSDV until safety and efficacy have been tested using challenge experiments (Personal communication, Tuppurainen ES, FAO and EuFMD on line meeting, 2015).

In a recent field study to investigate the adverse reactions that were reported after vaccine administration, 63 dairy cattle farms, with a total of 19,539 animals, were included in the study (Abutarbush et al. 2014). Of those, 56 farms reported adverse clinical signs after vaccine administration, while the rest did not. The duration between vaccine administration and appearance of adverse clinical signs ranged from 1 to 20 days (mean = 10.3, SD ± 3.9). Reported clinical signs were similar to those observed in natural cases affected with LSD. Reported clinical signs were mainly fever, decreased feed intake, decreased milk production, and variable-sized cutaneous nodules (a few millimeters to around 2 cm in diameter) that were seen anywhere on the body (head, neck, trunk, perineum), udder, and/or teats. Nodules were raised and firmed initially and then formed dry scabs that could be peeled off the skin. The characteristic deep "sitfast" appearance was rarely seen, and most lesions were superficial. Some cattle had swollen lymph nodes, while a few pregnant animals aborted. The percentage of affected cattle ranged from 0.3 to 25 % (mean = 8, SD ± 5.1). Fever, decreased feed intake, and decreased milk production were seen in 83.9, 85.7, and 94.6 % in cattle on the affected farms, respectively. All affected cattle displayed skin nodules over their entire bodies, while 33.9 and 7.1 % of the affected farms reported nodular lesions present on the udder and teats, respectively. No mortalities were reported due to vaccine adverse reactions. Duration (course) of clinical signs ranged from 3 to 20 days (mean = 13.7, SD ± 4.1). In the same study, two types of LSD vaccines were used by the farmers: sheep pox virus (SPPV) vaccine derived from the RM65 isolate and a strain of LSDV vaccine with unknown source (Abutarbush et al. 2014).

Vaccination of cattle with sheep pox virus may result in a small percentage of cattle that develop granulomatous local reactions, but there is no spread of the sheep pox to sheep running with the cattle (Radostits et al. 2007). Vaccination of a herd at the start of an outbreak has limited efficacy as most animals will already be incubating the disease and poor needle hygiene in these circumstances may spread the disease (Radostits et al. 2007). However, in a recent study, a small number of farms were vaccinated after the onset of the LSD clinical signs (Abutarbush et al. 2014). In those, morbidity and mortality rates, percentage of decrease in milk production, duration of illness (days), and the total cost of treatment and losses per animal in the holding (£) were all lower than those reported for unvaccinated farms. This suggests that using vaccine after the appearance of clinical signs may have some benefits. However, the case fatality rate was similar in the two groups of farms, and abortion rate was almost six times higher in the vaccinated group after clinical signs compared with the unvaccinated group (Abutarbush et al. 2014).

Slaughter of affected and in-contact animals and destruction of contaminated hides, coupled with vaccination of at-risk animals, are used when the disease gains access to a previously free country (Radostits et al. 2007).

References

Abutarbush SM (2015) Haematological and serum biochemical findings associated with clinical cases, naturally infected with lumpy skin disease in cattle. J Infect Dev Ctries 9(3):283–288

Abutarbush SM (2014) Efficacy of vaccination against lumpy skin disease in Jordanian cattle. Vet Rec. pii: vetrec-2013-102271. doi:10.1136/vr.102271

Abutarbush SM, Hananeh WM, Ramadan W et al (2014) Adverse reactions to field vaccination against lumpy skin disease in Jordan. Transbound Emerg Dis. doi:10.1111/tbed.12257

Abutarbush SM, Ababneh MM, Al Zoubi IG et al (2013) Lumpy skin disease in Jordan: disease emergence, clinical signs, complications and preliminary-associated economic losses. J Transbound Emerg Dis. doi:10.1111/tbed.12177

Alexander RA, Plowright W, Haig DA (1957) Cytopathogenic agents associated with lumpy-skin disease of cattle. Bull Epizoot Dis Afr 5:489–492

Ali AA, Esmat M, Attia H et al (1990) Clinical and pathological studies on lumpy skin disease in Egypt. Vet Rec 127:549–550

Al-Salihi KA, Hassan IQ (2015) Lumpy skin disease in Iraq: study of the disease emergence. Transbound Emerg Dis 62(5):457–462

Annandale CH, Holm DE, Ebersohn K, Venter EH et al (2014) Seminal transmission of lumpy skin disease virus in heifers. Transbound Emerg Dis 61(5):443–448

Awadin W, Hussein H, Elseady Y et al (2006) Detection of lumpy skin disease virus antigen and genomic DNA in formalin-fixed Paraffln-embedded tissues from an Egyptian outbreak in 2006. Transbound Emerg Dis. doi:10.1111/j. 1865-1682

Ayelet G, Abate Y, Sisay T et al (2013) Lumpy skin disease: preliminary vaccine efficacy assessment and overview on outbreak impact in dairy cattle at DebreZeit, central Ethiopia. Antiviral Res 98:261–265

Babiuk S, Bowden TR, Parkyn G et al (2008) Quantification of lumpy skin disease virus following experimental infection in cattle. Transbound Emerg Dis 55:299–307

Ben-Gera J, Klement E, Khinich E et al (2015) Comparison of the efficacy of Neethling lumpy skin disease virus and x10RM65 sheep-pox live attenuated vaccines for the prevention of lumpy skin disease -the results of a randomized controlled field study. Vaccine 33(38):4837–4842

Brenner J, Bellaiche M, Gross E et al (2009) Appearance of skin lesions in cattle populations vaccinated against lumpy skin disease:statutory challenge. Vaccine 27:1500–1503

Buller RM, Arif BM, Black DN et al (2005) Poxviridae. In: Fauquet, CM, Mayo MA, et al (eds) Virus taxonomy: eight report of the International Committee on the Taxonomy of Viruses, Elsevier Academic Press, Oxford, pp 117–133

Burdin ML (1959) The use of histopathological examinations of skin materialfor the diagnosis of lumpy skin disease in Kenya. Bull Epizoot Dis Afr 7:27–36

Carn VM, Kitching RP (1995) An investigation of possible routes of transmission of lumpy skin disease virus (Neethling). Epidemiol Infect 114:219–226
Chihota CM, Rennie LF et al (2003) Attempted mechanical transmission of lumpy skin disease virus by biting insects. Med Vet Entomol 17:294–300
Chihota CM, Rennie LF, Kitching RP et al (2001) Mechanical transmission of lumpy skin disease virus by Aedes aegypti (Diptera.: Culicidae). Epidemiol Infect 126:317–321
Coetzer JAW (2004) Lumpy skin disease. In: Coetzer JAW, RC Tustin (eds) Infectious diseases of livestock, 2nd edn. University Press Southern Africa, Oxford, pp 1268–1276
Davies FG, Atema C (1981) Relationships of Capripox viruses found in Kenya with two Middle Eastern strains and some Orthopox viruses. Res Vet Sci 31:253–255
Davies FG (1991) Lumpy skin disease of cattle: a growing problem in Africa and the Near East. World Anim Rev 68:37–42
Davies FG (1976) Characteristics of a virus causing a pox disease in sheep and goatsin Kenya, with observations on the epidemiology and control. J Hyg 76:163–171
Davies FG, Otema C (1978) The antibody response in sheep infected with Kenyan sheep and goat pox virus. J Comp Pathol 88:205–210
Davies FG (1982) Observations on the epidemiology of lumpy skin disease in Kenya. J Hyg 88:95–102
Haig DA (1957) Lumpy skin disease. Bull Epizoot Dis Afr 5:421–430
Hunter P, Wallace D (2001) Lumpy skin disease in southern Africa: a review of the disease and aspects of control. J S Afr Vet Assoc 72:68–71
Lubinga JC, Tuppurainen ES, Mahlare R et al (2015) Evidence of transstadial and mechanical transmission of lumpy skin disease virus by Amblyommahebraeum ticks. Transbound Emerg Dis 62(2):174–182
Lubinga JC, Tuppurainen ES, Stoltsz WH et al (2013) Detection of lumpy skin disease virus in saliva of ticks fed on lumpy skin disease virus-infected cattle. Exp Appl Acarol 61(1):129–138
Gari G, Bonnet P, Roger F et al (2011) Epidemiological aspects and financial impact of lumpy skin disease in Ethiopia. Prev Vet Med 102:274–283
Gari G, Biteau-Coroller F, LeGoff C et al (2008) Evaluation of indirect fluorescent antibody test (IFAT) for the diagnosis and screening of lumpy skin diseaseusing Bayesian method. Vet Microbiol 129:269–280
Hailu B, Tolosa T, Gari G et al (2014) Estimated prevalence and risk factors associated with clinical Lumpy skin disease in north-eastern Ethiopia. Prev Vet Med 115(1–2):64–68
Kitching RP (1983) Progress towards sheep and goat pox vaccines. Vaccine 1:4–9
Kitching RP, Hammond JM, Taylor WP (1987) A single vaccine for the control of capripox infection in sheep and goats. Res Vet Sci 42:53–60
Kitching R (2003) Vaccines for lumpy skin disease, sheep pox and goat pox. Dev Biol (Basel) 114:161–167
Le Goff C, Lamien CE, Fakhfakh E et al (2009) Capripoxvirus G-protein-coupled chemokine receptor: a host-rangegene suitable for virus animal origin discrimination. J Gen Virol 90:1967–1977
MacDonald RAS (1931) Pseudo-urticaria of cattle. Annual Report for 1930. Department of Animal Health, Northern Rhodesia, pp. 20–21
Magori-Cohen R, Louzoun Y, Herziger Y et al (2012) Mathematical modelling and evaluation of the different routes of transmission of lumpy skin disease virus. Vet Res 43:1
MarshallM(2006)Availableat:http://www.promedmail.org/pls/apex/f?p=2400:1001:::NO::F2400_P1001_BACK_PAGE_F2400_P1001_PUB_MAIL_ID:1000_33506
Ngichabe CK, Wamwayi HM, Ndungu EK et al (2002) Long term immunity in African cattle vaccinated with a recombinant capripox-rinderpest virus vaccine. Epidemiol Infect 128:343–349
Orlova ES, Shcherbakov AV et al (2006) Differentiation of capripoxvirus species and strains by polymerase chain reaction. Mol Biol (NY) 40:139–145
Plowright W, Ferris RD (1959) Ether sensitivity of some mammalian poxviruses. Virology 7:357

Prozesky L, Barnard JH (1982) A study of the pathology of lumpy skin disease in cattle. Onderstepoort J Vet Res 49:167–175

Radostits OM, Gay CC et al (2007) Veterinary medicine. A textbook of the diseases of cattle, horses, sheep, pigs and goats, 10th edn. Saunders, Philadelphia, pp 1424–1426

Salib FA, Osman AH (2011) Incidence of lumpy skin disease among Egyptian cattle in Giza Governorate, Egypt. Vet World 4:162–167

Thomas AD, Mare CVE (1945) Knopvelsiekte. J S Afr Vet Assoc 16:36–43

Tulman E, Alfonso C, Lu Z et al (2002) The genomes of sheep pox and goat pox viruses. J Virol 76:6054–6061

Tuppurainen ESM, Lubinga JC, Stoltsz WH et al (2012) Mechanical transmission of lumpy skin disease virus by Rhipicephalus appendiculatus male ticks. Epidemiol Infect 141:425–430

Tuppurainen ESM, Stoltsz WH, Troskie M et al (2011) A potential role for ixodid (hard) tick vectors in the transmission of lumpy skin disease virus in cattle. Transbound Emerg Dis 58:93–104

Tuppurainen ES, Oura CA (2012) Review: lumpy skin disease: an emerging threat to Europe, the Middle East and Asia. Transbound Emerg Dis 59:40–48

Tuppurainen ESM, Venter EH, Coetzer JAW (2005) The detection of lumpy skin disease virus in samples of experimentally infected cattle using different diagnostic techniques. Onderstepoort J Vet Res 72:153–164

Tuppurainen ESM, Stoltsz WH, Troskie M, Wallace DB, Oura CAL, Mellor PS, Coetzer JAW, Venter EH (2011) A potential role for ixodid (hard) tick vectors in the transmission of lumpy skin disease virus in cattle. Transbound Emerg Dis 58:93–104

Yeruham I, Nir O, Braverman Y et al (1995) Spread of lumpy skin disease in Israeli dairy herds. Vet Rec 137(4):91–93

Yeruham I, Perl S, Nyska A et al (1994) Adverse reactions in cattle to a capripox vaccine. Vet Rec 135:330–332

Wainwrigh S, El Idrissi A, Mattioli R et al (2013) Emergence of lumpy skin disease in the Eastern Mediterranean Basin countries. EMPRES Watch. © FAO 2013. Available at http://www.fao.org/ag/empres.html

Weiss KE (1968a) Lumpy skin disease virus. Virol Monogr 3:111–131

Weiss WE (1968b) Lumpy skin disease. In emerging diseases of animals. FAO Agric Stud Bull 61:179–201

Zheng M, Liu Q, Jin NY et al (2007) A duplex PCR assay for simultaneous detection and differentiation of Capripoxvirus and Orf virus. Mol Cell Probes 21:276–281

Hepatitis E in Livestock

15

Marcelo Alves Pinto, Jaqueline Mendes de Oliveira, and Debora Regina Lopes dos Santos

15.1 Introduction

Hepatitis E, the main cause of enterically transmitted hepatitis in many developing countries, is now recognized as an emerging zoonotic disease worldwide (Meng 2013). The causative agent, hepatitis E virus (HEV), is a small, non-enveloped virus, with a single-stranded positive-sense RNA genome of approximately 7.2 kb organized in three open reading frames (ORF) (Tam et al. 1991). ORF1 at the 5' end of the genome encodes nonstructural proteins including a methyltransferase, a papain-like protease, a helicase, and an RNA-dependent RNA polymerase. ORF2 at the 3' end of the genome encodes the major viral capsid protein. ORF3, partially overlapping with ORF1 and ORF2, encodes a small regulatory phosphoprotein (Chandra et al. 2008).

HEV was first recognized as a new etiological agent of enterically transmitted hepatitis in the early 1980s and was associated to waterborne hepatitis epidemics that occurred in New Delhi (Viswanathan 1957) and in the Kashmir Valley of India (Khuroo 1980, 2011; Khuroo et al. 1983). Virus particles were first visualized by immune electron microscopy in the feces of an experimentally infected volunteer (Balayan et al. 1983). The virus was successfully transmitted to cynomolgus monkeys, and its genome was cloned and characterized in 1991 (Emerson and Purcell 2003; Tam et al. 1991).

M. Alves Pinto (✉) • J.M. de Oliveira
Laboratory of Technological Development in Virology,
Oswaldo Cruz Institute, Fiocruz, Rio de Janeiro, Brazil
e-mail: marcelop@ioc.fiocruz.br

D.R.L.d. Santos
Laboratory of Veterinary Viruses, Department of Microbiology and Immunology,
Veterinary Institute, Rural Federal University of Rio de Janeiro, Rio de Janeiro, Brazil

J. Bayry (ed.), *Emerging and Re-emerging Infectious Diseases of Livestock*,
DOI 10.1007/978-3-319-47426-7_15

Subsequent studies confirmed the existence of genetically related, but epidemiologically distinct, HEV variants. In 1995, HEV-specific antibodies were demonstrated in naturally infected swine from Nepal (Clayson et al. 1995). The first genetically characterized strain was from a swine in the USA (Meng et al. 1997). Until then, only human isolates from endemic areas have been characterized and classified into genotypes 1 and 2 (Reyes et al. 1990).

The discovery of swine as potential reservoirs for HEV led to a new perspective over hepatitis E epidemiological pattern, especially where the outbreaks occurred only as imported cases. Various studies developed in different endemic and non-endemic areas showed the enzootic pattern of HEV among swine populations, especially from commercial herds (Meng 2010; Thiry et al. 2014).

HEV was also detected and characterized from other mammals such as wild boars, cervids, cattle, camels, rabbits, rodents, musks, ferrets, mongooses, foxes, and bats (Bodewes et al. 2013; Drexler et al. 2012; Feng et al. 2011; Guan et al. 2013; Hu and Ma 2010; Johne et al. 2010; Li et al. 2014; Nakamura et al. 2006; Rutjes et al. 2010; Sonoda et al. 2004; Woo et al. 2014; Zhao et al. 2009). Also, the cutthroat trout virus (CTV), isolated in 1988 from trouts, was shown to be genetically related to HEV (Batts et al. 2011).

Additionally, HEV-related virus has been isolated from chickens and turkeys. A significant prevalence has been observed in chicken flocks in Europe, the USA, and Asia (Hsu and Tsai 2014; Peralta et al. 2009). In chickens, HEV has been associated to big liver and spleen disease (BLS) or hepatitis-splenomegaly syndrome (HS), ovarian regression, diarrhea, and other pathogeneses (Haqshenas et al. 2001). Phylogenetic analysis has shown that avian HEV is genetically related to, but distinct from, other known HEV strains described (Sun et al. 2004). Indeed, avian HEV infection was not successfully reproduced in nonhuman primates (Sun et al. 2004) (Table 15.1).

The discovery of new HEV or HEV-related viruses led to progressive increase in the number of genotypes and consequently consensus classification (Smith et al. 2014). According to the International Committee on Taxonomy of Viruses (ICTV) *Hepeviridae* Study Group, all HEV are classified into family *Hepeviridae* with two genera: *Orthohepevirus* with four species (A–D) and *Piscihepevirus* with a single species (A). In *Orthohepevirus* A, classified strains were found in humans, pigs, wild boar, rabbit, cervids, mongoose, and camels; *Orthohepevirus* B contains strains from chickens; *Orthohepevirus* C encompasses strains found in rats (HEV-C1) and ferrets (HEV-C2); and *Orthohepevirus* D contains bat strain. Also *Orthohepevirus* A is classified into seven genotypes. Strains in genotypes HEV-1 and HEV-2 are identified exclusively in humans, while genotypes HEV-3 and HEV-4 have been isolated from both humans and different animal species associated to zoonotic transmission. Genotypes HEV-5 and HEV-6 have been found in wild boar in Japan and genotype HEV-7 in dromedary camels in Dubai (Smith et al. 2014; Thiry et al. 2015). Recently, other isolates have been obtained from moose, fox, and mink and could represent new members of the *Hepeviridae* family.

Table 15.1 Distribution of HEV infection among livestock species

Species	Liver enzymes	Prevalence	Histological finds	Genotype	Country	Reference
Swine	Not done	20–100 % of growing and adult pigs	Not done	3	USA	Meng et al. (1997, 2002)
	Not done	15 %	Not done	3	Korea	Choi et al. (2003)
	Not done	75 %	Not done	3	France	Bouquet et al. (2011), Colson et al. (2010), Walachowski et al. (2014)
	Not done	Not done	Twenty-three HEV-infected pigs had diarrhea (51.1 %)	3	Italy	Bartolo et al. (2015), Monini et al. (2015)
	Not done	Not done	Not done	3	Democratic Republic of the Congo	Kaba et al. (2010)
	Not done	15 %	Not done	3	Japan	Takahashi et al. (2003a, b)
	Not done	54.6–74.4 %	Not done	4	India(Arankalle et al. 2001, 2006; Chobe et al. 2006)	
	Not done	7–15 %	Not done	3–4	The Netherlands	Bouwknegt et al. (2007), Hakze-van der Honing et al. (2011)
	Not done	Not done	Not done	3–4	Belgium	Bouamra et al. (2014), Hakze-van der Honing et al. (2011)
	Not done	10–32 %	Not done	2	Portugal	Berto et al. (2012a, b)
	Not done	25–32 % and 48.4 %	Mild to moderate hepatitis	Close relationship with human HEV strains/3	Spanish	Casas et al. (2009), de Deus et al. (2007), Jiménez de Oya et al. (2011), Pina et al. (2000)
	Not done	88.4 %	Necroinflammatory liver lesions, focal inflammation	3	Brazil	dos Santos et al. (2009)
	Not done	82.2 %	Not done	4	China	Geng et al. (2010), Wang et al. (2002)
	Not done	4–58 %.	Not done	3	Argentina	Munné et al. (2006)

(continued)

Table 15.1 (continued)

Species	Liver enzymes	Prevalence	Histological finds	Genotype	Country	Reference
Wild boar	Not done	8.1 %	Not done	3, 4, and unrecognized genotypes	Japan	Sato et al. (2011), Takahashi and Okamoto (2014)
	Not done	26 %	Not done	3	Portugal	Boadella et al. (2012), Mesquita et al. (2016)
	Not done	30.63–55.65 %	Not done	3	Spanish	de Deus et al. (2008b)
Bovine	Not done	10.4 % and 4.4–6.9 %	Not done	4	China	Arankalle et al. (2001), Hu and Ma (2010)
	Not done	1.42 %	Not done	Not done	Brazil	Vitral et al. (2005)
Sheep	Not done	35.20 %	Not done	4	China	Wang and Ma (2010), Wu et al. (2015)
Goat	Not done	16 %	Not done	HEV-related agent	USA	Sanford et al. (2013)
	Not done	6.3 and 28.2 %	Not done	Not detected	China	Geng et al. (2010), Shao et al. (2009), Wang et al. (2002)
Rabbits	ALT and AST elevation	not done	Multifocal lymphohistiocytic infiltrates and local hepatocellular necrosis	Rabbit HEV genotype	China	Ma et al. (2010)
	Not done	15.4 %	Not done	Rabbit HEV genotype	China	Geng et al. (2011a)
				Rabbit HEV genotype (distant member of the zoonotic genotype 3 HEV)	USA	Cossaboom et al. (2012)
Horse	AST elevation	13 %	Not done	1 (Accidental hosts for human HEV?)	Egypt	Saad et al. (2007)
Deer	Not done	absence	Not done	Not detected	Italy	Serracca et al. (2015)
	Not done	5.43 %	Not done	Not done	China	Zhang et al. (2015)
	Not done	13.9 %	Not done	HEV genotype 3	Italy	Di Bartolo et al. (2015)

Chickens	Not tested	71 % of chicken flocks and 30 % of chickens	Hepatitis-splenomegaly syndrome	Avian HEV genotype 2	USA	Haqshenas et al. (2001)
	Not tested	Not done	Normal animals	Avian HEV grouped together with European and Chinese isolates, and the aHEV18198 with Canadian ones	Russia	(Sprygin et al. 2012)
	Not done	Not done	Hepatitis-splenomegaly syndrome	Avian HEV genotype 1	Korea and Australia	Bilic et al. (2009)
	Not done	Not done	Hepatitis-splenomegaly syndrome	Avian HEV 1 and 3	Australia and Europe	Bilic et al. (2009), Marek et al. (2010)
	Not done	Not done	Hepatitis-splenomegaly syndrome	Avian HEV	USA	Billam et al. (2005)
	Not done	Not done	Hepatitis-splenomegaly syndrome	Avian HEV	Spain	Peralta et al. (2009)
	Not done	Not done	Hepatitis-splenomegaly syndrome	Avian HEV	Canada	Agunos et al. (2006)
	Not done	35.9–54.62 %	Not done	Avian HEV	China	Geng et al. (2011b), Zhao et al. (2013)
	Not done	Not done	Hepatitis-splenomegaly syndrome	Avian HEV	Hungary	Morrow et al. (2008)

15.2 Comparative Hepatitis E Virus Infection (Hepatitis E Virus Infection and Clinical Features)

In humans, acute HEV infection is clinically indistinguishable from hepatitis A and exhibits a wide spectrum of clinical manifestations that may vary from asymptomatic or subclinical to a fulminant outcome. However, in contrast with hepatitis A, some patients can be persistently infected and develop chronic viral hepatitis progressing to cirrhosis, as it has been shown in transplant recipients and in other immunosuppressive conditions such as human immunodeficiency syndrome (AIDS) and leukemia. (Halac et al. 2012; Kamar et al. 2010).

The potential for xenogeneic transmission of HEV from animals to humans via organ, tissue, or cellular transplantation or via ex vivo exposure of humans to porcine biologic products for medical purposes is of concern for the public health (Abrahante et al. 2011; Denner 2015; Meng 2003). Xenotransplantation allows viruses to bypass the normal immunological defense mechanisms of the recipient. Furthermore, the use of immunosuppressive drugs following transplantation may facilitate the xenogeneic transmission of swine HEV that does not cause any clinical symptoms in the natural host but can infect humans and cause hepatitis (Yoo and Giulivi 2000). Additionally, immunocompromised patients have higher likelihood of developing chronic hepatitis E and cirrhosis (Echevarría et al. 2015; Kamar et al. 2010; Neukam et al. 2013; Passos-Castilho et al. 2014).

15.3 Natural History of Hepatitis E in Swine

The presence of HEV throughout swine production farms worldwide is of concern to livestock industry, since HEV can be transmitted, either to humans or to other animals through direct contact with contaminated secretes, excretes, and organs and through the swine food chain. Indeed, HEV has been detected in several animal species (mainly swine), which are considered potentially natural hosts of this virus. However, there are very few studies evidencing clinical signs of viral hepatitis in naturally or experimentally HEV-infected pigs. Moreover, it has been proven, at least in experimental setting, that HEV can persistently circulate within a closed animal population (Van der Poel 2014). Additionally, urine and meat were identified as possible HEV sources for pig-to-pig and pig-to-human HEV transmission (Bouwknegt et al. 2009).

Regarding vertical transmission of HEV in pigs, it is known that colostrum-deprived piglets, born from anti-HEV-positive sows, did not show detectable serum markers of HEV infection (dos Santos et al. 2009) (Fig. 15.1a) or any evidence of vertical transmission of HEV to fetuses or clinical disease in the gilts or their offspring (Kasorndorkbua et al. 2003). In humans, on the contrary, HEV infection during pregnancy may induce respiratory distress syndrome in neonates, followed by preterm birth and signs of sepsis and hepatosplenomegaly (El Sayed Zaki et al. 2013) and fatal cases (Bonney et al. 2012; Khuroo and Kamili 2003). In pigs, the absence of anti-HEV IgG is probably due to the natural poor transplacental

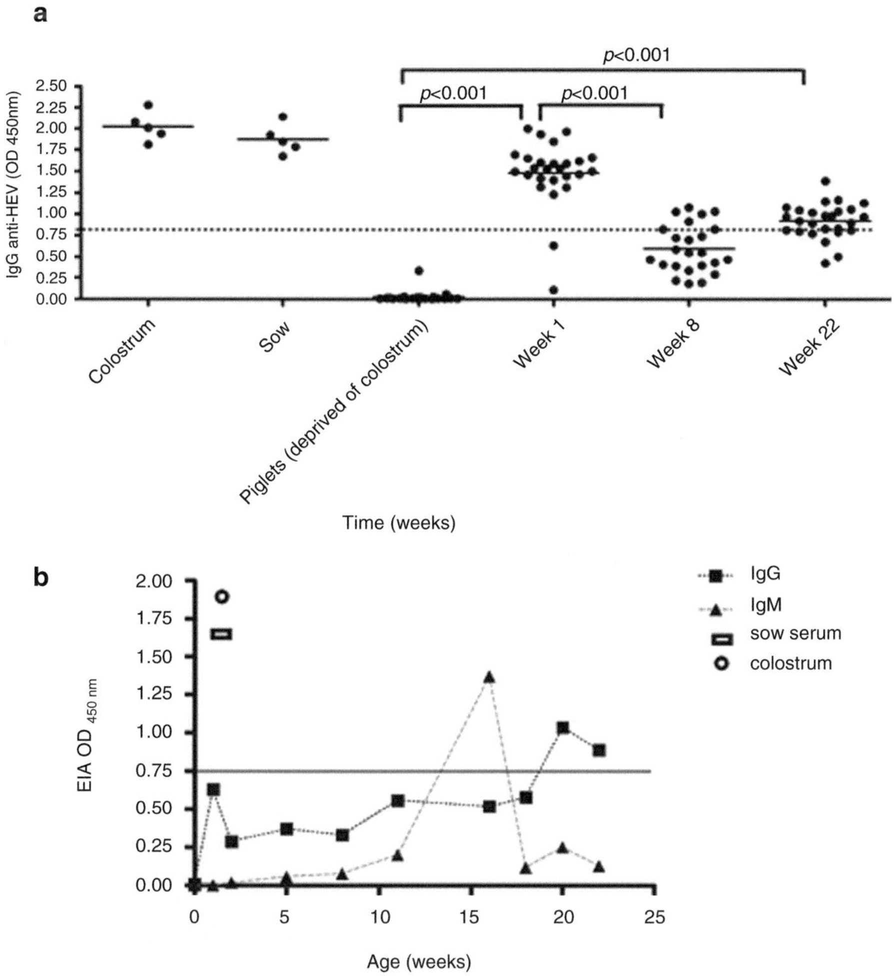

Fig. 15.1 (**a**) Distribution of OD values for the detection of HEV-specific IgG by EIA in colostrum and serum samples collected from each of five lactating sows before breastfeeding, as well as in serum samples of 26 colostrum-deprived piglets at postpartum week 1, and after colostrum intake at 8 and 22 weeks of age in study 1. Samples presenting OD_{450} values greater than 0.757 (*dashed line*) were considered reactive for anti-HEV. P-values under value 0.001 are statistically significant. Week 1 samples were collected 24 h after birth and colostrum uptake. (**b**) HEV-specific IgM/IgG kinetics by EIA in piglet #12. The *solid line* indicates the cutoff value. IgM OD value above the cutoff indicates recent HEV infection. IgG OD values from Piglet #12 respective sow and colostrum are indicated

antibody transfer during sow pregnancy (Salmon et al. 2009). The intact antibodies are transferred into the circulation. Enteric absorptive cells take up antibodies by endocytosis and transport across the epithelium in vacuoles (Kraehenbuhl and Campiche 1969). Finally, newborns are very susceptible to HEV infection (Andraud et al. 2014) and other viruses soon after delivery. Immediately after colostrum intake, elevated anti-HEV IgG levels are detected in the blood, because newborns present elevated enteric permeability at 24–36 h after birth and are detectable in serum for 4–5 weeks (Salmon et al. 2009; Wagstrom et al. 2000). Other authors have suggested that maternal antibodies delay the onset of genotype 3 HEV viremia and seroconversion in piglets (Kanai et al. 2010).

After weaning, between 10 and 13 weeks of age, the HEV antibody levels are decreased in the peripheral blood, and the piglets become susceptible to HEV infection again. Immunologically mature animals are naturally infected by pig-to-pig contact in herd and become highly reactive to HEV (Fig. 15.1a) (Andraud et al. 2013; Bouwknegt et al. 2009). Increased IgM titers are detectable between 11 and 17 weeks (Fig. 15.1b). The prevalence of HEV reaches 97.3 % in animals older than 25 weeks (slaughterhouse age) (dos Santos et al. 2009; Feng et al. 2011). Despite elevated serum titers of anti-HEV IgG at slaughterhouse age, HEV Ag or RNA has been currently detected in liver samples by other authors (Ahn et al. 2005; Ha and Chae 2004; Kanai et al. 2010) including liver samples sold as food in butcher shops and grocery stores (Gutiérrez-Vergara et al. 2015; Okano et al. 2014a, b; Wenzel et al. 2011). The elevated titers of anti-HEV IgG may reduce the viremia as demonstrated by Feng et al. (2011). However, there is evidence that circulating antibodies may not clear HEV Ag from liver parenchyma. A partial protection has been observed in HEV vaccine studies in pigs (Sanford et al. 2012), even after postinfection challenge with genotype 3 HEV (Sanford et al. 2011). The possibility of multiple reinfections has not been discarded in humans from hyperendemic areas (WHO 2015). The absence of clinical signs of HEV infection in pigs can be described as limitation of HEV antibody-mediated protection in pigs (de Deus et al. 2007; Halbur et al. 2001; Krawczynski et al. 2001).

Despite developing subclinical disease, pigs experimentally inoculated with human HEV genotype 3 can show severe and persistent hepatic lesions (multifocal lymphoplasmacytic hepatitis and focal hepatocellular necrosis, Fig. 15.2) than those infected with the swine HEV (dos Santos et al. 2009; Halbur et al. 2001). A single evidence of apparent clinical disease was reported in only one HEV-infected pig presenting mild diarrhea at 3 day postinfection (dpi), and a gross lesion observed was mildly enlarged mesenteric lymph nodes at 50 dpi. Histopathologically, mild multifocal lymphoplasmacytic infiltration, foci of hepatocellular vacuolation, and rare focal hepatocellular necrotic lesions were found in HEV-infected pigs (Halbur et al. 2001; Lee et al. 2009a, b; Meng et al. 1997). In addition, swine HEV cDNA was detected in normal hepatocytes but not degenerative hepatocytes suggesting that liver damage induced by HEV infection may be due to the host immune response; liver injury may not be a direct cause of viral replication in hepatocytes (Prabhu et al. 2011). Using in situ hybridization, HEV RNA was already detected in hepatocytes, Kupffer cells, bile epithelial cells, and interstitial lymphocytes (Lee et al. 2009a). Pigs naturally infected with HEV present characteristic features of necroinflammatory activity: degeneration and apoptosis of hepatocytes, focal and zonal hepatocyte necrosis, erosion of the limiting plate and portal inflammation consisting of a mixed population of lymphocytes, macrophages, and mastocytes (dos Santos et al. 2009) with detectable HEV Ag in hepatocytes (de Carvalho et al. 2013). Other authors described extrahepatic sites of HEV replication in the small

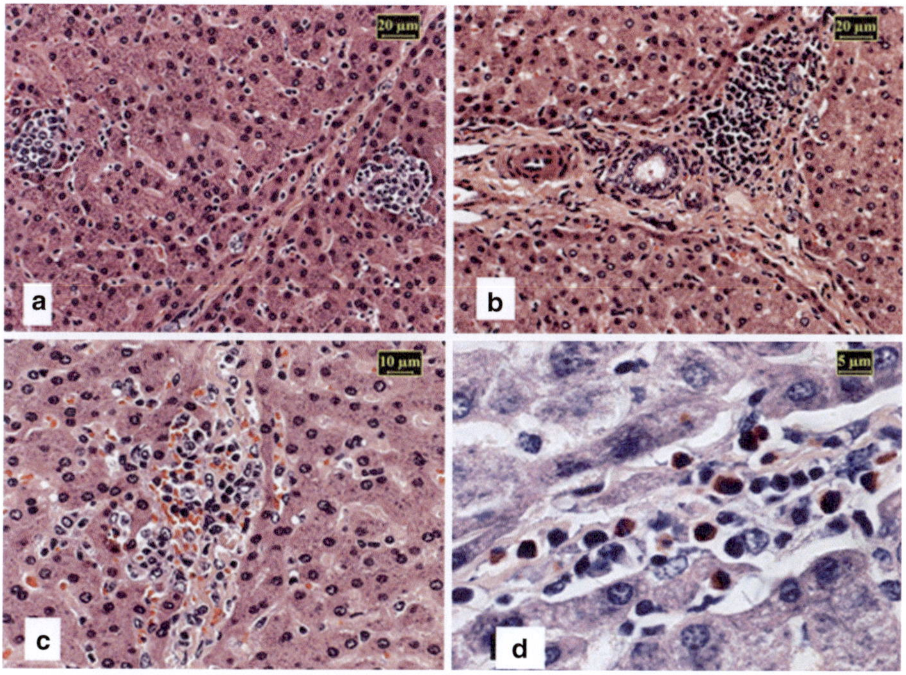

Fig. 15.2 Photomicrographs of inflammatory liver lesions in pigs naturally infected with HEV. In (**a**), lymphohistiocytic infiltrates within the liver parenchyma (H&E); in (**b**), portal inflammation with mixed mononuclear inflammatory infiltrate (H&E); in (**c**), moderate interface hepatitis involving lymphocyte, macrophage, and mastocyte infiltration (H&E); in (**d**), better visualization of periportal infiltration of mastocytes using specific staining (Giemsa)

intestine, lymph nodes, and colon of naturally infected pigs (de Deus et al. 2008a; Williams et al. 2001). The mechanism of HEV-induced liver injury is immune-mediated; however, the severity of HEV infection may be dose-dependent for experimentally infected pigs, as suggested by other authors (Bouwknegt et al. 2007). It was also suggested that the severity of HEV infection might be worse by coinfections with other pig pathogens, particularly, with porcine circovirus type 2 (PCV2), swine TTV, and porcine reproductive and respiratory syndrome virus (PRRSV) (Gagnon et al. 2007; Mao et al. 2013; Savic et al. 2010). HEV and PCV2 coinfection in piglets may produce severe pathological changes and death (Yang et al. 2015). Additionally, transplacental coinfection with PCV2 may occur in sow, causing reproductive failure (Hosmillo et al. 2010).

15.4 Diagnostic Tools for Characterization of HEV Infection in Livestock

Zoonotic transmission of HEV (genotypes 3 and 4) from animal reservoirs to humans has been confirmed by several authors. Domestic pigs are considered the major reservoirs (Meng 2010a), and virus spreads mainly by the fecal-oral route.

The role of foodborne transmission of HEV-related viruses by other livestock products, such as game meat (e.g., wild boar and deer), cattle, sheep, and seafood, has also been described (Colson et al. 2010; Krog et al. 2014; Moal et al. 2012; Shao et al. 2008). As the infection in pigs is subclinical, the diagnosis relies exclusively on laboratory tests, namely, on the detection of viral RNA or specific anti-HEV antibodies (Martin-Latil et al. 2014; Pezzoni et al. 2014; van der Poel et al. 2014). To date, there are no well-established diagnostic criteria for HEV detection in livestock.

A gold standard for anti-HEV testing in swine is not available. Hence HEV seroprevalence rates might be either underestimated or overestimated in most studies that employed commercial enzyme-linked immunosorbent assay (ELISA) kits designed for the detection of human HEV-specific antibodies (Vitral et al. 2005). These studies were performed based on the concept of cross reactivity between swine IgG/IgM antibodies and the antihuman IgG-/IgM-labeled antibodies (Meng et al. 2002). If not, some authors have employed modified commercial ELISAs in order to detect anti-HEV in swine by using anti-swine IgG obtained from a heterologous animal species as a secondary antibody (Crossan et al. 2015; Feng et al. 2011; Krumbholz et al. 2013; Wang et al. 2014).

Several HEV antibody assays have been developed for swine serology. However, most of them used recombinant antigens or synthetic peptides based on the human HEV genotype 1 strains. Since zoonotic HEV infections are caused by genotypes 3 and 4, and given the high variability among HEV strains, homologous HEV antibody assays may have a better performance (van der Poel et al. 2014). Indeed, seroprevalence studies using genotype 1-based ELISAs have shown markedly lower degree of agreement as compared to other tests including genotype 3 or genotype 1 and genotype 3-based assays. On the other hand, compared to genotype 3-based ELISAs, genotype 1-based assays have shown to give a higher degree of anti-HEV reactivity. However, such discordant results (positive in genotype 1-based but negative in genotype 3-based assays) need a further evaluation, for instance, by using a supplementary test, such as a strip immunoassay based on HEV genotype 1 and 3 recombinant antigens (Krumbholz et al. 2013).

A recombinant protein-based HEV immunoassay for the detection of HEV-specific IgG in pig sera was recently developed (van der Poel et al. 2014). This in-house ELISA used HEV recombinant antigen (ORF2, genotype 3) produced in a baculovirus expression system as a capture protein. The proposed immunoassay was compared with two other assays, a commercially available ELISA, which is able to detect HEV-specific antibodies independent of host species and immunoglobulin class, and an immunoassay developed in-house, both assays being based on HEV genotype 1. By using a Bayesian analysis approach, the newly developed HEV genotype 3 immunoassay was considered to be accurately validated and showed similar sensitivity and specificity (0.84 and 0.89 vs. 0.93 and 0.89), as compared to commercial ELISA, and higher sensitivity compared to the genotype 1-based immunoassay. In the absence of a gold standard, different approaches have been used to validate the accuracy of immunoassays for detection of swine HEV-specific antibodies. Comparative studies (genotype 1- vs. genotype 3-based assays) have shown excellent correlation in results and interdependency between assays for true seropositives.

15.5 Vaccine and Other Preventive Measures to Interrupt HEV Infection in Commercial Swine Farm

The hypothesis of massive human exposure to zoonotic HEV has been raised by the high incidence of HEV among swine at slaughter age. Thus, it is important to determine measures to reduce HEV spread and infection of slaughter-age pig within the farms. However, much data is not available about HEV epidemiology in pig farms worldwide. Studies on HEV transmission among pigs have suggested that the age of HEV infection is not strictly dependent on the proportion of piglets with colostrum intake but is also linked to farm-specific husbandry (mixing of piglets after weaning) and hygiene practices (Andraud et al. 2013, 2014). In France, the risk of HEV-positive livers was increased by early slaughter, genetic background, lack of hygiene measures, and surface origin of drinking water (Walachowski et al. 2014). Besides hygiene measures, herd vaccination has been recognized as an efficient method to reduce the proportion of infectious animals at slaughter age, transmission rate parameter, susceptibility, mean infectious period, and/or a combination of these parameters (Backer et al. 2012). Whether hepatitis E vaccine provides long-term protection against heterogenous HEV infection is still unsolved due to human cases of HEV reinfections in hyperendemic countries (Baylis et al. 2015). Indeed, decrease in HEV IgG titers (acquired after natural HEV infection or vaccination) over time has already been reported in urban and rural children of North India (Mathur et al. 2001; Zhu et al. 2010). Efforts to improve immune response to HEV vaccination have been made, and capsid-designed vaccines are the most promising candidates. Tsarev and colleagues described a cross protection between Pakistani and Mexican HEV strains in nonhuman primates vaccinated twice with a 50 microgram dose of the recombinant capsid protein (ORF2). The animals were protected from hepatitis after heterologous genotype challenge with the Mexican strain (Tsarev et al. 1997). In fact, an efficacious anti-HEV antibody response was observed against conformational epitopes located in a 459–606 of HEV pORF2. All reported neutralization epitopes are present on the dimer domain constructed by this peptide (Tang et al. 2015). So far, two recombinant vaccines to HEV are available. The rHEV, based on the 56 kd capsid protein, have had short-term efficacy in clinical trials. The other one, HEV 239 (Hecolin-Xiamen Innovax Biotech Co., Ltd., China), is licensed in China, for humans, since December 2011. HEV 239 vaccine is described as highly immunogenic, although the seroconversion have occurred only after three doses (0, 1, and 6 months). A protective response to HEV infection was evaluated by comparison with seronegative placebo subjects (evaded naturally acquired immunity). HEV 239 vaccine-induced immunity significantly reduced the risk of infection in the vaccinated group. The risk of reinfection observed in the placebo group, which had evaded naturally acquired immunity (0.52 %, 95 % CI 0.30–0.83), was similar to the risk of breakthrough infection observed in the vaccine group, which had evaded vaccine-induced immunity (0.30 %, 95 % CI 0.19–0.44), although geometric mean concentration (GMC) of the vaccine group was substantially higher (Purcell et al. 2003). Despite many reports on commercial HEV vaccines in humans, clinical vaccine trials are scarce in swine. Recently, one study showed that rabbits vaccinated with 20 μg of the HEV p179 produced high titers of anti-HEV antibodies (1:10^4–1:10^5) and the animals were completely protected against HEV infection. In this study, when rabbits were vaccinated with 10 μg anti-HEV, animals produced lower titers (1:10^3–1:10^4). Importantly, two of the five rabbits showed fecal virus shedding upon challenge (Cheng et al. 2012).

In summary, some efforts are required to reduce HEV infection in livestock and consequent prevention of human infection: improvements in sanitary conditions of pigs in breeding farms and slaughterhouses, including the wastewater and sewage systems treatment; identification of HEV in different livestock-derived products in order to discover the source of HEV contamination in production chain of pork meat and derived products; and eventually, the, development of a safe and effective hepatitis E vaccine for pigs.

References

Abrahante JE, Martins K, Papas KK, Hering BJ, Schuurman HJ, Murtaugh MP (2011) Microbiological safety of porcine islets: comparison with source pig. Xenotransplantation 18(2):88–93

Agunos AC, Yoo D, Youssef SA, Ran D, Binnington B, Hunter DB (2006) Avian hepatitis E virus in an outbreak of hepatitis–splenomegaly syndrome and fatty liver haemorrhage syndrome in two flaxseed-fed layer flocks in Ontario. Avian Pathol 35(5):404–412

Ahn JM, Kang SG, Lee DY, Shin SJ, Yoo HS (2005) Identification of novel human hepatitis E virus (HEV) isolates and determination of the seroprevalence of HEV in Korea. J Clin Microbiol 43(7):3042–3048

Andraud M, Casas M, Pavio N, Rose N (2014) Early-life hepatitis e infection in pigs: the importance of maternally-derived antibodies. PLoS One 9(8):e105527

Andraud M, Dumarest M, Cariolet R, Aylaj B, Barnaud E, Eono F, Pavio N, Rose N (2013) Direct contact and environmental contaminations are responsible for HEV transmission in pigs. Vet Res 44:102

Arankalle VA, Chobe LP, Chadha MS (2006) Type-IV Indian swine HEV infects rhesus monkeys. J Viral Hepat 13(11):742–745

Arankalle VA, Joshi MV, Kulkarni AM, Gandhe SS, Chobe LP, Rautmare SS, Mishra AC, Padbidri VS (2001) Prevalence of anti-hepatitis E virus antibodies in different Indian animal species. J Viral Hepat 8(3):223–227

Backer JA, Berto A, McCreary C, Martelli F, van der Poel WH (2012) Transmission dynamics of hepatitis E virus in pigs: estimation from field data and effect of vaccination. Epidemics 4(2):86–92

Balayan MS, Andjaparidze AG, Savinskaya SS, Ketiladze ES, Braginsky DM, Savinov AP, Poleschuk VF (1983) Evidence for a virus in non-A, non-B hepatitis transmitted via the fecal-oral route. Intervirology 20(1):23–31

Batts W, Yun S, Hedrick R, Winton J (2011) A novel member of the family Hepeviridae from cutthroat trout (*Oncorhynchus clarkii*). Virus Res 158(1–2):116–123

Baylis SA, Crossan C, Corman VM, Blümel J, Scobie L, Dalton HR (2015) Unusual serological response to hepatitis E virus in plasma donors consistent with re-infection. Vox Sang 109(4):406–409

Berto A, Backer JA, Mesquita JR, Nascimento MS, Banks M, Martelli F, Ostanello F, Angeloni G, Di Bartolo I, Ruggeri FM, Vasickova P, Diez-Valcarce M, Hernandez M, Rodriguez-Lazaro D, van der Poel WH (2012a) Prevalence and transmission of hepatitis E virus in domestic swine populations in different European countries. BMC Res Notes 5:190

Berto A, Mesquita JR, Hakze-van der Honing R, Nascimento MS, van der Poel WH (2012b) Detection and characterization of hepatitis E virus in domestic pigs of different ages in Portugal. Zoonoses Public Health 59(7):477–481

Bilic I, Jaskulska B, Basic A, Morrow CJ, Hess M (2009) Sequence analysis and comparison of avian hepatitis E viruses from Australia and Europe indicate the existence of different genotypes. J Gen Virol 90(Pt 4):863–873

Billam P, Huang FF, Sun ZF, Pierson FW, Duncan RB, Elvinger F, Guenette DK, Toth TE, Meng XJ (2005) Systematic pathogenesis and replication of avian hepatitis E virus in specific-pathogen-free adult chickens. J Virol 79(6):3429–3437

Boadella M, Ruiz-Fons JF, Vicente J, Martín M, Segalés J, Gortazar C (2012) Seroprevalence evolution of selected pathogens in Iberian wild boar. Transbound Emerg Dis 59(5):395–404

Bodewes R, van der Giessen J, Haagmans BL, Osterhaus AD, Smits SL (2013) Identification of multiple novel viruses, including a parvovirus and a hepevirus, in feces of red foxes. J Virol 87(13):7758–7764

Bonney JH, Kwame-Aryee RA, Obed S, Tamatey AA, Barnor JS, Armah NB, Oppong SA, Osei-Kwesi M (2012) Fatal hepatitis E viral infection in pregnant women in Ghana: a case series. BMC Res Notes 5:478

Bouamra Y, Gérolami R, Arzouni JP, Grimaud JC, Lafforgue P, Nelli M, Tivoli N, Ferretti A, Motte A, Colson P (2014) Emergence of autochthonous infections with hepatitis E virus of genotype 4 in Europe. Intervirology 57(1):43–48

Bouquet J, Tessé S, Lunazzi A, Eloit M, Rose N, Nicand E, Pavio N (2011) Close similarity between sequences of hepatitis E virus recovered from humans and swine, France, 2008–2009. Emerg Infect Dis 17(11):2018–2025

Bouwknegt M, Lodder-Verschoor F, van der Poel WH, Rutjes SA, de Roda Husman AM (2007) Hepatitis E virus RNA in commercial porcine livers in The Netherlands. J Food Prot 70(12):2889–2895

Bouwknegt M, Rutjes SA, Reusken CB, Stockhofe-Zurwieden N, Frankena K, de Jong MC, de Roda Husman AM, Poel WH (2009) The course of hepatitis E virus infection in pigs after contact-infection and intravenous inoculation. BMC Vet Res 5:7

Casas M, Pujols J, Rosell R, de Deus N, Peralta B, Pina S, Casal J, Martín M (2009) Retrospective serological study on hepatitis E infection in pigs from 1985 to 1997 in Spain. Vet Microbiol 135(3–4):248–252

Chandra V, Kar-Roy A, Kumari S, Mayor S, Jameel S (2008) The hepatitis E virus ORF3 protein modulates epidermal growth factor receptor trafficking, STAT3 translocation, and the acute-phase response. J Virol 82(14):7100–7110

Cheng X, Wang S, Dai X, Shi C, Wen Y, Zhu M, Zhan S, Meng J (2012) Rabbit as a novel animal model for hepatitis E virus infection and vaccine evaluation. PLoS One 7(12):e51616

Chobe LP, Lole KS, Arankalle VA (2006) Full genome sequence and analysis of Indian swine hepatitis E virus isolate of genotype 4. Vet Microbiol 114(3–4):240–251

Choi IS, Kwon HJ, Shin NR, Yoo HS (2003) Identification of swine hepatitis E virus (HEV) and prevalence of anti-HEV antibodies in swine and human populations in Korea. J Clin Microbiol 41(8):3602–3608

Clayson ET, Innis BL, Myint KS, Narupiti S, Vaughn DW, Giri S, Ranabhat P, Shrestha MP (1995) Detection of hepatitis E virus infections among domestic swine in the Kathmandu Valley of Nepal. Am J Trop Med Hyg 53(3):228–232

Colson P, Borentain P, Queyriaux B, Kaba M, Moal V, Gallian P, Heyries L, Raoult D, Gerolami R (2010) Pig liver sausage as a source of hepatitis E virus transmission to humans. J Infect Dis 202(6):825–834

Cossaboom CM, Córdoba L, Cao D, Ni YY, Meng XJ (2012) Complete genome sequence of hepatitis E virus from rabbits in the United States. J Virol 86(23):13124–13125

Crossan C, Grierson S, Thomson J, Ward A, Nunez-Garcia J, Banks M, Scobie L (2015) Prevalence of hepatitis E virus in slaughter-age pigs in Scotland. Epidemiol Infect 143(10):2237–2240

de Carvalho LG, Marchevsky RS, dos Santos DR, de Oliveira JM, de Paula VS, Lopes LM, Van der Poel WH, González JE, Munné MS, Moran J, Cajaraville AC, Pelajo-Machado M, Cruz

OG, Pinto MA (2013) Infection by Brazilian and Dutch swine hepatitis E virus strains induces haematological changes in *Macaca fascicularis*. BMC Infect Dis 13:495
de Deus N, Casas M, Peralta B, Nofrarías M, Pina S, Martín M, Segalés J (2008a) Hepatitis E virus infection dynamics and organic distribution in naturally infected pigs in a farrow-to-finish farm. Vet Microbiol 132(1–2):19–28
de Deus N, Peralta B, Pina S, Allepuz A, Mateu E, Vidal D, Ruiz-Fons F, Martín M, Gortázar C, Segalés J (2008b) Epidemiological study of hepatitis E virus infection in European wild boars (*Sus scrofa*) in Spain. Vet Microbiol 129(1–2):163–170
de Deus N, Seminati C, Pina S, Mateu E, Martín M, Segalés J (2007) Detection of hepatitis E virus in liver, mesenteric lymph node, serum, bile and faeces of naturally infected pigs affected by different pathological conditions. Vet Microbiol 119(2–4):105–114
Denner J (2015) Xenotransplantation and Hepatitis E virus. Xenotransplantation 22(3):167–173
Di Bartolo I, Angeloni G, Ponterio E, Ostanello F, Ruggeri FM (2015). Detection of hepatitis E virus in pork liver sausages. Int J Food Microbiol 16(193):29–33
Di Bartolo I, Ponterio E, Angeloni G, Morandi F, Ostanello F, Nicoloso S, Ruggeri FM (2015) Presence of Hepatitis E virus in a RED deer (*Cervus elaphus*) population in Central Italy. Transbound Emerg Dis. doi:10.1111/tbed.12353
dos Santos DR, Vitral CL, de Paula VS, Marchevsky RS, Lopes JF, Gaspar AM, Saddi TM, Júnior NC, Guimarães Fde R, Júnior JG, Ximenes LL, Souto FJ, Pinto MA (2009) Serological and molecular evidence of hepatitis E virus in swine in Brazil. Vet J 182(3):474–480
Drexler JF, Seelen A, Corman VM, Fumie Tateno A, Cottontail V, Melim Zerbinati R, Gloza-Rausch F, Klose SM, Adu-Sarkodie Y, Oppong SK, Kalko EK, Osterman A, Rasche A, Adam A, Müller MA, Ulrich RG, Leroy EM, Lukashev AN, Drosten C (2012) Bats worldwide carry hepatitis E virus-related viruses that form a putative novel genus within the family Hepeviridae. J Virol 86(17):9134–9147
Echevarría JM, Fogeda M, Avellón A (2015) Epidemiology of hepatitis E virus infection in Spain. Enferm Infecc Microbiol Clin 33(4):281–286
El Sayed Zaki M, El Aal AA, Badawy A, El-Deeb DR, El-Kheir NY (2013) Clinicolaboratory study of mother-to-neonate transmission of hepatitis E virus in Egypt. Am J Clin Pathol 140(5):721–726
Emerson SU, Purcell RH (2003) Hepatitis E virus. Rev Med Virol 13(3):145–154
Feng R, Zhao C, Li M, Harrison TJ, Qiao Z, Feng Y, Ma Z, Wang Y (2011) Infection dynamics of hepatitis E virus in naturally infected pigs in a Chinese farrow-to-finish farm. Infect Genet Evol 11(7):1727–1731
Gagnon CA, Tremblay D, Tijssen P, Venne MH, Houde A, Elahi SM (2007) The emergence of porcine circovirus 2b genotype (PCV-2b) in swine in Canada. Can Vet J 48(8):811–819
Geng J, Wang L, Wang X, Fu H, Bu Q, Liu P, Zhu Y, Wang M, Sui Y, Zhuang H (2011a) Potential risk of zoonotic transmission from young swine to human: seroepidemiological and genetic characterization of hepatitis E virus in human and various animals in Beijing, China. J Viral Hepat 18(10):e583–e590
Geng J, Wang L, Wang X, Fu H, Bu Q, Zhu Y, Zhuang H (2011b) Study on prevalence and genotype of hepatitis E virus isolated from Rex Rabbits in Beijing, China. J Viral Hepat 18(9):661–667
Geng Y, Wang C, Zhao C, Yu X, Harrison TJ, Tian K, Wang Y (2010) Serological prevalence of hepatitis E virus in domestic animals and diversity of genotype 4 hepatitis E virus in China. Vector Borne Zoonotic Dis 10(8):765–770
Guan D, Li W, Su J, Fang L, Takeda N, Wakita T, Li TC, Ke C (2013) Asian musk shrew as a reservoir of rat hepatitis E virus, China. Emerg Infect Dis 19(8):1341–1343
Gutiérrez-Vergara C, Quintero J, Duarte JF, Suescún JP, López-Herrera A (2015) Detection of hepatitis E virus genome in pig livers in Antioquia, Colombia. Genet Mol Res 14(1):2890–2899
Ha SK, Chae C (2004) Immunohistochemistry for the detection of swine hepatitis E virus in the liver. J Viral Hepat 11(3):263–267

Hakze-van der Honing RW, van Coillie E, Antonis AF, van der Poel WH (2011) First isolation of hepatitis E virus genotype 4 in Europe through swine surveillance in the Netherlands and Belgium. PLoS One 6(8):e22673

Halac U, Béland K, Lapierre P, Patey N, Ward P, Brassard J, Houde A, Alvarez F (2012) Cirrhosis due to chronic hepatitis E infection in a child post-bone marrow transplant. J Pediatr 160(5):871–4.e1

Halbur PG, Kasorndorkbua C, Gilbert C, Guenette D, Potters MB, Purcell RH, Emerson SU, Toth TE, Meng XJ (2001) Comparative pathogenesis of infection of pigs with hepatitis E viruses recovered from a pig and a human. J Clin Microbiol 39(3):918–923

Haqshenas G, Shivaprasad HL, Woolcock PR, Read DH, Meng XJ (2001) Genetic identification and characterization of a novel virus related to human hepatitis E virus from chickens with hepatitis-splenomegaly syndrome in the United States. J Gen Virol 82(Pt 10):2449–2462

Hosmillo M, Jeong YJ, Kim HJ, Park JG, Nayak MK, Alfajaro MM, Collantes TM, Park SJ, Ikuta K, Yunoki M, Kang MI, Park SI, Cho KO (2010) Molecular detection of genotype 3 porcine hepatitis E virus in aborted fetuses and their sows. Arch Virol 155(7):1157–1161

Hsu IW, Tsai HJ (2014) Avian hepatitis E virus in chickens, Taiwan, 2013. Emerg Infect Dis 20(1):149–151

Hu GD, Ma X (2010) Detection and sequences analysis of bovine hepatitis E virus RNA in Xinjiang Autonomous Region. Bing Du Xue Bao 26(1):27–32

Jiménez de Oya N, de Blas I, Blázquez AB, Martín-Acebes MA, Halaihel N, Gironés O, Saiz JC, Escribano-Romero E (2011) Widespread distribution of hepatitis E virus in Spanish pig herds. BMC Res Notes 4:412

Johne R, Plenge-Bönig A, Hess M, Ulrich RG, Reetz J, Schielke A (2010) Detection of a novel hepatitis E-like virus in faeces of wild rats using a nested broad-spectrum RT-PCR. J Gen Virol 91(Pt 3):750–758

Kaba M, Colson P, Musongela JP, Tshilolo L, Davoust B (2010) Detection of hepatitis E virus of genotype 3 in a farm pig in Kinshasa (Democratic Republic of the Congo). Infect Genet Evol 10(1):154–157

Kamar N, Abravanel F, Mansuy JM, Peron JM, Izopet J, Rostaing L (2010) Hepatitis E infection in dialysis and after transplantation. Nephrol Ther 6(2):83–87

Kanai Y, Tsujikawa M, Yunoki M, Nishiyama S, Ikuta K, Hagiwara K (2010) Long-term shedding of hepatitis E virus in the feces of pigs infected naturally, born to sows with and without maternal antibodies. J Med Virol 82(1):69–76

Kasorndorkbua C, Thacker BJ, Halbur PG, Guenette DK, Buitenwerf RM, Royer RL, Meng XJ (2003) Experimental infection of pregnant gilts with swine hepatitis E virus. Can J Vet Res 67(4):303–306

Khuroo MS (1980) Study of an epidemic of non-A, non-B hepatitis. Possibility of another human hepatitis virus distinct from post-transfusion non-A, non-B type. Am J Med 68(6):818–824

Khuroo MS (2011) Discovery of hepatitis E: the epidemic non-A, non-B hepatitis 30 years down the memory lane. Virus Res 161(1):3–14

Khuroo MS, Duermeyer W, Zargar SA, Ahanger MA, Shah MA (1983) Acute sporadic non-A, non-B hepatitis in India. Am J Epidemiol 118(3):360–364

Khuroo MS, Kamili S (2003) Aetiology, clinical course and outcome of sporadic acute viral hepatitis in pregnancy. J Viral Hepat 10(1):61–69

Kraehenbuhl JP, Campiche MA (1969) Early stages of intestinal absorption of specific antibiodies in the newborn. An ultrastructural, cytochemical, and immunological study in the pig, rat, and rabbit. J Cell Biol 42(2):345–365

Krawczynski K, Kamili S, Aggarwal R (2001) Global epidemiology and medical aspects of hepatitis E. Forum (Genova) 11(2):166–179

Krog JS, Larsen LE, Schultz AC (2014) Enteric porcine viruses in farmed shellfish in Denmark. Int J Food Microbiol 186:105–109

Krumbholz A, Joel S, Neubert A, Dremsek P, Dürrwald R, Johne R, Hlinak A, Walther M, Lange J, Wutzler P, Sauerbrei A, Ulrich RG, Zell R (2013) Age-related and regional differences in the prevalence of hepatitis E virus-specific antibodies in pigs in Germany. Vet Microbiol 167(3–4):394–402

Lee YH, Ha Y, Ahn KK, Chae C (2009a) Localisation of swine hepatitis E virus in experimentally infected pigs. Vet J 179(3):417–421

Lee YH, Ha Y, Ahn KK, Cho KD, Lee BH, Kim SH, Chae C (2009b) Comparison of a new synthetic, peptide-derived, polyclonal antibody-based, immunohistochemical test with in situ hybridisation for the detection of swine hepatitis E virus in formalin-fixed, paraffin-embedded tissues. Vet J 182(1):131–135

Li TC, Yang T, Ami Y, Suzaki Y, Shirakura M, Kishida N, Asanuma H, Takeda N, Takaji W (2014) Complete genome of hepatitis E virus from laboratory ferrets. Emerg Infect Dis 20(4):709–712

Ma H, Zheng L, Liu Y, Zhao C, Harrison TJ, Ma Y, Sun S, Zhang J, Wang Y (2010) Experimental infection of rabbits with rabbit and genotypes 1 and 4 hepatitis E viruses. PLoS One 5(2):e9160

Mao J, Zhao Y, She R, Xiao P, Tian J, Chen J (2013) One case of swine hepatitis E virus and porcine reproductive and respiratory syndrome virus co-infection in weaned pigs. Virol J 10:341

Marek A, Bilic I, Prokofieva I, Hess M (2010) Phylogenetic analysis of avian hepatitis E virus samples from European and Australian chicken flocks supports the existence of a different genus within the Hepeviridae comprising at least three different genotypes. Vet Microbiol 145(1–2):54–61

Martin-Latil S, Hennechart-Collette C, Guillier L, Perelle S (2014) Method for HEV detection in raw pig liver products and its implementation for naturally contaminated food. Int J Food Microbiol 176:1–8

Mathur P, Arora NK, Panda SK, Kapoor SK, Jailkhani BL, Irshad M (2001) Sero-epidemiology of hepatitis E virus (HEV) in urban and rural children of North India. Indian Pediatr 38(5):461–475

Meng XJ (2003) Swine hepatitis E virus: cross-species infection and risk in xenotransplantation. Curr Top Microbiol Immunol 278:185–216

Meng XJ (2010) Hepatitis E virus: animal reservoirs and zoonotic risk. Vet Microbiol 140(3–4):256–265

Meng XJ (2013) Zoonotic and foodborne transmission of hepatitis E virus. Semin Liver Dis 33(1):41–49

Meng XJ, Purcell RH, Halbur PG, Lehman JR, Webb DM, Tsareva TS, Haynes JS, Thacker BJ, Emerson SU (1997) A novel virus in swine is closely related to the human hepatitis E virus. Proc Natl Acad Sci U S A 94(18):9860–9865

Meng XJ, Wiseman B, Elvinger F, Guenette DK, Toth TE, Engle RE, Emerson SU, Purcell RH (2002) Prevalence of antibodies to hepatitis E virus in veterinarians working with swine and in normal blood donors in the United States and other countries. J Clin Microbiol 40(1):117–122

Mesquita JR, Oliveira RM, Coelho C, Vieira-Pinto M, Nascimento MS (2016) Hepatitis E virus in sylvatic and captive wild boar from Portugal. Transbound Emerg Dis 63(5):574–578

Moal V, Gerolami R, Colson P (2012) First human case of co-infection with two different subtypes of hepatitis E virus. Intervirology 55(6):484–487

Monini M, Di Bartolo I, Ianiro G, Angeloni G, Magistrali CF, Ostanello F, Ruggeri FM (2015) Detection and molecular characterization of zoonotic viruses in swine fecal samples in Italian pig herds. Arch Virol 160(10):2547–2556

Morrow CJ, Samu G, Mátrai E, Klausz A, Wood AM, Richter S, Jaskulska B, Hess M (2008) Avian hepatitis E virus infection and possible associated clinical disease in broiler breeder flocks in Hungary. Avian Pathol 37(5):527–535

Munné MS, Vladimirsky S, Otegui L, Castro R, Brajterman L, Soto S, Guarnera E, Molina V, Monfellano M, Schlauder GG, González JE (2006) Identification of the first strain of swine hepatitis E virus in South America and prevalence of anti-HEV antibodies in swine in Argentina. J Med Virol 78(12):1579–1583

Nakamura M, Takahashi K, Taira K, Taira M, Ohno A, Sakugawa H, Arai M, Mishiro S (2006) Hepatitis E virus infection in wild mongooses of Okinawa, Japan: demonstration of anti-HEV antibodies and a full-genome nucleotide sequence. Hepatol Res 34(3):137–140

Neukam K, Barreiro P, Macías J, Avellón A, Cifuentes C, Martín-Carbonero L, Echevarría JM, Vargas J, Soriano V, Pineda JA (2013) Chronic hepatitis E in HIV patients: rapid progression to cirrhosis and response to oral ribavirin. Clin Infect Dis 57(3):465–468

Okano H, Nakano T, Sugimoto K, Takahashi K, Nagashima S, Takahashi M, Arai M, Okamoto H (2014a) High genomic similarity between European type hepatitis E virus subgenotype 3e strains isolated from an acute hepatitis patient and a wild boar in Mie, Japan. Hepatol Res 44(6):694–699

Okano H, Takahashi M, Isono Y, Tanaka H, Nakano T, Oya Y, Sugimoto K, Ito K, Ohmori S, Maegawa T, Kobayashi M, Nagashima S, Nishizawa T, Okamoto H (2014b) Characterization of sporadic acute hepatitis E and comparison of hepatitis E virus genomes in acute hepatitis patients and pig liver sold as food in Mie, Japan. Hepatol Res 44(10):E63–E76

Passos-Castilho AM, Porta G, Miura IK, Pugliese RP, Danesi VL, Porta A, Guimarães T, Seda J, Antunes E, Granato CF (2014) Chronic hepatitis E virus infection in a pediatric female liver transplant recipient. J Clin Microbiol 52(12):4425–4427

Peralta B, Biarnés M, Ordóñez G, Porta R, Martín M, Mateu E, Pina S, Meng XJ (2009) Evidence of widespread infection of avian hepatitis E virus (avian HEV) in chickens from Spain. Vet Microbiol 137(1–2):31–36

Pezzoni G, Caminiti A, Stercoli L, Grazioli S, Galletti G, Santi A, Tamba M, Brocchi E (2014) Comparison of three in-house ELISAs for the detection of hepatitis E virus infection in pigs under field conditions. J Virol Methods 207:95–103

Pina S, Buti M, Cotrina M, Piella J, Girones R (2000) HEV identified in serum from humans with acute hepatitis and in sewage of animal origin in Spain. J Hepatol 33(5):826–833

Prabhu SB, Gupta P, Durgapal H, Rath S, Gupta SD, Acharya SK, Panda SK (2011) Study of cellular immune response against Hepatitis E virus (HEV). J Viral Hepat 18(8):587–594

Purcell RH, Nguyen H, Shapiro M, Engle RE, Govindarajan S, Blackwelder WC, Wong DC, Prieels JP, Emerson SU (2003) Pre-clinical immunogenicity and efficacy trial of a recombinant hepatitis E vaccine. Vaccine 21(19–20):2607–2615

Reyes GR, Purdy MA, Kim JP, Luk KC, Young LM, Fry KE, Bradley DW (1990) Isolation of a cDNA from the virus responsible for enterically transmitted non-A, non-B hepatitis. Science 247(4948):1335–1339

Rutjes SA, Lodder-Verschoor F, Lodder WJ, van der Giessen J, Reesink H, Bouwknegt M, de Roda Husman AM (2010) Seroprevalence and molecular detection of hepatitis E virus in wild boar and red deer in The Netherlands. J Virol Methods 168(1–2):197–206

Saad MD, Hussein HA, Bashandy MM, Kamel HH, Earhart KC, Fryauff DJ, Younan M, Mohamed AH (2007) Hepatitis E virus infection in work horses in Egypt. Infect Genet Evol 7(3):368–373

Salmon H, Berri M, Gerdts V, Meurens F (2009) Humoral and cellular factors of maternal immunity in swine. Dev Comp Immunol 33(3):384–393

Sanford BJ, Dryman BA, Huang YW, Feagins AR, Leroith T, Meng XJ (2011) Prior infection of pigs with a genotype 3 swine hepatitis E virus (HEV) protects against subsequent challenges with homologous and heterologous genotypes 3 and 4 human HEV. Virus Res 159(1):17–22

Sanford BJ, Emerson SU, Purcell RH, Engle RE, Dryman BA, Cecere TE, Buechner-Maxwell V, Sponenberg DP, Meng XJ (2013) Serological evidence for a hepatitis e virus-related agent in goats in the United States. Transbound Emerg Dis 60(6):538–545

Sanford BJ, Opriessnig T, Kenney SP, Dryman BA, Córdoba L, Meng XJ (2012) Assessment of the cross-protective capability of recombinant capsid proteins derived from pig, rat, and avian hepatitis E viruses (HEV) against challenge with a genotype 3 HEV in pigs. Vaccine 30(44):6249–6255

Sato Y, Sato H, Naka K, Furuya S, Tsukiji H, Kitagawa K, Sonoda Y, Usui T, Sakamoto H, Yoshino S, Shimizu Y, Takahashi M, Nagashima S, Jirintai Nishizawa T, Okamoto H (2011) A nationwide survey of hepatitis E virus (HEV) infection in wild boars in Japan: identification of boar HEV strains of genotypes 3 and 4 and unrecognized genotypes. Arch Virol 156(8):1345–1358

Savic B, Milicevic V, Bojkovski J, Kureljusic B, Ivetic V, Pavlovic I (2010) Detection rates of the swine torque teno viruses (TTVs), porcine circovirus type 2 (PCV2) and hepatitis E virus (HEV) in the livers of pigs with hepatitis. Vet Res Commun 34(7):641–648

Serracca L, Battistini R, Rossini I, Mignone W, Peletto S, Boin C, Pistone G, Ercolini R, Ercolini C (2015) Molecular investigation on the presence of Hepatitis E Virus (HEV) in wild game in North-Western Italy. Food Environ Virol 7(3):206–212

Shao ZJ, Li JH, Zheng YJ, Zhang JX, Ma YH, Ma WT, Jiang QW, Dang RL (2009) Epidemiological screening for hepatitis E virus in bile specimens from livestock in northwest China. J Clin Microbiol 47(3):814–816

Shao ZJ, Zheng YJ, Zhang JX, Xiong CL, Lu YH, Ma WT, Xu DZ, Jiang QW (2008) An investigation on hepatitis E virus infection status among livestock in Xi'an area. Zhonghua Liu Xing Bing Xue Za Zhi 29(2):158–160

Smith DB, Simmonds P, Jameel S, Emerson SU, Harrison TJ, Meng XJ, Okamoto H, Van der Poel WH, Purdy MA, International Committee on Taxonomy of Viruses Hepeviridae Study Group (2014) Consensus proposals for classification of the family Hepeviridae. J Gen Virol 95(Pt 10):2223–2232

Sonoda H, Abe M, Sugimoto T, Sato Y, Bando M, Fukui E, Mizuo H, Takahashi M, Nishizawa T, Okamoto H (2004) Prevalence of hepatitis E virus (HEV) Infection in wild boars and deer and genetic identification of a genotype 3 HEV from a boar in Japan. J Clin Microbiol 42(11):5371–5374

Sprygin AV, Nikonova ZB, Zinyakov NG (2012) Avian hepatitis E virus identified in Russian chicken flocks exhibits high genetic divergence based on the ORF2 capsid gene. Avian Pathol 41(5):459–463

Sun ZF, Larsen CT, Huang FF, Billam P, Pierson FW, Toth TE, Meng XJ (2004) Generation and infectivity titration of an infectious stock of avian hepatitis E virus (HEV) in chickens and cross-species infection of turkeys with avian HEV. J Clin Microbiol 42(6):2658–2662

Takahashi M, Nishizawa T, Miyajima H, Gotanda Y, Iita T, Tsuda F, Okamoto H (2003a) Swine hepatitis E virus strains in Japan form four phylogenetic clusters comparable with those of Japanese isolates of human hepatitis E virus. J Gen Virol 84(Pt 4):851–862

Takahashi M, Nishizawa T, Okamoto H (2003b) Identification of a genotype III swine hepatitis E virus that was isolated from a Japanese pig born in 1990 and that is most closely related to Japanese isolates of human hepatitis E virus. J Clin Microbiol 41(3):1342–1343

Takahashi M, Okamoto H (2014) Features of hepatitis E virus infection in humans and animals in Japan. Hepatol Res 44(1):43–58

Tam AW, Smith MM, Guerra ME, Huang CC, Bradley DW, Fry KE, Reyes GR (1991) Hepatitis E virus (HEV): molecular cloning and sequencing of the full-length viral genome. Virology 185(1):120–131

Tang ZM, Tang M, Zhao M, Wen GP, Yang F, Cai W, Wang SL, Zheng ZZ, Xia NS (2015) A novel linear neutralizing epitope of hepatitis E virus. Vaccine 33(30):3504–3511

Thiry D, Mauroy A, Saegerman C, Thomas I, Wautier M, Miry C, Czaplicki G, Berkvens D, Praet N, van der Poel W, Cariolet R, Brochier B, Thiry E (2014) Estimation of hepatitis E virus (HEV) pig seroprevalence using ELISA and Western blot and comparison between human and pig HEV sequences in Belgium. Vet Microbiol 172(3–4):407–414

Thiry D, Mauroy A, Pavio N, Purdy MA, Rose N, Thiry E, de Oliveira-Filho EF (2015) Hepatitis E virus and related viruses in animals. Transbound Emerg Dis. doi:10.1111/tbed.12351

Tsarev SA, Tsareva TS, Emerson SU, Govindarajan S, Shapiro M, Gerin JL, Purcell RH (1997) Recombinant vaccine against hepatitis E: dose response and protection against heterologous challenge. Vaccine 15(17–18):1834–1838

Van der Poel WH (2014) Food and environmental routes of Hepatitis E virus transmission. Curr Opin Virol 4:91–96

van der Poel WH, Pavio N, van der Goot J, van Es M, Martin M, Engel B (2014) Development and validation of a genotype 3 recombinant protein-based immunoassay for hepatitis E virus serology in swine. Braz J Med Biol Res 47(4):334–339

Viswanathan R (1957) A review of the literature on the epidemiology of infectious hepatitis. Indian J Med Res 45(Suppl):145–155

Vitral CL, Pinto MA, Lewis-Ximenez LL, Khudyakov YE, dos Santos DR, Gaspar AM (2005) Serological evidence of hepatitis E virus infection in different animal species from the Southeast of Brazil. Mem Inst Oswaldo Cruz 100(2):117–122

Wagstrom EA, Yoon KJ, Zimmerman JJ (2000) Immune components in porcine mammary secretions. Viral Immunol 13(3):383–397
Walachowski S, Dorenlor V, Lefevre J, Lunazzi A, Eono F, Merbah T, Eveno E, Pavio N, Rose N (2014) Risk factors associated with the presence of hepatitis E virus in livers and seroprevalence in slaughter-age pigs: a retrospective study of 90 swine farms in France. Epidemiol Infect 142(9):1934–1944
Wang XJ, Zhao Q, Jiang FL, Liu BY, Zhao JN, Dang L, Sun YN, Mu Y, Xiao SQ, Wang CB, Hsu WH, Liu L, Widén F, Zhou EM (2014) Genetic characterization and serological prevalence of swine hepatitis E virus in Shandong province, China. Vet Microbiol 172(3–4):415–424
Wang Y, Ma X (2010) Detection and sequences analysis of sheep hepatitis E virus RNA in Xinjiang autonomous region. Wei Sheng Wu Xue Bao 50(7):937–941
Wang YC, Zhang HY, Xia NS, Peng G, Lan HY, Zhuang H, Zhu YH, Li SW, Tian KG, Gu WJ, Lin JX, Wu X, Li HM, Harrison TJ (2002) Prevalence, isolation, and partial sequence analysis of hepatitis E virus from domestic animals in China. J Med Virol 67(4):516–521
Wenzel JJ, Preiss J, Schemmerer M, Huber B, Plentz A, Jilg W (2011) Detection of hepatitis E virus (HEV) from porcine livers in Southeastern Germany and high sequence homology to human HEV isolates. J Clin Virol 52(1):50–54
Weekly Epidemiological Record (2015) No. 18,185–200
Williams TP, Kasorndorkbua C, Halbur PG, Haqshenas G, Guenette DK, Toth TE, Meng XJ (2001) Evidence of extrahepatic sites of replication of the hepatitis E virus in a swine model. J Clin Microbiol 39(9):3040–3046
Woo PC, Lau SK, Teng JL, Tsang AK, Joseph M, Wong EY, Tang Y, Sivakumar S, Xie J, Bai R, Wernery R, Wernery U, Yuen KY (2014) New hepatitis E virus genotype in camels, the Middle East. Emerg Infect Dis 20(6):1044–1048
Wu J, Si F, Jiang C, Li T, Jin M (2015) Molecular detection of hepatitis E virus in sheep from southern Xinjiang, China. Virus Genes 50(3):410–417
Yang Y, Shi R, She R, Mao J, Zhao Y, Du F, Liu C, Liu J, Cheng M, Zhu R, Li W, Wang X, Soomro MH (2015) Fatal disease associated with Swine Hepatitis E virus and Porcine circovirus 2 co-infection in four weaned pigs in China. BMC Vet Res 11:77
Yoo D, Giulivi A (2000) Xenotransplantation and the potential risk of xenogeneic transmission of porcine viruses. Can J Vet Res 64(4):193–203
Zhang XX, Qin SY, Zhang Y, Meng QF, Jiang J, Yang GL, Zhao Q, Zhu XQ (2015) First report of hepatitis E virus infection in sika deer in China. Biomed Res Int 2015:502846
Zhao C, Ma Z, Harrison TJ, Feng R, Zhang C, Qiao Z, Fan J, Ma H, Li M, Song A, Wang Y (2009) A novel genotype of hepatitis E virus prevalent among farmed rabbits in China. J Med Virol 81(8):1371–1379
Zhao Q, Sun Y, Zhao J, Hu S, Zhao F, Chen F, Clavijo A, Zhou EM, Xiao Y (2013) Development and application of an indirect ELISA for detection of antibodies against avian hepatitis E virus. J Virol Methods 187(1):32–36
Zhu FC, Zhang J, Zhang XF, Zhou C, Wang ZZ, Huang SJ, Wang H, Yang CL, Jiang HM, Cai JP, Wang YJ, Ai X, Hu YM, Tang Q, Yao X, Yan Q, Xian YL, Wu T, Li YM, Miao J, Ng MH, Shih JW, Xia NS (2010) Efficacy and safety of a recombinant hepatitis E vaccine in healthy adults: a large-scale, randomised, double-blind placebo-controlled, phase 3 trial. Lancet 376(9744):895–902

Malignant Catarrhal Fever

16

Richa Sood, Naveen Kumar, and Sandeep Bhatia

16.1 Introduction

Malignant catarrhal fever (MCF) is a generalized lymphoproliferative and fatal disease, particularly of cattle and a variety of other ruminants, including captive and farmed species of antelope, deer, bison and water buffalo, and has been recently reported in domestic pigs (Schultheiss et al. 2000; Sood et al. 2013). The disease is sporadic and has a worldwide distribution (Plowright 1990). Clinically, in susceptible species, MCF is manifested by high fever, profuse nasal discharge, leukopenia, ophthalmitis, corneal opacity, generalized lymphadenopathy, erosions of the upper respiratory tract and alimentary tract and occasionally diarrhoea and neurologic signs. Histologically, the disease is characterized by hyperplasia of lymphoid organs, accumulation of lymphocytes in many tissues and generalized lymphocytic vasculitis. Although ten different forms of MCF have been identified, two are very similar and well-known forms of MCF, i.e. wildebeest-derived malignant catarrhal fever (WD-MCF) and sheep-associated malignant catarrhal fever (SA-MCF). These two forms are clinico-pathologically almost indistinguishable, but are epidemiologically and aetiologically different. WD-MCF was named because of its first outbreak documentation in African cattle acquired from wildebeest (*Connochaetes taurinus*), which acts as reservoir of this virus (Plowright et al. 1960). The other common form of the disease, SA-MCF, was initially reported from Europe and has been diagnosed in different species like sheep, cattle, captive moose, pigs and water buffalo and is distributed worldwide. The main source of transmission is through the nasal secretions of young carrier species (wildebeest calves and lambs). The disease is clinically indistinguishable from various other diseases of cattle and buffalo like mucosal disease, foot-and-mouth disease, lumpy skin disease and infectious bovine rhinotracheitis. Though the histological lesions are

R. Sood (✉) • N. Kumar • S. Bhatia
National Institute of High Security Animal Diseases (NIHSAD),
Bhopal, Madhya Pradesh, India
e-mail: richa.bhatia0609@gmail.com

J. Bayry (ed.), *Emerging and Re-emerging Infectious Diseases of Livestock*,
DOI 10.1007/978-3-319-47426-7_16

characteristic to MCF, genomic detection by PCR remains the confirmatory diagnostic test for MCF. The prognosis of the disease is usually grave, and there is no specific treatment of the disease in the susceptible animal. There is no effective vaccine against MCF though attempts to develop one have been long initiated. Therefore, the major step in controlling the disease relies only on the segregation of the susceptible host species from the reservoir hosts.

16.2 Aetiology, Classification and Genomic Structure of MCF Viruses

Malignant catarrhal fever (MCF) is caused by the viruses within the genus *Macavirus* of the family *Herpesviridae* (subfamily *Gammaherpesvirinae*). The genus *Macavirus* comprises of ten different viruses that are known to cause the disease. The MCF groups of viruses are related to but distinct from other ruminant gammaherpesviruses, including non-MCF lymphotropic herpesviruses and bovine herpesvirus-6, and infect members of the order Artiodactyla in the families Bovidae, Giraffidae and Cervidae. The first MCF virus identified under the genus *Macavirus* was alcelaphine herpesvirus-1 (AlHV-1) whose common reservoir host is the African wildebeest (*Connochaetes* spp.), and the susceptible animal is the domestic cattle (*Bos taurus*). Alcelaphine herpesvirus-2 (AlHV-2) causes MCF in Barbary red deer (*Cervus elaphus barbarus*), while hartebeest (*Alcelaphus buselaphus*) and topi (*Damaliscus korrigum*) are the reservoir hosts. Another well-known virus recognized to cause MCF worldwide is ovine herpesvirus-2 (OvHV-2). OvHV-2 is enzootic worldwide in the domestic sheep and causes sheep-associated MCF (SA-MCF) in various domestic and wild ruminants, i.e. domestic cattle (*B. taurus*), multiple cervid species, moose (*Alces alces*), bison (*Bison bison*), domestic pigs (*Sus scrofa domesticus*), water buffalo (*Bubalus bubalis*), banteng (*Bos javanicus*) and domestic goats (*Capra aegagrus hircus*). Hippotragine herpesvirus-1 (HiHV-1) has been detected in the roan antelope (*Hippotragus equinus*) and oryx (*Oryx gazelle*). Caprine herpesvirus-2 (CpHV-2) is enzootic in domestic goats and can cause MCF disease in several ruminants, including sika deer (*Cervus nippon*), white-tailed deer (*Odocoileus virginianus*), moose (*Alces alces*) and roe deer (*Capreolus capreolus*). Another MCFV of unknown origin MCF-WTD causes MCF in white-tailed deer. These six viruses are known to cause the MCF disease, while other three members (gemsbok-MCFV, muskox-MCFV and aoudad-MCFV) are not associated with disease. The remaining virus hippotragine herpesvirus-1 (HiHV-1) causes MCF in rabbits experimentally, but naturally transmitted disease has not been reported. Phylogenetic classification based on the conserved region of the polymerase gene divides the members causing MCF into two groups: group 1 (Alcelaphinae/Hippotraginae group), AlHV-1, AlHV-2, hippotragine herpesvirus-1 (HiHV-1) and the MCFV carried by oryx and group 2 (Caprinae group), OvHV-2, CpHV-2, MCFV-WTD and the MCFVs carried by ibex, musk ox and aoudad (Fig. 16.1).

Among these MCF groups of viruses, only two of them are well characterized, and their genomes have been sequenced. Both these virus genomes share unique segments of about 130 kbp, which is bounded by terminal repeats of 1.1 kbp in AlHV-1 and 4.2 kbp in OvHV-2. Total open reading frames (ORFs) identified in the

Order	: *Herpesvirales*	
Family	: *Herpesviridae*	
Subfamily	: *Gammaherpesvirinae*	
Genus	: *Macavirus*	
Species	: **Alcelaphine herpesvirus 1 (AlHV-1) (Type species)**	**Alcelaphinae/Hippotraginae group** (also includes MCFV-oryx)
	Alcelaphineherpesvirus 2 (AlHV-2)	
	Hippotragine herpesvirus 1 (HiHV-1)	
	Bovine herpesvirus 6	
	Caprine herpesvirus 2 (CpHV-2)	**Caprinae group** (also includes MCFV-WTD, MCFV-ibex, MCFV-muskox and MCFV-aoudad)
	Ovine herpesvirus 2 (OvHV-2)	
	Suid herpesvirus 3	
	Suid herpesvirus 4	
	Suid herpesvirus 5	

Fig. 16.1 Classification of MCFVs in the genus *Macavirus*

OvHV-2 and AlHV-1 are 73 and 71, respectively. To date, ten unique genes have been described in AlHV-1 (A1–A10), eight of which are homologues in OvHV-2. There are no equivalents of A1 or A4 in OvHV-2, but OvHV-2 encodes four additional unique genes, i.e. Ov2.5, Ov3.5, Ov4.5 and Ov8.5. The Ov2.5 protein has typical cellular IL-10 functions. The Ov4.5 functions similar to BALF1 gene product of Epstein-Barr virus and Bcl-2 family of apoptosis-related protein and is involved in the regulation of cell death.

16.3 Epidemiology

OvHV-2 which is the causative agent of SA-MCF is globally distributed, whereas AlHV-1 which causes WD-MCF is prevalent in the natural habitats of antelopes in Africa, and zoological collections where mixed populations of members of Artiodactyla, including wildebeest, are kept. There are however reports of OvHV-2 infection in South Africa also. Besides these two, there are incidences of CpHV-2 in buffaloes in Switzerland. Serological and molecular studies have been conducted to know the prevalence of MCF viruses in an array of domestic and wild animals worldwide (Table 16.1). Certain species like sheep, goat, muskox and oryx had shown a high frequency of seropositivity, almost more than 50 %. Conversely, MCF-susceptible species like cattle, bison, deer, caribou, elk, llama, reindeer, moose, caribou, pronghorn antelope and different species of deer had a low degree of seropositivity, less than 50 %.

16.4 Pathogenesis

Despite years of research on MCF viruses, the pathogenesis is still not fully understood. Major factors that had impeded progress in the pathogenesis studies include the absence of a permissive cell culture system, especially for OvHV-2, and lack of

Table 16.1 Prevalence studies of MCFVs in both domestic and wild animals

S. no.	Countries	Species	Viruses tested	Prevalence	Methods employed	References
1.	Kenya	Sheep	MCFV	97 % (162/167)	Indirect immunofluorescence test	Rossiter (1981)
2.	Kenya	Oryx	MCFV	100 % (50/50)	Neutralization test	Mushi et al. (1981)
3.	USA	Sheep	SA-MCF	99 % (143/144)	CI-ELISA	Li et al. (1995)
				94 % (136/144)	PCR	
4.	USA	Bison	MCFV	2 % (2/103)	CI-ELISA	Li et al. (1996)
		Elk	MCFV	9 % (30/323)		
		Mule deer	MCFV	2 % (2/101)		
		White-tailed deer	MCFV	3 % (2/63)		
		Pronghorn antelope	MCFV	25 % (20/80)		
		Bighorn sheep	MCFV	37 % (124/339)		
		Muskox	MCFV	40 % (8/20)		
		Mouflon sheep	MCFV	62 % (13/21)		
		Cattle	MCFV	7.8 % (31/395)		
		Domestic sheep	MCFV	56.3 % (547/971)		
5.	Germany	Fallow deer	MCFV	9.3 % (10/108)	CI-ELISA	Frolich et al. (1998)
		Sika deer	MCFV	7 % (1/14)		
		Domestic sheep	MCFV	72 % (36/50)		
		Domestic sheep	OvHV-2	100 % (20/20)	PCR	
6.	USA	Cattle	OvHV-2	5.3 % (10/190)	PCR	Collins et al. (2000)

Table 16.1 (continued)

S. no.	Countries	Species	Viruses tested	Prevalence	Methods employed	References
7.	USA	Muskox	MCFV	96 % (100/104)	CI-ELISA	Zarnke et al. (2002)
		Dall sheep	MCFV	95 % (212/222)		
		Elk	MCFV	27 % (14/51)		
		Bison	MCFV	17 % (34/197)		
		Caribou	MCFV	4 % (9/232)		
		Sitka black-tailed deer	MCFV	2 % (1/49)		
		Moose	MCFV	1 % (3/219)		
8.	USA	Sika deer	MCFV	27.27 % (3/11)	CI-ELISA	Li et al. (2003)
		White-tailed deer	MCFV	33.33 % (3/9)	CI-ELISA	
			CpHV-2	22.22 % (2/9)	PCR	
		Fallow deer	MCFV	87.5 % (7/8)	CI-ELISA	
		Pygmy goat	MCFV	66.66 % (4/6)	CI-ELISA	
			CpHV-2	50 % (3/6)	PCR	
		Domestic goat	MCFV	100 % (9/9)	CI-ELISA	
			CpHV-2	100 % (9/9)	PCR	
9.	Poland	Sheep	SA-MCF	6 % (3/50)	PCR	Dullin et al. (2005)
10.	Norway	Red deer	MCFV	5 % (13/260)	CI-ELISA	Vikøren et al. (2006)
		Reindeer	MCFV	4 % (9/225)		
		Roe deer	MCFV	2 % (5/250)		
		Moose	MCFV	0.4 % (1/250)		
11.	India	Sheep	OvHV-2	85 % (28/33)	PCR	Wani et al. (2006)
		Goat	OvHV-2	61 % (16/26)		
		Cattle	OvHV-2	31 % (17/55)		

(continued)

Table 16.1 (continued)

S. no.	Countries	Species	Viruses tested	Prevalence	Methods employed	References
12.	Turkey	Sheep	MCFV	97.5 % (195/200)	CI-ELISA	Yeşlbağ (2007)
		Goat	MCFV	96.0 % (192/200)		
		Cattle	MCFV	15.0 % (30/200)		
13.	India	Sheep and goat	OvHV-2	2.77 % (17/612)	PCR	Banumathi et al. (2008)
14.	South Africa	Sheep	OvHV-2	76.5 % (65/85)	PCR	Bremer (2010)
15.	Turkey	Cameroon sheep	MCFV	36.3 % (4/11)	CI-ELISA	Yeşilbağ et al. (2011)
		Pygmy goats	MCFV	83.3 % (15/18)		
		Mountain goats	MCFV	90.0 % (9/10)		
		Llama	MCFV	14.2 % (1/7)		
16.	Japan	Sheep and goat	OvHV-2	37.66 % (58/154)	Neutralization test	Giangaspero et al. (2013)
17.	India	Sheep	OvHV-2	24.44 % (87/356)	PCR	Premkrishnan et al. (2015)

established study model. However, the rabbit has been suggested as a suitable model to study pathogenesis of OvHV-2 recently (Cunha et al. 2013). In the case of AlHV-1, generation of a bacterial artificial chromosome (BAC) clone possessing the complete genome further facilitated in understanding the pathogenesis of MCF.

In general, gammaherpesviruses target either B or T lymphocytes, but MCFVs characteristically infect T cells. The infected large granular lymphocyte (LGL) cells derived from MCF-affected animals have a phenotype similar to lymphokine-activated killer cells, and these cells are responsible for the pathology observed in the affected animals (Anderson et al. 2008). The studies carried out in rabbits demonstrated a progressive T cell hyperplasia in both the lymphoid and non-lymphoid organs. Further, extensive vasculitis and the dysregulated cytotoxic lymphocytes lead to tissue destruction. The striking difference in the pathology caused by AlHV-1 and OvHV-2 viruses is the lesions developed in peripheral lymph nodes in the former and visceral lymphoid tissue in the latter. In addition, the lesions produced by OvHV-2 viruses contained more necrotic areas compared to AlHV-1 viruses. The other lesions include epithelial erosions, kerato-conjunctivitis, enlarged lymph nodes and synovitis in the tibiotarsal joints. Thus, the pathology induced by both the viruses, i.e. AlHV-1 and OvHV-2, is mainly an immune-mediated aetiology, associated with infiltration of lymphocytes in the multiple tissues. In cattle and bison, both CD8+ and CD4+ cells predominate within the vascular infiltrates. The infiltrate from the experimentally induced sheep-associated MCF in bison consists of CD8(+)/perforin(+) WC1(−) gammadelta T cells, CD4(+)/perforin(−) alphabeta T cells and B cells. Recently, eight miRNAs encoded by OvHV-2 have been identified in an OvHV-2 bovine lymphocyte cell line. The miRNAs encoded by herpesviruses are effective regulators of both cellular and viral gene expression. However, the role of these miRNAs in inducing MCF specified pathological lesions has not been fully elucidated.

16.5 Transmission of Malignant Catarrhal Fever Viruses

16.5.1 Transmission of AlHV-1

In general, AlHV-1 is not transmitted from one clinically susceptible host to another via natural methods, as the virus secreted from the non-host animals is cell associated. Therefore, sick animals may be housed with healthy animals without fear of horizontal disease transmission. However, cell-free AlHV-1 is known to retain its infectivity for extended periods of time, especially in the high relative humidity environmental conditions.

The cell-free AlHV-1 has been detected in the wildebeest calf nasal secretions. AlHV-1 can spread to cattle through aerosolize droplets generated from the nasal secretions of wildebeest calves. The main source of infection of AlHV-1 is the young wildebeest of less than 3 months of age. In them, the virus has been isolated from the ocular and nasal secretions, and almost all animals get infected by 4 months of age. By the time wildebeest calves attain the age of 6 months, there is very low level of virus in their nasal and ocular secretions. The viral shedding in adult wildebeest is quite low and occurs primarily during periods of stress or parturition. AlHV-1 DNA has been detected in the urine of 3-month-old wildebeest calves by DNA hybridization, raising the possibility that urine might also act as a source of AlHV-1 infection to cattle. However, there are no reports that wildebeest urine is infectious to cattle. All ages of cattle are susceptible to the disease, but the adults particularly peri-parturient females are more prone to the disease.

16.5.2 Transmission of OvHV-2

The pattern of transmission of OvHV-2 among sheep is markedly different from that of AlHV-1 in wildebeest. Although nearly all of the adult populations of both the reservoir species are infected with their respective MCF viruses, the viraemic status varies between these two species. Only a few adult wildebeest are viraemic with AlHV-1 as demonstrated in the cell culture experiments, but almost all adult sheep have OvHV-2 DNA in their peripheral blood leucocytes, as determined by PCR assays. In contrast to wildebeest, pregnant ewes rarely transmit OvHV-2 to their lambs in utero, nor do lambs become infected with OvHV-2 within the first few months of life. Also, sheep foetuses and their lymph nodes or peripheral blood leucocytes are free from OvHV-2-specific antibodies and viral DNA, respectively. Even though OvHV-2 viral DNA is present in the colostrum and milk of ewes, OvHV-2 is not transmitted in utero or by the consumption of colostrum and milk in lambs. Lambs weaned at 2–2.5 months of age and reared in isolation from other sheep remain uninfected with OvHV-2, even though their dams are infected. OvHV-2 infections are not established in lambs until they reach the age of 3–3.5 months. OvHV-2 from ovine nasal secretions infects naive sheep and also can induce MCF in both the cattle and bison. Cattle has low susceptibility to MCF infection and may require even 1000-fold higher viral doses by intranasal nebulization to get infected compared to sheep and bison. The MCF-susceptible species generally are thought to be dead-end hosts because they do not transmit virus to other animals. This strategy is pivotal in limiting the spread of disease during outbreaks. Failure of natural transmission of OvHV-2 among cattle is due to the cell-bound nature of OvHV-2 in infected cattle and the inability of cattle to produce and shed cell-free OvHV-2 in a large enough quantity to be infectious to other cattle. It has been postulated that OvHV-2 is cell bound because the viral DNA of OvHV-2 becomes incorporated in the cellular genome of infected cells or exists as an episome in infected cells (Metzler 1991).

16.6 Clinical Signs and Histopathological Lesions

The clinical and pathological changes associated with either AlHV-1 or OvHV-2 in the susceptible animal species are almost similar and cannot be unfailingly differentiated. The incubation period of the disease is highly variable ranging from 2 weeks–9 months.

16.6.1 Clinical Forms

MCF is a multisystemic syndrome and described basically in four forms: (1) peracute form, (2) alimentary tract form, (3) common 'head and eye' form and (4) mild form. All these forms/gradations are based on the appearance of clinical signs which are more evident in the protracted form of the disease (Radostits et al. 2010).

16.6.1.1 Peracute Form

The peracute form is characterized by rapid onset of depression and high fever, followed by the development of diarrhoea, which in some cases, may be haemorrhagic, with death occurring in 12–24 h after its onset.

16.6.1.2 Alimentary Tract Form

In this form, there is marked diarrhoea in SA-MCF, whereas there is constipation in WD-MCF with only minor eye changes consisting of conjunctivitis rather than ophthalmia as seen in SA-MCF.

Fig. 16.2 Typical symptoms of corneal opacity, lacrimation and nasal discharge in cattle suffering from MCF

16.6.1.3 Common 'Head and Eye' Form

This form is the most common manifestation of MCF and is typically characterized by a sudden onset of extreme dejection, anorexia, agalactia, high fever (rectal temperature of 40–43 °C), rapid pulse rate (100–120 bpm) and serous nasal discharge which later becomes profuse mucopurulent. Due to obstruction of the nasal cavities with exudate, severe dyspnoea with stertor is also observed resulting in the head and neck being held extended and a foul odour accompanying the nasal discharges. There is lacrimation and ocular discharge with variable degrees of oedema of the eyelids. Sclera becomes reddened due to congestion of vessels, and the cornea becomes progressively opaque, to the extent that the animal becomes blind (Fig. 16.2). The progression of corneal oedema correlates well with disease outcome and that non-improvement of uveitis is a bad prognostic sign. There is swelling of lymph nodes especially in the head and neck region. Superficial necrosis is seen in the anterior nasal and buccal mucosa. This begins as a diffuse reddening of the mucosa and is a regular finding about day 19 or 20 after infection. Distinct local areas of necrosis are seen on the hard palate, gums and commissures of the mouth, inside the lips and gingivae. In lactating animals, the milk yield drops. Lesions may occur at the skin-hoof junction of the feet, especially at the back of the pastern. The skin of the teats, vulva and scrotum in acute cases may slough off entirely upon touch or become covered with dry, tenacious scabs.

16.6.1.4 Mild Form

The mild form is generally observed in experimental animals. There are transient fever and mild erosions on the oral and nasal mucosae. This form may also be followed by complete recovery.

16.6.2 Pathological Findings

The most common gross pathological changes associated with MCF are the erosions throughout the gastrointestinal tract and haemorrhagic cystitis which can typically be seen in bison (Sood et al. 2012). There are congestion and oedema

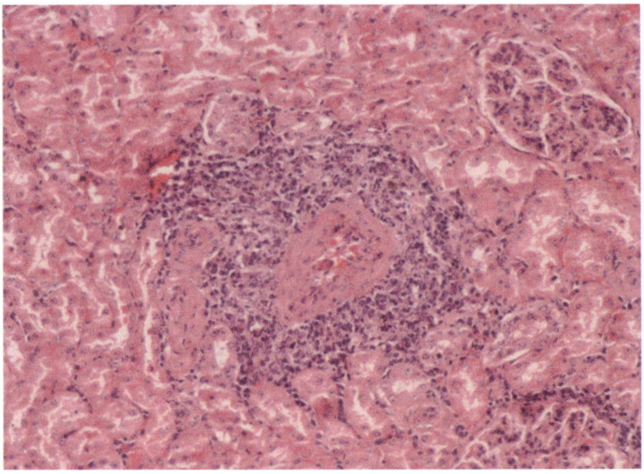

Fig. 16.3 Vasculitis and infiltration of lymphocytes around the blood vessels – 'perivascular cuffing'

of the lungs, erosions of the tracheal and bronchial mucosa, focal white lesions in the kidney, erythema of the turbinate mucosa and lymph node enlargement. The gross pathology is often subtle and inconsistent among cases, making confirmation of MCF difficult based on gross pathology or symptoms alone. Histopathologically, lymphocytic infiltrations and lymphocytic vasculitis are seen throughout many tissues and organs. Perivascular cuffing was earlier considered gold standard for the confirmatory diagnosis of the disease and is still significant for the tentative diagnosis (Fig. 16.3). Other lesions include bronchopneumonia in lungs and extensive haemorrhage in the heart. A non-purulent encephalitis, necrosis of vessel walls and gliosis in the central nervous system are also seen sometimes.

16.7 Diagnosis of MCF

Diagnosis of MCF in field is challenging as the disease syndrome resembles symptomatically to many other diseases like bovine viral diarrhoea, infectious bovine rhinotracheitis, bluetongue, rinderpest, salmonellosis, coccidiosis, besnoitiosis, etc. A history of proximity with sheep or wildebeest or grazing in pastures recently by either of these species is a support for the tentative diagnosis. Clinically, a tentative diagnosis of the disease can be done, but laboratory confirmation is required to arrive to a definitive diagnosis. Probably, MCF is the only disease where the World Organisation for Animal Health (OIE) recognized histopathology as a definitive diagnostic test for MCF, but other approaches such as indirect immunofluorescence and PCR assays are routinely used in the laboratories for quick confirmatory diagnosis. The confirmatory diagnosis for AlHV-1 is by virus isolation, but isolation of OvHV-2 has not been successful to date. Therefore, molecular diagnostic tests help in the early confirmatory diagnosis for this disease. Various methods used in the diagnosis of MCF are enumerated below.

16.7.1 Virological and Serological Detection of MCF

16.7.1.1 Virus Isolation

This test is used for isolation of AlHV-1 only. This virus can only be recovered from the viable cells; therefore, collection of tissue samples should be done either before or shortly after the death of the animal. Tissue samples should also not be frozen as virus cannot be recovered from dead cells. AlHV-1 recovery can be achieved by cocultivating either peripheral blood leucocytes or cells from affected tissues with monolayer cell cultures of bovine or ovine origin. The typical cytopathic effect of multinucleate foci is visible only after 7–21 days of infection. In case of SA-MCF, lymphoblastoid cells infected with OvHV-2 have been propagated from cattle, deer and rabbits with sheep-associated malignant catarrhal fever, but cell-free virus has not yet been cultivated despite numerous attempts to date.

16.7.1.2 Virus Neutralization Assay

This test can be performed using either of three strains (WC11, hartebeest isolate and C500) and is considered to be specific for A1HV-1. It does not detect antibodies against other bovine herpes viruses or OvHV-2. The test is carried out by inoculating primary or secondary cell cultures of bovine kidney, bovine thyroid, low-passage bovine testis or another permissive cell type with a mixture of serum from a test animal and live A1HV-1, and the serum is positive if it inhibited growth of AlHV-1 in cell culture. This test, however, does not detect the latent infection in the ruminants and cannot be used as a diagnostic test in clinically affected animals, as these animals are not able to produce virus neutralizing antibodies (OIE 2008). The difficulty in maintaining the cell culture and more time requirement to perform this test in the laboratory led to search for alternative tests, which could be more easily adapted to diagnostic laboratory environments.

16.7.1.3 Indirect Immunofluorescence Assay (IFA)

This test is not very specific, as other herpesviruses such as bovine herpesvirus-4 and infectious bovine rhinotracheitis virus show cross-reaction. However, it is useful in detecting antibodies in AlHV-1-infected cell monolayers. This assay also cannot be reliably utilized to detect SA-MCF (OIE 2008).

16.7.1.4 Competitive Inhibition-Enzyme Linked Immunosorbent Assay (CI-ELISA)

The competitive inhibition-ELISA (CI-ELISA) test was developed to detect antibodies against OvHV-2 (Li et al. 1994, 2001). This test uses a MAb (15-A) targeting an epitope on a complex of glycoproteins and is conserved among all MCF viruses. The test has been used to detect antibody in the serum of both wild and domestic ruminants. Antibodies to AlHV-1, AlHV-2, OvHV-2, CpHV-2, MCFV-oryx, MCFV-WTD and MCFV-muskox have also been detected using this test. The CI-ELISA is faster and more efficient than the IFA. The limitation of this test is that it gives false-positive reaction in the serum of uninfected lambs of less than 4 months of age because of the presence of maternal antibodies. To overcome this limitation, paired sampling is required for conclusive identification because suboptimal virus doses of infection takes almost 4 weeks for antibody production. Also, the test could not differentiate among different MCFVs.

16.7.2 Genomic Detection of MCF

Nucleic acid-based detection of MCF is the currently accepted method for diagnosis worldwide. The hemi-nested PCR test, which is an OIE-approved diagnostic test for the detection of OvHV-2 infection, was developed by Baxter and co-workers (Baxter et al. 1993). In this test, two sets of primers, i.e. the outer set of primers (556 and 755) and the nested primer set (556 and 555), amplify a 422 bp and a 238 bp fragment of the ORF75 (code for tegument protein), respectively (Table 16.2). Both these amplicons can be cleaved by the restriction endonuclease *RsaI* into fragments that are consistent with their predicted size. These primers are very specific and do not amplify DNA from bovine herpesvirus-1 (BoHV-1), bovine herpesvirus-2 (BoHV-2), bovine herpesvirus-4 (BoHV-4) and AlHV-1. This PCR test is currently being used for detection of OvHV-2 DNA in infected sheep as well as in animals with clinical MCF and considered as gold standard in most of the veterinary institutes worldwide (Sood et al. 2014). Since the virus cannot be grown in cell culture, this method was useful for qualitative detection of the infection. This test however needs extra care, while execution as potential for carry over amplicon contamination is very common leading to false-positive results. Modification in this PCR with fluorogenic probes was developed using the same region of the tegument gene to decrease the chances of contamination (Hussy et al. 2001). In this PCR, replication of OvHV-2 could be quantified, and thus, the infection could be measured along the time axis of a natural infection. To detect different MCFV infections, different PCRs need to be run which consume a lot of time, effort and money, and thus a multiplex PCR was developed targeting the polymerase gene which could simultaneously differentiate different MCFV infections. This multiplex real-time PCR uses one pair of primers in conjunction with fluorescently labelled probes which are specific for OvHV-2, CpHV-2, MCFV-WTD, MCFV-ibex and AlHV-1 (Cunha et al. 2009).

It is recommended to first identify the target population in which MCF causative agents have to be identified. Like in case of the reservoir host such as sheep, goat and wildebeest, serology is useful and an efficient method except in very young animals, where maternal antibodies would interfere with correct interpretation. Hemi-nested PCR is a preferred test where MCF-free status in the reservoir host is to be identified. In the case of clinically susceptible animals, quantitative PCR and hemi-nested PCR are tests of choice for the diagnosis of MCF. The samples from mixed species infection can best diagnosed with multiplex PCR.

16.8 Treatment and Control

No effective treatment of the disease is available, and the prognosis of the disease is usually grave in clinical cases. The symptomatic treatment may only prolong the course of disease. Prophylactic measures were initiated as early as 1975, and

Table 16.2 Commonly used PCR tests for identification and differentiation of MCFVs

Genomic detection methods	Primer name	Oligonucleotide sequences (5' to 3')	Amplicon size	Remarks	References
Hemi-nested PCR to detect OvHV-2 DNA	556	AGTCTGGGTATATGAATCCAGATGGCTCTC	422 bp	First round PCR	Baxter et al. (1993)
	755	AAGATAAGCACCAGTTATGCATCTGATAAA			
	556	AGTCTGGGTATATGAATCCAGATGGCTCTC	238 bp	Second round PCR	
	555	TTCTGGGGTAGTGGCGAGCGAAGGCTTC			
Hemi-nested PCR to distinguish AlHV-1 and OvHV-2 DNA	Primer POL1	GGC(CT)CA(CT)AA(CT)CTATGCTACTCCAC	386 bp	First round PCR	Flach et al. (2002)
	Primer POL2	ATT(AG)TCCACAAACTGTTTTGT			
	Primer OHVPol OR Primer AHVPol	AAAAACTCAGGGCCATTCTG (OvHV-2) OR CCAAAATGAAGACCATCTTA (AlHV-1)	172 bp	Second round PCR	
	Primer POL2	ATT(AG)TCCACAAACTGTTTTGT			
Real-time PCR for OvHV-2	Forward primer	TGGTAGGAGCAGGCTACCGT	131 bp	–	Hussy et al. (2001)
	Reverse primer	ATCATGCTGACCCCTTGCAG			
	FAM probe	TCCACGCCGTCCGCACTGTAAGA			
Pan-herpesvirus PCR	DFA	GAYTTYGCNAGYYTNTAYCC	–	First round PCR	VanDevanter et al. (1996)
	ILK	TCCTGGACAAGCAGCARNYSGCNMTNAA			
	KG1	GTCTTGCTCACC AGNTCNACNCCYTT			
	TGV	TGTAACTCGGTGTAYGGNTTYACNGGNGT	215–315 bp	Second round PCR	
	IYG	CACAGAGTCCGTRTCNCCRTADAT			

attempts to make a vaccine using live or attenuated strains of AlHV-1 were made. Serum neutralizing antibodies were developed; however, the vaccination was unsuccessful both in experimental and natural virus challenge (Plowright et al. 1975). Subsequently in 2008, a combination of prime and boost strategy for immunization with attenuated AlHV-1 emulsified with Freund's adjuvant was used for vaccination purpose (Haig et al. 2008). The vaccination in the upper neck region and challenge with the low-passage cell-free virus in the vaccinated calves led to development of high-titre neutralizing antibodies in the nasal secretions as well as in the blood. The neutralizing antibodies in the blood, however, do not play a significant role in the protection as much as mucosal antibodies which form a mucosal barrier and develop host immunity. The only effective control measure at present is the segregation of the susceptible host from the reservoir hosts. The grazing areas should be separate for the flocks kept under mixed farming. Different grazing hours and timings can also be defined for sheep and cattle. For example, if sheep is grazed in the morning hours in the field, the cattle can be grazed in the evening hours on the same field. Allowing this time gap between grazing will help in the inactivation of labile MCFVs in the sunlight. It is critical to separate cattle from the potential reservoir hosts such as sheep, goats, and wildebeest, especially during lambing, kidding or calving seasons, respectively. If lambs are removed from contact with infected sheep prior to that age, they remain uninfected and can be raised virus-free.

In an outbreak situation, immediate separation of cattle or the susceptible host from sheep and goats in the case of the SA-MCF and the susceptible host from alcelaphine or wild ruminants in the case of WD-MCF is imperative. Minimizing the stress can help in prevention of this disease in subclinically or mildly affected animals.

References

Anderson IE, Deane D, Swa S et al (2008) Production and utilization of interleukin-15 in malignant catarrhal fever. J Comp Pathol 138(2–3):131–144

Banumathi N, Sood R, Patil SS et al (2008) Genomic detection of Ovine herpes virus-2 in sheep and goat in states of southern India. Ind J Anim Sci 78:13–16

Baxter SI, Pow I, Bridgen A et al (1993) PCR detection of the sheep-associated agent of malignant catarrhal fever. Arch Virol 132:145–159

Bremer CW (2010) The prevalence of ovine herpesvirus-2 in 4 sheep breeds from different regions in South Africa. J S Afr Vet Assoc 81:93–96

Collins JK, Bruns C, Vermedahl TL et al (2000) Malignant catarrhal fever: polymerase chain reaction survey for ovine herpesvirus 2 and other persistent herpesvirus and retrovirus infections of dairy cattle and bison. J Vet Diagn Invest 12(5):406–411

Cunha CW, O'Toole D, Taus NS et al (2013) Are rabbits a suitable model to study sheep-associated malignant catarrhal fever in susceptible hosts? Vet Microbiol 163(3–4):358–363

Cunha CW, Otto L, Taus NS et al (2009) Development of a multiplex real-time PCR for detection and differentiation of malignant catarrhal fever viruses in clinical samples. J Clin Microbiol 47:2586–2589

Dullin P, Galbas M, Mikowska A et al (2005) Comparison of the prevalence of sheep-associated malignant catarrhal fever virus (SA-MCF) in two ovine herds of the Wielkopolska region. Med Weter 61(9):1046–1048

Flach EJ, Reid H, Pow I et al (2002) Gamma herpesvirus carrier status of captive artiodactyls. Res Vet Sci 73(1):93–99

Frölich K, Li H, Müller-Doblies U (1998) Serosurvey for antibodies to malignant catarrhal fever-associated viruses in free-living and captive cervids in Germany. J Wildl Dis 34(4):777–782
Giangaspero M, Orusa R, Savini G et al (2013) Serological survey to determine the occurrence of blue tongue virus, bovine leukemia virus and herpesvirus infections in the Japanese small ruminant population from northern districts. Clin Microbiol 2:1–6
Haig DM, Grant D, Deane D et al (2008) An immunisation strategy for the protection of cattle against alcelaphine herpesvirus-1-induced malignant catarrhal fever. Vaccine 26(35):4461–4468
Hussy D, Stauber N, Leutenegger CM et al (2001) Quantitative fluorogenic PCR assay for measuring ovine herpesvirus 2 replication in sheep. Clin Diagn Lab Immunol 8:123–128
Li H, McGuire TC, Müller-Doblies U et al (2001) A simpler, more sensitive competitive inhibition enzyme-linked immunosorbent assay for detection of antibody to malignant catarrhal fever viruses. J Vet Diagn Invest 13(4):361–364
Li H, Shen DT, Jessup DA et al (1996) Prevalence of antibody to malignant catarrhal fever virus in wild and domestic ruminants by competitive-inhibition ELISA. J Wildl Dis 32(3):437–443
Li H, Shen DT, O'Toole D et al (1995) Investigation of sheep-associated malignant catarrhal fever virus infection in ruminants by PCR and competitive inhibition enzyme-linked immunosorbent assay. J Clin Microbiol 33(8):2048–2053
Li H, Wunschmann A, Keller J et al (2003) Caprine herpesvirus-2-associated malignant catarrhal fever in white-tailed deer (*Odocoileus virginianus*). J Vet Diagn Invest 15(1):46–49
Li H, Shen DT, Knowles DP et al (1994) Competitive inhibition enzyme-linked immunosorbent assay for antibody in sheep and other ruminants to a conserved epitope of malignant catarrhal fever virus. Clin Microbiol 32:1674–1679
Metzler AE (1991) The malignant catarrhal fever complex. Comp Immunol Microbiol Infect Dis 14:107–124
Mushi EZ, Rossiter PB, Jessett D et al (1981) Isolation and characterization of a herpesvirus from topi (*Damaliscus korrigum, Ogilby*). J Comp Pathol 91:63–68
OIE (2008) OIE manual of diagnostic tests and vaccines for terrestrial animals, OIE, ed. Paris
Plowright W, Ferris RD, Scott GR (1960) Blue wildebeest and the aetiological agent of bovine malignant catarrhal fever. Nature 188:1167–1169
Plowright W, Herniman KA, Jessett DM et al (1975) Immunisation of cattle against the herpesvirus of malignant catarrhal fever: failure of inactivated culture vaccines with adjuvant. Res Vet Sci 19:159–166
Plowright W (1990) Malignant catarrhal fever virus. In: Dinter Z, Morein B (eds) Virus infections of ruminants. Elsevier Science, New York, pp 123–150
Premkrishnan GN, Sood R, Hemadri D et al (2015) Cross-sectional study indicates nearly a quarter of sheep population in Karnataka state of India is infected with ovine herpesvirus 2. Virusdisease 26(3):180–188
Radostits OM, Gay CC, Hinchcliff KW et al (2010) Veterinary medicine: a text book of diseases of cattle, sheep goat pigs and horses, 10th edn. Saunders, Philadelphia, p 1245
Rossiter PB (1981) Antibodies to malignant catarrhal fever virus in sheep sera. J Comp Pathol 91:303–311
Schultheiss PC, Collins JK, Spraker TR et al (2000) Epizootic malignant catarrhal fever in three bison herds: differences from cattle and association with ovine herpesvirus- 2. J Vet Diagn Invest 12:497–502
Sood R, Hemadri D, Bhatia S (2013) Sheep associated malignant catarrhal fever: an emerging disease of bovids in India. Indian J Virol 24(3):321–331
Sood R, Khandia R, Bhatia S et al (2014) Detection and molecular characterization of naturally transmitted sheep associated malignant catarrhal fever in cattle in India. Tropl Anim Health Prod 46(6):1037–1043
Sood R, Kumar M, Bhatia S et al (2012) Transboundary diseases: ovine herpesvirus type 2 infection in captive bison in India. Vet Rec 170:654
VanDevanter DR, Warrener P, Bennett L et al (1996) Detection and analysis of diverse herpesviral species by consensus primer PCR. J Clin Microbiol 34(7):1666–1671
Vikøren T, Li H, Lillehaug A et al (2006) Malignant catarrhal fever in free-ranging cervids associated with OvHV-2 and CpHV-2 DNA. J Wildl Dis 42:797–807

Wani SA, Samanta I, Pandit F et al (2006) Molecular epidemiology of ovine herpesvirus type 2 infection in Kashmir India. Vet Rec 159:587–590

Yeşilbağ K, Alpay G, Karakuzulu H (2011) A serologic survey of viral infections in captive ungulates in Turkish zoos. J Zoo Wildl Med 42(1):44–48

Yeşlbağ K (2007) Seroprevalence of malignant catarrhal fever-related gammaherpesviruses in domestic ruminants in Turkey. Tropl Anim Health Prod 39(5):363–368

Zarnke RL, Li H, Crawford TB (2002) Serum antibody prevalence of malignant catarrhal fever viruses in seven wildlife species from Alaska. J Wildl Dis 38(3):500–504

Part II

Emerging Rickettsial Diseases of Livestock

Ehrlichiosis

17

Daniel Moura de Aguiar

17.1 Definition

The genus *Ehrlichia* consists of tick-transmitted bacteria that infect leukocytes and endothelial cells in mammals and different tissues in ticks. Ehrlichiosis is considered an emerging infectious disease in both humans and animals. Bovine ehrlichiosis in the Americas is caused by an obligate intracellular organism of the genus *Ehrlichia* closely related to *Ehrlichia canis*, the etiologic agent of canine monocytic ehrlichiosis (CME). The proposed name for this species is *Ehrlichia minasensis* (Cabezas-Cruz et al. 2016).

17.2 Etiology

The genus *Ehrlichia* belongs to the family *Anaplasmataceae* and consists of six recognized species: *E. canis*, *E. chaffeensis*, *E. muris*, *E. ewingii*, *E. ruminantium*, and *Ehrlichia minasensis* (Dumler et al. 2001; Cabezas-Cruz et al. 2016). Unnamed species isolated from *Ixodes ovatus* ticks (designated *Ixodes ovatus Ehrlichia*) (Shibata et al. 2000), mice (designated *E. muris*-like agent – EMLA) (Pritt et al. 2011), and non-volant small mammals (designated "*Candidatus Ehrlichia khabarensis*") (Rar et al. 2015) have been reported. In addition, a novel *Ehrlichia* genotype known as Panola Mountain *Ehrlichia*, which is very closely related to *E. ruminantium*, has been identified in the USA and is associated with infections in goats, dogs, and humans (Loftis et al. 2008; Reeves et al. 2008; Qurollo et al. 2013).

In North and South America, different genotypes of *Ehrlichia minasensis* were reported in dairy cattle and mule deer (*Odocoileus hemionus*) in Canada

D.M. de Aguiar
Laboratory of Virology and Rickettsiosis, Veterinary Hospital of the Federal University of Mato Grosso, Brazil Avenida Fernando Correa, 2367, CEP78060 -900 Cuiabá, MT, Brazil
e-mail: danmoura@ufmt.br

J. Bayry (ed.), *Emerging and Re-emerging Infectious Diseases of Livestock*,
DOI 10.1007/978-3-319-47426-7_17

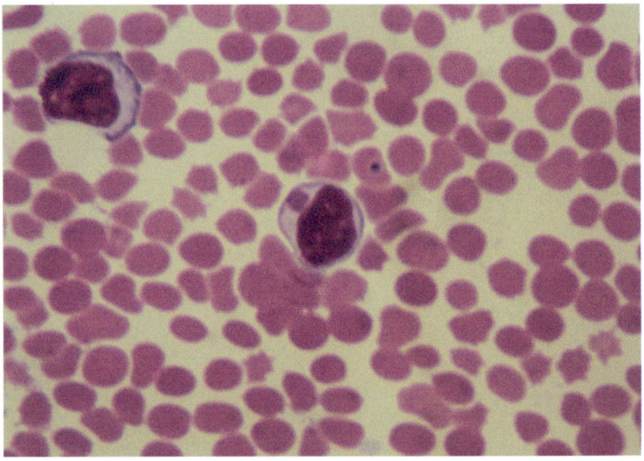

Fig. 17.1 *Ehrlichia minasensis* UFMT-BV morula in the cytoplasm of a monocyte of an experimentally infected calf, as seen in blood smears (originally photographed at 1000 × magnification) (Reproduced from Aguiar et al. 2014)

(genotype BOV2010) and in dairy and beef cattle in midwestern Brazil (genotype UFMT-BV) (Gajadhar et al. 2010; Lobanov et al. 2012; Aguiar et al. 2014). The type strain, *E. minasensis* UFMG-EV, was isolated from a partially engorged *R. microplus* female collected in Minas Gerais, Brazil, 2010. This fact suggests that this ehrlichial agent in cattle appears to be widespread throughout the North and South American continents (Aguiar et al. 2014). Interestingly, this species in bovines may be related to the previously described ehrlichial agent involved in Brazilian dairy cattle and referred to as *Ehrlichia bovis* (Massard and Massard 1982). This organism has been referred to as *Anaplasma bovis*, but its molecular identity has not been confirmed in later scientific literature (Santos and Carvalho 2006; Melo et al. 2010). Moreover, there are no data confirming the presence of *A. bovis* in Brazil.

Ehrlichia are small, Gram-negative, tick-transmitted obligate intracellular bacteria that form microcolonies within membrane-bound cytoplasmic vacuoles, called morulae (Latin *morum* = mulberry) (Popov et al. 1998). *Ehrlichia* infects primarily leukocytes (monocytes, macrophages, granulocytes) (Fig. 17.1) and endothelial cells in mammals and salivary glands, intestinal epithelium, and hemolymph cells in ticks (Groves et al. 1975; Cohn 2003; Allsopp 2010).

Electron microscopic examinations of *Ehrlichia* reveal two morphologically distinct forms, a large (0.4–0.6 μm × 0.7–1.9 μm) pleomorphic form with dispersed nucleoid filaments (reticulate) and a small form (0.4 μm, dense cored) with condensed nucleoid filaments (Popov et al. 1998). The organisms are round and oval shaped, and morulae have typical tri-layered cytoplasmic and outer membranes that, in some cases, may be rippled (Fig. 17.2) (Cabezas-Cruz et al. 2013).

Ehrlichia can be cultured in vitro in different cell lineages. A canine histiocytic cell line (DH82) has been routinely used for the isolation of *E. canis*, *E. chaffeensis*, *E. ruminantium*, *E. muris*, and *E. minasensis* (Fig. 17.3) (Zweygarth et al. 2013). In addition, the tick cell line IDE8 (*Ixodes scapularis*) has been used for the in vitro culture of *E. minasensis* in Brazil (Cabezas-Cruz et al. 2012).

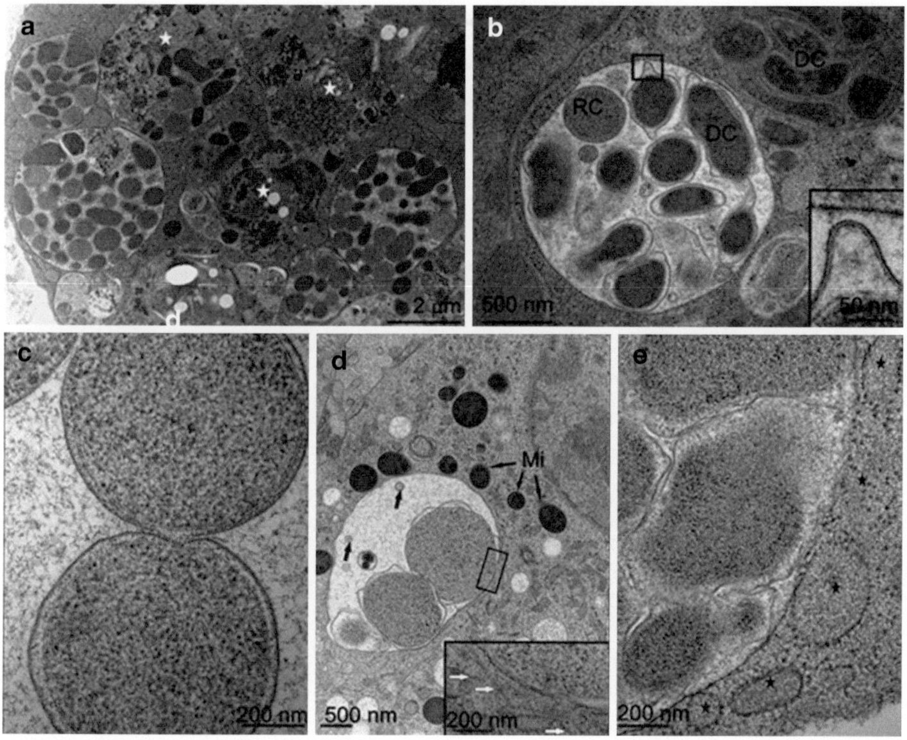

Fig. 17.2 SEM micrograph of *Ehrlichia minasensis* – infected IDE8 cells. (**a**) Cells containing phagolysosomes/secondary lysosomes (*white asterisk*) and numerous bacteria-containing vacuoles. (**b**) Electron-dense bodies and small vesicles inside membrane-lined vacuoles. The inset provides a detailed view of the rippled membrane. (**c**) Reticulate cells undergoing binary fission. (**d**) Vacuole containing reticulate bodies with ruffled outer membranes and small vesicles (*black arrows*) surrounded by mitochondria (Mi) and microtubules (detail in inset, *white arrows*). (**e**) A cisterna of endoplasmic reticulum in close contact with the membrane of the morulae (Reproduced from Cabezas-Cruz et al. 2013)

17.3 Epidemiology

Ehrlichia are transmitted to mammals by way of vectors. Ehrlichial infections are transmitted through the salivary secretions of an attached tick during feeding (Groves et al. 1975; Cabezas-Cruz et al. 2012). Several different tick species are able to transmit diseases horizontally via infection from vectors to animals (Cohn 2003). Furthermore, different *Ehrlichia* species appear to be adapted to distinct tick species, which in turn are adapted to different animal species (Table 17.1) that are responsible for maintaining the bacteria in nature. For instance, dogs are primary hosts for *Rhipicephalus sanguineus*, the brown dog tick (Stich et al. 2008), which is the main vector of *E. canis* transmission to dogs. On the other hand, dogs, which are considered the main reservoir of *E. canis*, can survive for months while

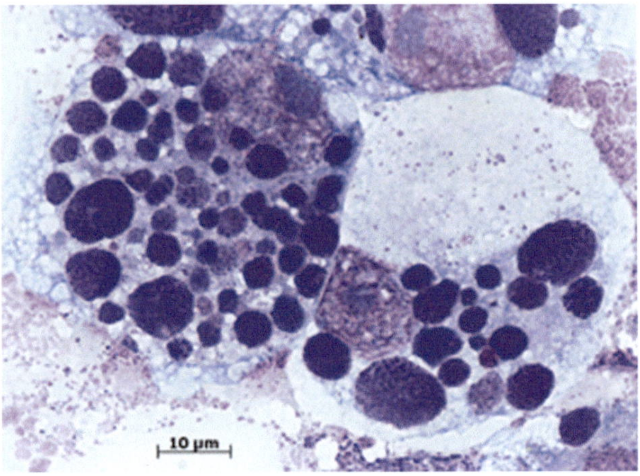

Fig. 17.3 Morulae of *Ehrlichia minasensis* with DH82 culture cells, visualized under light microscopy (originally photographed at 400 × magnification) (Reproduced from Zweygarth et al. 2013)

Table 17.1 *Ehrlichia* species that infect domestic and wild animals and ticks

Species	Geographic distribution	Vector	Reservoir
Ehrlichia canis	Worldwide	*Rhipicephalus sanguineus*	Dog
Ehrlichia chaffeensis	USA	*Amblyomma americanum*	White-tailed deer, coyotes
Ehrlichia muris	Japan	*Haemaphysalis flava*	Wild mouse (*Eothenomys kageus*)
Ehrlichia ewingii	USA	*Amblyomma americanum*	White-tailed deer
Ehrlichia ruminantium	Africa and Caribbean	*Amblyomma hebraeum*, *A. variegatum*	Bovine, dogs, wild ungulates
Ehrlichia minasensis	Brazil and Canada	*Rhipicephalus microplus*, *Amblyomma cajennense* s.l. *Dermacentor* spp., *Ixodes* sp.	Bovine, mule deer (*Odocoileus hemionus*)
Ixodes ovatus Ehrlichia	Japan	*Ixodes ovatus*	
Ehrlichia muris-like	USA	*Ixodes scapularis*	White-footed mouse (*Peromyscus leucopus*)
Panola Mountain Ehrlichia	USA	*Amblyomma americanum*	White-tailed deer, goats
"*Candidatus* Ehrlichia khabarensis"	Russia	*Haemaphysalis* spp..*Ixodes* spp.	Red-backed voles (*Myodes rutilus*), gray red-backed voles (*Myodes rufocanus*)

supporting cyclic bacteremia at different levels that can infect new populations of *R. sanguineus* ticks (Harrus et al. 1998; Stich et al. 2008). The mode of transmission is transstadial, whereby the infection is transmitted to subsequent tick developmental stages. On the other hand, given that ticks cannot transmit *Ehrlichia* via the transovarial route, the tick itself cannot serve as a reservoir of disease (Aguiar et al. 2007).

Not coincidentally, *E. canis* is a species closely related to the ehrlichial agent detected in cattle in the Americas. The new species *E. minasensis* (genotype UFMG-EV) was first isolated from the salivary secretion of *R. microplus* ticks, the probable main tick vector in the tropical region of the New World (Cabezas-Cruz et al. 2012). In Brazil, this tick is considered the most widespread ectoparasite of livestock (Estrada-Peña et al. 2006). Similarly to the brown dog tick, *R. microplus* is a one-host tick naturally adapted to bovines, especially present in dairy cattle herds. However, this tick molts all its phases in a single animal, and transmission between animals should occur due to the high herd density typical in dairy cattle farms, enabling *R. microplus* to be transferred between animals (Aguiar et al. 2014).

Ehrlichiosis may also be spread mechanically through infected hypodermic needles, castration and dehorning instruments, and blood transfusions. The infection may also be transmitted during immunization against babesiosis, when blood from an infected donor is used.

Prevalence rates based on Giemsa-stained blood smears in dairy cattle range from 3 to 5.5 % (Santos and Carvalho 2006; Melo et al. 2010), while a higher frequency of PCR positivity was observed in dairy cattle (14 %) than in beef cattle, which presented a lower prevalence (3.3 %). In fact, the susceptibility of dairy cattle to *R. microplus* may be an important factor contributing to the higher transmission rates in these herds (Aguiar et al. 2014). Calves appear to be prone to developing clinical symptoms, whereas after the acute stage, older cattle maintain a lifelong persistent infection, serving as reservoirs of the pathogen for continued transmission to susceptible hosts (Lobanov et al. 2012).

In Canada, the genotype of *E. minasensis*, namely, the genotype BOV2010, was also detected in mule deer, suggesting that other tick species could also be involved in the transmission of this *Ehrlichia* species in the related area (Lobanov et al. 2012). Ticks that are common in British Columbia, Canada, and that have been found in cattle include *Dermacentor albipictus*, *Dermacentor andersoni*, and different *Ixodes* species (Gregson 1956; Gajadhar et al. 2010).

In Brazil, *Amblyomma cajennense* sensu lato (s.l.), a three-host tick species, feeds chiefly on horses, capybaras (*Hydrochoerus hydrochaeris*), and tapirs (*Tapirus terrestris*) (Labruna 2009), but also commonly feeds on cattle when pastures are close to forests or sylvatic environments (Aguiar et al. 2014). A study in Brazil found that *A. cajennense* s.l. was an experimental vector of the so-called *E. bovis* transmitted to dairy cattle (Massard 1984). In this scenario, in addition to *R. microplus*, the presence of *A. cajennense* could be considered in the transmission of this *Ehrlichia* species to cattle in Brazil.

17.4 Pathogenesis and Clinical Signs

Despite the fact that reports of bovine ehrlichiosis in Brazil date back to the 1980s, several aspects of its pathogeny remain unclear. However, in view of its close relationship with *E. canis* and the similarities observed between canine and bovine ehrlichiosis, this section describes the pathogenesis of *E. canis* and the species of the genera, which suggests that they may present various similarities.

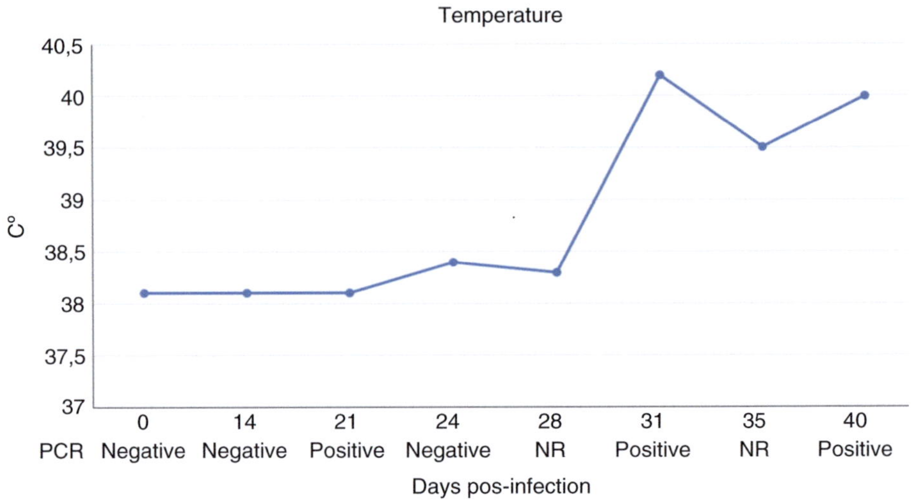

Fig. 17.4 Increase in the body temperature (°C) of a calf during the course of infection with *Ehrlichia minasensis* UFMT-BV (Reproduced from Aguiar, DM. personal archive)

Clinical infection in dogs is manifested in three stages: acute, subclinical, and chronic infection. Acute infection, which occurs 1–3 weeks after a dog is bitten by an infected tick, is typically mild. The organism invades the host's mononuclear cells, where it replicates, forming morulae or bacterial colonies bound within vacuolar membranes. The most consistent hematological change during acute infection is thrombocytopenia, which results from vascular endothelial inflammation with resulting platelet consumption, immunologically mediated destruction of platelets and splenic sequestration of platelets (Cohn 2003). Detectable *E. canis* DNA has been observed 10–14 days postinfection (dpi) (Harrus and Waner 2011). On the other hand, calves experimentally infected with *E. minasensis* BOV2010 showed positive PCR results beginning 12–23 dpi (Gajadhar et al. 2010). In Brazil, a calf experimentally infected with the *E. minasensis* UFMT-BV showed *Ehrlichia* DNA detectable by PCR 21 dpi. On day 28 and 35 dpi, ehrlichial morulae were observed in the cytoplasm of monocytes in peripheral blood smears (Fig. 17.1). The clinical signs observed in the calf were similar to those observed in the acute phase of canine monocytic ehrlichiosis. Anorexia, fever (rectal temperature > 40.0 °C) (Fig. 17.4), and lethargy were the first signs observed at 28 dpi, and bleeding at the venipuncture site was also observed at this time (Fig. 17.5), as well as the appearance of morulae in cytology blood smears (Aguiar et al. 2014).

The clinical findings reported in experimentally infected calves are consistent with those of the first report of *E. bovis* in Brazilian cattle in the 1980s, which described organic debility, anemia, and morulae in the cytoplasm of peripheral blood mononuclear cells. According to the first report, the disease in cattle has also been associated with hyperacute, acute, subacute, and chronic forms. In the hyperacute form, neurological disorders followed by a sudden death may occur after 12 h. In the other forms, hyperplasia of lymph nodes, anorexia, incoordination, fever, and

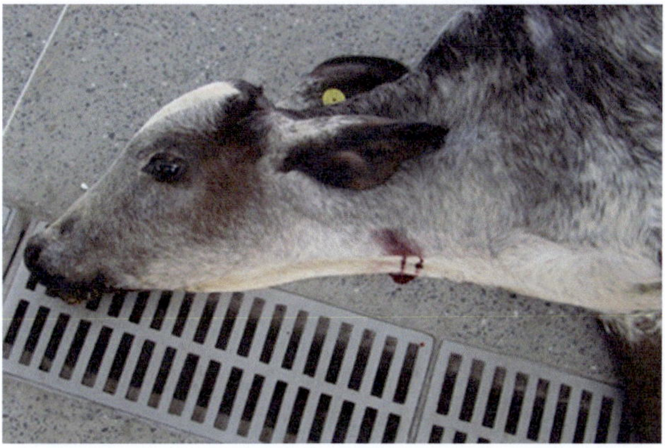

Fig. 17.5 Cutaneous bleeding in a calf due to *Ehrlichia minasensis* UFMT-BV infection (Reproduced from Aguiar, DM. personal archive)

constipation have been observed (Massard and Massard 1982). No other symptoms have been described for this ehrlichiosis in cattle.

17.5 Diagnosis

The diagnosis is made based on a combination of the animal's history (i.e., living in an endemic area, tick infestation, age), clinical and hematological indicators, serologic evidence, and molecular confirmation. The most common clinical signs are nonspecific (anorexia, fever, and lethargy) in calves. In older animals, infection is usually asymptomatic. Due to the nonspecificity of signs, coinfection with other tick-borne pathogens transmitted by the same tick vector should be considered and investigated (i.e., *Anaplasma marginale*, *Babesia* spp.).

Laboratory abnormalities observed in the experimentally infected calf included anemia (4.6×10^6 cells/ mm^3) on day 31, leukopenia (3.7×10^3 cells/ mm^3) on day 35, and thrombocytopenia (116×10^3, 50×10^3 and 124×10^3 platelets/ mm^3) on days 28, 31, and 35, respectively (Figs. 17.6 and 17.7) (Aguiar et al. 2014).

PCR has been shown to be a sensitive method for detecting acute experimental *E. minasensis* UFMT-BV in a calf and in naturally infected cows. A negative PCR result denotes that no target DNA was detected, but does not necessarily prove that no DNA was present in the sample (false-negative result). PCR assays are usually based on the target genes 16S rRNA and *dsb* for diagnosis.

The demonstration of typical cytoplasmic *Ehrlichia* morulae (Fig. 17.1) in monocytes in blood smears by light microscopy strongly supports a diagnosis of ehrlichiosis in cattle. Unfortunately, the search for morulae is difficult and time-consuming. In addition, sensitivity values for cytological evaluation are questionable, since it has been estimated that only about 4 % of cases have been successful in canine monocytic ehrlichiosis (Harrus and Waner 2011).

Traditionally, indirect immunofluorescence (IFA) has been the serological test of choice for ehrlichiosis. The interpretation of indirect FA results must take into account the history, clinical signs, and laboratory findings. A positive result must be

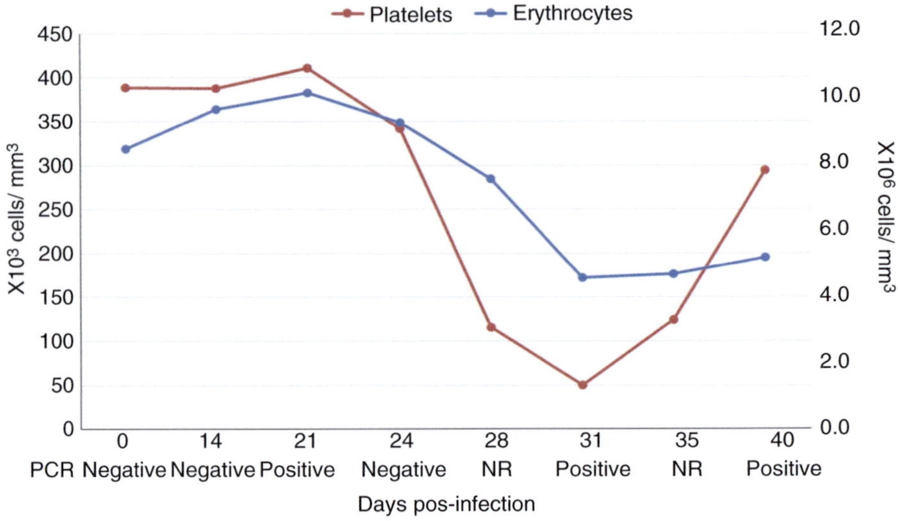

Fig. 17.6 Decreased percentage of erythrocyte and platelets in a calf during the course of infection with *Ehrlichia minasensis* UFMT-BV infection (Reproduced from Aguiar, DM. personal archive)

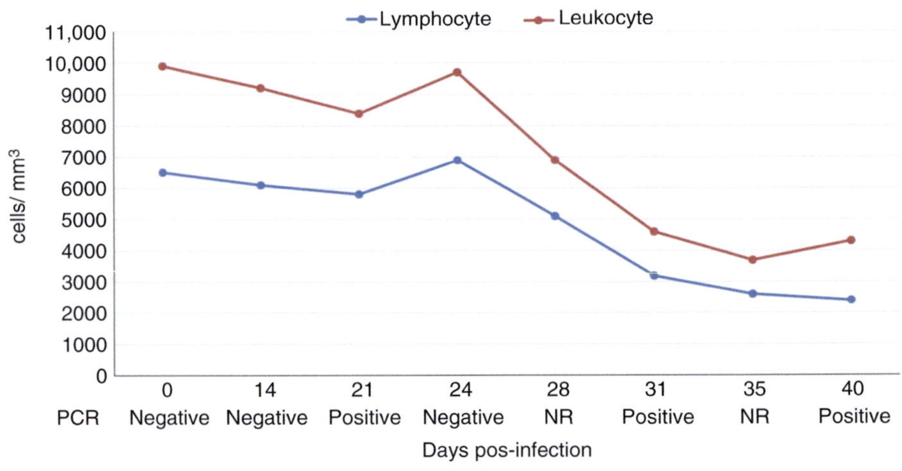

Fig. 17.7 Decreased percentage of leukocytes and lymphocytes in a calf during the course of infection with *Ehrlichia minasensis* UFMT-BV infection (Reproduced from Aguiar, DM. personal archive)

interpreted with caution, as it may represent current infection, resolved infection, or merely exposure. In this case, a second evaluation after 15 days should be considered and may be helpful in the interpretation of serologic results in these circumstances. Unfortunately, the results of studies based solely on serological diagnostic methods are difficult to interpret because of cross-reactivity between *Anaplasma* spp. and *Ehrlichia* spp. when crude antigens are utilized.

Macroscopic and histopathological findings include lymph node enlargement and hyperplasia, periportal mononuclear infiltrates in the liver, and hypercellularity of bone marrow during acute phase of ehrlichiosis in a calf (Aguiar et al. 2014). All reported findings were consistent with ehrlichial infection caused by *E. canis* in dogs (Castro et al. 2004). Lymph node hyperplasia and periportal mononuclear infiltrates in the liver may be associated with the presence of the organism and with stimulation of immune and nonspecific inflammatory mechanisms (Dumler and Bakken 1995; Castro et al. 2004). Bone marrow hypercellularity in the acute phase of ehrlichiosis may be involved with the response against erythrocyte and platelet destruction (Aguiar et al. 2014).

17.6 Treatment

The literature contains no treatment protocols for American ehrlichiosis in cattle. However, treatments against *Ehrlichia* in other species involve the use of antibacterial agents and support. Efficacious drugs include the class of antibiotics known as tetracyclines. The disease can be clinically treated with oxytetracycline, 5–20 mg/kg, at dosing intervals of 12–24 h, via intravenous (IV) or intramuscular (IM) injection for 14 days. Chloramphenicol is not recommended for cattle; however, florfenicol can be applied in doses of 20 mg/kg, at a dosing interval of 48 h, via IM for 14 days. The treatment duration of 14 days is based on the therapeutic protocols used in the treatment of other types of ehrlichiosis. With lower or uncertain effectiveness (Eddlestone et al. 2006), but adjuvant in the treatment of coinfection with *A. marginale* (de Vos et al. 1987), a single dose (SD) of 1.2 mg/kg of imidocarb dipropionate can be administered subcutaneously (SC). Antibacterial agents that are ineffective against monocytotropic *Ehrlichia* spp. in humans are erythromycin, new macrolides (azithromycin, clarithromycin, and telithromycin), penicillins, and aminoglycosides. The effectiveness of quinolones varies according to the drug used and the organism involved. Enrofloxacin has been evaluated for the treatment of experimentally induced *E. canis* infection and found to be ineffective (Neer et al. 1999). Blood transfusion is indicated in animals with a PCV of less than 15 %.

17.7 Prophylactic Measures

Similarly to anaplasmosis in cattle, the eradication of ehrlichiosis is impracticable because of the widespread tick vector that is able to carry the disease among herds and the long period the disease remains in the carrier animal (reservoir).

The introduction of the disease into herds by carrier animals should be prevented by prior serological testing or PCR. Special attention should focus on the prevention of iatrogenic transmission. Intensive control of ticks with acaricides should be carried out in tropical regions, including strategic control, by measuring susceptibility to acaricides to prevent the development of tick resistance (Pereira et al. 2008).

Commercial vaccines are not available for bovine ehrlichiosis. So far, a vaccination protocol using an attenuated strain of *E. canis* in the prophylaxis of canine ehrlichiosis has been tested and has shown promising results (Rudoler et al. 2012).

References

Aguiar DM, Ziliani TF, Zhang X, Melo ALT, Braga IA, Witter R et al (2014) A novel *Ehrlichia* genotype strain distinguished by the TRP36 gene naturally infects cattle in Brazil and causes clinical manifestations associated with ehrlichiosis. Ticks Tick Borne Dis 5:537–544. doi:10.1016/j.ttbdis.2014.03.010

Aguiar DM, Cavalcante GT, Pinter A, Gennari SM, Camargo LM, Labruna MB (2007) Prevalence of Ehrlichia canis (Rickettsiales: Anaplasmataceae) in dogs and Rhipicephalus sanguineus (Acari: Ixodidae) ticks from Brazil. J Med Entomol 44:126-132. doi:http://dx.doi.org/10.1093/jmedent/41.5.126

Allsopp BA (2010) Natural history of *Ehrlichia ruminantium*. Vet Parasitol 167:123–135. doi:10.1016/j.vetpar.2009.09.014

Cabezas-Cruz A, Zweygarth E, Ribeiro MF, da Silveira JA, de la Fuente J, Grubhoffer L et al (2012) New species of *Ehrlichia* isolated from *Rhipicephalus* (*Boophilus*) *microplus* shows an ortholog of the *E. canis* major immunogenic glycoprotein gp36 with a new sequence of tandem repeats. Parasit Vectors 5:291. doi:10.1186/1756-3305-5-291

Cabezas-Cruz A, Vancová M, Zweygarth E, Ribeiro MF, Grubhoffer L, Passos LM (2013) Ultrastructure of *Ehrlichia mineirensis*, a new member of the *Ehrlichia* genus. Vet Microbiol 167:455–458. doi:10.1016/j.vetmic.2013.08.001

Cabezas-Cruz A, Zweygarth E, Vancová M, Broniszewska M, Grubhoffer L, Passos LM et al (2016) *Ehrlichia minasensis* sp. nov., a new species within the genus *Ehrlichia* isolated from the tick *Rhipicephalus microplus*. Int J Syst Evol Microbiol. doi:10.1099/ijsem.0.000895

Castro MB, Machado RZ, de Aquino LP, Alessi AC, Costa MT (2004) Experimental acute canine monocytic ehrlichiosis: clinicopathological and immunopathological findings. Vet Parasitol 119:73–86

Cohn LA (2003) Ehrlichiosis and related infections. Vet Clin North Am Small Anim Pract 33:863–884

de Vos AJ, McGregor W, Stewart NP, Dalgliesh RJ, Barry DN (1987) Influence of imidocarb prophylaxis on the pathogenicity and immunogenicity of *Anaplasma marginale* infections in cattle. Aust Vet J 64:83–85

Dumler JS, Bakken JS (1995) Ehrlichial diseases of humans: emerging tick-borne infections. Clin Infect Dis 20:1102–1110

Dumler JS, Barbet AF, Bekker CP, Dasch GA, Palmer GH, Ray SC et al (2001) Reorganization of genera in the families Rickettsiaceae and Anaplasmataceae in the order Rickettsiales: unification of some species of *Ehrlichia* with *Anaplasma*, *Cowdria* with *Ehrlichia* and *Ehrlichia* with *Neorickettsia*, descriptions of six new species combinations and designation of *Ehrlichia equi* and 'HGE agent' as subjective synonyms of *Ehrlichia phagocytophila*. Int J Syst Evol Microbiol 51:2145–2165

Eddlestone SM, Neer TM, Gaunt SD, Corstvet R, Gill A, Hosgood G et al (2006) Failure of imidocarb dipropionate to clear experimentally induced *Ehrlichia canis* infection in dogs. J Vet Intern Med 20:840–844

Estrada-Peña A, Bouattour A, Camicas JL, Guglielmone A, Horak I, Jongejan F et al (2006) The known distribution and ecological preferences of the tick subgenus *Boophilus* (Acari: Ixodidae) in Africa and Latin America. Exp Appl Acarol 38:219–235

Gajadhar AA, Lobanov V, Scandrett WB, Campbell J, Al-Adhami B (2010) A novel *Ehrlichia* genotype detected in naturally infected cattle in North America. Vet Parasitol 173:324–329

Gregson JD (1956) The Ixodoidea of Canada. Science Service, Entomology Division of Canada Department of Agriculture. Queen's Printer, Ottawa. doi:10.5962/bhl.title.58947

Groves MG, Dennis GL, Amyx HL, Huxsoll DL (1975) Transmission of *Ehrlichia canis* to dogs by ticks (*Rhipicephalus sanguineus*). Am J Vet Res 36:937–940

Harrus S, Waner T (2011) Diagnosis of canine monocytotropic ehrlichiosis (*Ehrlichia canis*): an overview. Vet J 187:292–296. doi:10.1016/j.tvjl.2010.02.001

Harrus S, Waner T, Aizenberg I, Foley JE, Poland AM, Bark H (1998) Amplification of ehrlichial DNA from dogs 34 months after infection with *Ehrlichia canis*. J Clin Microbiol 36:73–76

Labruna MB (2009) Ecology of rickettsia in South America. Ann N Y Acad Sci 1166:156–166. doi:10.1111/j.1749-6632.2009.04516.x

Lobanov VA, Gajadhar AA, Al-Adhami B, Schwantje HM (2012) Molecular study of free-ranging mule deer and white-tailed deer from British Columbia, Canada, for evidence of *Anaplasma* spp. and *Ehrlichia* spp. Transbound Emerg Dis 59:233–243

Loftis AD, Levin ML, Spurlock JP (2008) Two USA *Ehrlichia* spp. cause febrile illness in goats. Vet Microbiol 130:398–402

Massard CA (1984) Ehrlichia bovis (Donatien & Letosquard, 1936) diagnostico, cultivo in vitro, e aspectos epidemiológicos em bovinos no Brasil. http://r1.ufrrj.br/wp/ppgcv/wp-content/themes/PPGCV/pdf/R060.pdf. Accessed 20 Oct 2015

Massard CA, Massard CL (1982) *Ehrlichia bovis* (Rickettsiales: Rickettsiaceae) em gado de leite no Brasil. Arq Univ Fed Rur Rio de J 5:237–239

Melo OA Jr, Albernaz AP, Thiebaut JTL, Miranda FJB, Machado JA, Silva ACF (2010) Ocorrência de *Anaplasma bovis* (Donatien & Lestoquard, 1936, Dumler et al. 2001) na região de Bom Jesus do Itabapoana, RJ. Ci Anim Bras 11:888–891. doi:10.5216/cab.v11i4.5575

Neer TM, Eddlestone SM, Gaunt SD, Corstvet R (1999) Efficacy of enrofloxacin for the treatment of experimentally induced *Ehrlichia canis* infection. J Vet Intern Med 13:501–504

Pereira MC, Labruna MB, Szabó MPJ, Klafe GM (2008) Rhipicephalus (Boophilus) microplus: biologia, controle e resistência, 1st edn. Medvet Livros, São Paulo

Popov VL, Han VC, Chen SM, Dumler JS, Feng HM, Andreadis TG et al (1998) Ultrastructural differentiation of the genogroups in the genus *Ehrlichia*. J Med Microbiol 47:235–251

Pritt BS, Sloan LM, Johnson DK, Munderloh UG, Paskewitz SM, McElroy KM et al (2011) Emergence of a new pathogenic *Ehrlichia* species, Wisconsin and Minnesota, 2009. N Engl J Med 365:422–429

Qurollo BA, Davenport AC, Sherbert BM, Grindem CB, Birkenheuer AJ, Breitschwerdt EB (2013) Infection with Panola Mountain *Ehrlichia* sp. in a dog with atypical lymphocytes and clonal T-cell expansion. J Vet Intern Med 5:1251–1255

Rar VA, Pukhovskaya NM, Ryabchikova EI, Vysochina NP, Bakhmetyeva SV, Zdanovskaia NI et al (2015) Molecular-genetic and ultrastructural characteristics of '*Candidatus* Ehrlichia khabarensis', a new member of the *Ehrlichia* genus. Ticks Tick Borne Dis 6:658–667. doi:10.1016/j.ttbdis.2015.05.012

Reeves WK, Loftis AD, Nicholson WL, Czarkowski AG (2008) The first report of human illness associated with the Panola Mountain *Ehrlichia* species: a case report. J Med Case Reports 2:139. doi:10.1186/1752-1947-2-139

Rudoler N, Baneth G, Eyal O, van Straten M, Harrus S (2012) Evaluation of an attenuated strain of *Ehrlichia canis* as a vaccine for canine monocytic ehrlichiosis. Vaccine 31:226–233. doi:10.1016/j.vaccine.2012.10.003

Santos CF, Carvalho CB (2006) Primeiro relato de *Anaplasma bovis* (Donatien & Lestoquard, 1936, Dumler et al. 2001) na microrregião de Campos dos Goytacazes, Estado do Rio de Janeiro, Brasil: nota de pesquisa. Rev Bras Parasit Vet 15:126–127

Shibata S, Kawahara M, Rikihisa Y, Fujita H, Watanabe Y, Suto C et al (2000) New *Ehrlichia* species closely related to *Ehrlichia chaffeensis* isolated from *Ixodes ovatus* ticks in Japan. J Clin Microbiol 38:1331–1338

Stich RW, Schaefer JJ, Bremer WG, Needham GR, Jittapalapong S (2008) Host surveys, ixodid tick biology and transmission scenarios as related to the tick-borne pathogen Ehrlichia canis. Vet Parasitol 158:256–273. doi:10.1016/j.vetpar.2008.09.013

Zweygarth E, Schöl H, Lis K, Cabezas-Cruz A, Thiel C, Silaghi C et al (2013) In vitro culture of a novel genotype of *Ehrlichia* sp. from Brazil. Transbound Emerg Dis 60 Suppl 2:86–92. doi:10.1111/tbed.12122

Part III

Emerging Parasitic Diseases of Livestock

Bovine Trypanosomiasis in Brazil

18

Solange de Araújo Melo, Renata Mondêgo de Oliveira, and Ana Lucia Abreu-Silva

The infection caused by *Trypanosoma vivax* in South America was first reported in 1919, in French Guiana (Silva et al. 2002). However, Lavier (1943) pointed that *T. vivax* was introduced in Latin America in the nineteenth century, through the importation of infected cattle herds.

The parasite is autochthone from the African continent, where it causes great economic losses (Silva et al. 1996; Jones and Dávila 2001). In this continent, tripanosomiasis is more prevalent in areas of the occurrence of tsetse fly (Levine 1973). *T. vivax* is a protozoan belonging to the order Kinetoplastida, family Trypanosomatidae, subgenus Duttonella (Vickerman 1976; Losos 1986; Dagnachew and Bezie 2015). It parasites domestic and wild ruminants, as well as horses (Maudlin et al. 2004). This parasite presents two basic morphological forms: the epimastigote form, found in the tissues, and the trypomastigotes, found in the bloodstream (Brener 1979; Silva et al. 2002) (Fig. 18.1).

T. vivax significantly impacts the livestock systems, since it affects animals of economic importance, such as ruminants (Desquesnes and Gardiner 1993; Wells and Betancourt 1977). It causes disease not only in cattle but also in buffaloes, and it is able to infect a large number of both domestic and wild species. However, literature suggests that swines and canines are resistant to infection (Silva et al. 2003). In Africa, the parasite also causes trypanosomiasis in horses, small ruminants, and camelids. Although the detection of antibodies is common in cattle, sheep, and buffaloes, most of the infected animals do not show an apparent infection (Paiva et al. 2000; Garcia et al. 2006).

In Brazil, the first register of *T. vivax* was made by Boulhosa in 1946, in cattles of the northeast mesoregion of Pará State. However, the official record was made by Shaw and Lainson in 1972, in buffaloes herd. Later, Serra-Freire reported the occurrence of the parasite in sheeps and cattles in the Amapá State, in 1981, and in

S. de Araújo Melo • R.M. de Oliveira • A.L. Abreu-Silva (✉)
Universidade Estadual do Maranhão, São Luís, Maranhão, Brazil
e-mail: abreusilva.ana@gmail.com

J. Bayry (ed.), *Emerging and Re-emerging Infectious Diseases of Livestock*,
DOI 10.1007/978-3-319-47426-7_18

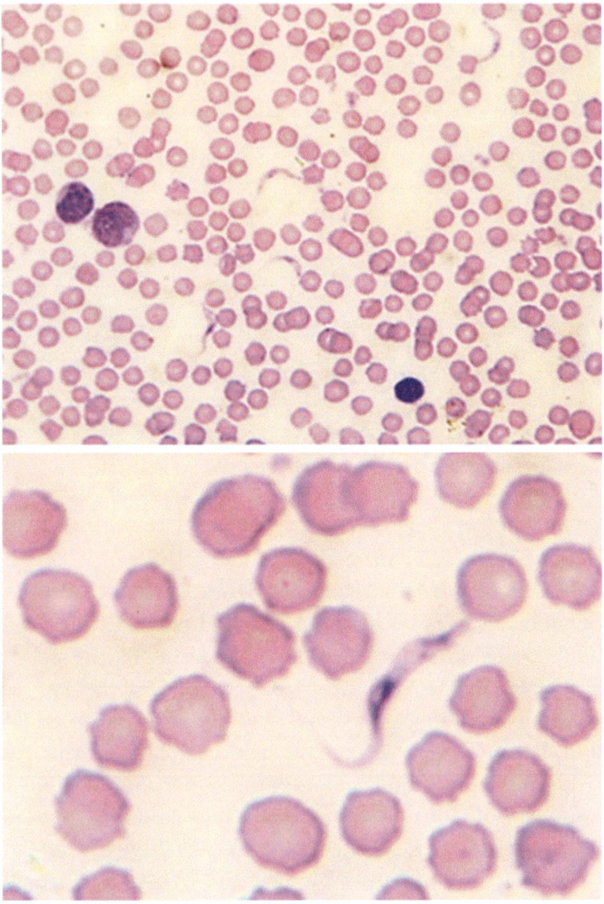

Fig. 18.1 *T. vivax* tripomastigote form isolated from a cattle from Maranhão State, Brazil

buffaloes, in 1983. Posteriorly, it was reported in cattles in Poconé and Miranda, both towns in the Pantanal regions of Mato Grosso and Mato Grosso do Sul. In Northeast Brazil, the parasite was found in Paraiba State (Batista et al. 2007), Maranhão State (Guerra et al. 2008; Melo et al. 2011), and Pernambuco State (Pimentel et al. 2012; Guerra et al. 2013) (Fig. 18.2).

Based on the way the parasite is transmitted to the host, the genus *Trypanosoma* is classified into two groups: Salivaria and Stercoraria (Hoare 1972; Losos 1986). In the first one, the parasite is present in the insect salivary gland and it is subsequently inoculated after a blood meal, whereas, in the *stercoraria* group, it is mechanically transmitted through feces (Silva et al. 2002).

T. vivax is mainly transmitted in a cyclic manner by the tsetse fly bite, in areas where this vector is present (Silva et al. 2002). In the American continent, due to the absence of the vector, the parasite has adapted to the mechanical transmission, via bloodsucking insects such as *Stomoxys calcitrans* and tabanids (Silva et al. 1996; Paiva et al. 2000; Levine 1973). Mechanical transmission is less efficient than the

Fig. 18.2 Areas of occurrence of *T. vivax* in Brazil

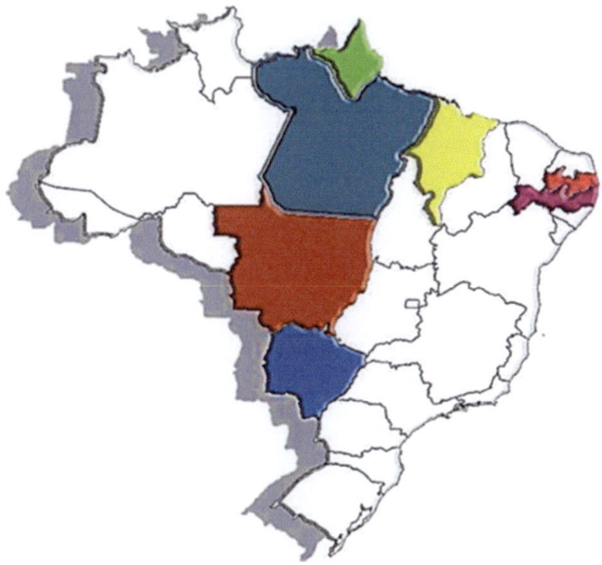

vectorial one, since the parasite remains viable in the fly mouthparts for a short period of time (Hoare 1972; Molyneux and Ashford 1983; Gardiner 1989). But, due to this adaptation, *T. vivax* expanded to the Americas. Iatrogenic transmission can also occur, through contaminated needles and blood transfusion, as already demonstrated in experimental infections (Silva et al. 1996).

The weather conditions, such as a marked dry season and an intense rainy season, as it happens in some Brazilian regions such as the city of Pedreiras, Maranhão State, also favor increased disease transmission (Mamoudou et al. 2016). Melo et al. (2011) observed that the clinical signs in infected animals disappear in three to four months between the dry and rainy season, which indicates that some areas of Brazil have a disease stabilization capacity. It is believed that this happens due to the transmission type and the parasite adaptation skills.

In contrast to Africa, in Latin America, there are reports of trypanosomiasis in newborn calves, which become more resistant than adults (Ogwu and Nuru 1981; Dwinger and Hall 2000; Silva et al. 2013). However, an experimental infection demonstrated that *T. vivax* is able to cross the placental barrier and cause abortion, perinatal mortality, and low birth weight (Meléndez et al. 1993).

In the areas where there are only mechanical route of transmission, as in Latin America, bovine trypanosomiasis occurs in epidemic outbreaks. Thus, it is not unusual to have multiple outbreaks in the same area. Thereafter, the infection can be eliminated either by treatment as self-cure (Desquesnes 1997). The parasite detection in the bloodstream depends on the course of the infection and usually can be found in the first month post-infection. From this moment, anti-*T. vivax* antibodies are found in the sera. Animals can remain asymptomatic for many years and, due to a possible immunosuppression, have a high parasitemia, featuring an epizootic outbreak. This phenomenon is called "epidemiological wave," typical of mechanical transmissions, which occurs in countries in South America such as Guyana, Colombia, and Venezuela (Meléndez et al. 1993; Desquesnes 1997, 2004).

In cattle, the incubation period ranges from 9 to 14 days, for virulent isolates, and 9–59 days for the less pathogenic isolates (Hoare 1972). The amount of circulating parasites undergoes fluctuations throughout the day, so it may be observed high levels in the morning and lower levels in the afternoon (Hoof et al. 1948). Generally, the parasitemia lasts several months, followed by a period where no parasites can be found in the bloodstream. Depending on the host immunological status, relapse can occur (Hoare 1972; Batista et al. 2006).

T. vivax has peculiar characteristics to overcome host immune response leading to resistance toward commonly used drugs and, consequently, its massive proliferation (Fairbairn 1953). This fact is due to its antigenic variation that allows the emergence of phenotypically different parasites, which influences its pathogenic potential (Walker 1964). The morphometric characteristics of *T. vivax* evidenced in many studies show that more elongated parasites provide the appearance of chronic course of illness, while those with shorter forms provide the acute course of trypanosomiasis (Fairbairn 1953).

In Brazil and Bolivia, shorter forms of *T. vivax* are recorded than those found in West Africa. However clinical symptoms recorded in these two regions of South America would be better defined as an acute disease (Shaw and Lainson 1972; Silva et al. 1996, 1998; D'Ávila et al. 1997).

The parasites circulate among the herd, and during stress condition, such as drought and abundance of bloodsucking insects, the disease manifests itself clinically. Thus the parasite alternates between silent periods, where parasite cannot be detected in the bloodstream, followed by a suddenly onset of the disease. That is why it occurs frequently in the form of multiple foci (Desquesnes 2004; Batista et al. 2008).

The main clinical signs of the infection are fever, anemia, miscarriage, appetite loss, lethargy, progressive weight loss, and dysentery. Experimental studies in cattle, sheep, goats, pigs, and horses showed that the most frequent clinical manifestations are intermittent fever, lacrimation, nasal discharge, anemia, edema, and progressive weight loss (Souza 1980). The most common reproductive changes are follicular degeneration (Silva et al. 2013), low semen quality, and infertility (Sekoni 1994, Almeida et al. 2010). Abortion may occur, depending on the stage in which the female is infected (Weems et al. 2007), indicating that it is associated with metabolic demands and immunological, biochemical, hematological, and hormonal changes (Silva et al. 2013).

After parasitemia, trypanosomes can be found in several organs, especially in lymphatic tissues, eyes, and cerebrospinal fluid, causing lesions at different degrees (Valli and Forsberg 1979; Van Den Ingh and Neijs-Baker 1979; Whitelaw et al. 1988; Kimeto et al. 1990; Tuntasuvan et al. 1997; Batista et al. 2006; Almeida et al. 2010).

In bovine trypanosomiasis, anemia is the major clinical sign and its persistence can lead to congestive heart failure and consequently, animal death. New studies in naturally infected herds showed that *T. vivax* does not usually cause disease in

Brazil, leading to a chronic and asymptomatic infection. However, in cattle, buffaloes, and sheep, parasites have been detected in the blood, as well as serum antibodies (Murray et al. 1979; Gardiner 1989). During the course of infection, hematocrit values undergo significant changes. Low hematocrit is associated with more severe stages of trypanosomiasis, which usually evolve to death (Ventura et al. 2001; Dávila et al. 2003; Silva et al. 2013).

The clinical manifestations of trypanosomiasis in the acute febrile phase are similar to those of other diseases and hence require differential diagnosis, especially babesiosis and anaplasmosis. Furthermore, differential diagnosis should be made with malnutrition, helminth infections, and diseases that cause neurological disorders, since the parasite can be lodged in cerebrospinal fluid and cause choroids, multifocal mononuclear meningitis, and encephalitis (Whitelaw et al. 1988, Osório et al. 2008, Silva et al. 2013).

The detection of antibodies against *T. vivax* in asymptomatic cattle suggests the existence of subclinical disease, which has been associated with low efficiency of mechanical transmission, animal resistance, or low virulence of the isolated parasite. The frequent inoculation of small amounts of *T. vivax* by mechanical vectors in endemic areas contributes to the development of protective antibodies and the coexistence of parasites in the herd without causing the disease (Batista et al. 2006).

The debilitating characteristics of trypanosomiasis predispose to secondary infections, hindering the diagnosis and determining the most severe cases (Paiva et al. 2000). Direct diagnosis can be made by methods of lymph node aspiration, thick smear, thin section, or microhematocrit method or Woo technique, which consists of direct observation of blood forms. Besides these, there are serological diagnostic methods such as indirect immunofluorescence, direct agglutination test, ELISA, tripanolise test, and molecular methods (Silva et al. 2002, 2013).

According to Pillay et al. (2013), the antigen TvGM6 could be used in indirect ELISA for the diagnosis of *T. vivax* in cattle, but no test based on this antigen is available until now. Regarding molecular techniques, the polymerase chain reaction is not very sensitive for the identification of parasite, especially some ITS1 regions, mainly because of the higher content of guanine-cytosine, and hence it is hard to find a good marker, as described by Fikru et al. (2016).

The disease can be controlled by chemical treatment, particularly during early periods of acute disease. However, in cattle with nervous symptoms, the treatment is not efficient because there is recurrence of the disease and death of the animals (Batista et al. 2007). Treatment should be combined with the restricted circulation of sick animals, epidemiological surveillance, and the elimination or control of the vector (Dwinger and Hall 2000). The satisfactory treatment of trypanosomiasis requires more than a correct administration of trypanocidal drug, and the rate of recovery is largely determined by the nutrition plan, the amount of exercise during convalescence, and duration of the disease (Kinabo 1993, Van Den Bossche et al. 2000).

For bovine trypanosomiasis, trypanocidal drugs are used such as homidium chloride, isometamidium chloride, and diminazene aceturate while for horses and buffaloes, suramin, quinapyramina, and melarsomine are the election drugs (Table 18.1).

Table 18.1 Currently available trypanocide drugs for the treatment of *T. vivax* infection

Active principle	Dosage	Species
Diminazene aceturate	3.5–7 mg/kg, intramuscularly	Large and small ruminants
Homidium chloride	1 mg/kg, intramuscularly	Pigs, small ruminants, and horses
Isometamidium chloride	0.25–0.5 mg/kg or 0.5–1 mg/kg, intramuscularly	Cattle and small ruminants
Quinapyramine dimethylsulphate/chloride	3–5 mg/kg, subcutaneously	Equines

Primarily, they are used to minimize the economic loss, but they can be used to clear the parasites of the hosts. In other words, depending on the dosage administered, these drugs could be used for curative and prophylactic purposes. In the latter case, the drugs are stored in the tissue and are gradually released into the bloodstream, maintaining the concentration of active compound sufficient to control trypanosomes (Dagnachew and Bezie 2015; Tsegaye et al. 2015).

Clinical and parasitological relapses observed after the treatment of animals with nervous signs have been associated with the location of the parasite in the nervous tissue, cerebrospinal fluid, and aqueous humor, where trypanosomes are protected against trypanocidal drugs (Whitelaw et al. 1988; Galiza et al. 2011). These extravascular foci are indicative of a worse prognosis and are likely to be important in the maintenance of trypanosomiasis. In humans, infected with *T. brucei*, recurrences occur because trypanocidal drugs do not cross the blood-brain barrier (Brun et al. 2001). This suggests that mass treatment of the herd is not necessary. Besides being an uneconomic measure, it predisposes to the resistance to trypanocidal drugs (Schunefeld et al. 1987). The speed upon which the trypanosomes develop resistance to these drugs has complicated the treatment and control of trypanosomiasis (Mutugi et al. 1995).

If the treatment is done at an early stage, when the parasitemia is always higher, there is a quick prevention at the mechanical transmission by insects (Stephen 1986). In order to prevent resistance to the drugs used, the administration of trypanocidal drug restricted only to infected animals is suggested, avoiding massive drug application in the herd (Vargas and Arellano 1997).

References

Almeida KS, Freitas FLC, Tebaldi JH, Alessi AC, Machado RZ, Nascimento AA (2010) Alterações clínicas, histopatológicas e enzimáticas em ovinos infectados experimentalmente por *Trypanosoma vivax*. Ciênc Anim Bras 11:669–676

Batista JS, Bezerra FSB, Lira RA, Carvalho JRG, Rosado Neto AM, Petri AA, Teixeira MMG (2008) Clinical, epidemiological and pathological signs of natural infection in cattle by *Trypanosoma vivax* in Paraíba Brazil. Pesq Vet Bras 28:63–69

Batista JS, Correa FR, Barbosa RC, Guerra JL (2006) Infecção experimental por *Trypanosoma vivax* em ovinos. Pesq. Vet. Bras 26:31–37

Batista JS, Riet-Correa F, Teixeira MMG, Madruga CR, Simões SDV, Maia TF (2007) Trypanosomiasis by *Trypanosoma vivax* in cattle in the Brazilian semiarid: description of an outbreak and lesions in the nervous system. Vet Parasitol 143:174–181

Boulhosa J (1946) Informação Científica, Boletim Técnico Ministério da Agricultura, pp 21–26

Brener Z (1979) O parasito: Relações hospedeiroparasito. In: Brener Z, Andrade Z (eds) *Trypanosoma cruzi* e doença de Chagas, Guanabara Koogan, Rio de Janeiro, pp 1–41

Brun R, Shucmancer R, Schmide K, Brurri C (2001) The phenomenon of treatment failures in human African trypanosomosis. Trop Med Int Health 6:906–914

D'avila AMR, Ramirez L, Silva RAMS (1997) Morphological and biometrical differences among Trypanosoma vivax isolates from Brazil and Bolivia. Mem Inst Oswaldo Cruz, Rio de Janeiro 92:357–358

Dagnachew SD, Bezie M (2015) Review on *Trypanosoma vivax*. Afr J Basic Appl Sci 7:41–64

D'ávila AM, Herrera HM, Schlebinger T, Souza SS,Traub Cseko YM (2003) Using PCR for unraveling the cryptic epizootiology of livestock trypanosomosis in the Pantanal. Braz. Vet Parasitol 117:1–13

Desquesnes M (1997) Evaluation of a simple PCR technique for the diagnosis of *Trypanosoma vivax* infection in the serum of cattle in comparison to parasitological techniques and antigen-enzyme-linked immunosorbent assay. Acta Trop 65:139–148

Desquesnes M (2004) Livestock trypanosomoses and their vectors in Latin America, OIE & CIRAD, Paris, p 190

Desquesnes MT, Gardiner PR (1993) Épidémiologie de la trypanosome bovine (Trypanosoma vivax) en Guyane Française. Rev Elev Med Vet Pays Trop 46:463–470

Dwinger RH, Hall MJR (2000) Trypanosomiasis due to *Trypanosoma vivax* in ruminants in Latin America – a review. In:51–58 International Atomic Energy Agency, Vienna, (eds) Animal trypanosomiasis: diagnosis and epidemiology, Backhuys Publishers, The Netherlandsz, pp 50–55.

Fairbairn H (1953) Studies on *Trypanosoma vivax*. IX Morphological differences in strains and their relation to pathogenicity. Ann Trop Med Parasitol 47:394–405

Fikru R, Matetovici I, Rogé S, Merga B, Goddeeris BM, Büscher P, Van Reet N (2016) Ribosomal DNA analysis of tsetse and non-tsetse transmitted Ethiopian Trypanosoma vivax strains in view of improved molecular diagnosis. Vet Parasitol 15:15–22

Galiza GJN, Garcia HA, Assis ACO, Oliveira DM, Pimentel LA, Dantas AFM, Simões SVD, Teixeira MMG, Riet-Correa F (2011) High mortality and lesions of the central nervous system in Trypanosomosis by *Trypanosoma vivax* in Brazilian hair sheep. Vet Parasitol 182:359–363

Garcia H, Garcia ME, Perez G, Bethencourt A, Zerpa E, Perez H, Leon AM (2006) Trypanosomiasis in Venezuelan water buffaloes: association of packed-cell volumes with seroprevalence and current trypanosomes infection. Ann Trop Med Parasitol 100:297–305

Gardiner PR (1989) Recent studies of the biology of *Trypanosoma vivax*. Adv Parasitol 28:229–317

Guerra RMSNC, Feitosa Jr AB, Santos HP, Abreu-Silva AL, Santos ACG (2008) Biometry of *Trypanosoma vivax* found in a calf in the state of Maranhão, Brazil. Ciência Rural, Santa Maria, 38, pp 833–835

Guerra NR, Monteiro MFM, Sandes HMM, Cruz NLN, Ramos CAN, Santana VLA, Souza MMA, Alves LC (2013) Detection of IgG antibodies against *Trypanosoma vivax* in cattle by indirect immunofluorescence test. Pesq. Vet. Bras 33:1423–1426

Hoare CA (1972)The trypanosomes of mammals: a zoological monograph. Oxford, Blackwell, p 749

Hoof L, Van Henrard C, Peel E (1948) Quelques observations sur les trypanosomes del grands mammifères au Congo Belge. Acta Trop 5:327

Jones TW, D'ávila AMR (2001) *Trypanosoma vivax* - out of Africa. Trends Parasitol 17:99–101

Kimeto BA, Mugera GM, Nyaga PN (1990) Haemorrhagic pancarditis in cattle infected with *Trypanosoma vivax*. Vet Parasitol 34:295–301

Kinabo LDB (1993) Pharmacology of existing drugs for animal trypanosomiasis. Acta Trop 54:169–183

Lavier G (1943) L'evolution de la morphologie dans le genre Trypanosoma. Ann Parasitol 19:168

Levine ND (1973) The hemoflagellates. In: Levine ND (ed) Protozoan parasites of domestic animals and of man, 2nd ed, Burgess Publishing, Minneapolis, pp 36–78

Losos GJ (1986) Infectious tropical diseases of domestic animals. Longman Scientific and Technical, Harlow Essex, p 938
Mamoudou A, Njanloga A, Hayatou A, Suh AF, Achukwi MD (2016) Animal trypanosomosis in clinically healthy cattle of north Cameroon: epidemiological implications. Parasit Vectors 9. doi:10.1186/s13071-016-1498-1
Maudlin I, Holmes PH, Miles MA (2004) The trypanosomiasis. CABI Publishing, Wallingford, pp 1–613
Meléndez RD, Forlano M, Figueiroa W (1993) Perinatal infection with *Trypanosoma vivax* in a calf in Venezuela. J Parasitol 79:293–294
Melo SA, Barros AC, Costa FB, De Carvalho Neta AV, Guerra RMC, Abreu-Silva AL (2011) Bovine trypanosomiasis an emerging disease in Maranhão State-Brazil. Vector Borne Zoonotic Dis 7:853–856
Molyneux DH, Ashford RW (1983) The parasites. In: Molyneux DH, Ashford RW. The biology of *Trypanosoma* and *Leishmania*, parasites of man and domestic animals, 1st ed, Taylor & Francis, London, pp 3–62
Murray M, Clifford DJ, MCintyre WIM (1979) Diagnosis of African trypanosomiasis in the bovine. Trans R Soc Trop Med Hyg 73:120–121
Mutugi MW et al (1995) Differences in cloning and subcloning sucess rates in four stocks of *Trypanosoma evansi* and variation in suramin resistance of the clones. Vet Parasitol 60:213–220
Ogwu D, Nuru S (1981) Transplacental transmission of trypanosomes in animals and man. Vet Bull 51:381–384
Osório ALAR, Madruga CR, Desquenes M, Soares CO, Ribeiro LRR, Costa SCG (2008) *Trypanosoma (Duttonela) vivax*: its biology, epidemiology, pathogenesis and introduction in the New World – a Review. Mem Inst Oswaldo Cruz 103:1–13
Paiva F, Lemos RAA, Nakazato L, Mori AE, Brum KE, Bernardo KCA (2000) *Trypanosoma vivax* em bovinos no Pantanal do Mato Grosso do Sul, Brazil: I. Acompanhamento clínico, laboratorial e anatomopatológico de rebanhos infectados. Rev Bras Parasitol Vet 9:135–141
Pillay D, Izotte J, Fikru R, Büscher P, Mucache H, Neves L, Boulangé A, Seck MT, Bouyer J, Napier GB, Chevtzoff C, Coustou V, Baltz T (2013) *Trypanosoma vivax* GM6 antigen: a candidate antigen for diagnosis of African animal trypanosomosis in cattle. PLoS One 25. doi:10.1371/journal.pone.0078565
Pimentel DS, Ramos CAN, Ramos RAN, Araújo FR, Borba ML, Faustino MAG, Alves LC (2012) First report and molecular characterization of *Trypanosome vivax* in cattle from state of Pernambuco Brazil. Vet Prasitol 185:286–289
Schunefeld A, Ruttcher D, Moloo SK (1987) The sensitivity to trypanocidal drugs of *Trypanosoma vivax* isolated in Kenya and Somalia. Trop Med Parasitol 3:177–180
Sekoni VO (1994) Reproductive disorders caused by animal trypanosomiases: a review. Theriogenology 42:557–570
Serra Freire NM (1981) Oiapoque: outro foco de Trypanosoma vivax no Brazil. Rev Braz Med Vet 4:30–31
Serra-Freire NM (1983) Reavaliação dos focos de Trypanosoma vivax em bovinos e bubalinos do Território Federal do Amapá (TFA). Revista da Faculdade de Veterinária da UFF 1:41–45
Shaw JJ, Lainson R (1972) *Trypanosoma vivax* in Brazil. Ann Trop Med Parasitol 66:25–32
Silva RAMS, Da Silva JA, Schneider RC, De Freitas J, Mesquita D, Mesquita T, Ramirez L, Dávila AMR, Pereira MEB (1996) Outbreak of trypanosomiasis due to *Trypanosoma vivax* (Ziemann, 1905) in bovines of the Pantanal Brazil. Mem Inst Oswaldo Cruz 91:561–562
Silva RAMS et al (1998) *Trypanosoma vivax*: biologia, diagnóstico e controle. In: Kessler RH, Schenk MAM (ed) Carrapato, tristeza parasitária e tripanossomose dos bovinos. Embrapa Gado de Corte, Campo Grande, MS, pp 123–125
Silva RAMS, Seidl A, Ramirez L, Dávila AMR (2002) *Trypanosoma evansi* e *Trypanosoma vivax*: biologia, diagnóstico e controle. Embrapa Pantanal, Corumbá, 1st, p 141
Silva RAMS, Sanchez V, Dávila AMR (2003) Métodos de diagnósticos parasitológicos das tripanosomoses bovinas e eqüinas, Circular Técnica, vol 41. Embrapa Pantanal, Corumbá, MS

Silva TM, Olinda RG, Rodrigues CM, Câmara AC, Lopes FC, Coelho WA, Ribeiro MF, Freitas CI, Teixeira MMG, Batista JS (2013) Pathogenesis of reproductive failure induced by *Trypanosoma vivax* in experimentally infected pregnant ewes. Vet Res 44:1

Souza MC (1980) Estudo de infecções experimentais por *Trypanosoma (Duttonella) vivax* Ziemann 1905, em animais domésticos, MSc Thesis, Universidade Federal Rural do Rio de Janeiro, Seropédica, p 110

Stephen LE (1986) Trypanosomiasis: a veterinary perspective. Pergamon Press, New York, p 533

Tsegaye B, Dagnachew S, Terefe G (2015) Review on drug resistant animal trypanosomes in africa and overseas. Afr J Basic Appl Sci 7:73–83

Tuntasuvan D, Sarataphan N, NISHIKAWA H (1997) Cerebral trypanosomiasis in native cattle. Vet Parasitol 31:357–363

Valli VEO, Forsberg CM (1979) The pathogenesis of *Trypanosoma congolense Quantitative histological changes*. Vet Pathol 16:334–368

Van Den Bossche P, Shumba W, Makhambera P (2000) The distribution and epidemiology of bovine trypasonomosis in Malawi. Vet Parasitol 88:163–176

Van Den Ingh TS, Neijs-Bakker MH (1979) Pancarditis in *Trypanosoma vivax* infection in cattle. Tropenmed Parasitol 30:239–243

Vargas TM, Arellano SC (1997) La tripanosomiasis bovina en América Latina y el Caribe. Veterinaria Montevideo 33:17–21

Ventura RM, Paiva F, Silva RAMS, Takeda GF, Buck GA, Teixeira MMG (2001) *Trypanosoma vivax:* characterization of the spliced-leader gene for a Brazilian stock and species-specific detection by PCR amplification of an intergenic space sequence. Exp Parasitol 99:37–48

Vickerman K (1976) The diversity of the kinetoplastid flagellates. In: Lumsden WHR, Evans DA (eds) Biology of the kinetoplastida. Academic Press, London, pp 1–34

Walker PJ (1964) Reproduction and heredity in trypanosomes. Int Rev Cytol 17:51

Weems YS, Kim L, Tsuda V, Yin C, Weems CW (2007) What regulates placental steroidogenesis in 90-day pregnant ewes? Prostaglandins Other Lipid Mediat 84:54–65

Wells EA, Betancourt A, Ramirez LE (1977) The geographic distribution of Trypanosoma vivax in the New World. Trans R Soc Trop Med Hyg 71:448

Whitelaw DD, Gardiner PR, Murray M (1988) Extravascular foci of Trypanosoma vivax in goats: the central nervous system and aqueous humor of the eye as potential sources of relapse infections after chemotherapy. Parasitology 97:51–61

Part IV

Emerging Mycotic Diseases of Livestock

Sporotrichosis

19

Anderson Messias Rodrigues, Geisa Ferreira Fernandes, and Zoilo Pires de Camargo

19.1 Emerging Fungal Diseases

The first fossils of terrestrial plants and fungi appeared between 480 and 460 million years ago (Heckman et al. 2001). Since then, fungal species have evolved and adapted to live in a wide range of environments and ecological niches and at present constitute a highly diverse group of organisms. Slightly more than 100,000 species have been identified; however, this is only a small percentage of the total fungal biodiversity. It is estimated that there are at least 1.5–5 million fungal species on Earth (Sullivan et al. 2005; O'Brien et al. 2005). Approximately 600 species can cause human diseases, which are commonly known as mycoses (de Hog et al. 2000).

These infections occur because the fungal species have adapted, to some extent, their pathogenic repertoire for the invasion of other hosts, including plants (Giraud et al. 2010), insects (Vega et al. 2009), and even single-celled organisms such as amoebae (Steenbergen et al. 2004), before infecting humans. The limited number of fungal diseases in mammals compared with other groups of pathogens is remarkable. The resistance of mammals to fungal infections is due to the basic physiological mechanisms acquired during evolution, including endothermy and innate immunity (Romani 2011; Casadevall 2012). Indeed, the evolution of endothermy in birds and mammals was one of the most important events in vertebrate evolution (Lovegrove 2012). Robert and Casadevall (2009) suggest that the development of endothermy and homeothermy by mammals and insects, respectively, is a potent nonspecific defense against most fungi. These physiological adaptations were intended to maintain body temperature and, although costly to warm-blooded hosts, may have favored the evolutionary survival of these organisms against the establishment of infections.

A.M. Rodrigues • G.F. Fernandes • Z.P. de Camargo (✉)
Department of Microbiology, Immunology and Parasitology, Cell Biology Division, Federal University of São Paulo, São Paulo 04023-062, Brazil
e-mail: zpcamargo1@gmail.com

J. Bayry (ed.), *Emerging and Re-emerging Infectious Diseases of Livestock*, DOI 10.1007/978-3-319-47426-7_19

Because only a few fungal species are essentially pathogenic, the mechanism of pathogenesis tends to be highly complex and results from preexisting adaptations of organisms with nonparasitic lifestyles (van Burik and Magee 2001). Despite the relatively small number of human pathogens, some fungal species have become a threat to public health due to the expanding population of vulnerable patients with immune deficiencies and increased environmental exposure to these organisms. The late twentieth century witnessed the emergence of numerous infectious diseases caused by microorganisms that rarely cause disease in immunocompetent hosts (Casadevall 2003), including *Cryptococcus neoformans* (the agent of cryptococcosis), *Aspergillus fumigatus* (agent of aspergillosis), *Candida albicans* (agent of candidiasis), and *Rhizopus* spp. (agent of mucormycosis).

Emerging infectious diseases are caused by pathogens for which the incidence, geographical distribution, host range, and virulence are increasing (Fisher et al. 2012; Daszak et al. 2000). The similarities of emerging infectious diseases between humans and wild animals extend to the beginning of human colonization and the dissemination of exotic pathogens (Gortazar et al. 2014; Taylor et al. 2001). The perception of the emergence of pathogenic fungi as a threat to human health has been changing rapidly. This is due to the frequent occurrence of adverse interactions between fungi and the vertebrate host. Fungal infections, which are similar to certain bacterial and viral infections, can be highly lethal to naïve hosts, with mortality rates approaching 100 % (Fisher et al. 2012). In this scenario, the study of two emerging pathogens demonstrates the risk of loss of biodiversity among vertebrates in the wild.

Amphibians have little resistance to infection by *Batrachochytrium dendrobatidis* (Chytridiomycota: Rhizophydiales) (Daszak et al. 1999). By invading the superficial layers of the skin of anurans (toads and frogs) and urodeles (salamanders and newts), *B. dendrobatidis* promotes the thickening of the keratin layer (hyperplasia). In adults, this skin alteration limits the development of physiological processes, including osmoregulation and respiration, causing the death of thousands of amphibians. This emerging phenomenon is not geographically restricted, and *B. dendrobatidis* has been responsible for declines in amphibian populations in North America, South America, Europe, and Africa (Kilpatrick et al. 2010).

Another example of the emergence of fungi in wildlife is *Pseudogymnoascus destructans* (formerly *Geomyces destructans*, Ascomycota: Pezizomycotina), which has been recently described as the causative agent of white nose syndrome in bats (Gargas et al. 2009). Photographic records indicate that the disease first appeared in 2006 in hibernating bats in caves near Albany, New York (Blehert et al. 2009). Since then, *P. destructans* has devastated bat populations hibernating in caves in the USA, with mortality rates between 75 and 95 % (Warnecke et al. 2012).

The comparison between other models of emerging diseases and the *Sporothrix*-sporotrichosis system is important. We have recently observed the emergence of an intriguing phenomenon in *Sporothrix brasiliensis* (Ascomycota: Pezizomycotina) among domestic cats (*Felis catus*) in Brazil. In recent decades (1998–2012), the zoonotic transmission of *Sporothrix* via scratches and bites or contact with sick cats led to the infection of more than 4000 people and 4124 cats that were diagnosed at the Evandro Chagas Institute of Clinical Research (Instituto de Pesquisa Clínica Evandro Chagas – IPEC) at the Oswaldo Cruz Foundation (Fundação Oswaldo Cruz – Fiocruz) in Rio de Janeiro, Brazil (Gremião et al. 2015; Pereira et al. 2014). Epidemiological data reveal the importance of this mycosis as a public health problem (Barros et al. 2010).

It is of note that zoonotic pathogens represent approximately 60 % of all known pathogens that infect humans (Taylor et al. 2001). These human infections are a result of the animal-human interface, which is defined as the close contact between humans and animals, their environment, and their products (Hubálek and Rudolf 2011; Gortazar et al. 2014). The recent outbreak of *S. brasiliensis* in cats, in contrast to the animal diseases that occur in amphibians and bats caused by *B. dendrobatidis* and *P. destructans*, respectively, had an augmenting factor: the zoonotic transmission to humans. However, similar to the geographical expansion described for *B. dendrobatidis* and *P. destructans*, it is inevitable that the area of occurrence of the disease remains limited to the initial foci in the southern and southeastern regions of Brazil. Therefore, the biogeographical distribution of *S. brasiliensis* will eventually increase.

19.2 *Sporothrix schenckii* Hektoen and Perkins 1900

The most important human and animal pathogen in the order Ophiostomatales is the thermal dimorphic fungus *Sporothrix schenckii*, which is the etiological agent of sporotrichosis (Guarro et al. 1999). The first case that unquestionably presented as a clinical manifestation of sporotrichosis was reported by Benjamin Robinson Schenck (1873–1920) in 1898 at Johns Hopkins Hospital in Baltimore, Maryland, USA (Schenck 1898). The genus name *Sporothrix* was proposed in 1900 by Hektoen and Perkins for the fungus isolated by B.R. Schenck 2 years prior (Hektoen and Perkins 1900). Sporotrichosis is caused by a complex of phylogenetically related cryptic species (de Beer et al. 2003; Marimon et al. 2006). Historically, several synonyms are found in the literature, making the taxonomy of *Sporothrix* confusing (Table 19.1)

Sporothrix is not a monotypic taxon, i.e., formed by a single species, and the genus has a remarkable genetic and ecological diversity, which is reflected in the different associations between organisms. *Sporothrix* is commonly found on decaying plant material, and its isolation from soil samples has been reported in endemic areas (Mackinnon et al. 1969; Mendoza et al. 2007; Mehta et al. 2007). However, the distribution of clinically relevant species in the soil and the conditions that favor them have been the subject of speculation. *Sporothrix* belongs to the order Ophiostomatales, which occupies many ecological niches. These microorganisms are adapted for dispersal by insects (Coleoptera: Scolytinae), are associated with the plants of the genus *Protea*, and are widely distributed in the soil, where they grow on plant debris; in addition, they are morphologically similar and generally share similar niches based on their biological and ecological characteristics.

Table 19.1 *Sporothrix schenckii*, Hektoen and Perkins, 1900. Synonyms

Sporotrichum schenckii, De Beurmann and Gougerot, 1906
Sporotrichum sp., Smith, 1898
Sporotrichum beurmanni, Matruchot and Ramond, 1905
Sporotrichum asteroids, Splendore, 1908
Trichosporium beurmanni, Lutz and Splendore, 1909
Sporotrichum equi, Carougeau, 1909
Sporotrichum jeanselmei, Brumpt and Langeron, 1910
Sporotrichum councilmani, Wolbach, 1971
Sporotrichum fonsecai, Pereira, 1929
Sporotrichum pereirai, Miranda, 1935
Sporotrichum beurmanni var. *schenckii*, Ciferri and Redaelli, 1942
Sporotrichum beurmanni var. *councilmani*, Redaelli and Ciferri, 1942
Sporotrichum beurmanni var. *jeanselmei*, Redaelli and Ciferri, 1942
Sporotrichum beurmanni var. *carougeaui*, Redaelli and Ciferri, 1942
Sporotrichum beurmanni var. *anglicum*, Redaelli and Ciferri, 1942

Many ascomycetes in the order Ophiostomatales are considered phytopathogenic and cause a variety of diseases related to vascular wilt in their hosts. However, despite their relation to plant pathogens, there is no corroborative evidence that *S. schenckii* and other species of clinical importance are plant pathogens. Another important ecological aspect of Ophiostomatales involves the relationship between fungi and beetles (Coleoptera: Scolytinae). These ecological relationships involving *Sporothrix* are found worldwide and range from commensalism and mutualism to parasitism. The literature reports that beetles are an important route of dissemination of environmental species of *Sporothrix* and *Ophiostoma* (Zipfel et al. 2006; Romon et al. 2014). However, we cannot extrapolate that these ecological relationships are the agents of animal and human infections. With vast genetic, ecological, and biological diversity, the taxonomy of *Sporothrix* is paradoxical.

For more than a century (1898–2007), *S. schenckii* was described as the only etiological agent of sporotrichosis (Schenck 1898; Marimon et al. 2006). However, with the advent and refinement of molecular biology techniques and their direct application to fungal taxonomy, it has been shown that the classic agent *S. schenckii* is actually a complex of phylogenetically related cryptic species (Marimon et al. 2006). The recognition and recent introduction in clinical practice of the new etiological agents *Sporothrix brasiliensis*, *Sporothrix globosa*, and *Sporothrix luriei* revealed important aspects of their epidemiology, pathogenicity, virulence, ecology, and genetics, as well as their sensitivity to the main antifungal drugs.

19.3 The *S. schenckii* Complex

Previous studies on different clinical isolates of *Sporothrix* have reported phenotypic variations among isolates (Marimon et al. 2007; Fernandes et al. 2009a). Such differences include changes in colony morphology, growth on different culture media, thermal tolerance, virulence in murine models, susceptibility to antifungal drugs, protein secretion, genomic architecture, and melanin synthesis (Serena et al. 2008b; Fernandes et al. 2009a; Arrillaga-Moncrieff et al. 2009; Sasaki et al. 2014; Almeida-Paes et al. 2015).

The first molecular evidence that the classical agent *S. schenckii* was represented by two or more species was based on the sequencing and phylogenetic analysis of the internal transcribed spacer (ITS) region located in the ribosomal DNA (rDNA) operon of these microorganisms. According to de Beer et al. (2003), phylogenetic analysis indicated two groups of *Sporothrix*. Organisms belonging to group I have been isolated from patients infected with *Sporothrix* in the Americas and South Africa, whereas organisms from group II have been isolated directly from the soil or decomposing plant material, with few clinical cases. This confirmed previous observations that indicated morphological and physiological differences among isolates of *S. schenckii s.l.* from clinical and environmental origins. Historically, *Sporothrix* isolated from wood and soil also tend to be less pathogenic in mice (Howard and Orr 1963; Arrillaga-Moncrieff et al. 2009; Rodrigues et al. 2015a).

A few years later, Marimon et al. (2006) reinforced the studies on genetic diversity conducted by de Beer et al. (2003). The combined phylogenetic analysis based on the DNA sequence of three protein-coding loci (chitin synthase, β-tubulin, and calmodulin) from 60 *Sporothrix* isolates from various geographical regions revealed the presence of 14 genetic groups with high statistical evidence (100 %), and six of these genetic groups represented six probable phylogenetic species.

Phylogenetic analyses suggest that not only existing populations of *S. schenckii* are undergoing a divergence process but that all the resulting isolates are undergoing speciation (Marimon et al. 2006). This is corroborated by a high degree of geographic isolation inherent to different phylogenetic groups. The analysis of the three loci revealed the presence of two main groups: one group includes the European isolates, and another group includes isolates from South America and South Africa. In the latter group, two clades were identified, one composed of 29 Brazilian isolates and the other composed of the remaining isolates (Marimon et al. 2006).

On a global scale, Marimon et al. (2007) characterized 127 isolates of *Sporothrix* previously classified as *S. schenckii* using phenotypic and genotypic methods. The results suggest that in addition to *S. schenckii s. str.*, three new species are causative agents of sporotrichosis, on the basis of their assimilation of carbon sources (sucrose, raffinose, and ribitol); average diameter of colonies after culture at 20 °C, 30 °C, 35 °C, and 37 °C; morphology of conidia; and partial sequence of the calmodulin gene: *S. brasiliensis*, *S. globosa*, and *S. mexicana*.

Ajello and Kaplan (1969) described an *S. schenckii* variety termed *S. schenckii* var. *luriei*. This sample was isolated from a native resident of South Africa presenting with sporotrichosis. In culture, this variety is not distinct from other species of *Sporothrix*. However, in tissues, it forms small budding yeast cells as well as large, spherical, thick-walled cells. In 2008, Marimon et al. (2008a) proposed that *S. schenckii* var. *luriei* was elevated to the level of species under the name *Sporothrix luriei*.

Therefore, the *S. schenckii* complex is formed by the species *S. brasiliensis* (clade I), *S. schenckii s. str.* (clades IIa and IIb), *S. globosa* (clade III), and *S. luriei* (clade VI) (Zhou et al. 2014). The phylogeny of the ITS region revealed a clear separation between the clinically relevant and the environmentally relevant species, corroborating the studies conducted by de Beer et al. (2003). The isolates of human and animal origin are located in the clades *S. brasiliensis*, *S. schenckii s. str.*, *S. globosa*, and *S. luriei*. The environmental species of *Sporothrix* are phylogenetically located at a relatively large distance away and are flanked by the environmental species of *Ophiostoma* (Fig. 19.1).

The clinically relevant *Sporothrix* species that often infect vertebrate hosts form a monophyletic group (Marimon et al. 2007; Roets et al. 2008); however, other genetically distant species in Ophiostomatales can also occasionally serve as accidental pathogens, including *Sporothrix pallida*, *Sporothrix mexicana*, *Sporothrix chilensis*, *Ophiostoma stenoceras*, and *Ophiostoma piceae* (Bommer et al. 2009; Morrison et al. 2013; Rodrigues et al. 2015a).

19.4 Ecology of *Sporothrix*

The ecology of medically relevant fungi is a neglected topic in mycology and generally does not follow clinical, biological, and molecular advancements. Small outbreaks of sporotrichosis have been recorded, including an outbreak among mine workers in Witwatersrand, South Africa, that was described by Pijper and Pullinger (1927). Studies on the epidemic of sporotrichosis involving miners in South Africa revealed important aspects of the ecology of *Sporothrix*. It was observed that the fungus can grow at temperatures between 26 and 27 °C and relative humidity levels between 92 and 100 %. Interestingly, the original report of the South African epidemic indicated that *S. schenckii s.l.* did not develop on the plant material in the gold mines, where the wooden beams that had supported the excavations were infected with lignocellulosic basidiomycetes, including *Poria* sp. (*Basidiomycota*: *Agaricomycotina*). In this situation, the miners were free of infection. Furthermore, it was found that the infection was not prevalent outside the gold mines (Rippon 1988). Recently, a lymphocutaneous sporotrichosis outbreak of lesser magnitude was described in a gold mine in South Africa (Govender et al. 2015), and *S. schenckii s. str.* was identified as the etiological agent. Although genetically distinct species were isolated from clinical and environmental sources, it is likely that the contaminated soil and untreated wood were the sources of the infection.

The species of *Sporothrix* are not plant pathogens; however, the incidence of sapronotic outbreaks is dependent on specific and rare conditions that occur in

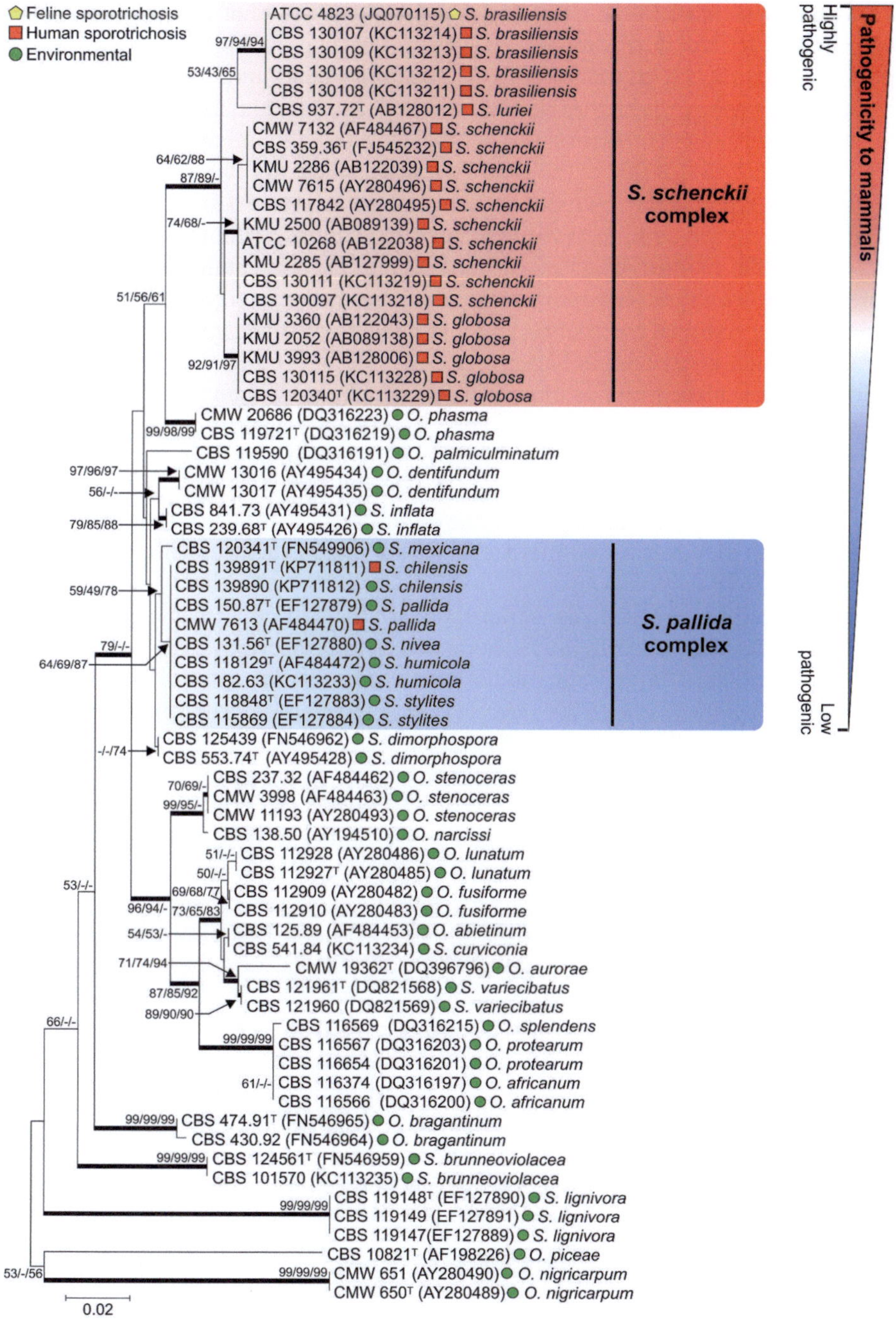

Fig. 19.1 Phylogenetic relationships among the clinical and environmental members of the clade *Sporothrix*, based on the ribosomal DNA (rDNA) operon sequences IT1, 5.8S, and ITS2. Method, neighbor joining. The numbers next to the branches (neighbor joining/maximum likelihood/maximum parsimony) refer to percentages of re-sampling (1000 bootstrap)

decaying plant material. Zhang et al. (2015) postulate that a certain degree of decomposition and fermentation of plant material would promote excessive growth of *Sporothrix* in nature. High temperature and humidity associated with metabolic disorders (induction of the respiratory system) and oxidative stress (due to decomposition and fermentation) can induce a morphological transition in species of *Sporothrix* that favors the growth of yeasts. The "fermentation hypothesis" is based on experimental evidence demonstrating the growth of the fungus on corn (*Zea mays*) in vitro combined with epidemiological observations that confirm the increase in the number of cases of human sporotrichosis in areas where maize crops are grown in northeastern China. It has been suggested that *Sporothrix* can grow exponentially on corn debris and function as a possible source of inoculation for rural workers involved in harvesting. In China, it is common for corn harvest workers to lower their bodies to collect small bundles of plant material deposited on the soil and transport these bundles close to their face. It is of note that in northeastern China, skin lesions are manifested predominantly on the face, which is the body region closest to the source of inoculation. After the infectious material was removed from the crops, the rate of human infection decreased over time, corroborating the observation that most cases in Jilin Province, China, occur during winter (Song et al. 2013).

Occasionally, sporotrichosis is transmitted to animals other than cats, including not only warm-blooded vertebrates but also arthropods (Lacaz et al. 2002). In animals such as horses, dogs, cats, rabbits, and mice, spontaneous sporotrichosis can occur (Fishburn and Kelley 1967; Crothers et al. 2009). Lutz and Splendore (1907) isolated pathogenic microorganisms of the genus *Sporothrix* from the buccal mucosa of rats. Some authors have reported cases of sporotrichosis after bites by mosquitoes, squirrels, bats, ants, and spiders (Zhang et al. 2015). The fungus was also isolated from plant material found in armadillo burrows (Mackinnon et al. 1969; Rodrigues et al. 2014a). Disseminated sporotrichosis can also occur in armadillos (*Dasypus novemcinctus*), as reported by Kaplan et al. (1982). The plant decomposition conditions and temperature in the burrows of armadillos may be suitable for the development of *Sporothrix* and subsequent dispersal by the host (Rodrigues et al. 2014a). Therefore, despite the prevalence of feline vectors, animal and plant host species can vary. Zhang et al. (2015) propose that wild animals occasionally provide favorable conditions similar to those found in fermented plant material. Cats collect the fungal spores from the soil during excavation and easily transfer the spores to the mouth by licking. The physiological conditions in the saliva of cats are associated with their body temperature (normal range between 37.7 and 39.1 °C) and can stimulate the yeast-mycelium transition of *Sporothrix*. The saliva of cats has a pH between 7.5 and 8.0, which is similar to the pH obtained during the fermentation of corn (approximately 8.0) and is the optimum pH for yeast-mycelium conversion in vitro (Travassos and Lloyd 1980). The hypothesis of conditional similarities between the plant fermentation, the digestive tract of animals, and the migration of *Sporothrix* from plant to animal hosts becomes understandable.

19.5 Epidemiology of Sporotrichosis: A Global and Neglected Mycosis

The incidence of human sporotrichosis varies widely worldwide based on the analysis of data and case reports (Pappas 2003; Chakrabarti et al. 2015). The global epidemiology of sporotrichosis indicates that the endemic countries with the highest incidence of the disease are Brazil (5814 cases), China (3299 cases), and South Africa (3154 cases), whereas the countries with the lowest incidence include Japan, Australia, India, and parts of South America (Zhang et al. 2015). Although a few hundred human cases were described at the beginning of the last century in France by Beurmann and Gougerot (1912), sporotrichosis is rare in this country and throughout Europe in general.

Sporothrix species have been described by many mycologists across the globe (Chakrabarti et al. 2015); however, there is an intriguing difference in the incidence and geographical distribution of the etiologic agents (Fig. 19.2). In almost all regions, there is a predominance of a particular species. In Asia, particularly in China, it is estimated that *S. globosa* is the etiological agent in 99.3 % of the human cases of sporotrichosis. In other endemic areas, including Australia and South Africa (94 %), North America, and parts of South America (89 %), *S. schenckii s. str.* is the prevalent species. In the south and southeast of Brazil, *S. brasiliensis* is the main etiological agent of human and animal sporotrichosis and is responsible for 88 % of cases (Fig. 19.2) (Rodrigues et al. 2013b, 2014d; Zhang et al. 2015).

Animal sporotrichosis was first described in São Paulo, Brazil, in 1907 by Lutz and Splendore (1907) in naturally infected rats. Subsequently, Freitas et al. (1956, 1965) reported a series of cases in cats and dogs in the city of São Paulo. Marques et al. (1993) documented feline sporotrichosis associated with human cases. Schubach et al. (2001) isolated *S. schenckii s.l.* from cat nails during an outbreak of sporotrichosis that affected these animals and from humans in the city of Rio de Janeiro, Brazil. Animal-animal and animal-human transmissions occur via scratches and bites by animals contaminated with fungal spores, resulting in the implantation of the pathogen in the skin of the host (Gremião et al. 2015).

Therefore, sporotrichosis has traditionally been an obscure disease from an epidemiological point of view. In the past few decades (1998–2015), a major zoonotic outbreak has occurred in Brazil (Schubach et al. 2004, 2005a, b). It is difficult to estimate the true magnitude of the epidemic because compulsory notification of public health agencies is not required for sporotrichosis. Rodrigues et al. (2014d) reported human cases of sporotrichosis in 14 of the 26 states in Brazil and indicated the main endemic areas and clinical forms of the disease. The southern and southeastern regions of Brazil have a very high incidence of human cases, which is directly associated with major epidemics of sporotrichosis transmitted by cats (Barros et al. 2010; Gremião et al. 2015).

Sporothrix brasiliensis infection, including human and animal cases, occurs in four of the five Brazilian regions. Phylogenetic analyses based on rDNA operon sequences (ITS1, 5.8S, and ITS2) and on the protein-coding loci of proteins such as

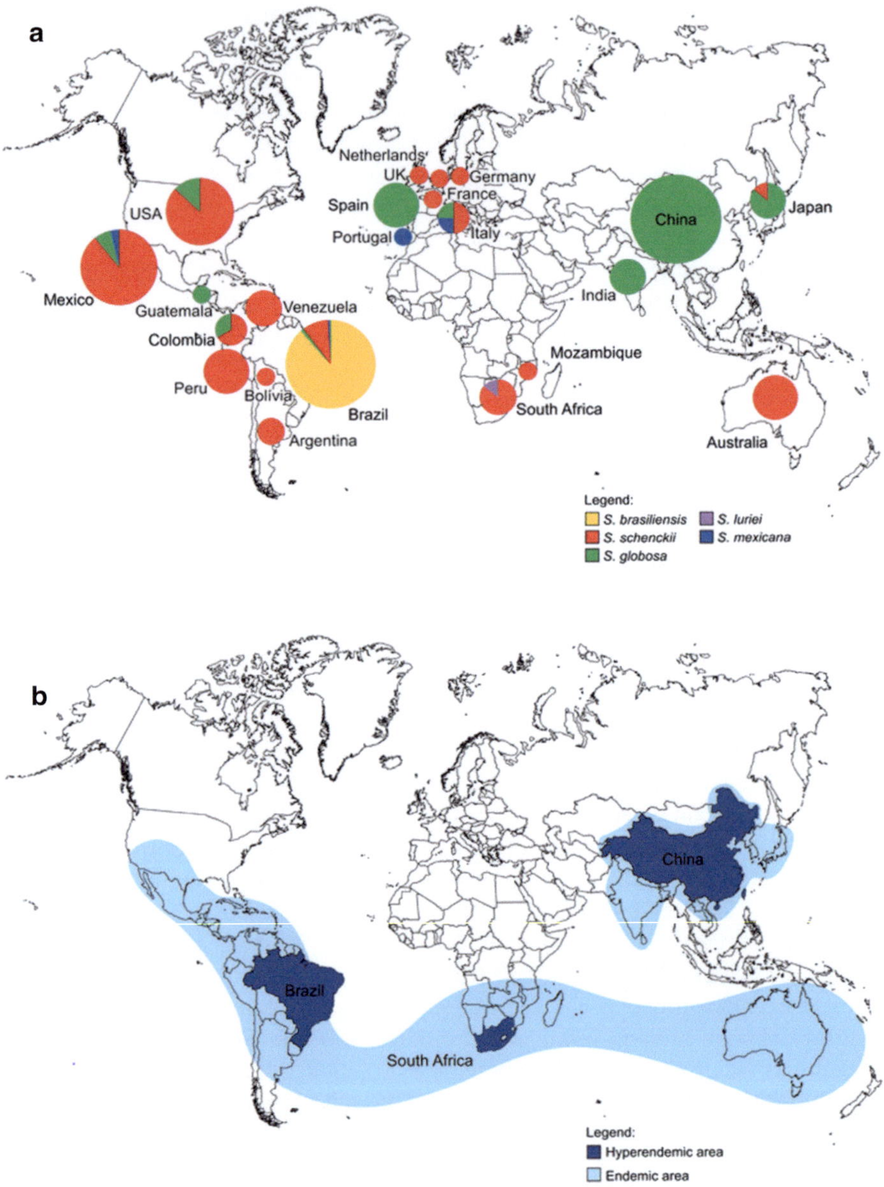

Fig. 19.2 Distribution of the clinically relevant species worldwide. (**a**) Note that *S. brasiliensis* is geographically restricted to Brazil, whereas *S. schenckii* and *S. globosa* have a transoceanic distribution. The graphics are proportional to the number of isolates identified by DNA sequencing. (**b**) The main endemic areas of human sporotrichosis are present in tropical and subtropical areas. Brazil, South Africa, and China documented the highest number of infected individuals (Zhang et al. 2015)

calmodulin and β-tubulin support the monophyly of this species. To date, there have been no reports of *S. brasiliensis* outside Brazil. Rodrigues et al. (2013b) reported the high prevalence of *S. brasiliensis* (96.9 %) in cats with sporotrichosis in southern and southeastern Brazil. The *S. brasiliensis* genotype that infects cats is identical to that found in humans in the same geographical area, and the zoonotic transmission of the disease has been confirmed by molecular analysis. *S. brasiliensis* is associated with atypical (de Macedo et al. 2015b) and more severe forms of the disease, including disseminated skin infection (without underlying disease), hypersensitivity reactions, and infection of the mucosa (Almeida-Paes et al. 2014).

The dangerous combination of late diagnosis and late treatment of infected cats leads to the rapid transmission of the disease between animals and humans. Socioeconomic factors such as the abandonment of sick animals on the streets and the improper disposal of animal carcasses, which are often buried in the yard or left in vacant lots, combined with the absence of public policies to contain the epidemic allowed the transmission to reach epidemic proportions. The increase in the number of cases in cats is usually followed by an increase in the number of cases in humans, and this situation is a serious public health problem. Epidemiological studies indicate that *S. brasiliensis* depends on the feline host for its development in southern and southeastern Brazil (Rodrigues et al. 2013b, 2014d). The increase in the number of human cases in the metropolitan region of Rio de Janeiro since 1998 is due to the success of zoonotic transmission (Barros et al. 2004, 2008; Schubach et al. 2008; Gremião et al. 2015).

Sporothrix schenckii has a greater genetic diversity and homogeneous distribution throughout the Brazilian territory compared with *S. brasiliensis* and includes several genetic groups. Most of the Brazilian isolates of *S. schenckii* were isolated from human cases of sporotrichosis, and this causative agent is prevalent in areas that are free from outbreaks of feline sporotrichosis. In Brazil, there is a low incidence of human sporotrichosis caused by *S. globosa* (Madrid et al. 2009; de Oliveira et al. 2010; Rodrigues et al. 2014d). To date, clinical isolates of *S. globosa* have been identified in the United Kingdom, Portugal, Spain, Italy, China, Japan, the USA, India, and Brazil (Madrid et al. 2009; Rodrigues et al. 2013a).

From the epidemiological and ecological points of view, the low number of cases due to *S. schenckii s. str.* in Brazil suggests that the infection of patients occurs via the classic route, i.e., direct contact with contaminated soil and plant material. Compared with the alternative route (i.e., feline hosts, as in *S. brasiliensis*), the classic route of infection may be less effective, leading to cases of sporotrichosis in groups of patients with specific occupations (gardeners, farm workers, etc.). However, outbreaks also occur via the classic route (Rodrigues et al. 2014d; Zhang et al. 2015; Govender et al. 2015). The most relevant cases of sapronosis in France, the USA, South Africa, and China indicate that very specific ecological conditions promote the spread of the pathogen via decaying plant material (Zhang et al. 2015). In the alternative route, transmission by the scratches and bites of cats is highly effective, and a larger number of individuals are at risk of contracting sporotrichosis.

In this respect, different transmission routes require different public policies for the containment of the sporotrichosis epidemic. In the classical route of sporotrichosis, the use of appropriate tools for the manipulation of plant material and the removal of environmental foci are essential for the containment of the outbreak. In contrast, in regions where horizontal animal transmission (between cats) and zoonotic transmission (from cats to humans) are predominant, containment measures should be directed at educating cat owners in endemic areas, castrating animals in areas of outbreaks, and implementing appropriate antifungal therapy.

19.6 The *Sporothrix*-Sporotrichosis System

The clinical manifestations of sporotrichosis are directly influenced by several factors, including the inoculum size, morphology of the inoculated cells, virulence of the etiologic agents, and immune status of the host. Sporotrichosis is characterized by the presence of suppuration, fistulas, and ulceration. In sporotrichosis, a primary lesion frequently occurs, followed by the development of nodular lymphangitis. In other cases, the lesion is restricted, without satellite lymphadenitis (Lacaz et al. 2002).

Fixed skin sporotrichosis is characterized by the presence of a single lesion, which can appear on the face, neck, trunk, or arms, in the form of ulcerative, verrucous, or papular wounds. The morphology can be vegetative, verrucous, infiltrated plate, and tuberous, without lymphatic involvement (Barros et al. 2011). The clinical manifestation can occur in the form of reinfection in individuals who have previously been sensitized to the fungus, and the etiological agent is often unable to grow at temperatures above 35 °C. Fixed sporotrichosis can be a challenge for clinical diagnosis because it resembles several other skin diseases, including lupus, sarcoidosis, and leishmaniasis, and therefore, differential diagnosis is required.

The lymphocutaneous form of sporotrichosis is characterized by a primary lesion usually located on the extremities, particularly the hands and forearms, which are the sites that are most often exposed to infection. The initial injury may have the appearance of an ulcer or a small subcutaneous nodule, and the skin color varies from blackish to reddish and subsequently acquires a dark color. Lymphocutaneous lesions are typical of sporotrichosis. Together with the fixed cutaneous form of the disease, these lesions account for more than 95 % of the cases of sporotrichosis (Bonifaz and Vázquez-González 2013). Lymphocutaneous lesions exert pressure under the skin, causing ischemia under the epidermis; progress to exudation and ulceration; and produce a purulent discharge. As the disease progresses, secondary lesions appear along the regional lymph channels, providing a sporotrichoid aspect to the infection. Lymph node involvement or the presence of systemic symptoms is unusual. Untreated cases of lymphocutaneous sporotrichosis can become chronic, but spontaneous cures sometimes occur (Barros et al. 2011; Lacaz et al. 2002).

Sporotrichosis can affect any organ, manifesting itself as a disseminated disease. Hematogenic dissemination is not very common but can occur, especially in immunocompromised patients (Silva-Vergara et al. 2012), and can be associated with transplants and other diseases, including cancer, diabetes, and particularly HIV/AIDS (Al-Tawfiq and Wools 1998). Disseminated sporotrichosis is characterized by multiple lesions, which initially appear as subcutaneous nodules and may progress into papules, pustules, ulcers, or ulcerative lesions. In cases of coinfection with HIV/AIDS, the mortality rate can reach 30 %, and there is a correlation between the involvement of the central nervous system and death (Moreira et al. 2015).

The mucosal form of sporotrichosis can be a variant of the cutaneous form and involves the nasal mucosa or the conjunctiva. Pulmonary sporotrichosis has also been reported (Aung et al. 2013). In general, untreated cases can become chronic, but spontaneous cures sometimes occur. The extracutaneous form is rare and difficult to diagnose. This type of manifestation is often associated with HIV/AIDS. The skin and bones are commonly affected, and osteoarticular manifestations can occur; tenosynovitis and bursitis have also been reported. Monoarthritis can occur in immunocompetent patients (Morris-Jones 2002).

Generally speaking, human sporotrichosis can assume different clinical forms, depending on the interaction between two key factors: the integrity of the host immune system and the biological characteristics of the microorganism. Owing to the distinct manifestations of the disease associated with the organic topology, number of lesions, and severity of lesions, various classifications are available in the literature. At present, the most widely adopted classification is based on a review published by Lopes-Bezerra et al. (2006):

(i) Cutaneous
  Lymphocutaneous
  Fixed
  Disseminated or multiple
(ii) Mucosal
  Ocular
  Nasal
  Other forms
(iii) Extracutaneous
  Pulmonary
  Osteoarticular
  Meningitis
  Generalized
(iv) Residual (sequel)
(v) Special forms
  Spontaneous regression
  Hypersensitivity (erythema nodosum, erythema multiforme)

In feline sporotrichosis, the distribution of skin lesions presents three patterns: L1 (skin lesions in one site), L2 (skin lesions in two nonadjacent sites), and L3 (skin lesions in three or more nonadjacent sites) (Schubach et al. 2004). These lesions vary from small papules to large areas of necrosis. In general, the lesions may exhibit a suppurative appearance (Fig. 19.3). Animals with the L1 pattern usually have well-formed granulomas. Animals with the L3 pattern present high fungal burden, and this may be associated with poorly formed granulomas (Miranda et al.

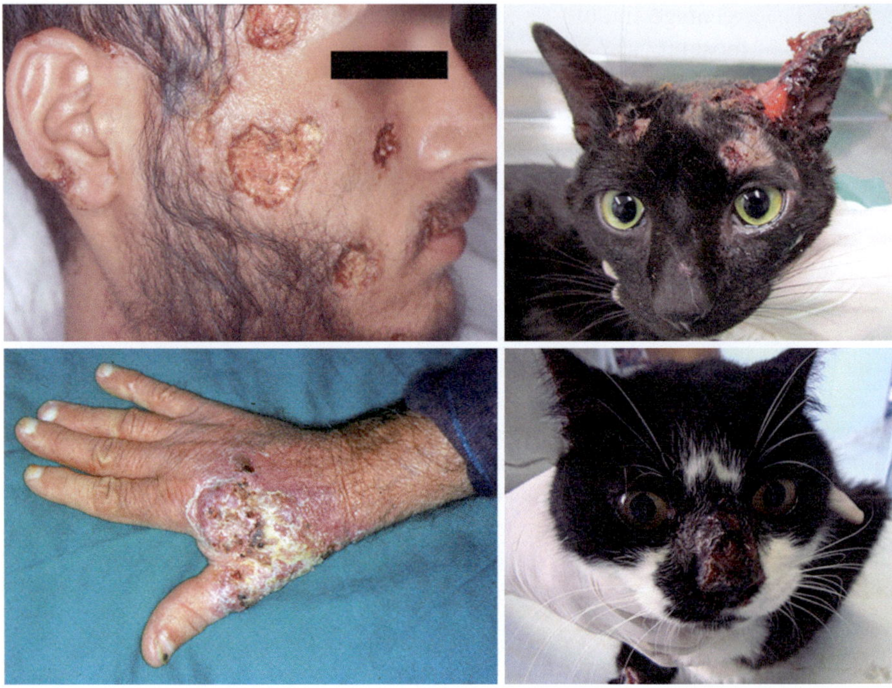

Fig. 19.3 Characteristics of human and animal sporotrichosis. Ulcerative and suppurative lesions occur in human and feline hosts during sporotrichosis outbreaks due to *Sporothrix* spp. The animals usually present ulcerative skin lesions in the cephalic region. The image of human sporotrichosis is courtesy of Dr. Flávio de Queiroz Telles Filho from the Federal University of Paraná, Brazil. The images of feline sporotrichosis are courtesy of Dr. Sandro Antonio Pereira from the Evandro Chagas National Institute of Infectious Diseases at the Oswaldo Cruz Foundation, Brazil, and Dr. Mário Augusto Ono from the State University of Londrina, Brazil

2013). Anemia, neutrophilic leukocytosis, hypoalbuminemia, and hypergammaglobulinemia are the main blood abnormalities present in infected cats (Schubach et al. 2004). Interestingly, in the case of feline sporotrichosis in southern and southeastern Brazil, there is a marked prevalence of *S. brasiliensis* in approximately 99 % of the cases (Rodrigues et al. 2013b), whereas in other regions of the world, the primary agent of animal sporotrichosis is *S. schenckii* (Kano et al. 2015).

19.7 Classic Diagnosis

The classic diagnosis of sporotrichosis is based on correlations among clinical, epidemiological, and laboratory data. The material aspirated from skin nodules, pus, exudate, curettage material, or swabs of open lesions is grown directly on Sabouraud agar medium and incubated at room temperature. Biological samples such as sputum, urine, blood, and cerebrospinal and synovial fluid are examined in cases of generalized manifestation. *Sporothrix* shows satisfactory growth in most culture media. The percentage of positive cultures in clinical cases of the disease is high, and this is the most reliable method for diagnosis. Growth normally occurs between 3 and 5 days of incubation. It is recommended that cultures are incubated for at least 4 weeks before discarding them as negative (Lacaz et al. 2002). The direct analysis of biological samples is performed using 10 % KOH (Barros et al. 2011). Although the observation of "cigar-shaped" yeast cells suggests infection by *Sporothrix*, in general, these structures are scarce in human biological samples, and their small size limits their identification.

By histopathological diagnosis, *S. schenckii* can be detected in tissues after staining with hematoxylin and eosin (H&E), Gomori methenamine silver (GMS), or periodic acid-Schiff (PAS) (Morris-Jones 2002; Larone 2011). The Splendore-Hoeppli reaction, which is characterized by asteroid bodies around *Sporothrix* yeasts, is frequently observed; however, its presence is not sufficient to diagnose sporotrichosis because this event occurs in several other diseases caused by parasites, bacteria, and other fungi. These structures are among the evasive strategies used by *Sporothrix* to avoid the host immune response (Daniel Da Rosa et al. 2008). The suppurative and granulomatous inflammatory reaction in the dermis and subcutaneous tissue is often accompanied by microabscesses and fibrosis. Skin infections may also present hyperkeratosis, parakeratosis, and pseudoepitheliomatous hyperplasia (Larone 2011).

Although delayed hypersensitivity is not used as a diagnostic tool, this approach aims to detect the cellular immune response and is generally useful in epidemiological studies, providing positive rates of up to 90 %. A positive result indicates active infection or previous contact with the fungus. This diagnostic test involves the application of sporotrichin on the dermis. At present, the most significant inconsistency of this tool is related to the preparation of the antigen, which varies between laboratories. Furthermore, the existence of different etiological agents limits diagnosis at the species level. The sporotrichin test remains positive after the clinical cure of the infection (Lacaz et al. 2002).

The gold standard for the diagnosis of classical sporotrichosis is the use of cultures, which allows the isolation and identification of the agent (Barros et al. 2011). The isolation of fungal cells from biological samples is usually performed in Sabouraud agar with chloramphenicol and in media containing cycloheximide, such as Mycobiotic agar, at 25 °C. The colonies grown on Sabouraud agar at room temperature are white to cream colored and leathery, with grooved surfaces. Some samples retain the cream color, whereas others become dark brown to black (Fig. 19.4) (Lacaz et al. 2002). The morphology of the colonies during the early growth of *Sporothrix* is relatively variable. The color is unstable, and isolates present considerable intra- and interspecies variation. The same isolate may change color during successive samplings, and this phenomenon is known as phenotypic plasticity.

At 37 °C, organisms grown in brain heart infusion agar with or without blood become yeast colonies, with a soft consistency, white to cream color, and uneven surfaces (Fig. 19.4) (Lacaz et al. 2002). The addition of thiamine and biotin to the cultures and the low CO_2 content facilitate the conversion of the mycelial form to the yeast form (Travassos and Lloyd 1980). The demonstration of dimorphism in

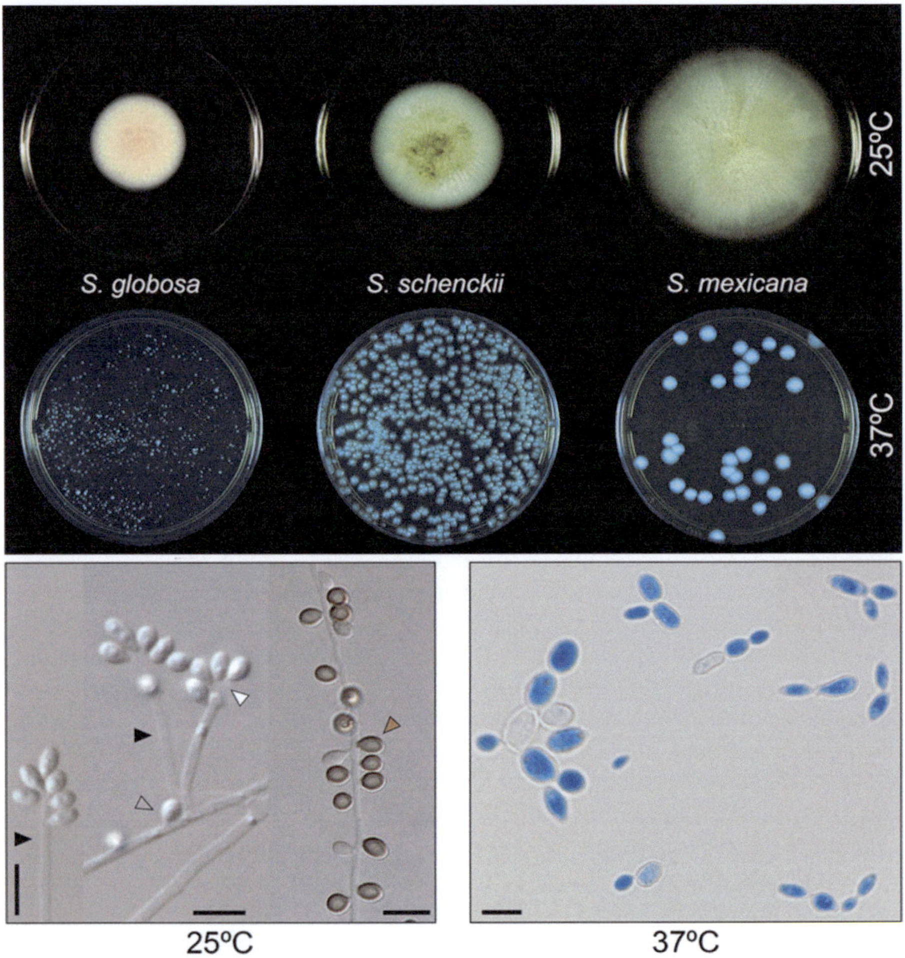

Fig. 19.4 Thermal dimorphism in isolates of *Sporothrix* spp. At room temperature (25 °C), the giant colonies, when grown in Sabouraud agar, are initially white to cream colored and leathery, with a grooved surface. When the isolates are grown at 37 °C in brain heart agar infusion (BHI), the fungi develop into small yeast colonies. Microscopically, at 25 °C, hyaline, thin, and septate hyphae, with erect conidiophores (*black arrow*) and primary hyaline conidia arranged sympodially (*white arrow*), are observed. Numerous lateral and secondary conidia with a *brown color* (*brown arrow*) or hyaline conidia can occur along the undifferentiated hyphae (*empty arrow*). At 37 °C, spherical or oval yeast cells, sometimes with elongated buds, are seen. *Bar* = 5 μm

culture is important for the specific characterization of the agents from the *S. schenckii* complex. The conversion to yeast is sometimes restricted to the peripheral areas of the colonies (Lacaz et al. 2002).

The ultrastructure of *Sporothrix* is characterized by the development of hyaline, thin, and septate hyphae, with erect conidiophores and primary hyaline conidia arranged sympodially. Numerous lateral and secondary conidia, which are usually brownish, may occur together with the undifferentiated hyphae. The primary conidia are usually hyaline, with dimensions of 2–3 µm by 3–6 µm. The arrangement resembles a palm tree or a daisy. Sometimes, conidia are angular or obovate. Over time, conidiogenesis increases, and conidia are formed along the conidiophores and eventually along undifferentiated hyphae. Long chains of conidia are observed in old cultures. Free conidia have an annular marking, which indicates the position where they were connected to the conidiophore (Lacaz et al. 2002).

Secondary conidia are produced by some isolates and emerge from undifferentiated hyphae. According to de Beer and Wingfield (2013), the most pathogenic species and those originating in the soil also produce small and pigmented secondary conidia, which are unique features of the *S. schenckii-O. stenoceras* complex, in addition to primary hyaline conidia. These conidia are usually brown, triangular, and globose, with thick walls. However, it is important to note that the presence, shape, and size of secondary conidia exhibit intra- and interspecific variations (Fernandes et al. 2009b). The only other group that produces pigmented secondary conidia, although typically significantly larger, is the genus *Raffaelea* (*Ascomycota*: *Pezizomycotina*) (de Beer and Wingfield 2013). Secondary conidia may be more resistant to environmental factors than primary, hyaline, thin-walled conidia.

At 37°C, the yeast is oval or spherical shaped and of various sizes. On average, the dimensions of the yeast are 1–3 µm by 3–10 µm (Rodrigues et al. 2015e). The yeast may present one or more buds (Fig. 19.4). The microscopic examination of smear cultures stained by the Gram method reveals the presence of gram-positive, globular, oval, or fusiform (cigar-shaped) cells (Lacaz et al. 2002).

However, the morphology of the saprophytic phase can be observed in many species in Ophiostomatales, including *Sporothrix*, *Ophiostoma*, *Hyalorhinocladiella*, and *Pesotum* (Zipfel et al. 2006; Zhou et al. 2006, 2014; Roets et al. 2010; de Beer and Wingfield 2013; Zhang et al. 2015; Rodrigues et al. 2015a), and also in unrelated clinically relevant fungi, including the basidiomycete *Quambalaria* (de Hoog et al. 2000). For this reason, the environmental isolation and characterization of the specimen using only morphological characteristics can mislead an inexperienced taxonomist.

19.8 Molecular Diagnosis

The molecular diagnosis of sporotrichosis at the species level can be performed from samples isolated in culture or from biological samples, including tissues and secretions. According to Zhou et al. (2014), all the clinically relevant species of *Sporothrix* (clades I–III and VI) can be easily recognized by sequencing and analysis of the ITS region (ITS1 + 5.8S + ITS2) using the ITS1 and ITS4 primers (White et al. 1990). The overall sampling does not affect the success of phylogenetic characterization of the clinical clade on the basis of the ribosomal operon. Phylogenetic analyses based on protein-coding genes such as β-tubulin (de Meyer et al. 2008), calmodulin (Marimon et al. 2006, 2007), and elongation factor 1α (Rodrigues et al. 2013b) are also important for the molecular identification of the causative agent of human and animal sporotrichosis.

Some simple methods based on polymerase chain reaction (PCR) are available to identify the clinically relevant *Sporothrix* species with high sensitivity and specificity (Rodrigues et al. 2015b). Identification methods that use DNA obtained from strains isolated in culture include random-amplified polymorphic DNA (RAPD) with the primer T3B (de Oliveira et al. 2012), PCR-restriction fragment length polymorphism (PCR-RFLP) (Rodrigues et al. 2014b), rolling circle amplification (Rodrigues et al. 2015f), and species-specific PCR (Rodrigues et al. 2015b). For direct molecular diagnosis from complex samples, species-specific PCR is used for the selective detection of the DNA of *Sporothrix* spp. using primers for *S. brasiliensis*, *S. schenckii*, *S. globosa*, *S. mexicana*, *S. pallida*, and *Ophiostoma stenoceras*. The molecular diagnosis using species-specific PCR was validated using a murine model of disseminated sporotrichosis (BALB/c). Infections due to *S. brasiliensis* and *S. schenckii* have been successfully detected using species-specific PCR in different organs, including the spleen, lungs, liver, kidneys, heart, brain, and tail (Rodrigues et al. 2015b). Moreover, the utility of PCR for the detection of *Sporothrix* DNA was demonstrated in stool samples from infected animals, suggesting the application of species-specific PCR in epidemiological studies (Rodrigues et al. 2015b). Rolling circle amplification (RCA) based on species-specific padlock probes certainly provides a valuable tool for the detection of *Sporothrix* DNA in the tissue of warm-blooded host and environmental samples, including soil, *Sphagnum* moss, and *Rose* spp. (Rodrigues et al. 2015f).

19.9 Serological Diagnosis

Serological tests are considered to be rapid and noninvasive tools for the diagnosis of fungal diseases. In the case of sporotrichosis, serological tests may be an auxiliary tool, particularly for the diagnosis of clinical forms or atypical extracutaneous forms. Well-standardized tests offer high sensitivity and specificity and can be automated, allowing large-scale execution. Several methods with variable sensitivity and specificity have been described for the detection of antibodies in the serum of patients with sporotrichosis. Initially, precipitation and agglutination tests were used for serological diagnosis.

Double immunodiffusion tests for sporotrichosis are inexpensive and combine high sensitivity and high specificity. Typically, these tests do not have cross-reactivity with sera from patients with other infectious diseases with similar clinical manifestations, such as chromoblastomycosis or leishmaniasis (Barros et al. 2011). In the past, assays such as immunoelectrophoresis, tube agglutination, and latex agglutination were successfully used for the serological diagnosis of sporotrichosis for the detection of antibodies of the host (Karlin and Nielsen 1970; Barros et al. 2011).

The humoral response of patients with sporotrichosis was investigated by Scott and Muchmore (1989) using Western blotting. In approximately 100 % of the sera from these patients, it was possible to observe the presence of IgG antibodies that recognized the 40 and 70 kDa antigenic fractions isolated from *Sporothrix* spp.

Mendoza et al. (2002) reported that antibodies in the sera of patients with sporotrichosis recognize different molecules, including the 40 kDa, 55 kDa, 74 kDa, 90 kDa, and 147 kDa fractions, depending on the culture medium used for the production of the protein/glycoprotein extract. In an experimental murine model of sporotrichosis, antibodies primarily recognized the 67 kDa molecule (Carlos et al. 1998). At present, it is clear that the isolates and especially the species directly interfere with the effectiveness of the antigen. The main antigen of human and animal sporotrichosis used for serological diagnosis was recently identified by mass spectrometry as 3-carboxymuconate cyclase. This protein appears as isoforms (pI between 4.33 and 4.85) and glycoforms (55–73 kDa) in the proteomes of *S. brasiliensis*, *S. schenckii*, and *S. globosa* (Rodrigues et al. 2015c, d). These physicochemical variations appear to be the result of posttranslational modifications (Rodrigues et al. 2015d).

The peptido-rhamnomannan fraction (CWPR) of *S. schenckii* has been used for serological and intradermal tests (Lima and Bezerra 1997). Penha and Bezerra (2000) used the peptido-rhamnomannan fraction of the fungal cell wall as an antigen for enzyme-linked immunosorbent assays (ELISAs) and noted that ELISA using the *S. schenckii* Con-A binding fraction (SsCBF) purified by affinity chromatography bound to Con-A was highly specific for the diagnosis of sporotrichosis. Bernardes-Engemann et al. (2005) demonstrated that SsCBF has 90 % sensitivity, 80 % specificity, and an overall efficiency of 86 % in ELISAs using sera from patients with distinct clinical forms of the disease. Later, the same group reported a variation in the activity of SsCBF molecules obtained from different *Sporothrix* isolates (Bernardes-Engemann et al. 2009).

Almeida-Paes et al. (2007b) described the diagnosis of human sporotrichosis by ELISA using crude *Sporothrix* exoantigens in the mycelial phase. The test had a sensitivity of 97 %, specificity of 89 %, and an overall efficiency of 92 %. The ELISA assay using *S. schenckii* exoantigens is a highly sensitive diagnostic tool for the serological diagnosis of human sporotrichosis and can be used in conjunction with conventional diagnostic methods. Almeida-Paes et al. (2007a) evaluated the immune response for IgG, IgM, and IgA in 41 patients with sporotrichosis before treatment with itraconazole and in 35 patients during treatment with itraconazole and demonstrated that the combination of the results for IgA, IgG, and IgM yielded high sensitivity and specificity for the diagnosis of sporotrichosis.

The diagnosis of feline sporotrichosis using ELISA for the detection of anti-*Sporothrix*-specific antibodies has been proposed by Fernandes et al. (2011) and Rodrigues et al. (2015c). The early diagnosis of feline sporotrichosis is critical to identify outbreaks and control the spread of the disease among humans. Rodrigues et al. (2015c) explored the diversity of molecules expressed by *S. brasiliensis* and *S. schenckii* that are recognized by IgG present in the serum of cats naturally infected with *S. brasiliensis*. After infection, cats produce an IgG-mediated response against antigens expressed by *Sporothrix* that is similar to the humoral response observed in murine sporotrichosis (Fernandes et al. 2013) and human sporotrichosis (Rodrigues et al. 2015d). There is a significant cross-reactivity between antigens of *S. brasiliensis* and *S. schenckii*, supporting the hypothesis that antigenic epitopes may be phylogenetically conserved between related species. IgG antibodies against the enzyme 3-carboxymuconate cyclase were identified in all the animals evaluated, independent of their clinical manifestations. Therefore, understanding the immune response in feline sporotrichosis is critical to advance the development of novel techniques for serological diagnosis and vaccine development and to improve our understanding of the evolution of *Sporothrix* (Rodrigues et al. 2015c). Moreover, ELISA is a valuable screening tool for the detection of anti-*Sporothrix*-specific antibodies in cats with sporotrichosis.

19.10 Host-Pathogen Interaction

Sporothrix undergoes a morphological transition in response to temperature variation and develops into filamentous hyphae during its saprophytic stage in the soil (25 °C) or into yeast forms in the host tissue during the parasitic stage (37 °C). The overall efficiency of yeast-mycelium conversion is high in the species from the clinical clades *S. brasiliensis*, *S. schenckii*, *S. globosa*, and *S. luriei*, whereas the yeast-mycelium transitional capacity varies considerably in the environmentally relevant species *S. inflata*, *S. humicola*, *S. pallida*, *S. mexicana*, *S. chilensis*, *S. dimorphospora*, and *S. brunneoviolacea*, among others, which, in general, exhibit poor morphological transition, producing few yeast cells (Rodrigues et al. 2015d, e). This result can be due to the attenuated virulence of environmental species toward mammals (Howard and Orr 1963; Dixon et al. 1992; Arrillaga-Moncrieff et al. 2009). Interestingly, members of the complex *O. stenoceras-S. schenckii*, such as *S. chilensis* and *Ophiostoma bragantinum*, also produce cream-colored colonies with a large number of yeast cells (Madrid et al. 2010; Rodrigues et al. 2015a). The morphological transition induced by temperature in *Sporothrix* is an important morphological adaptation for the infection of warm-blooded animals, and this adaptation is shared with phylogenetically distant human pathogens of the orders *Onygenales* and *Eurotiales* (Sil and Andrianopoulos 2014).

Within the *S. schenckii* complex, the virulence profiles vary according to the characteristics of the pathogen (Fernandes et al. 2009b) and the immunity of the host (Rodrigues et al. 2015d). *Sporothrix brasiliensis* is the most virulent species of the complex and causes intense tissue invasion and death, whereas *S. schenckii s. str.* presents different levels of pathogenicity, ranging from high to low virulence, and *S. globosa* displays little or no virulence in murine models (Arrillaga-Moncrieff et al. 2009; Fernandes et al. 2013; Castro et al. 2013; Almeida et al. 2015). The environmental species of *Sporothrix* present attenuated virulence in murine models, with low invasive potential, and the host can control infection a few weeks after challenge (Howard and Orr 1963; Dixon et al. 1992; Arrillaga-Moncrieff et al. 2009; Rodrigues et al. 2015a). In other clinically relevant fungal species, specimens with a high degree of virulence and those without virulence are often grouped side by side in phylogenetic analyses, as what occurred with the melanized fungi *Cladophialophora bantiana* and *Cladophialophora psammophila* (Badali et al. 2011). This reinforces the idea that virulence is an emerging, changing, and highly complex characteristic (Casadevall and Pirofski 2003; Casadevall and Fang 2011; Pirofski and Casadevall 2015). In fact, the infectivity of a pathogen is often a result of adaptations of preexisting characteristics related to the nonparasitic lifestyle of these microorganisms (van Burik and Magee 2001).

The mechanisms involved in the pathogenesis and invasion of *Sporothrix* are still largely unknown, and many virulence factors have not been identified. The molecules involved in host-parasite interactions are often immunogenic, and the yeasts of *Sporothrix* express several antigens, which vary from proteins to cell wall carbohydrates (Alves et al. 1994; Lopes-Bezerra 2011). Such molecules may trigger distinct innate and adaptive immune mechanisms in the host. Cellular and humoral responses triggered by host-pathogen interactions play an important role in the development and severity of sporotrichosis. The analysis of the humoral response indicates that fungal diseases can generate protective and non-protective antibodies (Nascimento et al. 2008; Romani 2011; Fernandes et al. 2013) against molecules expressed during the host-parasite interaction. Previous studies have shown that *Sporothrix* antigens are involved in the induction of protective immunity and that the humoral immune response may be one of the mechanisms involved in the establishment of protection (Nascimento et al. 2008; Almeida 2012; de Lima Franco et al. 2012). Although scarce, initial studies on these antigenic molecules have discovered several potential markers for serological diagnosis and candidates for vaccine development and have expanded our understanding of the virulence factors associated with the pathogenesis of *Sporothrix*.

Cellular and humoral responses triggered by host-pathogen interactions play an important role in the development and severity of sporotrichosis. Rodrigues et al. (2015d) investigated the molecules expressed during this interaction that are capable of inducing a humoral response in human sporotrichosis. The characterization of specific antigens indicated a converging profile of humoral immune responses in human sporotrichosis that is critical for clinical diagnosis and immunotherapy. The cross-reactivity between *S. brasiliensis*, *S. schenckii*, and *S. globosa* is remarkable, and antibodies react strongly against a 60–70 kDa glycoprotein (gp60-70). This molecule has been successfully identified as 3-carboxymuconate cyclase and is present in pathogenic species (*S. brasiliensis*, *S. schenckii*, and *S. globosa*) but absent in the non-virulent ancestral species *S. mexicana*. This observation may be related to the attenuated virulence of *S. mexicana* in mammals. In feline sporotrichosis, all the animals infected with *S. brasiliensis* produced IgG antibodies against 3-carboxymuconate cyclase (Rodrigues et al. 2015c).

Although the primary function of the gp60-70 orthologs in the environment is the degradation of benzoate and aromatic polymers, it has been suggested that 3-carboxymuconate cyclase is important during pathogenesis and invasion in human and animal sporotrichosis (Fernandes et al. 2013). In addition, 3-carboxymuconate cyclase was initially described as an adhesin for fibronectin and laminin in *S. schenckii s. str.* and is primarily located in the cell wall (Nascimento et al. 2008; Teixeira et al. 2009; Castro et al. 2013; Ruiz-Baca et al. 2014). Rodrigues et al. (2015c, d) have shown that gp60-70 is immunogenic, on the basis of the bioinformatics prediction of linear epitopes for B cells, which is consistent with the in vitro and in vivo antigenic profile of this molecule. The results obtained by Fernandes et al. (2013) indicate different levels of virulence of *S. schenckii* and suggest a correlation between protein secretion, immunogenicity, genetic diversity, and virulence in the *S. schenckii* complex. In this scenario, BALB/c mice inoculated with virulent isolates of *S. brasiliensis* and *S. schenckii* exhibit anti-gp60-70 antibodies (Fernandes et al. 2013).

19.11 Treatment

Treatment options for sporotrichosis include local hyperthermia, saturated solutions of potassium iodide, azole drugs (ketoconazole, itraconazole, and fluconazole), amphotericin B, and the allylamine terbinafine (Bonifaz and Vázquez-González 2013). The average duration of treatment in immunocompetent patients, regardless of the oral drug administered, is 60 days if the patient has an intact immune system (Orofino-Costa et al. 2015). For decades, it has been known that hyperthermia decreases the size of cutaneous sporotrichosis lesions. Therefore, the use of specific devices that heat the tissue to 42–43 °C using infrared wavelengths is recommended.

A saturated solution of potassium iodide administered orally has been used for more than a century, in increasing dosages, depending on the tolerance of the patient (2–3 g daily for adults) and is still a good choice for the treatment of sporotrichosis (Costa et al. 2013; de Macedo et al. 2015a, b). However, itraconazole is currently recommended for the treatment of all forms of sporotrichosis (Mahajan 2014). For cutaneous and lymphocutaneous infections, the positive response rate is 90–100 % for itraconazole therapy but only 63–71 % for fluconazole therapy. The guidelines of the American Infectious Disease Society for the management of patients with fixed lymphocutaneous and cutaneous sporotrichosis recommend daily treatment of itraconazole (200 mg/day) for 2–4 weeks after the remission of the lesions, usually for a total of 3 to 6 months. Patients who do not respond to therapy should receive a higher dosage of itraconazole (200 mg/twice daily) (Kauffman et al. 2007).

De Lima Barros et al. (2011) reported successful treatment using the minimum dose of itraconazole. Among the 610 patients evaluated, 94.6 % were cured with 50–400 mg/day; of these, 547 patients were cured with 100 mg/day, 59 patients with 200–400 mg/day, and four children with 50 mg/day. The authors report a low incidence of adverse events and treatment failure.

Amphotericin B is initially used for the treatment of serious systemic disease during pregnancy and in immunosuppressed patients until recovery (Costa et al. 2011; Orofino-Costa et al. 2015), followed by the use of itraconazole for the rest of treatment (Silva-Vergara et al. 2012). In vitro results indicate that *Sporothrix* species differ in sensitivity to the main antifungal agents. Serena et al. (2008b) evaluated the efficacy of 12 antifungal agents in vitro using a reference method of microdilution. There were significant differences in the dose of antifungal agents for the inhibition of *S. brasiliensis*, *S. schenckii*, *S. globosa*, *S. pallida*, and *S. mexicana*. *Sporothrix brasiliensis* exhibited the best response to antifungal agents, whereas *S. mexicana* was the most tolerant to the drugs. Terbinafine is the most active drug in vitro, followed by ketoconazole and posaconazole.

Previous studies that correlated antifungal susceptibility and genetic diversity indicate that genetically homogeneous populations with little genetic variability, as observed for *S. brasiliensis*, have little variation in susceptibility. In contrast, an increase in genetic diversity, as observed in populations of *S. schenckii*, leads to a wider variation in susceptibility. Rodrigues et al. (2014c) evaluated the minimum inhibitory concentration (MIC) and minimum fungicidal contraction (MFC) in a group of genetically diverse isolates of *Sporothrix*. Itraconazole and posaconazole were moderately effective against *S. brasiliensis* (MIC_{90} values of 2 µg/ml for both) and against *S. schenckii* (MIC_{90} values of 4 µg/ml and 2 µg/ml, respectively). Interestingly, posaconazole showed a low MIC against *S. mexicana*, which is tolerant to most commercially available antifungal agents. Caspofungin, 5-fluorocytosine, and fluconazole had no antifungal activity in vitro against *S. brasiliensis*, *S. schenckii*, *S. globosa*, and *S. mexicana*. The results indicate significant differences between the MICs required to inhibit the growth of *Sporothrix* and the MIC required to decrease the number of colony-forming units, demonstrating the fungistatic and not fungicidal effect of most of the antifungal agents evaluated.

Borba-Santos et al. (2015) reported that isolates of *S. brasiliensis* of human origin recently recovered during the epidemic in Rio de Janeiro (2011–2012) were less susceptible to amphotericin B, itraconazole, posaconazole, voriconazole, and terbinafine compared with isolates recovered between 1998 and 2004. Terbinafine, followed by posaconazole, was more effective in vitro against the mycelial and yeast forms of *S. brasiliensis*. Terbinafine has been reported to be effective for the treatment of cutaneous sporotrichosis (Mahajan 2014). Several fungal isolates showed high MIC values for amphotericin B and/or itraconazole, which are currently used as the preferred agents for the treatment of various forms of sporotrichosis. The interaction of *S. brasiliensis* with amphotericin B, itraconazole, and terbinafine leads to morphological changes that range from an increase in the microfibrillar layer to changes in cell wall thickness (Borba-Santos et al. 2015).

Treatment of feline sporotrichosis presents a challenge owing to the limited number of oral antifungal agents that are associated with numerous side effects and high cost. Similar to the treatment options for human sporotrichosis, the treatment of feline sporotrichosis includes local hyperthermia, potassium iodide, azole drugs (ketoconazole and itraconazole), amphotericin B, and terbinafine. The azoles and the saturated solution of potassium iodide are the most commonly used drugs for the treatment of feline sporotrichosis (Pereira et al. 2010; Reis et al. 2012; Gremião et al. 2015), and clinical cure is achieved regardless of initial clinical findings or coinfection with FIV and/or FeLV. Treatment can take several weeks to several months (average duration of 4–9 months) and should be continued for at least 1 month after clinical cure. Recurrence may occur, indicating the possibility of reactivation of the lesions, despite the success of treatment (Pereira et al. 2010; Gremião et al. 2011, 2015; Chaves et al. 2013).

References

Ajello L, Kaplan W (1969) A new variant of *Sporothrix Schenckii*. Mycoses 12(11):633–644. doi:10.1111/j.1439-0507.1969.tb03423.x

Almeida SR (2012) Therapeutic monoclonal antibody for sporotrichosis. Front Microbiol 3:409. doi:10.3389/fmicb.2012.00409

Almeida-Paes R, Pimenta MA, Monteiro PC, Nosanchuk JD, Zancope-Oliveira RM (2007a) Immunoglobulins G, M, and A against *Sporothrix schenckii* exoantigens in patients with sporotrichosis before and during treatment with itraconazole. Clin Vaccine Immunol 14(9):1149–1157. doi:10.1128/cvi.00149-07

Almeida-Paes R, Pimenta MA, Pizzini CV, Monteiro PC, Peralta JM, Nosanchuk JD, Zancope-Oliveira RM (2007b) Use of mycelial-phase *Sporothrix schenckii* exoantigens in an enzyme-linked immunosorbent assay for diagnosis of sporotrichosis by antibody detection. Clin Vaccine Immunol 14(3):244–249. doi:10.1128/cvi.00430-06

Almeida-Paes R, de Oliveira MM, Freitas DF, do Valle AC, Zancope-Oliveira RM, Gutierrez-Galhardo MC (2014) Sporotrichosis in Rio de Janeiro, Brazil: *Sporothrix brasiliensis* is associated with atypical clinical presentations. PLoS Negl Trop Dis 8(9):e3094. doi:10.1371/journal.pntd.0003094

Almeida-Paes R, de Oliveira LC, Oliveira MME, Gutierrez-Galhardo MC, Nosanchuk JD, Zancopé Oliveira RM (2015) Phenotypic characteristics associated with virulence of clinical isolates from the *Sporothrix* complex. BioMed Res Int 2015: Article ID 212308. doi:10.1155/2015/212308

Al-Tawfiq JA, Wools KK (1998) Disseminated sporotrichosis and *Sporothrix schenckii* fungemia as the initial presentation of Human Immunodeficiency Virus infection. Clin Infect Dis 26(6):1403–1406. doi:10.1086/516356

Alves LL, Travassos LR, Previato JO, Mendonça-Previato L (1994) Novel antigenic determinants from peptidorhamnomannans of *Sporothrix schenckii*. Glycobiology 4(3):281–288. doi:10.1093/glycob/4.3.281

Arrillaga-Moncrieff I, Capilla J, Mayayo E, Marimon R, Mariné M, Gené J, Cano J, Guarro J (2009) Different virulence levels of the species of *Sporothrix* in a murine model. Clin Microbiol Infect 15(7):651–655. doi:10.1111/j.1469-0691.2009.02824.x

Aung AK, Teh BM, MCgrath C, Thompson PJ (2013) Pulmonary sporotrichosis: case series and systematic analysis of literature on clinico-radiological patterns and management outcomes. Med Mycol 51(5):534–544. doi:10.3109/13693786.2012.751643

Badali H, Prenafeta-Boldu FX, Guarro J, Klaassen CH, Meis JF, de Hoog GS (2011) *Cladophialophora psammophila*, a novel species of Chaetothyriales with a potential use in the bioremediation of volatile aromatic hydrocarbons. Fungal Biology 115(10):1019–1029. doi:10.1016/j.funbio.2011.04.005

Barros MB, Schubach Ade O, do Valle AC, Gutierrez Galhardo MC, Conceição-Silva F, Schubach TM, Reis RS, Wanke B, Marzochi KB, Conceição MJ (2004) Cat-transmitted sporotrichosis epidemic in Rio de Janeiro, Brazil: description of a series of cases. Clin Infect Dis 38(4):529–535. doi:10.1086/381200

Barros MBL, Schubach AO, Schubach TMP, Wanke B, Lambert-Passos SR (2008) An epidemic of sporotrichosis in Rio de Janeiro, Brazil: epidemiological aspects of a series of cases. Epidemiol Infect 136(09):1192–1196. doi:10.1017/S0950268807009727

Barros MB, Schubach TP, Coll JO, Gremião ID, Wanke B, Schubach A (2010) [Sporotrichosis: development and challenges of an epidemic] (Sporotrichosis: development and challenges of an epidemic). Rev Panam Salud Publica 27(6):455–460. doi:10.1590/S1020-49892010000600007

Barros MB, de Almeida Paes R, Schubach AO (2011) *Sporothrix schenckii* and sporotrichosis. Clin Microbiol Rev 24(4):633–654. doi:10.1128/cmr.00007-11

Bernardes-Engemann AR, Orofino Costa RC, Miguens BP, Penha CVL, Neves E, Pereira BAS, Dias CMP, Mattos M, Gutierrez MC, Schubach A, Oliveira Neto MP, Lazéra M, Lopes-Bezerra LM (2005) Development of an enzyme-linked immunosorbent assay for the serodiagnosis of several clinical forms of sporotrichosis. Med Mycol 43(6):487–493. doi:10.1080/13693780400019909

Bernardes-Engemann AR, Loureiro y Penha CV, Benvenuto F, Braga JU, Barros ML, Orofino-Costa R, Lopes-Bezerra LM (2009) A comparative serological study of the SsCBF antigenic fraction isolated from three *Sporothrix schenckii* strains. Med Mycol 47(8):874–878. doi:10.3109/13693780802695520

Beurmann L, Gougerot H (1912) Les Sporotrichose. Librairie Felix Alcan, Paris

Blehert DS, Hicks AC, Behr M, Meteyer CU, Berlowski-Zier BM, Buckles EL, Coleman JT, Darling SR, Gargas A, Niver R, Okoniewski JC, Rudd RJ, Stone WB (2009) Bat white-nose syndrome: an emerging fungal pathogen? Science (New York, NY) 323(5911):227. doi:10.1126/science.1163874

Bommer M, Hütter M-L, Stilgenbauer S, de Hoog GS, de Beer ZW, Wellinghausen N (2009) Fatal *Ophiostoma piceae* infection in a patient with acute lymphoblastic leukaemia. J Med Microbiol 58(3):381–385. doi:10.1099/jmm.0.005280-0

Bonifaz A, Vázquez-González D (2013) Diagnosis and treatment of lymphocutaneous sporotrichosis: What are the options? Curr Fungal Infect Rep 7(3):252–259. doi:10.1007/s12281-013-0140-3

Borba-Santos LP, Rodrigues AM, Gagini TB, Fernandes GF, Castro R, de Camargo ZP, Nucci M, Lopes-Bezerra LM, Ishida K, Rozental S (2015) Susceptibility of *Sporothrix brasiliensis* isolates to amphotericin B, azoles, and terbinafine. Med Mycol 53(2):178–188. doi:10.1093/mmy/myu056

Carlos IZ, Sgarbi DB, Placeres MC (1998) Host organism defense by a peptide-polysaccharide extracted from the fungus *Sporothrix schenckii*. Mycopathologia 144(1):9–14

Casadevall A (2012) Fungi and the rise of mammals. PLoS Pathog 8(8):e1002808. doi:10.1371/journal.ppat.1002808

Casadevall A, Fang FC, Pirofski LA (2011) Microbial virulence as an emergent property: Consequences and opportunities. PLoS Pathog 7(7):e1002136. doi:10.1371/journal.ppat.1002136

Casadevall A, Pirofski LA (2003) The damage-response framework of microbial pathogenesis. Nat Rev Micro 1(1):17–24

Castro RA, Kubitschek-Barreira PH, Teixeira PAC, Sanches GF, Teixeira MM, Quintella LP, Almeida SR, Costa RO, Camargo ZP, Felipe MSS, de Souza W, Lopes-Bezerra LM (2013) Differences in cell morphometry, cell wall topography and Gp70 expression correlate with the virulence of *Sporothrix brasiliensis* clinical isolates. PLoS ONE 8(10):e75656. doi:10.1371/journal.pone.0075656

Chakrabarti A, Bonifaz A, Gutierrez-Galhardo MC, Mochizuki T, Li S (2015) Global epidemiology of sporotrichosis. Med Mycol 53(1):3–14. doi:10.1093/mmy/myu062

Chaves AR, de Campos MP, Barros MBL, do Carmo CN, Gremião IDF, Pereira SA, Schubach TMP (2013) Treatment abandonment in feline sporotrichosis – study of 147 cases. Zoonoses Public Health 60(2):149–153. doi:10.1111/j.1863-2378.2012.01506.x

Costa RO, Bernardes-Engemann AR, Azulay-Abulafia L, Benvenuto F, Neves Mde L, Lopes-Bezerra LM (2011) Sporotrichosis in pregnancy: case reports of 5 patients in a zoonotic epidemic in Rio de Janeiro, Brazil. Anais Brasileiros de Dermatologia 86(5):995–998

Costa RO, Macedo PM, Carvalhal A, Bernardes-Engemann AR (2013) Use of potassium iodide in dermatology: updates on an old drug. Anais Brasileiros de Dermatologia 88(3):396–402. doi:10.1590/abd1806-4841.20132377

Crothers SL, White SD, Ihrke PJ, Affolter VK (2009) Sporotrichosis: a retrospective evaluation of 23 cases seen in northern California (1987-2007). Veterinary Dermatology 20(4):249–259. doi:10.1111/j.1365-3164.2009.00763.x

Daniel Da Rosa W, Gezuele E, Calegari L, Goni F (2008) Asteroid body in sporotrichosis. Yeast viability and biological significance within the host immune response. Med Mycol 46(5):443–448. doi:10.1080/13693780801914898

Daszak P, Berger L, Cunningham AA, Hyatt AD, Green DE, Speare R (1999) Emerging infectious diseases and amphibian population declines. Emerging Infectious Diseases 5(6):735–748. doi:10.3201/eid0506.990601

Daszak P, Cunningham AA, Hyatt AD (2000) Emerging infectious diseases of wildlife—threats to biodiversity and human health. Science (New York, NY) 287(5452):443–449 de Beer ZW, Wingfield MJ (2013) Emerging lineages in the Ophiostomatales. In: Seifert KA, de Beer ZW, Wingfield MJ (eds) The Ophiostomatoid Fungi: Expanding Frontiers, CBS biodiversity series, vol 12. CBS-KNAW Fungal Biodiversity Centre, Utrecht

de Beer ZW, Harrington TC, Vismer HF, Wingfield BD, Wingfield MJ (2003) Phylogeny of the *Ophiostoma stenoceras–Sporothrix schenckii* complex. Mycologia 95:434–441

de Hoog GS, Guarro J, Gené J, Figueras MJ (2000) Atlas of clinical fungi, 2nd edn. Centraalbureau voor Schimmelcultures, Utrecht

de Lima Barros MB, Schubach AO, de Vasconcellos Carvalhaes de Oliveira R, Martins EB, Teixeira JL, Wanke B (2011) Treatment of cutaneous sporotrichosis with Itraconazole – study of 645 patients. Clin Infect Dis 52(12):e200–e206. doi:10.1093/cid/cir245

de Lima Franco D, Nascimento RC, Ferreira KS, Almeida SR (2012) Antibodies against *Sporothrix schenckii* enhance TNF-α production and killing by macrophages. Scand J Immunol 75(2):142–146. doi:10.1111/j.1365-3083.2011.02636.x

de Macedo PM, Lopes-Bezerra LM, Bernardes-Engemann AR, Orofino-Costa R (2015a) New posology of potassium iodide for the treatment of cutaneous sporotrichosis: study of efficacy and safety in 102 patients. J Eur Acad Dermatol Venereol 29(4):719–724. doi:10.1111/jdv.12667

de Macedo PM, Sztajnbok DC, Camargo ZP, Rodrigues AM, Lopes-Bezerra LM, Bernardes-Engemann AR, Orofino-Costa R (2015b) Dacryocystitis due to *Sporothrix brasiliensis*: a case report of a successful clinical and serological outcome with low-dose potassium iodide treatment and oculoplastic surgery. Br J Dermatol 172(4):1116–1119. doi:10.1111/bjd.13378

de Meyer EM, de Beer ZW, Summerbell RC, Moharram AM, de Hoog GS, Vismer HF, Wingfield MJ (2008) Taxonomy and phylogeny of new wood- and soil-inhabiting *Sporothrix* species in the *Ophiostoma stenoceras-Sporothrix schenckii* complex. Mycologia 100(4):647–661. doi:10.3852/07-157r

de Oliveira MM, de Almeida-Paes R, de Medeiros MM, de Lima Barros MB, Galhardo MC, Zancope-Oliveira RM (2010) Sporotrichosis caused by *Sporothrix globosa* in Rio De Janeiro, Brazil: case report. Mycopathologia 169(5):359–363. doi:10.1007/s11046-010-9276-7

de Oliveira MME, Sampaio P, Almeida-Paes R, Pais C, Gutierrez-Galhardo MC, Zancope-Oliveira RM (2012) Rapid identification of *Sporothrix* species by T3B fingerprinting. J Clin Microbiol 50(6):2159–2162. doi:10.1128/JCM.00450-12

Dixon DM, Duncan RA, Hurd NJ (1992) Use of a mouse model to evaluate clinical and environmental isolates of *Sporothrix* spp. from the largest U.S. epidemic of sporotrichosis. J Clin Microbiol 30(4):951–954

Fernandes GF, Amaral CCD, Sasaki A, Godoy PM, De Camargo ZP (2009a) Heterogeneity of proteins expressed by Brazilian *Sporothrix schenckii* isolates. Med Mycol 47(8):855–861. doi:10.3109/13693780802713216

Fernandes GF, dos Santos PO, Amaral CC, Sasaki AA, Godoy-Martinez P, de Camargo ZP (2009b) Characteristics of 151 Brazilian *Sporothrix schenckii* isolates from 5 different geographic regions of Brazil: a forgotten and re-emergent pathogen. Open Mycol J 3(1):48–58. doi:10.2174/1874437000903010048

Fernandes GF, Lopes-Bezerra LM, Bernardes-Engemann AR, Schubach TM, Dias MA, Pereira SA, de Camargo ZP (2011) Serodiagnosis of sporotrichosis infection in cats by enzyme-linked immunosorbent assay using a specific antigen, SsCBF, and crude exoantigens. Vet Microbiol 147(3-4):445–449. doi:10.1016/j.vetmic.2010.07.007

Fernandes GF, dos Santos PO, Rodrigues AM, Sasaki AA, Burger E, de Camargo ZP (2013) Characterization of virulence profile, protein secretion and immunogenicity of different *Sporothrix schenckii sensu stricto* isolates compared with *S. globosa* and *S. brasiliensis* species. Virulence 4(3):241–249. doi:10.4161/viru.23112

Fishburn F, Kelley DC (1967) Sporotrichosis in a horse. Journal of the American Veterinary Medical Association 151(1):45–46

Fisher MC, Henk DA, Briggs CJ, Brownstein JS, Madoff LC, McCraw SL, Gurr SJ (2012) Emerging fungal threats to animal, plant and ecosystem health. Nature 484(7393):186–194. doi:10.1038/nature10947

Freitas DC, Migliano MF, Zani Neto L (1956) Sporotrichosis. Observation of spontaneous case in domestic cat (*Felis catus*). Rev Fac Med Vet Univ Sao Paulo 5:601–604

Freitas DC, Moreno G, Saliba AM, Botino JÁ, Mós EM (1965) Sporotrichosis in dogs and cats. Rev Fac Med Vet Univ Sao Paulo 7:381–387

Gargas A, Trest MT, Christensen M, Volk TJ, Blehert DS (2009) *Geomyces destructans* sp. nov. associated with bat white-nose syndrome. Mycotaxon 108(1):147–154. doi:10.5248/108.147

Giraud T, Gladieux P, Gavrilets S (2010) Linking the emergence of fungal plant diseases with ecological speciation. Trends in Ecology & Evolution 25(7):387–395. doi:10.1016/j.tree.2010.03.006

Gortazar C, Reperant LA, Kuiken T, de la Fuente J, Boadella M, Martinez-Lopez B, Ruiz-Fons F, Estrada-Pena A, Drosten C, Medley G, Ostfeld R, Peterson T, VerCauteren KC, Menge C, Artois M, Schultsz C, Delahay R, Serra-Cobo J, Poulin R, Keck F, Aguirre AA, Henttonen H, Dobson AP, Kutz S, Lubroth J, Mysterud A (2014) Crossing the interspecies barrier: opening the door to zoonotic pathogens. PLoS Pathog 10(6):e1004129. doi:10.1371/journal.ppat.1004129

Govender NP, Maphanga TG, Zulu TG, Patel J, Walaza S, Jacobs C, Ebonwu JI, Ntuli S, Naicker SD, Thomas J (2015) An outbreak of lymphocutaneous sporotrichosis among mine-workers in South Africa. PLoS Negl Trop Dis 9(9):e0004096. doi:10.1371/journal.pntd.0004096

Gremião I, Schubach T, Pereira S, Rodrigues A, Honse C, Barros M (2011) Treatment of refractory feline sporotrichosis with a combination of intralesional amphotericin B and oral itraconazole. Australian Veterinary Journal 89(9):346–351. doi:10.1111/j.1751-0813.2011.00804.x

Gremião ID, Menezes RC, Schubach TM, Figueiredo AB, Cavalcanti MC, Pereira SA (2015) Feline sporotrichosis: epidemiological and clinical aspects. Med Mycol 53(1):15–21. doi:10.1093/mmy/myu061

Guarro J, Gené J, Stchigel AM (1999) Developments in fungal taxonomy. Clin Microbiol Rev 12(3):454–500

Heckman DS, Geiser DM, Eidell BR, Stauffer RL, Kardos NL, Hedges SB (2001) Molecular evidence for the early colonization of land by fungi and plants. Science (New York, NY) 293(5532):1129–1133. doi:10.1126/science.1061457

Hektoen L, Perkins CF (1900) Refractory subcutaneous abscesses caused by *Sporothrix schenckii*: A new pathogenic fungus. The Journal of experimental medicine 5(1):77–89

Howard DH, Orr GF (1963) Comparison of strains of *Sporotrichum schenckii* isolated from nature. Journal of Bacteriology 85:816–821

Hubálek Z, Rudolf I (2011) The epidemic process in zoonoses and sapronoses. In: Hubálek Z, Rudolf I (eds) Microbial zoonoses and sapronoses. Springer, Dordrecht/Heidelberg/London/New York, pp 33–50. doi:10.1007/978-90-481-9657-9

Kano R, Okubo M, Siew HH, Kamata H, Hasegawa A (2015) Molecular typing of *Sporothrix schenckii* isolates from cats in Malaysia. Mycoses 58(4):220–224. doi:10.1111/myc.12302

Kaplan W, Broderson JR, Pacific JN (1982) Spontaneous systemic sporotrichosis in nine-banded armadillos (*Dasypus novemcinctus*). Sabouraudia 20(4):289–294

Karlin JV, Nielsen HS Jr (1970) Serologic aspects of sporotrichosis. J Infect Dis 121(3):316–327

Kauffman CA, Bustamante B, Chapman SW, Pappas PG (2007) Clinical practice guidelines for the management of sporotrichosis: 2007 update by the Infectious Diseases Society of America. Clin Infect Dis 45(10):1255–1265. doi:10.1086/522765

Kilpatrick AM, Briggs CJ, Daszak P (2010) The ecology and impact of chytridiomycosis: an emerging disease of amphibians. Trends in Ecology & Evolution 25(2):109–118. doi:10.1016/j.tree.2009.07.011

Lacaz CS, Porto E, Martins JEC, Heins-Vaccari EM, de Melo NT (2002) Tratado de Micologia Médica, 9 edn. Sarvier, São Paulo

Larone DH (2011) Medically important fungi: a guide to identification, 5th edn. ASM Press, Washington, DC

Lima OC, Bezerra LML (1997) Identification of a concanavalin A-binding antigen of the cell surface of *Sporothrix schenckii*. Journal of Medical and Veterinary Mycology 35(3):167–172. doi:10.1080/02681219780001101

Lopes-Bezerra LM (2011) *Sporothrix schenckii* cell wall peptidorhamnomannans. Front Microbiol 2:243. doi:10.3389/fmicb.2011.00243
Lopes-Bezerra LM, Schubach A, Costa RO (2006) *Sporothrix schenckii* and sporotrichosis. Anais da Academia Brasileira de Ciências 78(2):293–308. doi:10.1590/S0001-37652006000200009
Lovegrove BG (2012) The evolution of endothermy in Cenozoic mammals: a plesiomorphic-apomorphic continuum. Biol Rev Camb Philos Soc 87(1):128–162. doi:10.1111/j.1469-185X.2011.00188.x
Lutz A, Splendore A (1907) On a mycosis observed in men and mice: Contribution to the knowledge of the so-called sporotrichosis (About a mycosis observed in rats and men: Contribution to the knowledge of the so-called sporotrichosis). Revista Médica de São Paulo 21:443–450
Mackinnon JE, Conti-Díaz IA, Gezuele E, Civila E, Da Luz S (1969) Isolation of *Sporothrix schenckii* from nature and considerations on its pathogenicity and ecology. Med Mycol 7(1):38–45. doi:10.1080/00362177085190071
Madrid H, Cano J, Gene J, Bonifaz A, Toriello C, Guarro J (2009) *Sporothrix globosa*, a pathogenic fungus with widespread geographical distribution. Revista iberoamericana de micologia 26(3):218–222. doi:10.1016/j.riam.2009.02.005
Madrid H, Gené J, Cano J, Silvera C, Guarro J (2010) *Sporothrix brunneoviolacea* and *Sporothrix dimorphospora*, two new members of the *Ophiostoma stenoceras-Sporothrix schenckii* complex. Mycologia 102(5):1193–1203. doi:10.3852/09-320
Mahajan VK (2014) Sporotrichosis: an overview and therapeutic options. Dermatology Research and Practice 2014:272376. doi:10.1155/2014/272376
Marimon R, Gené J, Cano J, Trilles L, Dos SantosLazéra M, Guarro J (2006) Molecular phylogeny of *Sporothrix schenckii*. J Clin Microbiol 44(9):3251–3256. doi:10.1128/JCM.00081-06
Marimon R, Cano J, Gené J, Sutton DA, Kawasaki M, Guarro J (2007) *Sporothrix brasiliensis*, *S. globosa*, and *S. mexicana*, three new *Sporothrix* species of clinical interest. J Clin Microbiol 45(10):3198–3206. doi:10.1128/JCM.00808-07
Marimon R, Gené J, Cano J, Guarro J (2008a) *Sporothrix luriei*: a rare fungus from clinical origin. Med Mycol 46(6):621–625. doi:10.1080/13693780801992837
Marimon R, Serena C, Gené J, Cano J, Guarro J (2008b) *In vitro* antifungal susceptibilities of five species of *Sporothrix*. Antimicrob Agents Chemother 52(2):732–734. doi:10.1128/AAC.01012-07
Marques SA, Franco SR, de Camargo RM, Dias LD, Haddad Junior V, Fabris VE (1993) Sporotrichosis of the domestic cat (*Felis catus*): human transmission. Revista do Instituto de Medicina Tropical de São Paulo 35(4):327–330
Mehta KIS, Sharma NL, Kanga AK, Mahajan VK, Ranjan N (2007) Isolation of *Sporothrix schenckii* from the environmental sources of cutaneous sporotrichosis patients in Himachal Pradesh, India: results of a pilot study. Mycoses 50(6):496–501. doi:10.1111/j.1439-0507.2007.01411.x
Mendoza M, Diaz AM, Hung MB, Zambrano EA, Diaz E, De Albornoz MC (2002) Production of culture filtrates of *Sporothrix schenckii* in diverse culture media. Med Mycol 40(5):447–454
Mendoza M, Diaz E, Alvarado P, Romero E, Bastardo de Albornoz MC (2007) Isolation of *Sporothrix schenckii* from environmental samples in Venezuela. Revista Iberoamericana de Micologia 24(4):317–319
Miranda LH, Conceicao-Silva F, Quintella LP, Kuraiem BP, Pereira SA, Schubach TM (2013) Feline sporotrichosis: histopathological profile of cutaneous lesions and their correlation with clinical presentation. Comparative Immunology, Microbiology and Infectious Diseases 36(4):425–432. doi:10.1016/j.cimid.2013.03.005
Moreira JA, Freitas DF, Lamas CC (2015) The impact of sporotrichosis in HIV-infected patients: A systematic review. Infection 43(3):267–276. doi:10.1007/s15010-015-0746-1
Morris-Jones R (2002) Sporotrichosis. Clinical and experimental dermatology 27(6):427–431
Morrison AS, Lockhart SR, Bromley JG, Kim JY, Burd EM (2013) An environmental *Sporothrix* as a cause of corneal ulcer. Med Mycol Case Rep 2:88–90. doi:10.1016/j.mmcr.2013.03.002
Nascimento RC, Espíndola NM, Castro RA, Teixeira PAC, Loureiro y Penha CV, Lopes-Bezerra LM, Almeida SR (2008) Passive immunization with monoclonal antibody against a 70-kDa putative adhesin of *Sporothrix schenckii* induces protection in murine sporotrichosis. Eur J Immunol 38(11):3080–3089. doi:10.1002/eji.200838513

O'Brien HE, Parrent JL, Jackson JA, Moncalvo J-M, Vilgalys R (2005) Fungal community analysis by large-scale sequencing of environmental samples. Applied and Environmental Microbiology 71(9):5544–5550. doi:10.1128/aem.71.9.5544-5550.2005

Orofino-Costa R, de Macedo P, Bernardes-Engemann A (2015) Hyperendemia of sporotrichosis in the Brazilian Southeast: Learning from clinics and therapeutics. Curr Fungal Infect Rep:9:220. doi:10.1007/s12281-015-0235-0

Pappas PG (2003) Sporotrichosis. Clinical Mycology. Oxford University Press, New York

Penha CV, Bezerra LM (2000) Concanavalin A-binding cell wall antigens of *Sporothrix schenckii*: a serological study. Med Mycol 38(1):1–7

Pereira SA, Passos SR, Silva JN, Gremião ID, Figueiredo FB, Teixeira JL, Monteiro PC, Schubach TM (2010) Response to azolic antifungal agents for treating feline sporotrichosis. The Veterinary Record 166(10):290–294. doi:10.1136/vr.b4752

Pereira SA, Gremião ID, Kitada AA, Boechat JS, Viana PG, Schubach TM (2014) The epidemiological scenario of feline sporotrichosis in Rio de Janeiro, State of Rio de Janeiro, Brazil. Revista da Sociedade Brasileira de Medicina Tropical 47(3):392–393. doi:10.1590/0037-8682-0092-2013

Pijper A, Pullinger BD (1927) An outbreak of sporotrichosis among South African native miners. The Lancet 210(5435):914–916. doi:10.1016/S0140-6736(01)35176-0

Pirofski LA, Casadevall A (2015) What is infectiveness and how is it involved in infection and immunity? BMC Immunology 16:13. doi:10.1186/s12865-015-0076-1

Reis EG, Gremião ID, Kitada AA, Rocha RF, Castro VS, Barros MB, Menezes RC, Pereira SA, Schubach TM (2012) Potassium iodide capsule treatment of feline sporotrichosis. J Feline Med Surg 14(6):399–404. doi:10.1177/1098612x12441317

Rippon JW (1988) Medical Mycology– The pathogenic fungi and the pathogenic actinomycetes 3rd edn. W. B, Saunders Company, Philadelphia, PA

Robert VA, Casadevall A (2009) Vertebrate endothermy restricts most fungi as potential pathogens. J Infect Dis 200(10):1623–1626. doi:10.1086/644642

Rodrigues AM, de Hoog S, de Camargo ZP (2013a) Emergence of pathogenicity in the *Sporothrix schenckii* complex. Med Mycol 51(4):405–412. doi:10.3109/13693786.2012.719648

Rodrigues AM, de Melo TM, de Hoog GS, Schubach TMP, Pereira SA, Fernandes GF, Bezerra LML, Felipe MS, de Camargo ZP (2013b) Phylogenetic analysis reveals a high prevalence of *Sporothrix brasiliensis* in feline sporotrichosis outbreaks. PLoS Negl Trop Dis 7(6):e2281. doi:10.1371/journal.pntd.0002281

Rodrigues AM, Bagagli E, de Camargo ZP, Bosco SMG (2014a) *Sporothrix schenckii sensu stricto* isolated from soil in an armadillo's burrow. Mycopathologia 177:199–206. doi:10.1007/s11046-014-9734-8

Rodrigues AM, de Hoog GS, Camargo ZP (2014b) Genotyping species of the *Sporothrix schenckii* complex by PCR-RFLP of calmodulin. Diagn Microbiol Infect Dis 78(4):383–387. doi:10.1016/j.diagmicrobio.2014.01.004

Rodrigues AM, de Hoog GS, de Cassia PD, Brihante RSN, da Costa Sidrim JJ, Gadelha MF, Colombo AL, de Camargo ZP (2014c) Genetic diversity and antifungal susceptibility profiles in causative agents of sporotrichosis. BMC Infect Dis 14(1):219. doi:10.1186/1471-2334-14-219

Rodrigues AM, de Hoog GS, Zhang Y, Camargo ZP (2014d) Emerging sporotrichosis is driven by clonal and recombinant *Sporothrix* species. Emerg Microbes Infect 3(5):e32. doi:10.1038/emi.2014.33

Rodrigues AM, Cruz Choappa R, Fernandes GF, De Hoog GS, Camargo ZP (2015a) *Sporothrix chilensis* sp. nov. (Ascomycota: Ophiostomatales), a soil-borne agent of human sporotrichosis with mild-pathogenic potential to mammals. Fungal Biol 120(2):246–264. doi:10.1016/j.funbio.2015.05.006

Rodrigues AM, Kubitschek-Barreira PH, Fernandes GF, de Almeida SR, Lopes-Bezerra LM, de Camargo ZP (2015b) Immunoproteomic analysis reveals a convergent humoral response signature in the *Sporothrix schenckii* complex. Journal of Proteomics 115:8–22. doi:10.1016/j.jprot.2014.11.013

Rodrigues AM, de Hoog GS, Camargo ZP (2015c) Molecular diagnosis of pathogenic *Sporothrix* species. PLoS Negl Trop Dis 9(12):e0004190. doi:10.1371/journal.pntd.0004190

Rodrigues AM, Kubitschek-Barreira PH, Fernandes GF, de Almeida SR, Lopes-Bezerra LM, de Camargo ZP (2015d) Two-dimensional gel electrophoresis data for proteomic profiling of *Sporothrix* yeast cells. Data Brief 2:32–38. doi:10.1016/j.dib.2014.11.004

Rodrigues AM, Fernandes GF, Araujo LM, Della Terra PP, Dos Santos PO, Pereira SA, Schubach TM, Burger E, Lopes-Bezerra LM, de Camargo ZP (2015e) Proteomics-based characterization of the humoral immune response in sporotrichosis: Toward discovery of potential diagnostic and vaccine antigens. PLoS Negl Trop Dis 9(8):e0004016. doi:10.1371/journal.pntd.0004016

Rodrigues AM, Najafzadeh MJ, De Hoog GS, Camargo ZP (2015f) Rapid identification of emerging human-pathogenic *Sporothrix* species with rolling circle amplification. Front Microbiol 6:1385. doi: 10.3389/fmicb.2015.01385

Roets F, de Beer ZW, Wingfield MJ, Crous PW, Dreyer LL (2008) *Ophiostoma gemellus* and *Sporothrix variecibatus* from mites infesting *Protea* infructescences in South Africa. Mycologia 100(3):496–510

Roets F, Wingfield BD, de Beer ZW, Wingfield MJ, Dreyer LL (2010) Two new *Ophiostoma* species from *Protea caffra* in Zambia. Persoonia 24:18–28. doi:10.3767/003158510x490392

Romani L (2011) Immunity to fungal infections. Nat Rev Immunol 11(4):275–288

Romon P, de Beer ZW, Zhou X, Duong TA, Wingfield BD, Wingfield MJ (2014) Multigene phylogenies of Ophiostomataceae associated with Monterey pine bark beetles in Spain reveal three new fungal species. Mycologia 106(1):119–132. doi:10.3852/13-073

Ruiz-Baca E, Hernandez-Mendoza G, Cuellar-Cruz M, Toriello C, Lopez-Romero E, Gutierrez-Sanchez G (2014) Detection of 2 immunoreactive antigens in the cell wall of *Sporothrix brasiliensis* and *Sporothrix globosa*. Diagn Microbiol Infect Dis 79(3):328–330. doi:10.1016/j.diagmicrobio.2014.02.020

Sasaki AA, Fernandes GF, Rodrigues AM, Lima FM, Marini MM, dos S. Feitosa L, de Melo Teixeira M, Felipe MSS, da Silveira JF, de Camargo ZP (2014) Chromosomal polymorphism in the *Sporothrix schenckii* complex. PLoS One 9(1):e86819. doi:10.1371/journal.pone.0086819

Schenck BR (1898) On refractory subcutaneous abscesses caused by a fungus possibly related to the *Sporotricha*. Bull Johns Hopkins Hosp 9:286–290

Schubach TM, Valle AC, Gutierrez-Galhardo MC, Monteiro PC, Reis RS, Zancope-Oliveira RM, Marzochi KB, Schubach A (2001) Isolation of *Sporothrix schenckii* from the nails of domestic cats (*Felis catus*). Med Mycol 39(1):147–149

Schubach TM, Schubach A, Okamoto T, Barros MB, Figueiredo FB, Cuzzi T, Fialho-Monteiro PC, Reis RS, Perez MA, Wanke B (2004) Evaluation of an epidemic of sporotrichosis in cats: 347 cases (1998-2001). J Am Vet Med Assoc 224(10):1623–1629. doi:10.2460/javma.2004.224.1623

Schubach A, Schubach TM, Barros MB, Wanke B (2005a) Cat-transmitted sporotrichosis, Rio de Janeiro, Brazil. Emerg Infect Dis 11(12):1952–1954. doi:10.3201/eid1112.040891

Schubach AO, Schubach TM, Barros MB (2005b) Epidemic cat-transmitted sporotrichosis. N Engl J Med 353(11):1185–1186. doi:10.1056/NEJMc051680

Schubach A, Barros MB, Wanke B (2008) Epidemic sporotrichosis. Curr Opin Infect Dis 21(2):129–133. doi:10.1097/QCO.0b013e3282f44c52

Scott EN, Muchmore HG (1989) Immunoblot analysis of antibody responses to *Sporothrix schenckii*. J Clin Microbiol 27(2):300–304

Sil A, Andrianopoulos A (2014) Thermally dimorphic human fungal pathogens – Polyphyletic pathogens with a convergent pathogenicity trait. Cold Spring Harb Perspect Med. doi:10.1101/cshperspect.a019794

Silva-Vergara ML, de Camargo ZP, Silva PF, Abdalla MR, Sgarbieri RN, Rodrigues AM, dos Santos KC, Barata CH, Ferreira-Paim K (2012) Disseminated *Sporothrix brasiliensis* infection with endocardial and ocular involvement in an HIV-infected patient. AmJTrop Med Hyg 86(3):477–480. doi:10.4269/ajtmh.2012.11-0441

Song Y, Li SS, Zhong SX, Liu YY, Yao L, Huo SS (2013) Report of 457 sporotrichosis cases from Jilin province, northeast China, a serious endemic region. J Eur Acad Dermatol Venereol 27(3):313–318. doi:10.1111/j.1468-3083.2011.04389.x

Steenbergen JN, Nosanchuk JD, Malliaris SD, Casadevall A (2004) Interaction of *Blastomyces dermatitidis*, *Sporothrix schenckii*, and *Histoplasma capsulatum* with *Acanthamoeba castellanii*. Infect Immun 72(6):3478–3488. doi:10.1128/iai.72.6.3478-3488.2004

Sullivan D, Moran G, Coleman D (2005) Fungal diseases of humans. In: Kavanagh K (ed) Fungi: biology and applications. John Wiley & Sons, Chichester, pp 171–190

Taylor LH, Latham SM, Woolhouse MEJ (2001) Risk factors for human disease emergence. Philos Trans R Soc Lond Ser B Biol Sci 356(1411):983–989. doi:10.1098/rstb.2001.0888

Teixeira PAC, de Castro RA, Nascimento RC, Tronchin G, Pérez Torres A, Lazéra M, de Almeida SR, Bouchara JP, Loureiro y Penha CV, Lopes-Bezerra LM (2009) Cell surface expression of adhesins for fibronectin correlates with virulence in *Sporothrix schenckii*. Microbiology 155(11):3730–3738. doi:10.1099/mic.0.029439-0

Travassos LR, Lloyd KO (1980) *Sporothrix schenckii* and related species of *Ceratocystis*. Microbiol Rev 44(4):683–721

van Burik JA, Magee PT (2001) Aspects of fungal pathogenesis in humans. Annu Rev Microbiol 55:743–772. doi:10.1146/annurev.micro.55.1.743

Vega FE, Goettel MS, Blackwell M, Chandler D, Jackson MA, Keller S, Koike M, Maniania NK, Monzón A, Ownley BH, Pell JK, Rangel DEN, Roy HE (2009) Fungal entomopathogens: new insights on their ecology. Fungal Ecol 2(4):149–159. doi:10.1016/j.funeco.2009.05.001

Warnecke L, Turner JM, Bollinger TK, Lorch JM, Misra V, Cryan PM, Wibbelt G, Blehert DS, Willis CK (2012) Inoculation of bats with European *Geomyces destructans* supports the novel pathogen hypothesis for the origin of white-nose syndrome. Proc Natl Acad Sci U S A 109(18):6999–7003. doi:10.1073/pnas.1200374109

White TJ, Bruns T, Lee S, Taylor J (1990) Amplification and direct sequencing of fungal ribosomal RNA genes for phylogenetics. In: Innis M, Gelfand D, Shinsky J, White T (eds) PCR protocols: a guide to methods and applications. Academic Press, San Diego, pp 315–322

Zhang Y, Hagen F, Stielow B, Rodrigues AM, Samerpitak K, Zhou X, Feng P, Yang L, Chen M, Deng S, Li S, Liao W, Li R, Li F, Meis JF, Guarro J, Teixeira M, Al-Zahrani HS, Camargo ZP, Zhang L, de Hoog GS (2015) Phylogeography and evolutionary patterns in *Sporothrix* spanning more than 14000 human and animal case reports. Persoonia 35:1–20. doi:10.3767/003158515x687416

Zhou X, de Beer ZW, Wingfield MJ (2006) DNA sequence comparisons of *Ophiostoma* spp., including *Ophiostoma aurorae* sp. nov., associated with pine bark beetles in South Africa. Stud Mycol 55:269–277

Zhou X, Rodrigues AM, Feng P, Hoog GS (2014) Global ITS diversity in the *Sporothrix schenckii* complex. Fungal Divers 66(1):153–165. doi:10.1007/s13225-013-0220-2

Zipfel RD, de Beer ZW, Jacobs K, Wingfield BD, Wingfield MJ (2006) Multi-gene phylogenies define *Ceratocystiopsis* and *Grosmannia* distinct from *Ophiostoma*. Stud Mycol 55:75–97

Part V

Miscellaneous

20 Emerging Infectious Diseases in Camelids

Abdelmalik I. Khalafalla

20.1 Introduction

It is estimated that 65 % of fatalities in Old World camels (OWC, i.e., *Camelus dromedarius* and *C. bactrianus*) and 50 % in New World camelids/South American camelids (NWC/SAC, i.e., the domestic alpaca (*Vicugna pacos*) and llama (*Lama glama*)) are caused by infectious diseases (Wernery and Kaaden 2002). In the past, camels were used mainly for transportation beside their role as the main source of milk and meat for pastoralists. Dromedary camel in sub-Saharan Africa was traditionally known to be reared in the arid and semiarid lands. Due to aridity and desertification, they obliged to move to the higher rainfall areas side by side with other domestic livestock and wildlife. This change resulted in exposure of camels to diseases that were uncommon in their natural habitat such as dermatophilosis, tick paralysis, trypanosomosis, and brucellosis. The situation in Niger, Chad, and Sudan is an example where diseases like contagious ecthyma, trypanosomosis, and tick paralysis have become very serious with increased mortality rates due to the migration of camels south of their well-known camel belt. Drought in the Sahel and the Horn of Africa has also brought pastoralists closer to urban centers, and sales of camel milk became their main source of cash income. Due to an increased demand of urban populations of many countries, particularly in North Africa and the Middle East for camel milk, many dairy farms are established in intensive and semi-intensive systems. This development may be responsible for making camels more susceptible to certain disease. Brucellosis, enterotoxemia, and Johne's disease are examples of these diseases. A similar situation could be envisaged for an increased incidence of enterotoxemia in camels when raised in an intensive husbandry system as in the UAE (Wernery and Kaaden 2002) or Syria (Khalafalla AI 2015, personnel communication). Changes in animal husbandry related to increasing camel contacts

A.I. Khalafalla
Camel Research Center, King Faisal University, Al-Ahsa 31982, Saudi Arabia
e-mail: abdokhlf@yahoo.co.uk

J. Bayry (ed.), *Emerging and Re-emerging Infectious Diseases of Livestock*,
DOI 10.1007/978-3-319-47426-7_20

with other animal species, such as equids, may cause disease emergence. Examples of newly emerged diseases of camels resulting from sharing premises with equines are glanders (Wernery et al. 2011), melioidosis (Wernery et al. 1997), and *Rhodococcus equi* infection (Kinne et al. 2011). Another factor that may contribute to the emergence of camel diseases is the migration into new habitat that never was reached before by camels (Faye and Vias 2010).

20.2 Emerging Viral Diseases of Camels

20.2.1 Camelpox

Camelpox is the only camel disease included in the OIE's list of reportable diseases. A chapter on camelpox has been recently added to the OIE's manual of terrestrial animal diseases, following its endorsement by OIE's assembly during the general session of May 2014. A special research interest in camelpox has resulted in numerous publications on different aspects of the disease and the causative virus. This is mainly attributable to the resemblance of the CMLV to small poxvirus (Baxby 1974). Interestingly, the CMLV is recently becoming the subject for studies on antiviral therapies (Duraffour et al. 2014), cellular ion channel analysis, and apoptosis.

Camelpox is a highly contagious skin disease and the most frequent infectious viral disease of the camelids that occurs in almost every country in which camel husbandry is practiced (Fig.20.1). Outbreaks have been reported in Asia (Bahrain, Iran, Iraq, Oman, Saudi Arabia, the UAE, Yemen, Syria, Afghanistan, southern parts of Russia and India, and Pakistan) and in Africa (Algeria, Egypt, Ethiopia, Kenya, Mauritania, Morocco, Niger, Somalia, and Sudan). The disease is endemic in these countries, and a pattern of sporadic outbreaks occurs with a rise in the seasonal incidence usually during the rainy season (OIE 2008). The disease was recently reported from Saudi Arabia (Yousif 2011), India (Bhanuprakash et al. 2010; Bera et al. 2011), Ethiopia (Ayelet et al. 2013), and Iran (Mosadeghhesari et al. 2014).

20.2.1.1 The Etiology

Camelpox is caused by the camelpox virus (CMLV), which belongs to the genus *Orthopoxvirus* (OPXV) of the subfamily Poxvirinae in the family Poxviridae. Phylogenetic analysis of CMLV revealed that CMLV is most closely related to variola virus (VARV), sharing all genes involved in basic replicative functions and the majority of genes involved in other host-related functions (Afonso et al. 2002; Gubser and Smith 2002).

20.2.1.2 Clinical Picture

The disease is species specific and characterized by localized or generalized pox lesions that vary in severity in correlation with age of affected animals (Khalafalla and Mohamed 1996). Pox lesion of various stages may develop, particularly on the face, the neck, and under the tail (Fig. 20.2). Other symptoms include fever and lymph node enlargement. Abortion rates may reach 87 %, as observed by Al Zi'abi et al. (2007) in Syria.

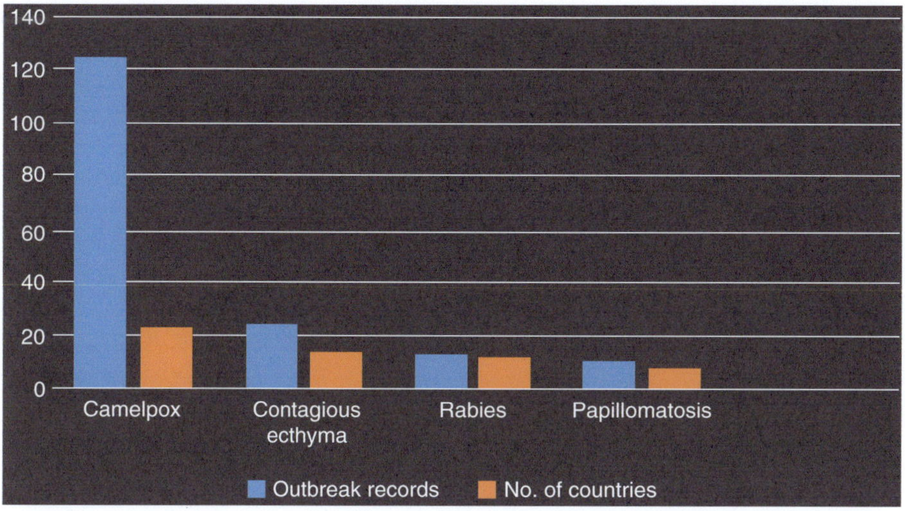

Fig. 20.1 Distribution of reported outbreaks of the important camelids diseases

Fig. 20.2 Early skin lesion of camelpox in dromedary camels

20.2.1.3 Diagnosis

The presumptive diagnosis of camelpox infection can be made based on clinical signs. However, infections of camels in the early clinical stages and in mild cases should be differentiated from contagious ecthyma, which is caused by a *Parapoxvirus* (PPV), papilloma virus infections, and insect bites (Khalafalla et al. 1998; Bhanuprakash et al. 2010). Various laboratory techniques are available for the diagnosis of camelpox including virus isolation, electron microscopy, serology, and polymerase chain reaction (PCR).

20.2.1.4 Zoonotic Potentiality

It has been over 106 years since camelpox was first described in Punjab, India, in 1909 (Wernery and Kaaden 2002), but the zoonotic nature of the CMLV remained a debate. According to Baxby (1972), CMLV is different from the VARV, the causative agent of smallpox, and is incapable of infecting man. Kritz in Somalia (Kritz 1982) described the first report of a case of human camelpox. The case was a 40-year-old camel herder who developed lesions resembling those of smallpox in June 1978. From the 1970s until recently, it has been well accepted that CMLV rarely infects humans (Duraffour et al. 2011). This is probably due to the cross-immunity induced via smallpox vaccination that ended in the late 1970s. It was therefore postulated that human camelpox may become more common as the immunity of the human population wanes (Duraffour et al. 2011). Human population more than three decades after cessation of the smallpox vaccination has lost protection against that deadly virus and all zoonotic infections caused by human and animal OPVs as well. The first conclusive evidence of zoonotic CMLV infection in humans, associated with outbreaks in dromedary camels, has been recently reported in India where three human cases of camelpox have been reported (Bera et al. 2011). They were detected in animal handlers during an outbreak of camelpox, and the lesions were confined to the hands and fingers of camel handlers and passed through all the stages of pox lesions until the formation of scabs. These are the first confirmed cases of zoonotic camelpox as infection was diagnosed by conventional PCR and seroconversion. Additional four cases of camelpox in humans (camel herders) were detected recently in 2015 in Showak area of Eastern Sudan (Khalafalla et al. 2016, Vector borne & Zoonotic Diseases, accepted) (Fig. 20.3).

20.2.2 Rift Valley Fever (RVF)

Rift Valley fever is an acute viral, mosquito-borne disease that affects domestic animals (such as sheep, cattle, and goats) and humans distributed in sub-Saharan African countries and the Arabian Peninsula.

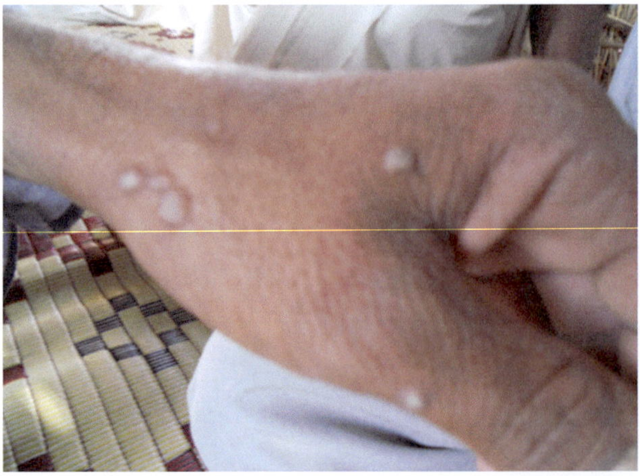

Fig. 20.3 Lesion of camelpox on hand of a camel herder, Sudan

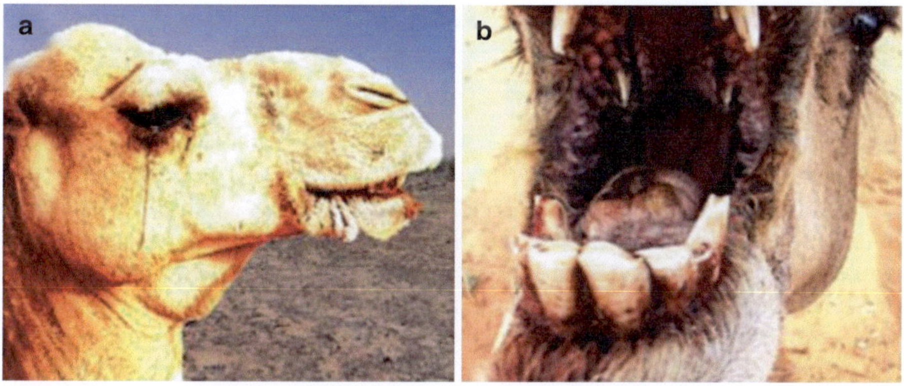

Fig. 20.4 Clinical symptoms of Rift Valley fever in dromedary camels during in Mauritania. (**a**) Conjunctivitis and ocular discharge, hemorrhages of the gums, and edema of the trough, (**b**) hemorrhages of gums and tongue (Courtesy of Dr. Ahmed El Mamy)

RVF is caused by an RNA virus of the family *Bunyaviridae*, genus *Phlebovirus*. Camels have been regularly involved in the RVF epidemics in East Africa and Egypt. However, clinical disease is not seen in adult camels, but abortion occurs and some early deaths have been observed (OIE 2008). Serological evidence of dromedary camel infection with RVF was documented. According to Davies et al. (1985), camel sera collected after an epizootic of Rift Valley fever in Kenya revealed positive sera with high titers of serum neutralizing antibody found in 22 % of camels at one of the seven sampling sites. Furthermore, the demonstration of specific IgG antibodies in camels (Nabeth et al. 2001) indicates that these animals are naturally infected. The disease emerged in Egypt in 1977, in Mauritania in 1998 (Nabeth et al. 2001), and in the Arabian Peninsula in 2000 (Abdo-Salem et al. 2006). In addition, the disease reappeared in Kenya in 2006/2007 (Bird et al. 2008). It also involved camels beside sheep, goats, and humans, again with abortion as the only clinical symptom.

Recently, El Mamy et al. (2014) have reported confirmatory evidence for a field camel infection with RVF. In September of 2010, an RVF outbreak occurred in northern Mauritania involving mass abortions in small ruminants and camels (*Camelus dromedarius*) and at least 63 human clinical cases, including 13 deaths. In camels, serological prevalence was 27.5–38.5 %, and for the first time, clinical signs other than abortions were reported in this species, including hemorrhagic septicemia and severe respiratory distress (Fig. 20.4). Phylogenetic analyses of the genome of isolates from camels suggested a shared ancestor between the Mauritania 2010 strain and strains from Zimbabwe, Kenya, South Africa, Uganda, and other strains linked to the 1987 outbreak of RVF in Mauritania.

20.2.3 Peste Des Petits Ruminants (PPR)

Peste des petits ruminants (PPR) is a highly contagious disease of sheep and goats which has recently reemerged and is now found widely distributed through large parts of Africa, the Middle East, and Asia. The disease is characterized by severe pyrexia, anorexia, ulcerative necrotic stomatitis, diarrhea due to purulent oculo-nasal discharge, and respiratory distress.

The etiology The disease is caused by the peste des petits ruminant virus (PPRV), which belongs to the *Morbillivirus* genus of the paramyxovirus family of viruses. PPRV is a non-segmented negative-strand RNA virus closely related to the rinderpest virus of cattle and buffaloes, the measles virus of humans, the distemper virus of dogs and some wild carnivores, and the morbilliviruses of aquatic mammals. To date, genetic characterization of PPR virus strains has allowed them to be categorized into four groups: three from Africa and one from Asia. This virus has a particular affinity for lymphoid tissues and epithelial tissue of the gastrointestinal (GI) and respiratory tracts, where it produces characteristic lesions.

PPR in camelids Camels were not reported as possible hosts to PPR until Ismail and coauthors (Ismail et al. 1992) in Egypt detected the infection through serology in Sudanese camels. The first documented outbreak of PPR in camels reported from Ethiopia in 1996, consisted of a highly contagious respiratory syndrome with elevated morbidity and low mortality rates (Roger et al. 2000, 2001). Consecutive outbreaks of a similar disease occurred in Eastern Sudan in 2004 and later in Somalia and Kenya. Though investigations in Somalia and Kenya were inconclusive and failed to identify the real causative agent, the disease in Sudan was diagnosed as a PPR virus (small ruminant virus) infection (Khalafalla et al. 2010). The incriminated virus belongs to the genus *Morbillivirus* in the family *Paramyxoviridae*—a group of RNA viruses characterized by continuous evolution and species jumping. Newly evolved viruses in this family include phocine distemper virus and Hendra virus, formerly called equine *Morbillivirus* (Earle et al. 2011).

The clinical and epidemiological picture of the previous PPR-suspected or PPR-confirmed outbreaks in northeast Africa is not similar. At least two forms can be identified: a per acute disease characterized by sudden death, abortion, and diarrhea with a high mortality rate occurred in the region during 2004–2007 (Gluecks and Younan 2010; Dawo 2010; Khalafalla et al. 2010), whereas the early outbreaks (1992–1996) showed an acute respiratory disease with low mortality rates (Roger et al. 2000). The presence of more than one form points to involvement of other pathogens as secondary invaders or to the genetic variations of the causative virus. The last hypothesis was verified by detecting lineage III of PPRV from camels in Ethiopia in 1997 and lineage IV from camels affected by PPRV in 2004 in Sudan (Kwiatek et al. 2011). Kwiatek and coworkers (2011) suggested that a virulent lineage IV strain might have been introduced in Africa during the 1990s, resulting in outbreaks in both camels and small ruminants.

Clinically the disease is characterized by sudden death of apparently healthy animals and yellowish and later bloody diarrhea and abortion (Fig. 20.5). Death has been always sudden and proceeded with colic and difficulty in respiration. Mortality rate ranges between 0 and 50 % and vary in accordance with the area with a mean of 7.4 %. More than 80 % of deaths were in pregnant and recently delivered she-camels.

Fig. 20.5 Carcasses of camels died of PPRV infection scattered in northern Butana, Sudan in 2004

20.2.4 Middle East Respiratory Syndrome Coronavirus Infection in Dromedary Camels

Middle East respiratory syndrome coronavirus (MERS-CoV) is an emerging pathogen associated with severe respiratory symptoms and renal failure in infected persons. Saudi Arabia is the country most severely affected by the virus and is where the first recognized case was reported in September 2012. The origin of MERS-CoV remains a mystery. Phylogenetic analyses of complete and partial genome sequences by Cotten et al. (2013) enabled estimates of the timing of introduction and geographic distribution of distinct virus clades. The authors proposed that MERS-CoV emerged in humans in 2011 and noted that sequence divergence among clades is consistent with multiple sporadic introductions of the virus into the human population, presumably from an animal reservoir (Alagaili et al. 2014).

Early observations that some MERS-CoV infected people had exposure to camels suggesting a possible role of this animal as an intermediate reservoir host (Milne-Price et al. 2014; Albarrak et al. 2012; Drosten et al. 2013). Serologic surveys subsequently conducted in several countries in the Arabian Peninsula and Africa have identified high rates of MERS-CoV-specific antibodies in camels (Hemida et al. 2013; Perera et al. 2013; Reusken et al. 2013; Meyer et al. 2014). Furthermore, researchers have found definitive proof that camels can be infected with MERS-CoV when viral genomic sequences were detected in nasal swabs collected in Qatar (Haagmans et al. 2014; Nowotny and Kolodziejek 2014), Saudi Arabia (Memish et al. 2014), and Egypt (Chu et al. 2014).

Infection in camels Susceptibility of dromedary camels to and infection with MERS-CoV have been documented by the detection of high levels of antibodies to the virus and viral sequences in nasal swabs and lung tissues. Nevertheless, infection remained subclinical with only mild upper respiratory tract symptoms (Fig. 20.6) observed in some camels (Khalafalla et al. 2015). Experimental infection of dromedary camels with a human isolate of MERS-CoV induced no systemic illness despite shedding of large quantities of virus from the upper respiratory tract (Adney et al. 2014). It is therefore concluded that MERS-CoV causes no well-defined disease in camels.

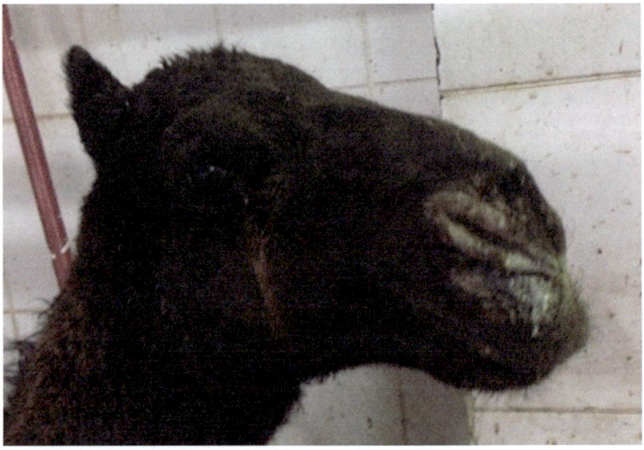

Fig. 20.6 Symptoms of MERS-CoV infection in a 10-month-old dromedary camel

20.3 Emerging Bacterial Infections

20.3.1 Brucellosis

Brucellosis is one of the most important worldwide zoonosis affecting livestock and humans. The disease is regarded as one of the most widespread diseases in the world by the Food and Agriculture Organization of the United Nations (FAO), the World Health Organization (WHO), and the World Animal Health Organization (OIE).

20.3.1.1 The Etiology

Camels of both species (*Camelus dromedarius* and *Camelus bactrianus*) are highly susceptible to brucellosis caused by *B. melitensis* and *B. abortus*. However, too few bacteriological surveys have been conducted to draw conclusions on the relative importance of either species of *Brucella* in the etiology of camel brucellosis in the respective countries (Abbas and Agab 2002). A recent report regarded brucellosis as a regionally emerging zoonotic disease in North Africa, the Middle East, and India (Gwida et al. 2010). Camels are infected by *Brucella abortus* and *B. melitensis*, which cause abortion and infertility (Wernery and Kaaden 2002). *B. melitensis* was isolated from camels in Iran, Libya, and Saudi Arabia; *B. abortus* was isolated in Sudan, Egypt, and Kuwait.

20.3.1.2 The Disease in Camelids

Serological surveys of camels conducted in many countries indicated that seroprevalence of *B abortus* ranged from 2 to15 %. However, recent reports showed a substantial increase in seroprevalence over the past few years (Ahmed et al. 2010). The increase reached 37.5 % in some areas in Sudan (Omer et al. 2010). Clinical disease was also reported (Musa et al. 2008; Al-Majali et al. 2008). Clinically, brucellosis in camelids induce symptoms similar to those in other livestock species. However, various researchers have noticed that abortion related to brucellosis is less in camels in comparison to other animals. Brucella infections in camels may cause stillborn calves, retained placenta, fetal death, mummification, reduced milk yield, delayed service age, and fertility (Wernery 2015).

20.3.2 Paratuberculosis (Johne's Disease)

Johne's disease (JD), also known as paratuberculosis, is widespread, chronic, and debilitating disease that affects mainly ruminants and causes severe economic loss. Once animals are infected, the disease gradually advances toward its chronic form, which is characterized by granulomatous enteritis, progressive weight loss with diarrhea, and finally death.

20.3.2.1 The Causative Agent

Mycobacterium avium subspecies *paratuberculosis* (MAP) causes Johne's disease in domestic and wild ruminant and in camelids. A recent study by Ghosh and coauthors based on gene typing of MAP isolates from Saudi Arabia indicated that all isolates belong to the sheep lineage of strains, suggesting a putative transmission from infected sheep herds (Ghosh et al. 2012).

20.3.2.2 Disease in Camelids

JD affects camels worldwide, causing characteristic clinical illness of severe diarrhea ending in death (Manefield and Tinson 1997; Wernery and Kaaden 2002) (Fig. 20.7). The course of disease is often more rapid than that in cattle (Higgins 1986). Reports from Saudi Arabia described many deaths due to this disease. According to Gameel et al. (1994), nine camels died after displaying

Fig. 20.7 A 2-year-old camel affected with Johne's disease

characteristic symptoms. The diagnosis was laboratory confirmed, and according to owners, 53 camels died in the field after showing typical symptoms. The disease continues to be reported as a serious and invariably fatal disease of the Arabian camel (Alharbi et al. 2012). According to several reports, Johne's disease is considered an important emerging disease in dromedary camels in the Saudi Arabia and Gulf states.

20.3.3 Glanders and Melioidosis

Glanders in equids caused by *Burkholderia mallei* recently reappeared in Pakistan and Brazil in 2008 and 2009, respectively, and appeared for the first time in Kuwait and Bahrain in 2010 (Wernery 2009; Roberts et al. 2010). Recently, an outbreak of glanders that killed three dromedaries out of six was diagnosed in Bahrain (Wernery et al. 2011). Melioidosis is a potentially fatal disease caused by the gram-negative bacterium *Burkholderia pseudomallei*. During 1990, seven out of 13 camels died from the disease in Queensland, Australia (Bergin and Torenbeck 1991). Since then, at least four incidents of melioidosis-related camel deaths have been diagnosed in the northern areas of the Northern Territory in Australia, and a single case was reported from the UAE (Wernery et al. 1997).

20.3.4 Dermatophilosis

Natural *Dermatophilus congolensis* infection of camels has been reported, for the first time, in Kenya, Sudan, the UAE, and Saudi Arabia in the mid-1990s (Bornstein 1995, Gitao et al. 1998a; Wernery and Ali 1990) (Fig. 20.8). According to camel owners in the Butana region of Sudan, this disease has never been observed before in their herds (Gitao et al. 1998b).

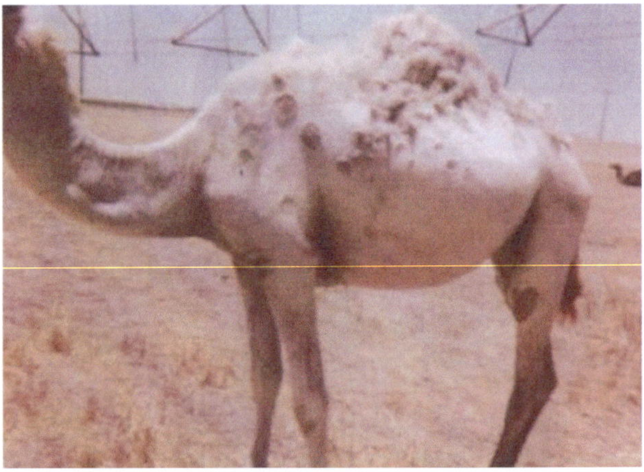

Fig. 20.8 Dermatophilosis lesion in a 3-year-old dromedary camel

20.4 Emerging Parasite Infections

20.4.1 Protozoal Infections

The most important and prevalent disease of camels, trypanosomiasis (Surra) caused by *Trypanosoma evansi*, has been reported from the Canary Islands, Spain, and recently from France (Molina et al. 1999; Gutierrez et al. 2010). *Eimeria* spp. have been introduced by importations of camels to Europe, e.g., *E. cameli* (Bornstein, personnel communication).

20.4.2 Ectoparasites

Camels harbor several tick species, and some are important vectors of pathogens; protozoal, virus, and bacteria spp. cause tick paralysis and toxicosis (see above). However, their role of transmitting these pathogens to camelids seems to be much less important than it is to other animals and humans (Bornstein 2002). Lately Alkhurma hemorrhagic fever virus (AHFV) was isolated in an *Ornithodoros savignyi*, the sand sampan, from Saudi Arabia (Charrel et al. 2007). This soft tick has a wide distribution in arid lands and may lay dormant for long periods during harsh conditions. The tick infests many different hosts including camels as well as humans. The AHFV causes an extremely severe hemorrhagic fever in humans with a case fatality rate of >30 %. About 20 human cases have been diagnosed with the infection in Saudi Arabia (Charrel et al. 2007). The authors associate these or some of these cases with tick bite history.

The above is one of many examples of emerging diseases transmitted by ticks and other arthropods (phylum Arthropoda including the class Insecta). The AHFV belong to tick-borne flaviviruses, which are among the medically most important arboviruses in Asia and Europe. Some are very important as human and animal pathogens; others have not yet been associated with human or animal diseases. Changes in human behavior, animal husbandry, land use, and/or climate may change the actual geographical distribution pattern and transmission intensity. This is already taking place, and it is often related to the present climate change. Tick-borne flaviviruses and other arboviruses may increase in medical and veterinary importance as, e.g., in the case with bluetongue virus (see below).

20.5 Emerging Viral Diseases of New World Camelids (NWC/ SAC)

Increasing numbers of South American camelids (SACs), particularly alpaca (*Vicugna pacos*) and llama (*Lama glama*), are being imported to various countries outside of South America, including Europe, for wool (fiber) production, breeding, and as pack and companion animals. These newly introduced species have proved to be susceptible to "old" and new pathogens in their new environments.

Three viral diseases can be regarded as emerging infections in NWCs.

20.5.1 Infections with Bovine Viral Diarrhea Virus (BVDV)

The BVDV is not limited to cattle, but may be detected in various species. There is documentation of infection of alpacas resulting in reproductive loss and illness. Kim et al. (2009) isolated BVDV from persistently infected alpacas and showed that unique genotypes of the virus may be maintained in the alpaca population even though camelids are susceptible to infection by other genotypes. There is clear evidence that the disease has been present in North American alpaca herds since at least 2001 and likely originated from BVDV-infected cows (van Amstel and Kennedy 2010).

20.5.2 Bluetongue (BT) Disease

The bluetongue virus (BTV) is an *Orbivirus* with 24 known sero-variants and is present worldwide. It exists in a broad band around the world covering large parts of the Americas, Africa, southern Asia and northern Australia (Mellor et al. 2008). Occasionally it has occurred on the southern borders of Europe. The BTV is transmitted by midges (*Culicoides* spp.) and naturally infects domestic and wild ruminants. The disease is considered to be one of the most important infections of domestic livestock. It is quite severe in sheep, less so in cattle. In June 2006, BTV appeared in northern Europe for the first time (Carpenter et al. 2009) and successfully wintered. Subsequently it caused substantial losses in livestock in 2007 and 2008. The virus has never previously occurred so far north anywhere in the world (Mellor et al. 2008). The explanation to this dramatic change in the epidemiology of BT is many, but one is linked to the recent extension in the distribution of its main vector(s) *Culicoides imicola* and perhaps new *Culicoides* vector(s) and to the ongoing climatic change (Wilson and Mellor 2009).

Alpacas and llamas are susceptible to BTV infections, but rarely show significant clinical signs. However, they remain infectious for several weeks and may thus serve as reservoirs of infection. Already Rivera et al. (1987) showed that camelids can be infected with BTV. About 100 sampled "healthy" alpacas were positive for specific antibodies against BVT. Although camelids are considered a "low-risk species," there are reports of cases, e.g., during the recent outbreak of BT in Europe 2007 (Henrich et al. 2007). One alpaca in the area that was affected in Germany during this outbreak was infected within a radius of 5 km from clinically BTV-infected sheep (with high mortality) and cattle. Ortega et al. (2010) described a fatal BTV infection in an alpaca (*Vicugna pacos*) in California. Meyer et al. (2009) described a lethal BTV infection in llamas. Vaccines are available and are used for ruminants. They are not licensed for SACs but may be used (Zanolari et al. 2010) and are recommended by many.

20.5.3 Coronavirus Infection

Recently, a novel coronavirus possibly associated with acute respiratory syndrome in alpacas in California, 2007, was reported (Crossley et al. 2010). Despite epidemiological evidence of BVDV, BT, and coronavirus infections in NWCs, current knowledge regarding the impact of these diseases is incomplete.

20.6 Emerging Parasitic Infections of NWCs

The small liver fluke or the lancet fluke (*Dicrocoelium dendriticum*) is rarely found in camelids. However, natural infections with this intriguing parasite have been reported from a few countries in alpacas in Europe, Switzerland, Germany (Wenker et al. 1998), and recently Sweden (de-Verdier et al. 2011). Infection rates are most probably rel. high leading to rel. few but significant mortalities. Llamas and alpacas may act as aberrant hosts to some parasites. In the USA, e.g., llamas cohabiting with the common white-tailed deer (*Odocoileus virginianus*) may be infected with a meningeal worm (*Parelaphostrongylus tenuis*), a nematode-causing neurological disease in the aberrant hosts (Fowler 1998).

References

Abbas B, Agab H (2002) A review of camel brucellosis. Prev Vet Med 55:47–56

Abdo-Salem S, Gerbier G, Bonnet P, Al-Qadasi M, Tran A, Thiry E, Al-Eryni G, Roger F (2006) Descriptive and spatial epidemiology of Rift valley fever outbreak in Yemen 2000-2001. Ann N Y Acad Sci 1081:240–242

Adney DR, van Doremalen N, Brown VR, Bushmaker T, Scott D, de Wit E, Bowen RA, Munster VJ (2014) Replication and shedding of MERS-CoV in upper respiratory tract of inoculated dromedary camels. Emerg Infect Dis 20(12):1999–2005

Afonso CL, Tulman ER, Lu Z, Zsak L, Sandybaev NT, Kerembekova UZ, Zaitsev VL, Kutish GF, Rock DL (2002) The genome of camelpox virus. Virology 295(1):1–9

Ahmed MO, Elmeshri SE, Abuzweda AR, Blauo M, Abouzeed YM, Ibrahim A, Salem H, Alzwam F, Abid S, Elfahem A, Elrais A (2010) Seroprevalence of brucellosis in animals and human populations in the western mountains region in Libya, December 2006-January (2008). Euro Surveill 15(30):1–3

Alagaili AN, Briese T, Mishra N, Kapoor V, Sameroff SC, de Wit E (2014) Middle East respiratory syndrome coronavirus infection in dromedary camels in Saudi Arabia. MBiol 5:e00884–e00814

Albarrak AM, Stephens GM, Hewson R, Memish ZA (2012) Recovery from severe novel coronavirus infection. Saudi Med J 33:1265–1269

Alharbi KB, Al-Swailem A, Al-Dubaib MA, Al-Yamani E, Al-Naeem A, Shehata M, Hashad ME, Albusadah KA, Mahmoud OM (2012) Pathology and molecular diagnosis of paratuberculosis of camels. Trop Anim Health Prod 44(1):173–177

Al-Majali AM, Al-Qudah KM, Al-Tarazi YH, Al-Rawashdeh OF (2008) Risk factors associated with camel brucellosis in Jordan. Trop Anim Health Prod 40(3):193–200

Al-Ziabi O, Nishikawa H, Meyer H (2007) The first outbreak of camelpox in Syria. J Vet Med Sci 69(5):541–543

Ayelet G, Jenberie S, Belay A, Mohammed A, Mola B, Gizaw Y et al (2013) The first isolation and molecular characterization of camelpox virus in Ethiopia. Antiviral Res 98(3):417–422

Baxby D (1972) Smallpox-like viruses from camels in Iran. Lancet 2:1063–1065

Baxby D (1974) Differentiation of smallpox and camelpox viruses in cultures of human and monkey cells. J Hyg (Lond) 72:251–254

Bera BC, Shanmugasundaram K, Barua S, Venkatesan G, Virmani N, Riyesh T, Gulati BR, Bhanuprakash V, Vaid RK, Kakker NK, Malik P, Bansal M, Gadvi S, Singh RV, Yadav V,

Nagarajan G, Balamurugan V, Hosamani M, Pathak KM, Singh RK (2011) Zoonotic cases of camelpox infection in India. Vet Microbiol 152(1–2):29–38

Bergin TJ, Torenbeck LR (1991) Melioidosis in camels. Austr Vet J 68:30–36

Bhanuprakash V, Balamurugan V, Hosamani M, Venkatesan G, Chauhan B, Srinivasan VA, Chauhan RS, Pathak KM, Singh RK (2010) Isolation and characterization of Indian isolates of camel pox virus. Trop Anim Health Prod 42(6):1271–1275

Biological Standard Commission. 2010. www.oie.int/doc/ged/D7690.PDF. Accessed 2/7/2011

Bird BH, Githinji JW, Macharia JM, Kasiiti JL, Muriithi RM, Gacheru SG, Musaa JO, Towner JS, Reeder SA, Oliver JB, Stevens TL, Erickson BR, Morgan LT, Khristova ML, Hartman AL, Comer JA, Rollin PE, Ksiazek TG, Nichol ST (2008) Multiple Virus Lineages sharing recent common ancestry were associated with a Large Rift Valley Fever outbreak among livestock in Kenya during 2006-2007. J Virol 82(22):11152–11166

Bornstein S (1995) Skin diseases of camels, in; Camel keeping in Kenya. Eds; Evans JO., Simpkin SP., Atkins DJ. Range Manage Handbook Kenya 3(8):7–13

Bornstein S (2002) Parasitic diseases. In: Wernery U, Kaaden OR (eds) Infectious diseases in Camelids. Blackwell Science, Berlin/Vienna, pp 267–387

Carpenter S, Wilson A, Mellor PS (2009) Culicoide and the emergence of bluetongue virus in northern Europe. Trends Microbiol 17(4):172–178

Charrel RN, Fagbo S, Moureau G, Algahtani MH, Temmam S, de Lamballerie X (2007) Alkhurma hemorrhagic fever virus in *Ornithodoros savignyi* ticks. Emerg Infect Dis 13(1):153–155

Chu DKW, Poon LLM, Gomaa MM, Shehata MM, Perera RAPM, Zeid DA, El Rifay AS, Siu LY, Guan Y, Webby RJ, Ali MA, Peiris M, Kayali G (2014) MERS coronaviruses in dromedary camels. Egypt Emerg Infect Dis 20(6):1049–1053

Cotten M, Watson SJ, Kellam P, Al-Rabeeah AA, Makhdoom HQ, Assiri A, Al-Tawfiq JA, Alhakeem RF, Madani H, AlRabiah FA, Al Hajjar S, Al-nassir WN, Albarrak A, Flemban H, Balkhy HH, Alsubaie S, Palser AL, Gall A, Bashford-Rogers R, Rambaut A, Zumla AI, Memish ZA (2013) Transmission and evolution of the Middle East respiratory syndrome coronavirus in Saudi Arabia: a descriptivegenomic study. Lancet 382(9909):1993–2002

Crossley BM, Barr BC, Magdesian KG, Ing M, Mora D, Jensen D, Loretti AP, McConnell T, Mock R (2010) Identification of a novel coronavirus possibly associated with acute respiratory syndrome in alpacas (*Vicugna pacos*) in California. 2007. J Vet Diagn Invest 22(1):94–97

Davies FG, Koros J, Mbugua H (1985) Rift Valley fever in Kenya: the presence of antibodies to the virus in camels (*Camelus dromedarius*). J Hyg (Lond) 94:241–244

Dawo F (2010) Mysterious mortality in camels (*Camelus dromedarius*) in Borana, Ethiopia: evidence of its association with reproductive age groups. Rev Sci Tech Off Int Epiz 29(3):621–628

de Verdier K, Sandros B, Christensson D, Osterman-Lind E, Lindqvist FK, Bornstein S (2011) Lilla leverflundran – första fallet hos alpacka i Sverige (The first case of the lancet fluke *(Dicrocoelium dendriticum)* in Swedish alpacas). Sv Vet Tidn 10(10):11–16

Drosten C, Seilmaier M, Corman VM, Hartmann W, Scheible G, Sack S (2013) Clinical features and virological analysis of a case of Middle East respiratory syndrome coronavirus infection. Lancet Infect Dis 13:745–751

Duraffour S, Meyer H, Andrei G, Snoeck R (2011) Camelpox virus. Antivir Res 92(2):167–186

Duraffour S, Drillien R, Haraguchi K, Balzarini J, Topalis D, van den Oord JJ, Andrei G, Snoeck R (2014) KAY-2-41, a novel nucleoside analogue inhibitor of orthopox viruses in vivo. Antimicrob Agents Chemother 58(1):27–37

Earle JA, Melia MM, Doherty NV, Nielsen O, Cosby SL (2011) Phocine distemper virus in seals, east coast, United States, 2006. Emerg Infect Dis 17(2):215–220

El Mamy AB, Lo MM, Thiongane Y, Diop M, Isselmou K, Doumbia B, Baba MO, El Arbi AS, Lancelot R, Kane Y, Albina E, Cêtre-Sossah C (2014 Dec) Comprehensive phylogenetic reconstructions of Rift Valley fever virus: the 2010 northern Mauritania outbreak in the *Camelus dromedarius* species. Vector Borne Zoonotic Dis 14(12):856–861

Faye N, Vias G (2010) The impact of climate change on camel farming system. Proc Int Camel Symp 'Camel science and development for sustainable livestock', Garissa, Kenya, KARI Ed, 10-15th June

Fowler M.E. 1998. Medicine and Surgery of South American Camelids: Llama, Alpaca, Vicuña, Guanaco, 2nd edn. Iowa State Press, Ames, p 549

Gameel AA, Ali AS, Razig SA, Brown J (1994) A clinic-pathological study on spontaneous Paratuberculosis in camels in Saudi Arabia. Pak Vet J 4(1):14–19

Ghosh P, Hsu C, Alyamani EJ, Shehata MM, Al-Dubaib MA, Al-Naeem A, Hashad M, Mahmoud OM, Alharbi KB, Al-Busadah K, Al-Swailem AM, Talaat AM (2012) Genome-wide analysis of the emerging infection with Mycobacterium avium subspecies paratuberculosis in the Arabian camels (*Camelus dromedarius*). PLoS One 7(2):e31947

Gitao CG, Agab H, Khalafalla AI (1998a) Outbreaks of *Dermatophilus congolensis* infection in camels (*Camelus dromedarius*) from the Butana region in eastern Sudan. Rev sci tech Off int Epiz 17(3):743–748

Gitao CG, Agab H, Khalafalla AI (1998b) Camel dermatophilosis in Kenya, Sudan and Saudi Arabia. Proceedings of the third annual meeting for animal production under arid conditions, vol 2. United Arab Emirates University Press, United Arab Emirates, pp 93–107

Gluecks IV, Younan M (2010) Camel sudden death syndrome: outbreak of an unknown camel disease in the Horn of Africa. http://www.mbali.info/doc560.htm. Accessed 1/7/2011

Gubser C, Smith GL (2002) The sequence of camelpox virus shows it is most closely related to variola virus, the cause of smallpox. J Gen Virol 83(Pt 4):855–872

Gutierrez C, Desquesnes M, Touratier L, Buscher P (2010) *Trypanosoma evansi* recent outbreaks in Europe. Vet Parasitol 174(1–2):26–29

Gwida M, Al DS, Melzer F, Rösler U, Neubauer H, Tomaso H (2010) Brucellosis – regionally emerging zoonotic disease? Croat Med J 51:289–295

Gwida M, El-Gohary A, Melzer F, Khan I, Rösler U, Neubauer H (2012) Brucellosis in camels. Res Vet Sci 92:351–355

Haagmans BL, Al Dhahiry SH, Reusken CB, Raj VS, Galiano M, Myers R et al (2014) Middle East respiratory syndrome coronavirus in dromedary camels: an outbreak investigation. Lancet Infect Dis 14(2):140–145

Hemida MG, Perera RA, Wang P, Alhammadi MA, Siu LY, Li M (2013) Middle East respiratory syndrome (MERS) coronavirus seroprevalence in domestic livestock in Saudi Arabia, 2010 to 2013. Euro Surveill 18:20659

Henrich M, Reinacher M, Hamann HP (2007) Lethal bluetongue virus infection in an alpaca. Vet Rec 161(22):764

Higgins AJ (1986) The camel in health and disease. Baillere Tindall, London, 104 p

Ismail TM, Hassan HB, Nawal MA, Rakha GM, Abd El-Halim MM, Fatebia MM (1992) Studies on prevalence of rinderpest and peste des petits ruminants antibodies in camel sera in Egypt. Vet Med J Giza 10(2):49–53

Jezek Z, Kriz B, Rothbauer V (1983) Camel pox and its risk to the human population. J Hyg Epidemiol Microbiol Immunol 27:29–42

Khalafalla AI (2000) Camel contagious ecthyma: risks in young calves. Rev Elev vet Pays Trop 53(2):173–176

Khalafalla AI, Abbas Z, Mohamed MEH (1998) Camel papillomatosis in the Sudan. J Camel Pract Res 5:157–159

Khalafalla AI, Mohamed MEH (1996) Clinical and epizootiological features of camelpox in eastern Sudan. J Cam Pract Res (2): 99–102

Khalafalla AI, Saeed IK, Ali YH, Abdurrahman MB, Kwiatek O, Libeau G, Obeida AA, Abbas Z (2010) An outbreak of peste des petits ruminants (PPR) in camels in the Sudan. Acta Trop 116(2):161–165

Khalafalla AI, Lu X, Al-Mubarak AI, Dalab AH, Al-Busadah KA, Erdman DD (2015) MERS-CoV in upper respiratory tract and lungs of dromedary camels, Saudi Arabia, 2013–2014. Emerg Infect Dis 21(7):1153–1158. doi:10.3201/eid2107.150070

Kim SG, Anderson RR, Yu JZ, Zylich NC, Kinde H, Carman S, Bedenice D, Dubovi EJ (2009) Genotyping and phylogenetic analysis of bovine viral diarrhea virus isolates from BVDV infected alpacas in North America. Vet Microbiol 136(3–4):209–216

Kinne J, Madarame H, Takai S, Jose S, Wernery U (2011) Disseminated *Rhodococcus equi* infection in dromedary camels (*Camelus dromedarius*). Vet Microbiol 149(1–2):269–272

Kritz B (1982) A study of camelpox in Somalia. J Comp Pathol 92:1–8

Kwiatek O, Ali YH, Saeed IK, Khalafalla AI, Mohamed OI, Abu OA, Abdelrahman MB, Osman HM, Taha KM, Abbas Z, El Harrak M, Lhor Y, Diallo A, Lancelot R, Albina E, Libeau G (2011) Asian lineage of peste des petits ruminants virus. Afr Emerg Infect Dis 17(7):1223–1231

Manefield GW, Tinson AH (1997) Camels – a compendium. Sydney Post Graduate Foundation, Australia, Vademecum Series C No 22

Mellor PS, Carpenter S, Harrup L, Baylis M, Mertens PP (2008) Bluetonge in Europe and the Mediterranean Basin: history of occurrence prior to 2006. Prev Vet Med 87(1–2):4–20

Memish ZA, Cotten M, Meyer B, Watson SJ, Alsahafi AJ, Al Rabeeah AA et al (2014) Human infection with MERS coronavirus after exposure to infected camels, Saudi Arabia, 2013. Emerg Infect Dis 20(6):1012–1015

Meyer G, Lacroux C, Léger S, Top S, Goyeau K, Deplanche M (2009) Lethal Bluetongue virus serotype 1 infection in llamas. Emerg Infect Dis 15(4):608–610

Meyer B, Müller MA, Corman VM, Reusken CBEM, Ritz D, Godeke GD (2014) Antibodies against MERS coronavirus in dromedary camels, United Arab Emirates, 2003 and 2013. Emerg Infect Dis 20:552–559

Milne-Price S, Miazgowicz KL, Munster VJ (2014) The emergence of the Middle East respiratory syndrome coronavirus. Pathog Dis 71(2):121–136

Molina JM, Ruiz A, Juste MC, Corbera JA, Amador R, Guitierrez C (1999) Seroprevalence of *Trypanosoma evansi* in dromedaries (*Camelus dromedarius*) from the Canary islands (Spain) using an Antibody Ab-ELISA. Prev Vet Med 47(1–2):53–59

Mosadeghhesari M, Oryan A, Zibaee S, Varshovi HR (2014) Molecular investigation and cultivation of camelpox virus in Iran. Arch Virol 159(11):3005–3011

Musa MT, Eisa MZ, El Sanousi EM, Abdel Wahab MB, Perrett L (2008) Brucellosis in camels (*Camelus dromedarius*) in Darfur, Western Sudan. J Comp Pathol 138(2–3):151–155

Nabeth P, Kane Y, Abdalahi MO, Diallo M, Ndiaye K, Schneegans M, Salam A, Mathiot C (2001) Rift Valley fever outbreak, Mauritania, 1998: seroepidemiologic, virologic, entomologic, and zoologic investigations. Emerg Infect Dis 7(6):1052–1054

Nowotny N, Kolodziejek J (2014) Middle East respiratory syndromecoronavirus (MERS-CoV) in dromedary camels, Oman, 2013. Euro Surveill 19(16). pii: 20781

OIE (2008) Manual of diagnostic tests and vaccines for terrestrial animals (terrestrial manual). Office international des épizooties, Paris

Omer MM, Musa MT, Bakhiet MR, Perrett L (2010) Brucellosis in camels, cattle and humans: associations and evaluation of serological tests used for diagnosis of the disease in certain nomadic localities in Sudan. Rev Sci Tech 29(3):663–669

Ortega J, Crossley B, Dechant JE, Drew CP, Maclachlan NJ (2010) Fatal Bluetongue virus infection in an alpaca (*Vicugna pacos*) in California. J Vet Diagn Investig 22(1):134–136

Perera RA, Wang P, Gomaa MR, ElShesheny R, Kandeil A, Bagato O (2013) Seroepidemiology for MERS coronavirus using microneutralisation and pseudoparticle virus neutralization assays reveal a high prevalence of antibody in dromedary camels in Egypt, June 2013. Euro Surveill 18:20574

Reusken CB, Ababneh M, Raj VS, Meyer B, Eljarah A, Abutarbush S (2013) Middle East respiratory syndrome coronavirus (MERS-CoV) serology in major livestock species in an affected region in Jordan, June to September 2013. Euro Surveill 18:20662

Rivera H, Madewell BR, Ameghino E (1987) Serological survey of viral antibodies in the Peruvian alpaca (*Lama pacos*). Am J Vet Res 48(2):189–191

Roberts H, Lopez M, Hancock R (2010) International disease monitoring. Vet Rec 167:192–195. doi:10.1136/vr.c5997

Roger F (2006) Descriptive and spatial epidemiology of Rift valley fever outbreak in Yemen 2000-2001. Ann N Y Acad Sci 1081:240–242

Roger F, Yigezu M, Hurard C, Libeau G, Mebratu GY, Diallo A, Faye B (2000) Investigations on a new pathological condition of camels in Ethiopia. J Camel Pract Res 7(2):163–165

Roger F, Guebre YM, Libeau G, Diallo A, Yigezu LM, Yilma T (2001) Detection of antibodies of rinderpest and peste des petits ruminants viruses (Paramyxoviridae, Morbillivirus) during a new epizootic disease in Ethiopian camels (*Camelus dromedarius*). Rev Méd Vét 152(3):265–268

van Amstel S, Kennedy M (2010) Bovine viral diarrhea infections in new world camelids – a review. Small Rumin Res 21:121–126

Wenker C, Hatt JM, Hertzberg H, Ossent P, Hänichen T, Brack A, Isenbuegel E (1998) Dicrocoeliosis in South American camelids. Tierärtzl Prax 26:355–361

Wernery U (2009) Glanders. In: Mair TS, Hutchinson RE (eds) Infectious diseases of the horse. Equine Veterinary Journal, Ltd., Fordham, pp 253–260

Wernery U (2015) Camelid brucellosis: a review. Rev Sci Tech Off Int Epiz 33(3):839–857

Wernery U, Ali M (1990) Dermatophilose in Renndromedaren – Fallbereicht. Tierärztl Umschau 45(3):209–210

Wernery U, Kaaden OR (2002) Infectious diseases of Camelids. Blackwell Science, Berlin

Wernery R, Kinne J, Hayden-Evans J, Ul HA (1997) Melioidosis in a seven year old camel. A new disease in the United Arab Emirates (UAE). J Camel Pract Res 4:141–143

Wernery U, Wernery R, Joseph M, Al-Salloom F, Johnson B, Kinne J, Jose J, Jose S, Tappendorf B, Hornstra H, Scholz HC (2011) Natural *Burkholderia mallei* infection in Dromedary, Bahrain. Emerg Infect Dis 17(7):1277–1279

Wilson AJ, Mellor PS (2009) Bluetongue in Europe: past, present and future. Philos Trans R Soc Lond Ser B Biol Sci 364(1530):2669–2681

Yousif AA, Al Naeem A (2011) Molecular characterization of enzootic camel pox in eastern province of Kingdom of Saudi Arabia. Int J Virol 7(4):135–136

Zanolari P, Bruckner L, Fricker R, Kaufmann C, Mudry M, Griot C, Meylan M (2010) Humoral response to 2 inactivated bluetongue virus serotype-8 vaccines in South American Camelids. J Vet Intern Med 24(4):956–959

Index

J. Bayry (ed.), *Emerging and Re-emerging Infectious Diseases of Livestock*,
DOI 10.1007/978-3-319-47426-7

MIX
Papier aus verantwortungsvollen Quellen
Paper from responsible sources
FSC® C105338

If you have any concerns about our products,
you can contact us on
ProductSafety@springernature.com

In case Publisher is established outside the EU,
the EU authorized representative is:
Springer Nature Customer Service Center GmbH
Europaplatz 3, 69115 Heidelberg, Germany

Printed by Libri Plureos GmbH
in Hamburg, Germany